The Evolution of the Crystalline Rocks

To
Kay and *Anne*

The Evolution of the Crystalline Rocks

Edited by

D.K. BAILEY, *Department of Geology, University of Reading, England*

and

R. MACDONALD, *Department of Environmental Sciences, University of Lancaster, England*

1976

ACADEMIC PRESS
London * New York * San Francisco
A Subsidiary of Harcourt Brace Jovanovich, Publishers

ACADEMIC PRESS INC. (LONDON) LTD.
24/28 Oval Road
London NW1

United States Edition published by
ACADEMIC PRESS INC.
111 Fifth Avenue
New York, New York 10003

Library of Congress Catalog Card Number: 75 46326
ISBN: 0 12 073450 8

Text set in 11/12 pt Monotype Bembo, printed by letterpress, and bound in Great Britain at The Pitman Press, Bath

Contributors

D.K. Bailey — Department of Geology,
University of Reading,
Whiteknights,
Reading RG6 2AB

W.S. Fyfe — Department of Geology,
Faculty of Science,
University of Western Ontario,
London 72,
Canada

H.J. Greenwood — Department of Geological Sciences,
University of British Columbia,
Vancouver 8,
Canada

W.C. Luth — Department of Geology,
School of Earth Sciences,
Stanford University,
California 94305,
U.S.A.

R.C. Newton — Department of the Geophysical Sciences,
University of Chicago,
5734, South Ellis Avenue,
Chicago,
Illinois 60637,
U.S.A.

W. Schreyer — Institut für Mineralogie,
Ruhr-Universität Bochum,
D-463 Bochum-Querenburg,
Universitatsstrasse 150,
Postfach 2148,
Germany

Preface

In 1928 a new landmark in petrology appeared, with the publication of Bowen's "The Evolution of the Igneous Rocks." Its aim was simple, to describe the techniques of phase equilibrium studies on silicates at high temperatures, and to couple the results with petrological observations, producing a unified picture of magmatic evolution. The achievement of this aim was so successful that it is safe to say that no later contributions to igneous petrology are uncoloured by Bowen's thinking. This demonstration of the power of experimental studies led to a gradual spread of experimentation and to the development of new techniques, particularly with the introduction, after the Second World War, of apparatus suitable for routine studies at high pressures. Since that time there has been an increasing need for a sequel to "The Evolution of the Igneous Rocks", especially to take account of the developments and the consequences of high pressure experimentation. Ideally, such a sequel should have been written by one person but the field is now so vast that the task would be almost impossible single-handed: certainly it is sufficiently daunting that no-one has yet tried. Meanwhile, the gap between Bowen's book and the developments of modern experimental petrology has been growing apace. The present book is a team effort to bridge that gap. The team was deliberately kept small so that each contributor could assume responsibility for a wide area in which the reader would find a unity of philosophy and style, in the spirit of Bowen's original. No attempt has been made at a complete coverage of the whole of experimental petrology because this would be impossible in one volume without superficial treatment of some important fields. Instead we have preferred to treat selected areas in depth, omitting others that appeared to be adequately covered in recent publications. In each section, the authors have tried to follow Bowen in producing a perspective view of the subject and the advances that have been made, relating the experiments to geological observations, and avoiding as far as possible unfamiliar terminology. Our joint aim has been to provide a basic text in applied experimental petrology, and thence a point of entry into the modern literature. This book is thus intended for earth scientists, from undergraduate level onwards. But the field of high temperature and high pressure equilibria is not the sole preserve of those concerned with the formation of crystalline rocks: it is common ground for the ceramics, glass, refractories, metals and mineral industries, so we hope that materials scientists generally may find here something of value.

The first part of the book deals with the methods of experimental petrology. Experimental equipment and techniques are described briefly so that the reader can appreciate the principles and understand the limitations. The main concern, however, is with the interpretation and uses of different types of phase diagrams by petrologists, attention being focused on low variance conditions, which are the prime concern of modern petrology.

The second part of the book comprises five chapters, in which the separate authors draw on their own experience to review the experimental petrology of selected groups of rocks. At the outset, the editors made but one stipulation, that the common theme should be the applications of experimental results to understanding rocks. The resulting variations on this theme, running through five chapters, provide the added bonus of a conspectus of the ways in which experimental petrologists can (and perhaps should) handle the information. Three of the chapters are devoted to the major groups of metamorphic rocks because this is the area where a sequel to Bowen's work is most sorely needed. Two final chapters deal with the experimental petrology of granites and alkaline rocks. Together they provide a much needed review of experimental developments in the realm of felsic igneous rocks, and at the same time they form a logical culmination to the preceding discussions of metamorphism. Some consideration is given to strongly undersaturated alkaline rocks, and the special conditions in the mantle that may be necessary to their formation, but the general theme of basic magmatism is too large to be encompassed in the scope of this book and, indeed, is more than adequately represented in recent literature. Each chapter in the book is independent and can be read as an entity. Some overlap is therefore unavoidable, but this is not without value because each contributor's approach is individual, and this gives added dimensions to important common areas.

In dealing with specific rock groups the order of discussion could have been decided on several grounds. One tempting way was to follow Bowen's general approach, working from highest to lowest temperatures, but because solid state reactions and the applications to metamorphic rocks are our main point of departure, we have chosen the reverse route: starting with low and proceeding to higher temperatures, culminating in igneous conditions.

There can be little doubt that part of Bowen's life-long success in communicating experimental results lay in his sparing use of specialist terminology. Chemical thermodynamic terms and concepts are not common currency among earth scientists, no matter how much we might wish otherwise. Even those who attain a facility for thermodynamics are commonly dismayed at how transient it is, a faculty soon atrophied if not in constant use. Many, too, are suspicious of the hiatus between classical thermodynamics, derived for ideal states, and the complex realities of rock systems. After all, experiments are necessary largely because geological systems are non-ideal. Consequently, we have tried to avoid lines of reasoning that rely heavily on mathematical thermodynamic proof, with all the special terms that this entails. We believe that Bowen understood the parodox that the application of experimental results to rocks could be portrayed more cogently (and in that sense, with greater accuracy) by avoiding the precision of thermodynamic terminology!

Since 1928, experimental petrology has explored more and more complex systems and conditions, so that many earth scientists are perhaps inclined to "give up when it comes to phase diagrams". Our hope is that readers of this book will be encouraged to make a more positive response. Understanding phase equilibria can give a new perception to all geologic observations, and has probably contributed more than any other factor to advancing the understanding of crystalline rocks.

Scientific progress is now so rapid that any carefully prepared book aimed at describing "current progress" must be out of date before publication. The present volume was conceived with a longer-term view. Altough the reader will find freshness and originality in the contributors' ideas, the book is primarily a distillation of their long experience in experimental petrology. Final manuscript revisions were made in 1974–1975, but our main aim has been to produce a survey that is factual and enduring. If we have provided a base from which the reader can participate in the excitement of new developments this aim is achieved. It is a pleasure to record that a major breakthrough in experimental technique has been made in the last year at the Geophysical Laboratory of Washington (see Appendix III). Temperatures in excess of 2000°C at pressures close to 500 Kbars can now be sustained in a newly-developed diamond-anvil, high pressure cell, and the prospect of investigating deep mantle conditions is suddenly revealed. Experimental petrology is on the threshold of a new era.

In the preparation of this book we have obviously had much help, and we would like to record our thanks to all our friends for their advice and encouragement. Particular thanks are due to Mrs. Kay Bailey for her painstaking compilation of the Index, and to Mrs. Angela Broughton, whose secretarial skills made the editors' work possible.

May, 1976

D.K. BAILEY

R. MACDONALD

"We're just trying to find out
how the Lord did it."

J. Frank Schairer

CONTENTS

List of contributors v
Preface vii

Part I Experimental Methods and the Uses of Phase Diagrams

Summary 1
Experimental Methods and the Uses of Phase Diagrams . . . 3

Part II Experimental Petrology

Section A Metamorphic Rocks

Summary 99
Chapter 1 High Pressure Metamorphism 101
Chapter 2 Metamorphism at Moderate Temperatures and Pressures . 187
Chapter 3 Experimental Metamorphic Petrology at Low Pressures and High Temperatures 261

Section B Igneous Rocks

Summary 333
Chapter 1 Granitic Rocks 335
Chapter 2 Applications of Experiments to Alkaline Rocks . . 419
Appendixes 470
Subject Index 475
Systems Index 483

Part I

Experimental Methods and the Uses of Phase Diagrams

SUMMARY

This first part of the book explains the methods of making experiments, constructing phase diagrams, and the ways in which these diagrams may be applied to petrological problems. It is divided into seven sections. The first four describe the equipment, materials and techniques of experimental petrology in sufficient detail for an appreciation of the method, but omitting technical data that would be needed only by an aspiring experimentalist. The aim is to let the reader know how the experiments are made, so that in the section which follows he will understand the limitations of each method, and the problems involved in the interpretation of experimental products. This leads naturally into the sixth section, which defines the phase rule and explains the methods by which phase diagrams are constructed from the experimental results.

Section seven, the major portion of this part of the book, is devoted to a description of the more commonly used types of phase diagram, their underlying rationale and the ways in which they may be interpreted. It starts with the now traditional discussion of phase equilibria involving liquids, considering first the various modes of melting exhibited by simple compounds, an appreciation of which is essential to understanding the more complex diagrams which follow. It is also traditional that solid $\rightleftharpoons$ liquid relations are illustrated in terms of heat loss and sequence of crystallisation, but here the emphasis is placed on the reverse process, heat gain and the initiation and progress of melting, because this goes immediately to low variance conditions. The difficulties of depicting polycomponent systems diagrammatically are described, and the special properties and potential ambiguities of various reduction methods, such as projections and pseudo-system diagrams, are examined. Isothermal sections are considered separately in an attempt to show why the experimentalist finds them so useful when the lay man frequently finds them uninformative. Isothermal sections give a lead into the topic of heating diagrams, which are outside the normal scope of phase equilibria. Space allows only a brief introduction to this complex subject incorporating the interplay of intensive and extensive variables. The intention is to show the important practical aspects, wherein the progress and the extent of isobaric melting are seen to be functions of mode of melting, bulk composition and heat input.

At temperatures below the special boundary condition represented by the onset of melting, experimental petrology is concerned with reactions among solids, in which the combined variation of pressure and temperature must be taken into account. For this purpose the pressure-temperature phase diagram becomes paramount, and consequently the characteristics and applications of PT diagrams must be examined in detail. Particular attention is given to the topological analysis of the univariant curves stemming from an invariant point in a PT diagram (Morey–Schreinemakers analysis), with the aim of high lighting the interpretive value of this method in metamorphic petrology. The discussion is rounded off with an examination of PT curves which constitute the upper limit of metamorphism, i.e. beginning of melting.

Finally, the influences of vapour phases on solid and solid–liquid relationships are examined, with special reference to water, oxygen and carbon dioxide.

Experimental Methods and the Uses of Phase Diagrams

D. K. Bailey

1 Introduction 4
(a) What is the purpose of experimental petrology? 4
(b) How is it possible? 4
(c) How is it done? 5
2 Equipment and Ranges of Physical Conditions 5
(a) High temperature experiments at atmospheric pressure 6
(b) Experiments at higher pressures 6
3 Range, Precision and Accuracy 11
4 Materials and Methods 11
(a) Sample compositions 11
(b) Advantages and drawbacks 12
(c) Size of Sample 14
(d) Sample Containers 14
(e) Length of Experiment 15
(f) Control of Temperature and Pressure 16
(g) Calibration 16
(h) Examining the Charge 16
5 Interpretation of Run Products and the Limitations 17
(a) Limitations inherent in the starting material 18
(b) Limitations inherent in the experimental methods 18
(c) Limitations in interpreting the phase relations 19
6 Underlying Principles and Methods of Constructing a Phase Diagram . 21
(a) Construction of a binary TX diagram 24
7 Presentation and Interpretation of Results 27
(a) TX diagrams involving liquids 28
(b) PX diagrams involving liquids 61
(c) Solid $\rightleftharpoons$ solid equilibria 61
(d) Characteristics of PT diagrams 73
(e) Mixed gas systems 91
(f) Variation in P_{O_2} 92
(g) Applications of PT diagrams to metamorphism 93

1. INTRODUCTION

Experimental studies of phase equilibria at high temperatures and pressures are now an integral part of modern geology, but, for those without first-hand experience, experimental petrology often seems wrapped in a veil of mystique. What the average petrologist needs most, to begin to penetrate this veil, are answers to the following questions:

(a) What is the purpose of experimental petrology?
(b) How is it possible?
(c) How is it done?

The aim of the first part of this book is to provide answers to these three questions and then to examine various types of phase diagrams and their petrological uses. Answers to the three questions are given first in general form and will then be elaborated in the sections which follow.

(a) WHAT IS THE PURPOSE OF EXPERIMENTAL PETROLOGY?

It is common petrological experience that most crystalline rocks are composed of a small number of phases (minerals) and that these phase assemblages have been continually repeated throughout space and time. Because the chemical variables in rocks are many and complex, these limited mineral assemblages suggest an approach to chemical equilibrium during rock crystallisation. If this is so, the rigorous and systematic exploration of the conditions of equilibrium in chemical systems pertaining to rocks must ultimately reveal the physical limits of crystalline rock formation. It is the knowledge of these limits which will enable us to understand the physical states, compositions, and processes within the Earth.

(b) HOW IS IT POSSIBLE?

Attainment of high temperatures and high pressures is not difficult. The main problems in conducting an experiment are connected with the control and calibration of temperature and pressure. But experimental petrology is *chiefly possible* because the reactions between silicates (and silicate liquids) are sluggish; therefore, if the sample is rapidly removed from the high temperature or high pressure condition there is insufficient time for retrograde reactions, and the phase assemblage of the experiment is preserved down to room temperature and pressure. The process is now commonly known as *quenching* the experiment. Quenching is successful because reaction rates fall dramatically with falling temperature, and it is also assisted by the fact that prograde reactions (in the experiment, as in nature) tend to increase the crystal grain size, which subsequently impedes re-equilibration during the period of falling temperature.

The exceedingly slow reaction rates in silicate assemblages are shown in nature by the preservation of mineral assemblages from the earliest pre-Cambrian on Earth, but perhaps even more spectacularly by the survival of glasses on the Moon for periods greater than 1.5 billion years.

(c) HOW IS IT DONE?

The vast majority of experiments take a fraction of a gram of the material to be studied, wrapped or sealed in an unreactive but ductile container (Pt, Ag, Au). This sample is heated in a furnace at the required temperature and pressure for the optimum period of time (usually hours or days). This simple statement has many ramifications, which will be explored at various levels in the following sections, but the time factor is one that troubles many geologists, especially those concerned with metamorphism because they sense a gross distortion of scale. Any responsible experimentalist will also be alert to this problem (especially as he relies in one way on sluggish reaction rates!). But it is easy to exaggerate the time scale distortion by neglecting the following facts:

(1) Very fine grained and therefore highly reactive materials are used as starting compositions, and even the products of the experiments are commonly fine grained so that reaction rates during the experiments are highly accelerated compared with the natural situation. The experimenter is not concerned with producing coarse grains; on the contrary, these will only present him with an alarming prospect of disequilibrium in his tiny sample.
(2) Where necessary (e.g. Al_2SiO_5) an experiment can be made in which the starting material contains all the possible products in known amounts and the *direction* of reaction can be discerned by changes in proportions, even though the experiment does not run to completion.
(3) The experimenter draws confidence about the time scale from the fact that most natural assemblages (though not the grain size) can be duplicated in laboratory times.
(4) The most important test of equilibrium that experimental petrologists apply is the ability to reproduce a phase change from either direction, e.g. up or down temperature. This is known as a *reversal*, and when demonstrated it is also a strong indication that the experimental times are realistic.

2. EQUIPMENT AND RANGES OF PHYSICAL CONDITIONS

In principle, the methods required for experimental petrology are simple. All that is needed is the means to achieve, measure and control temperature and pressure. In some cases it may be necessary to control specific vapour pressures, e.g. P_{O_2}, P_{H_2O} and P_{CO_2}, but these are essentially chemical controls on the composition and are additional to the normal procedures considered below. The main technological problems arise from the combination of high pressures and high temperatures because most materials are seriously weakened at high temperature and it therefore becomes increasingly difficult to maintain pressure.

This last factor, plus the need for measuring, controlling and recording temperatures and pressures accurately, may make experimental apparatus appear complex but this need not concern the petrologist. For comprehensive details of the equipment the reader is referred to Edgar (1973): the following description is designed to highlight the essential features of the more common types of apparatus used by experimental petrologists.

(a) HIGH TEMPERATURE EXPERIMENTS AT ATMOSPHERIC PRESSURE

Experiments in an open furnace are probably the most familiar, because most experimentation up to the late 1940's was of this type and it is the results of this work, largely carried out at the Geophysical Laboratory, which form the foundation of the subject and are still featured in most petrology textbooks.

Experiments are normally made with the sample suspended in the hot zone of a vertical tube electric furnace of about 1.5 cm bore (Faust, 1936; Edgar, 1973) shown diagrammatically in Fig. 1. This arrangement is traditionally known as a quenching furnace because it allows the sample to be instantaneously removed from the hot zone. The sample is suspended in the hot zone by a thin wire (inset to Fig. 1) which is fused at the close of the experiment by the passage of a heavy electric current. As the wire is fused the sample falls immediately through the bottom of the furnace where it is quenched to room temperature in a waiting dish of water or mercury. The temperature is measured by a sensor (usually a thermocouple) placed next to the charge. Temperature may be controlled by various means, such as incorporating the furnace winding as one arm of a Wheatstone Bridge and maintaining a constant resistance, or by using the signal from the sensor to actuate a control circuit on the power supply to the furnace.

With careful handling and continual calibration it is possible to make experiments to within $\pm 0.5°C$. Routine usage, with intermittent calibration, should give results within $\pm 3°C$. Quenching furnaces are also built in which it is possible to arrange a controlled atmosphere, e.g. nitrogen, argon, etc. by having adequate seals and supplying the gas to the furnace at a fixed pressure (usually 1 atmosphere).

(b) EXPERIMENTS AT HIGHER PRESSURES

The requirement here is that in addition to being able to heat the sample it must be possible to transmit a known pressure to it throughout the experiment. This may be done either hydrostatically, i.e. with a fluid pressure medium, or by squeezing the sample between solid surfaces.

(1) *Hydrostatic pressure*

(i) *Externally heated pressure vessel.* In this arrangement the sample is placed inside a pressure vessel (usually of test-tube form, see Fig. 2) and the pressure is

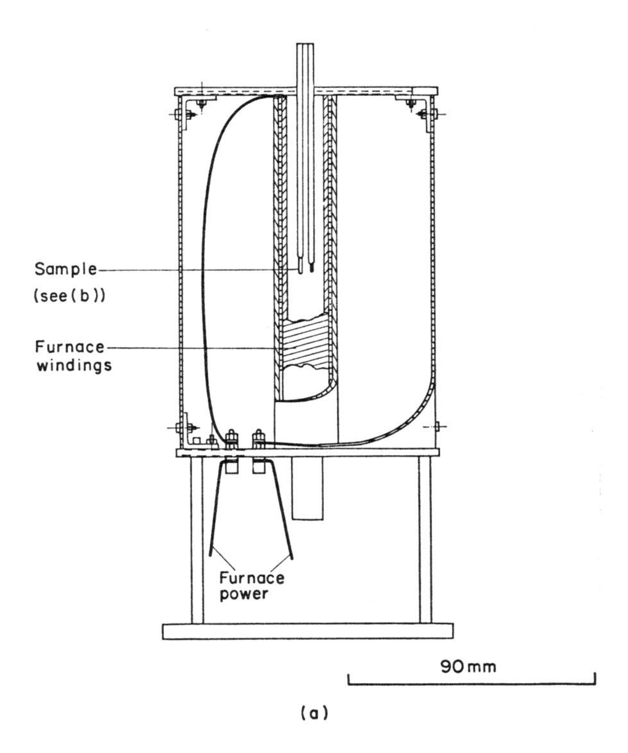

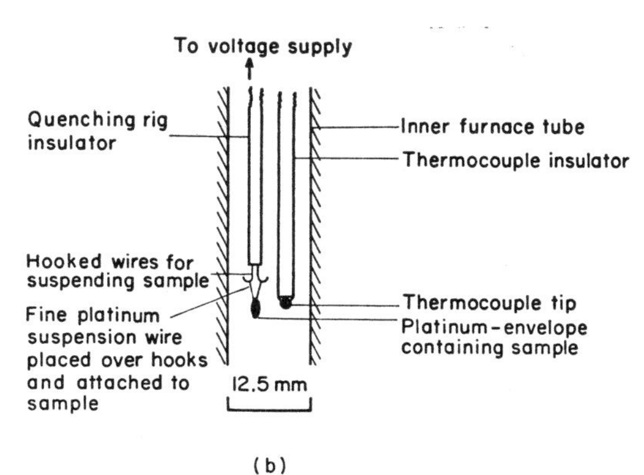

Fig. 1. Section of quenching furnace for use at atmospheric pressure: (a) quenching furnace; (b) details of quenching rig. (After Faust, 1936.)

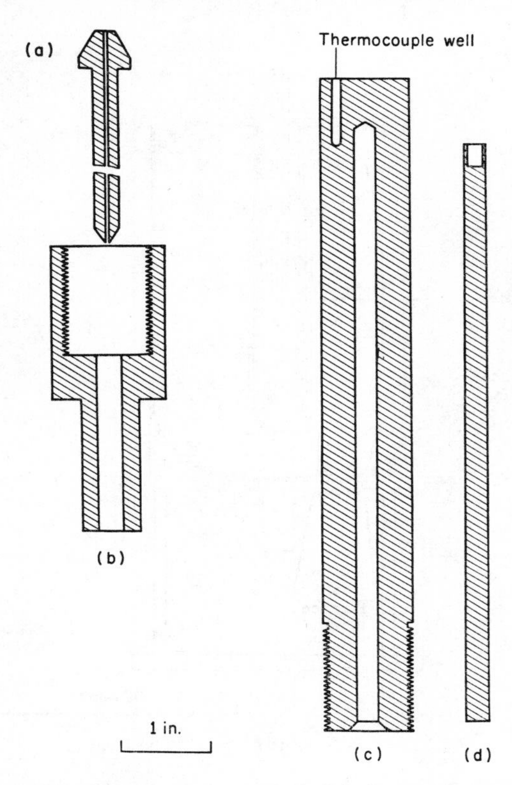

Fig. 2. Cross-section of cold-seal pressure vessel. (a) Cone-seal connection to pressure vessel (c). (b) Closure nut to seal (a) to (c). Charges are placed at the upper end of a support rod, such as (d), so that the centre of the charge is opposite the thermocouple junction. (After Tuttle and Bowen, 1958.)

applied by pumping the fluid pressure medium (commonly H_2O or argon) into the vessel. Pressure is normally measured directly and continuously with a Bourdon-tube pressure gauge. Alternatively, it may be measured by means of a pressure sensitive resistance winding immersed in the pressure medium; with this arrangement a continuous recording of the pressure is possible during the experiment. Heat is applied to the experiment by enclosing the pressure vessel in a furnace, sometimes a tube furnace, sometimes a hinged split furnace which closes to a tube and fits snugly round the vessel. Temperatures are measured and controlled by thermocouples embedded in the walls of the pressure vessel, as close as possible to the sample without weakening the vessel. The experiment is normally quenched by removing the furnace, and dousing the vessel with compressed air and water: the quench period is much longer than at 1 atm, and may take 30–90 seconds.

The connections from the pressure supply to the pressure vessel, and the seals, are normally outside the furnace and are therefore at a much lower temperature than the sample, (see Fig. 2). For this reason this apparatus is commonly referred to as a cold-seal, or a Tuttle cold-seal pressure vessel (Tuttle, 1949). It is affectionately known among experimentalists as the "Tuttle bomb", but this belies the safety of the method when carefully applied. It is the simplest and cheapest method of making experiments at high temperatures and pressures, and has made possible the now widespread studies in this field in universities throughout the world. The main limitation of this method is the liability to failure of the pressure vessel with increasing temperature and pressure; at 1 Kbar it is possible to conduct experiments up to 1000°C but at 10 Kbar 750°C is the normal upper temperature limit. To achieve higher temperatures in the pressure range up to 10 Kbar it is necessary to use a different method, in which the sample *and the furnace* are placed inside a large pressure vessel, the exterior of which is cooled by a water jacket.

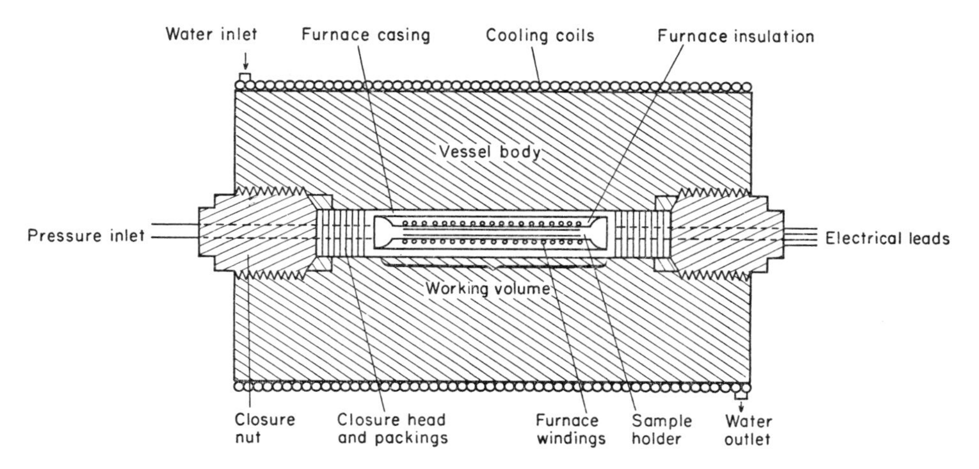

Fig. 3. Sketch of simplified internally heated pressure-vessel. (After Edgar, 1973.)

(ii) *Internally Heated Pressure Vessel.* The device of enclosing the sample and the furnace inside a large pressure vessel was employed at the Geophysical Laboratory in Washington between the two World Wars and was used by Goranson in his classic experiments on the melting of granite in the presence of water (1931). But this equipment was difficult to maintain, and routine hydrostatic experiments at high temperatures were not possible until Yoder (1950) produced the design which is the basis of most internally heated systems now in use (Fig. 3). With this type of apparatus temperature limits are set only by the melting temperature of the furnace winding e.g. with a Pt winding it is possible to run routinely at temperatures up to 1500°C. The pressure limits are set by the physical characteristics of the pressure medium; in the simple apparatus this is normally argon, which solidifies above 12 Kbar at room temperatures. Higher pressures are currently being achieved in some laboratories by heating the argon before the final compression to prevent solidification or, alternatively, by using nitrogen as a pressure medium. By these more sophisticated techniques hydrostatic experiments are now being conducted in the 10–20 Kbar range, which is probably not much lower than the eventual pressure limit in this type of experiment.

The main technical problems concerned with the internally heated pressure vessel are in creating effective pressure seals, especially at the points of entry of the power and thermocouple leads, which must be electrically insulated from the vessel. Apart from these, the experiment is essentially similar to the 1 atm experiments, and quenching is almost as rapid because as soon as the power to the furnace is cut off the whole apparatus is rapidly cooled by the external water jacket.

(2) *Solid media pressure apparatus*

For the range above 20 Kbar it is necessary to apply the pressure through solids acting on the sample, i.e. the sample is being mechanically crushed. This clearly creates problems in measuring temperature and pressure and also in minimising pressure gradients, which must always be inherently possible under these conditions. Several elegant methods have been devised to minimise these problems, the most widely used at the present time being the piston and cylinder device of Boyd and England (1963) with various modifications (Fig. 4). For

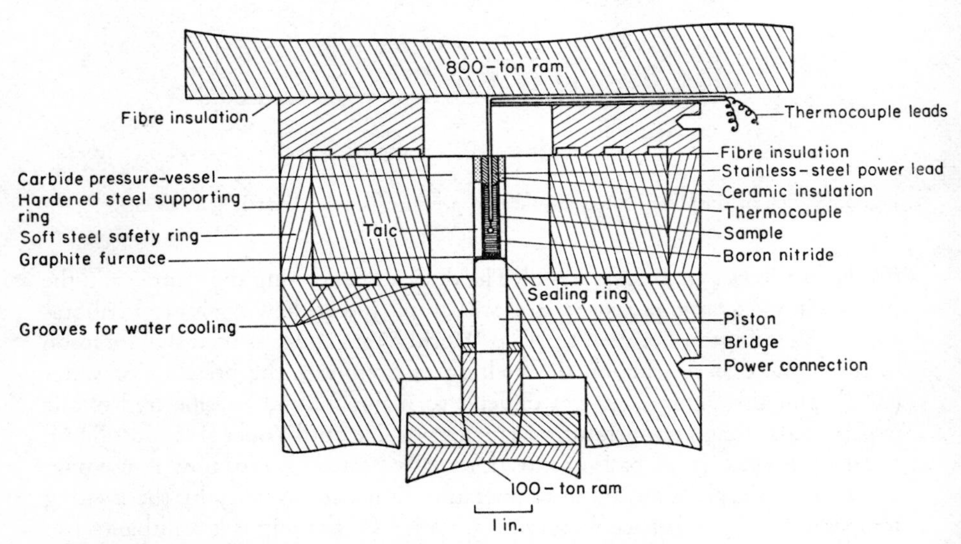

Fig. 4. Piston-cylinder apparatus. (After Boyd and England, 1963.)

further details of this and other equipment see Edgar (1973). The essence of the piston-cylinder apparatus is that the sample is enclosed in a tiny resistance furnace (usually graphite), the whole of which is then enclosed in crushable, heat resistant material such as pyrophyllite. This assembly together with power and thermocouple leads is then squeezed in a cylinder by a piston driven by a hydraulic ram. Transmitted pressure is normally obtained by calibrating the actuating pressure of the ram using the known inversion points of given substances. See Table 1 for ranges, accuracy and precision. Quenching is rapidly achieved by cutting the power to the furnace.

Solid state pressure equipment, although essentially simple in concept, is expensive to maintain and operate and requires continuous expert workshop support. Consequently, research in this field is restricted to a small number of centres.

(3) *Very high pressures*

Information on the physical properties of materials at pressures greater than 100 Kbar is obtained using explosion or shock wave techniques to generate the pressures required. Most of the information, however, is essentially a description of the physical state of materials under these conditions and would not normally qualify as phase equilibria studies. Possible exceptions to this might be simple materials such as one component systems, but even here the normal criteria of equilibrium cannot be applied because of the exceedingly short duration of the experiment. (But now see Appendix III.)

3. RANGE, PRECISION AND ACCURACY

The normal operating ranges, and the precision of temperature and pressure measurements are summarised in the following table, on top of p. 12.

Accuracy of the measurement for both temperature and pressure is a function of the skill and frequency of calibration. In the case of temperature, accuracy of the measurement is also strongly dependent on the configuration of the furnace, sample, and thermal sensor. Even in one atmosphere experiments, high accuracy requires painstaking and thorough tests of the equipment being used (see Biggar and O'Hara, 1969). In the more complex apparatus used for higher pressures, high degrees of accuracy are correspondingly harder to achieve—even though the degree of precision (or reproducibility) may be of a similar order to that in one atmosphere experiments.

4. MATERIALS AND METHODS

(a) SAMPLE COMPOSITIONS

The starting materials for phase equilibrium studies are generally prepared as synthetic glasses, or gels, or analysed rock or mineral powders. Some studies have been made using mixed powders, e.g. weighed proportions of separated

TABLE 1

Ranges and precision of common apparatus

Apparatus	Temperature			Pressure			Ref.
		Precision			Precision		
	Range	Best	Usual	Range	Best	Usual	
One atmosphere	0–1600°C	±0.5°C	±3°C	1 atm	—	—	1
Cold-seal	0–900°C	±1°C	±5°C	to 10 Kbar	±10 bars	±50 bars	2
Internally heated	0–1200°C	±2°C	±5°C	to 10 Kbar	±20 bars	±50 bars	5
Piston-cylinder	0–1750°C	±5°C	±15°C	to 60 Kbar	±100 bars	±5%	3, 4

1. Schairer, J. F., 1959; Biggar, G. M. and O'Hara, M. J., 1969.
2. Greenwood, H. J., 1961.
3. Boyd, F. R. and England, J. L., 1963.
4. Richardson, S. W., Gilbert, M. C. and Bell, P. M., 1969.
5. Yoder, H. S. and Tilley, C. E., 1962.

minerals, but in these cases there may always be a lurking doubt about the adequacy of the mixing and the final homogeneity of the sample. Homogeneity is always a source of concern to the experimenter, especially in the study of reactions in which there is no liquid and, even more so, no fluid present. The most homogeneous starting materials are glasses. These are prepared from carefully weighed amounts of pure chemicals, usually oxides or carbonates, by repeated fusion and crushing. Homogeneity is easily checked by examining a small amount of crushed glass in a matching refractive index liquid. Determination of refractive indices in a series of compositions also provides an additional check against composition errors because there should be a smooth variation throughout the composition range. When the glass is homogeneous it is customary to acclimatise it by holding at progressively lower temperatures, and eventually to crystallise part (typically half of a 10 g sample) as fully as possible. Gels are usually prepared by precipitating the required composition from solutions of organic silica compounds (e.g. tetraethyl ortho silicate) and nitrates and carbonates of the required cations. The dried precipitate is ground and fired at moderate temperature to produce an amorphous oxide mix. This product is normally used directly in the experiments but may itself be fused to make a glass. Rock and mineral samples are used directly as experimental materials, taking finely-ground analysed powders.

(b) ADVANTAGES AND DRAWBACKS

Glasses have the advantages that they are homogeneous, are less likely to produce persistent metastable phases during an experiment, and in experiments involving liquid they provide (when run simultaneously with their crystallised equivalents) a convenient method for constantly checking reversibility of reaction. They have the disadvantages that in systems containing alkalis there may be some losses during the high temperature fusions; they may be too sluggish in reaction (although this difficulty is largely obviated by prior crystallisation); and their preparation is time-consuming and requires expensive equipment. As an example of the sluggishness of some glasses it is worth noting

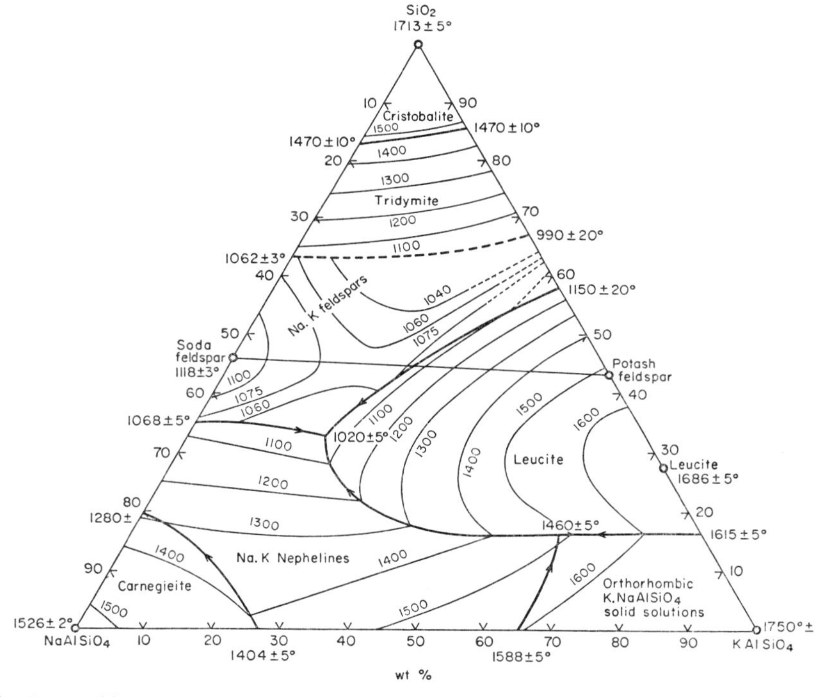

Fig. 5. Equilibrium diagram for the system $NaAlSiO_4$-$KAlSiO_4$-SiO_2, "Petrogeny's Residua System". (After Schairer, 1950.)

that the rhyolite minimum is wrongly depicted as being completely determined in most texts which figure Petrogeny's Residua System ($NaAlSiO_4$–$KAlSiO_4$–SiO_2: Fig. 5). In the original description of this system Schairer (1950) makes it quite clear that compositions in the synthetic rhyolite range could not be crystallised even after periods as long as 5 years at appropriate temperatures, and therefore "quenching data are entirely lacking for this region".

Gels have the advantage that they are easy to make and do not require expensive equipment. Control of the total composition of the gel is also accurately achieved: but it is less easy to control the homogeneity. Lack of homogeneity can sometimes be so serious as to make it impossible to achieve equilibrium in reasonable experiment time. Gels are finely divided and therefore highly reactive, but, because their structural state is so far removed from the required state in the general run of experimental conditions, there is a danger that they may run rapidly to metastable phases that are subsequently difficult to eliminate during the experiment.

Analysed minerals and rocks offer the advantages that they can be very close to the composition it is desired to study and are presumably close to the desired energy state and phase structure. But with rocks there are problems with homogenisation of the starting powder, the large number of components, and the constant question of how closely a crystalline rock represents the composition of the natural system in which it formed. In experiments on separated minerals the last problem is perhaps the greatest.

The choice of starting materials will clearly be determined by a variety of factors but it is quite common for experimental petrologists to try different starting materials in an attempt to produce identical run products under the same experimental conditions. The achievement of this convergence inspires confidence that the products represent equilibrium.

(c) SIZE OF SAMPLE

Most experimenters work with what seems to a non-experimenter to be an exceedingly small sample, most commonly a few tens of mgs. This small sample size is necessary because, (a) for a small sample, it is easier to produce a hot-zone in the furnace which is accurately controlled and free from temperature gradients, (b) the smaller the sample, the smaller the volume of the pressure vessel, making the equipment easier to produce and maintain, and making it easier to minimise pressure gradients, and (c) most experimental samples are contained in expensive noble metals. It is clearly an advantage to reduce the sample size to a minimum. The small sample size is possible because the starting materials and the run products in most experiments are very fine-grained and problems of inhomogeneity due to large crystal size are rarely encountered. The methods for examining the run products are described below.

(d) SAMPLE CONTAINERS

Samples are normally wrapped or sealed in silver, gold, platinum or silver-palladium alloy. In experiments with controlled atmospheres or in which reaction between charge and atmosphere can be neglected, the sample containers are unsealed, and in many of the classic one atmosphere experiments the sample was contained in a small envelope (about 1 cm square) made of platinum foil. In the vast majority of modern experiments it is desired to isolate the sample from the external pressure medium, and in these cases the container is typically made from a short length of metal tube (about 1½ cm long by 2 mm diameter) which is sealed by crimping and welding the ends. Such a thin-walled tube has the twin advantages of being non-reactive and ductile so that the external pressure is readily transmitted to the sample, which is otherwise isolated from external changes. Platinum and silver alloys have the additional property that they are permeable to hydrogen and this property is utilised in the oxygen buffer technique. In this technique, the sample (providing it is saturated with H_2O) can equilibrate in hydrogen activity with an external water-saturated environment. The hydrogen activity in the external environment is normally fixed by an oxidation equilibrium under water-saturated conditions, e.g. Ni $\rightleftharpoons$ NiO (see Fig. 6 and p. 92 for details).

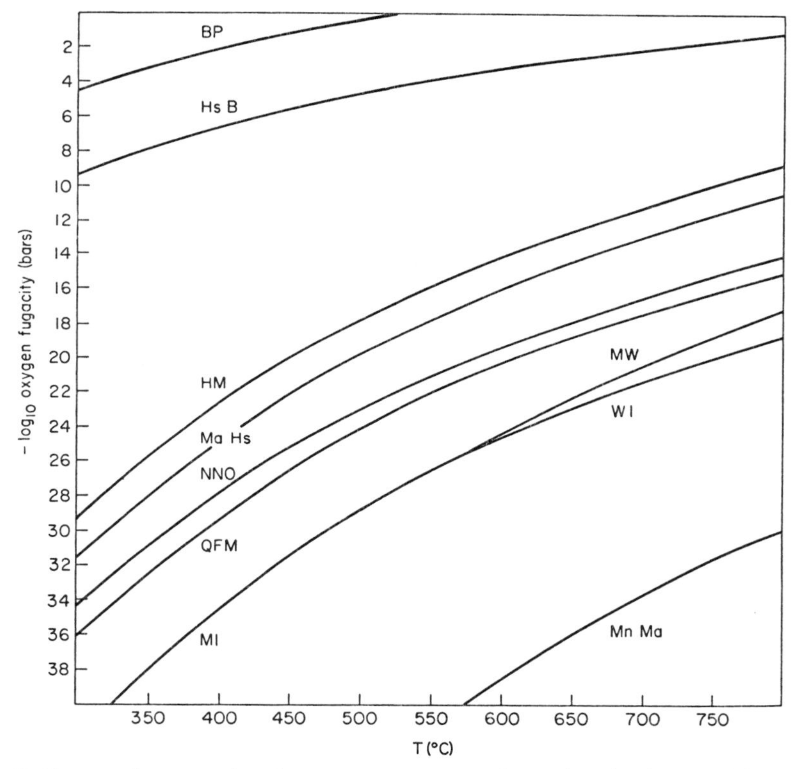

Fig. 6. Curves of oxygen fugacity versus temperature for various buffer assemblages at a total pressure of one bar. For abbreviations see Appendix.

(e) LENGTH OF EXPERIMENT

The ideal experimental time is the minimum period required to achieve an equilibrium assemblage but this of course varies greatly from system to system. The shorter the time the better because this improves the chances of maintaining accurate control of the physical conditions during the experiment. In some cases the optimum period may be measured only in minutes, in difficult cases such as that described by Schairer (1950) it may be impossible to reach equilibrium in any reasonable period. Normally the experimenter finds the optimum experimental period by trial and error, and in the majority of cases this is measured in hours or days.

(f) CONTROL OF TEMPERATURE AND PRESSURE

Temperature is normally controlled continuously during the experiment, most commonly by a thermo-electric sensor embedded in the furnace winding or placed next to the charge. The signal from the sensor actuates the control circuit which continuously operates on the power supply to the furnace. Such controllers are commercially available and their accuracy and efficiency may be very high, being effectively limited only by cost and the requirements of the experiment.

Pressure may be controlled continuously in an analogous manner to temperature using a pressure sensor, but continuous automatic control of pressure is not commonly employed because most pressure equipment is not liable to short-term fluctuation. Usually during an experiment, if there is any pressure change it is a slow decay and most experimenters achieve satisfactory results by regularly checking the pressure and restoring any small losses as they occur.

(g) CALIBRATION

Temperature calibration normally consists of calibrating the e.m.f. from the temperature measuring and controlling sensors against the known melting point of pure substances, the most commonly used being sodium chloride, gold and diopside. In hydrostatic experiments the pressure gauges are usually directly calibrated against a weight gauge. Solid media equipment needs to be calibrated for pressure using known phase transformations such as bismuth (BiI–BiII at 25°C) which constitute fixed points on a pressure scale: these known pressures in the experimental cell can then be directly related to the gauge pressure on the actuating hydraulic ram. Further details of control and calibration techniques are available in Edgar (1973).

(h) EXAMINING THE CHARGE

At the end of a run the experimenter is left with a small noble metal packet or capsule containing a few tens of mgs of (it is hoped!) the quenched equilibrium products at the conditions of the experiment. It is now his job to extract the maximum amount of information from this small specimen. The normal procedure is to open the charge with the maximum care so as to observe the form and distribution of the material in the container. In one atmosphere experiments this usually consists of carefully peeling the foil envelope from the charge, which, typically, is a small pellet. This is carefully examined for any signs of inhomogeneity in colour or crystallinity. Provided that the charge is homogeneous a small portion is gently crushed, made into an immersion mount in an appropriate refractive index liquid and examined with a polarising microscope. For many years, until the advent of simple X-ray powder diffraction equipment, this was the paramount technique for determining the run products and it is still the *essential* first step in the process today. In examining the charge the investigator seeks not only to identify the phases present but also to observe the size and form of the crystals, any textural relations, the relative amounts of the different phases and whether they are evenly distributed. After the optical examination it is now customary to use another small part of the charge to prepare an X-ray diffraction chart. In addition to confirming the optical identification this frequently allows the determination of a solid solution composition, and in the case of highly crystalline charges is the only method for the complete identification of all the phases present.

In experiments using sealed capsules the charge is normally removed by carefully cutting or filing the edges of the capsule until the sides can be folded back to reveal the charge. In experiments involving volatiles it is especially important to observe whether a separate volatile phase was present: it is customary to examine the charge with a binocular microscope for signs of vesicularity, the presence of a meniscus, and any evidence of precipitation from the vapour of crystals on the walls of the capsule, or liquid globules which are preserved in the vapour cavity as glass balls. After this preliminary examination on opening the capsule, the charge is examined with a polarising microscope, and by X-ray diffraction as in one atmosphere experiments. In experiments where losses or gains of volatiles due to capsule leakage might be critical, it is normal to weigh the capsule before and after the experiment. A further observation can also be made if the experiment involves a free vapour phase, in that during the initial opening, vapour may be heard to hiss from the capsule and if it is rich in H_2O, liquid may also be seen to escape from the first perforation. In experiments which generate a large amount of vapour the capsule may even be seen to be inflated at the end of the experiment. Interpretation of the run products and some of the limitations are considered more fully in the next section.

5. INTERPRETATION OF RUN PRODUCTS AND THE LIMITATIONS

An amazing feature of phase diagrams is the authority they assume when published. Petrologists who are otherwise sceptical, and acutely aware of the difficulties in interpreting the *phase relations in a rock*, seem to accept without question that their fellow experimentalists can either eliminate ambiguity in the experiments, or are gifted with infallibility in interpreting the charges. It is true that by studying the behaviour of one composition under different conditions much ambiguity can be eliminated, especially in a simple system, but in the end the starting material, the experimental conditions, and the interpretation of the charge depend on the human factor. Error or ambiguity can arise at any of these stages, which are discussed below. Even if these are all successfully eliminated, however, there is still the possibility of serious error or ambiguity in the construction of the phase diagram and its ultimate interpretation. These latter communication difficulties will be considered separately in succeeding sections on the structure of various types of phase diagrams.

(a) LIMITATIONS INHERENT IN THE STARTING MATERIAL

Some of the limitations imposed by different starting materials were referred to in the previous section and simply need amplification here.

(1) *Glass*

Alkali-bearing glasses always have the possibility that they have lost alkalis during high temperature preparation and if the subsequent experiments are also at high temperatures there is always the danger of further loss during the experiment. Some glass compositions are also sluggish in reaction. In general, circumspection is needed in any liquidus experiments (under which conditions all starting materials must go into the liquid state) because there is the danger that over-heating may invest the liquid with the ability to supercool due to nucleation failure (Gibb, 1974).

(2) *Gel*

With gel starting materials there is always the question of homogeneity and the constant possibility of rapid crystallisation of persistent metastable phases.

(3) *Rocks and minerals*

There is the ever present question of homogenisation, and in the case of coarsely crystalline samples there is a general sampling problem (familiar to anyone concerned with rock analyses) with the additional question for the experimenter, what is now missing from the original system?

Additional uncertainty may be brought into the experiment where additives of any sort are added to the starting material. This may be seemingly innocuous or even irrelevant, e.g. the addition of water in hydrothermal experiments on rock compositions, but a moment's thought will indicate the complications that this might introduce in the final results and their interpretation.

In some experiments substances have been added with the intention that they should function as catalysts or as fluxes, but it is usually extremely difficult, if not impossible, to define the reaction steps that these additives produce and we should not lose sight of this when trying to apply the results. J. F. Schairer summed up this somewhat culinary approach to phase equilibria in one of his better known aphorisms "if you start with horse manure, you end with horse manure".

(b) LIMITATIONS INHERENT IN THE EXPERIMENTAL METHOD

It is reasonable to assume that the experimenter will have done his utmost to ensure that the physical conditions of the experiment have been accurately measured and controlled, and that he will be aware of any limitations imposed by the configuration of the apparatus, e.g. temperature and pressure gradients.

Probably the most serious limitations involve the duration of the experiment and quenching. The good experimenter will have taken considerable pains to establish the optimum times for the achievement of equilibria, but even with the best intentions there is always the possibility that the phase structure of the starting material leads to a metastable equilibrium which gives the illusion of stability. Liquidus determinations using samples which have been over-heated, by whatever means, are a case in point. Perhaps the classic example of metastable persistence was found in the system MgO–SiO_2–H_2O (Bowen and Tuttle, 1949). These experimenters tried without success to produce anthophyllite in the appropriate composition range and concluded that it had no stability field in the conditions of the experiments. Other experimenters working with related compositions reached similar conclusions and it required a determined and painstakingly rigorous study by Greenwood (1963) to demonstrate that the stable formation of anthophyllite had been effectively masked by the metastable formation of talc, which persisted for long periods in the anthophyllite field.

The normal experimental method is totally dependent on the ability to quench the equilibrium at high temperature and pressure, so that this is preserved for examination at room temperature and pressure. In the vast majority of silicate systems this is readily possible due to the sluggishness of reaction, especially at low temperatures, so that once the charge has been plunged into the low temperature regime further reaction is impossible. In some systems, especially carbonate systems, re-equilibration during the quenching period may be extensive and real problems of interpretation arise because the resulting charge is a mixture of the high temperature equilibrium phases and those formed during the quench. With care, and the skilful use of the petrological microscope, it is possible to circumvent this problem. An example of how this may be done is described by Wyllie and Tuttle (1960) in their study of the system CaO–CO_2–H_2O, in which they were able to identify the presence of liquid by careful study of its crystallisation products, which formed during the quench.

In some systems it may not be possible to quench the high temperature-high pressure equilibrium at all. Such a case is described by Bell and England (1964) who found that from above a certain temperature it was impossible to quench aragonite, which always inverted to calcite during the quench. They were able to establish this by continuous differential thermal analysis which fixed the temperature of the calcite-aragonite boundary. This result would be undetectable by the standard quenching methods of experimental petrology and must stand as a salutary warning to all of us against complacency in interpreting run products.

(c) LIMITATIONS IN INTERPRETING THE PHASE RELATIONS

The methods used in examining the charge were outlined in the previous section but some mention must be made of the limitations in the available methods. When most petrologists see an experimental charge for the first time they are usually moved to wonder at the fineness of the grain size, and they are concerned about the identification of the phases present. In fact, the experience of the experimenter and the composition constraints imposed by the system make this a minor concern, provided the optical and X-ray examinations are carefully applied. The two forms of examination are especially powerful in combination but there are dangers if X-ray examinations or optical examinations are used exclusively. In densely crystalline charges, optical examination alone is always liable to miss some of the phases which may be present. But the fact that the X-ray examination appears to "see" all the phases should not seduce the experimenter away from optical methods, for the following reasons:

(i) Even in favourable cases it may be necessary for a phase to be present in amounts greater than 5% (in some cases 10%) before it is detectable by X-ray powder diffraction. When liquid is abundant it is possible to detect crystals optically at much lower concentrations, commonly below the 1% level.
(ii) X-rays cannot detect quench phases, corrosion, reaction rims, armouring of unreacted crystals, zoned crystals and uneven distribution of phases.

All these aspects can be resolved by optical examination but of course in densely crystalline charges this becomes increasingly difficult, and this is why the near-solidus and sub-solidus phase relations in some systems have been subject to different interpretations by different workers, and even later re-interpretation by the same workers.

Many modern studies are made in hydrothermal systems using cold seal pressure vessels and mention should therefore be made of the special problems that may arise with a free vapour (fluid) phase. Under these conditions total identification of all the phases may be difficult, if not impossible. There is little reason to suppose that the vapour will be congruent with any coexisting liquid and it is rarely possible to know the composition of the vapour, and consequently the total composition of the other phases present. These difficulties will be especially pronounced in systems containing alkalis or mixed gases, and the petrologist should treat the published results (which usually refer the compositions of the quenched liquid and solid phases to the anhydrous bulk composition of the starting material) with corresponding circumspection. It goes without saying that we should be especially wary of experiments that have become so difficult that the investigator acknowledges that all the phases could not be completely identified. It is worth getting into the habit, when examining published results, of checking that the number of phases present is permitted by the phase rule.

After reading the foregoing paragraphs, the reader may begin to feel that all experimental results should be viewed with suspicion, but this is merely to go to the opposite extreme from accepting all results without question. The moral to be drawn from the present section is that experimental results should be viewed realistically and with some knowledge of how they were obtained.

6. UNDERLYING PRINCIPLES AND METHODS OF CONSTRUCTING A PHASE DIAGRAM

As was stated at the outset, the observation that mineral assemblages are repeatedly produced in similar geologic circumstances suggests that they represent equilibrium or near-equilibrium conditions. This basic assumption allows us to apply Willard Gibbs' phase rule for heterogeneous equilibria, which states that a chemical equilibrium must conform to the following numerical relationship:

$$F = C + 2 - P$$

where F is the *variance* (number of degrees of freedom), C the number of independent chemical *components*, and P the number of *phases* in the *system*. This apparently simple equation embodies a number of concepts which many petrologists find difficult. The *system* is the entity of material under consideration, which can be regarded in isolation from the rest of the Universe: it is normally defined in terms of the minimum number of chemical components. The *components* are the formulae required to write the chemical composition of every phase that may appear in the range of conditions under consideration. The number of *phases* refers to the number of forms of matter in the system which may be mechanically separated from each other, e.g. the different minerals, liquids and gases that may coexist in the system. It may be seen therefore that the system, its phases and its chemical components are interdependent. Probably the concept that is most difficult for the non-chemist is that of *variance*, or *degrees of freedom*. It refers to the number of ways in which the variables in the system might be changed without changing the phase assemblage of any given equilibrium. The Gibbs' phase rule refers specifically to a closed system, i.e. one that may be considered in chemical isolation from the rest of the material Universe, and in this condition there are normally only two wholly independent physical variables, pressure and temperature, hence the "2" in the equation. Strictly speaking, other physical variables might also be considered, e.g. gravitational, magnetic and electrical fields, but for most equilibria these may be neglected. An equilibrium state in a chemical system may vary if either temperature or pressure is changed. Each separate chemical component represents an additional factor by which the equilibrium may vary. Thus the limiting number of variables to be considered at any equilibrium state of the system is $C + 2$. But each phase that is present at equilibrium places a constraint on the behaviour of the chemical components, so that each phase reduces the variability of the equilibrium by one degree of freedom. Hence at any equilibrium involving P phases, the number of ways in which the system may vary (without altering the number of phases) is given by the total number of variables ($C + 2$) minus the number of phases (P). When the number of phases equals the number of variables it is impossible for the chemical behaviour of the components, or the temperature, or the pressure, to change until a phase is eliminated—this is an invariant equilibrium ($F = 0$).

Pressure, temperature, and the chemical components are all factors that are independent of the size of the system, and are known as *intensive* variables. The phase rule applies regardless of the mass, volume, and heat content (which are *extensive*, or size-dependent properties). Consequently, the phases participating in an equilibrium may be present in any proportions.

Having made these simple definitions, the best way to understand the use of the phase rule is to consider some simple phase diagrams, their construction and their significance. Phase diagrams may be of many kinds and should be constructed to best illustrate the conditions in which the petrologist is interested. They can be considered in two major groups, (a) those in which the changes in the system are shown in terms of the independent physical variables, commonly PT diagrams, and (b) those in which compositional variations are plotted in terms of changing temperature or changing pressure. Figure 7 shows the PT diagram for the familiar substance water. In the range of conditions shown, water behaves as a single component, i.e. all the phases which appear can be described in terms of the one component H_2O. This would not be true however at very high temperatures when water dissociates, and in that range of conditions the system would have to be described in terms of two components, hydrogen and oxygen. Even in this simple system therefore the range of conditions must be carefully specified for which the phase diagram is relevant. In Fig. 7, PT space is divided into regions in which the various forms of water are stable. The stability field of liquid water is the range of temperature and pressure in which liquid is the *only* stable phase structure. Taking liquid water at a point (A) within this field it is possible to change *either* the pressure (to B) *or* the temperature (to D) without introducing another phase. At any point within the liquid field, therefore, the conditions are that the system consists of one component which exists as a single phase, and therefore the phase rule may be written as $F = 1 + 2 - 1 = 2$. The equilibrium is said to be divariant, i.e. pressure or temperature may be varied independently without changing the equilibrium phase assemblage, namely, liquid water. In any PT diagram the fields must have two degrees of freedom or more.

If in Fig. 7 the pressure or the temperature were changed until a boundary of the liquid field is encountered (e.g. C or E), a new phase assemblage must form if equilibrium is maintained. At E ice crystals must coexist in equilibrium with liquid and the phase rule may be written as $F = 1 + 2 - 2 = 1$. The equilibrium is univariant, i.e. if the pressure is changed the temperature must also change if the two phases are to coexist in equilibrium. If either pressure or temperature were to be changed independently, one of the phases (either ice or liquid) must be eliminated and the equilibrium becomes divariant once more. Thus in any PT diagram a line between two fields indicates a univariant equilibrium in which the number of phases exceeds the number of components by one ($P = C + 1$).

If, from point E in Fig. 7, the pressure and temperature were changed, maintaining equilibrium between ice and water, in each direction the line is found to terminate at a junction with two other univariant boundary curves (points H and G). These intersections indicate pressure and temperature conditions at which three phases coexist in equilibrium. The phase rule can therefore

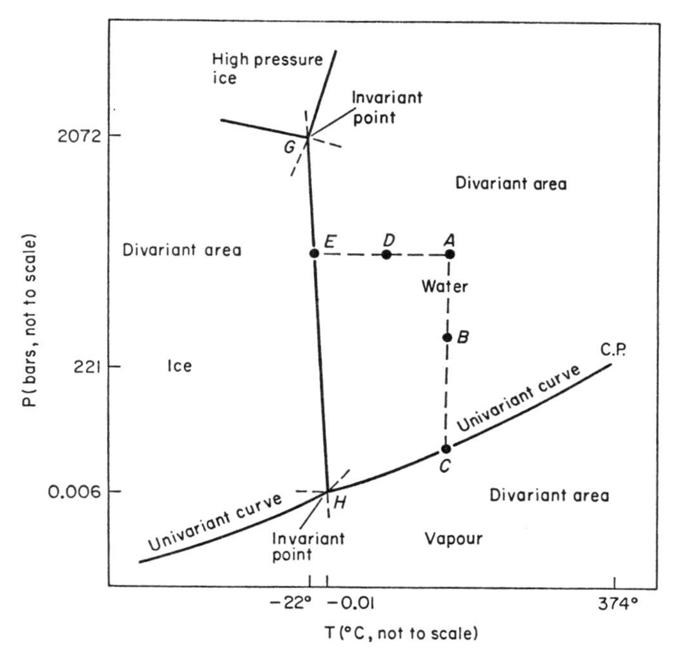

Fig. 7. Partial phase diagram for water. Four phases are shown. In the areas, where only one phase is present both pressure and temperature may be varied independently without changing the number of phases present. Where two phases are in equilibrium (along a line), neither pressure nor temperature may be varied *independently* without changing the number of phases. At the triple points, where three phases coexist, neither the pressure nor the temperature may be changed until one of the phases disappears. Metastable extensions of univariant lines are dashed. (Adapted from Frye, 1974.)

be written as $F = 1 + 2 - 3 = 0$ for such an equilibrium, which is said to be invariant. In the case of invariant point H, ice, liquid and water vapour are in equilibrium and if equilibrium is maintained the pressure and temperature of the system cannot change. Any attempt to lower the pressure (at constant temperature) at point H will result in the gradual reduction of ice and water and an increase in the amount of vapour until ice and water are eliminated, leaving only vapour, when the system is no longer invariant. Similarly, if heat is added to the system at point H there will be an increase in vapour, ultimately leading to the elimination of ice and water. At the instant when one or more of the three phases disappears, the system is no longer invariant and the addition of further heat will produce an increase in temperature. The requirement for an invariant equilibrium in any closed system is exemplified in point H and may be stated in general terms as, the number of phases exceeds the number of components by two ($P = C + 2$).

It may be seen from Fig. 7 that at any given pressure, say 100 bars, there is only one temperature at which ice may coexist in equilibrium with water. Expressed in another way, this says that if the pressure is made constant a degree of freedom has been removed from the system and the temperature at which ice and water coexist in equilibrium becomes invariant. Under isobaric conditions therefore the phase rule is effectively reduced to $F = C + 1 - P$, because the freedom for the pressure to change has been removed. The same is equally true if temperature is held constant and pressure is retained as the only wholly independent physical variable. The principle is widely applied in temperature-composition (TX) and pressure-composition (PX) phase diagrams, of which probably the most familiar are isobaric TX diagrams such as that shown in Fig. 8.

(a) CONSTRUCTION OF A BINARY TX DIAGRAM

Figure 8 shows the isobaric TX diagram for the two component (or binary) system albite–silica (Schairer and Bowen, 1956). It is useful in that it shows the compositions studied in the experiments (represented by circles) and can be conveniently used to describe, firstly, the mode of construction, and secondly, the underlying principles in a binary TX diagram. Each composition was made up from pure chemicals fused to a homogeneous glass and partly crystallised, as previously described. Small portions of the partly crystallised material were then held at a series of temperatures in order to bracket the temperature at which crystals first exist in stable equilibrium with liquid. For instance, in composition K the run data given by Schairer and Bowen (1956) show that at 1095°C after a ten day experiment the quenched sample consisted only of glass, but at 1090°C (after fourteen days) the quenched sample consisted of rare crystals of albite in glass. The beginning of crystallisation of albite in this composition is therefore bracketed between 1090°C and 1095°C. Determination of the bracket for all compositions defines the boundary of initial crystallisation in albite–silica at 1 atm pressure. This boundary is described as the *liquidus*. The condition at this boundary is that two phases coexist in equilibrium and the modified phase rule reduces to $F = 2 + 1 - 2 = 1$, i.e. the condition is isobaric univariant. Further lowering of the temperature in composition K must be accompanied by increasing crystallisation of albite, which in turn requires that the remaining liquid is depleted in this component and hence relatively enriched in silica. With cooling and crystallisation, therefore, the remaining liquid changes from K to I.

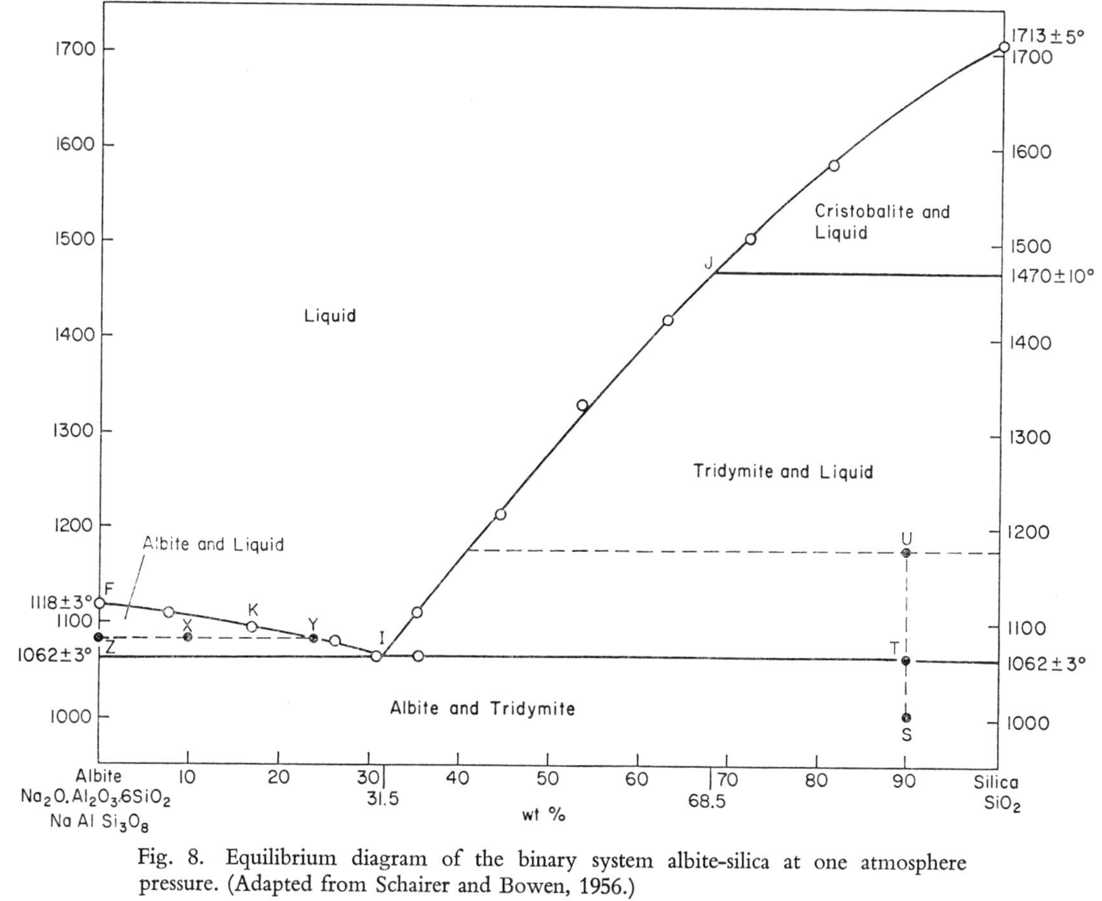

Fig. 8. Equilibrium diagram of the binary system albite-silica at one atmosphere pressure. (Adapted from Schairer and Bowen, 1956.)

The composition of the liquid coexisting with albite crystals at any given temperature is given by the horizontal temperature intercept on the albite liquidus boundary. When the temperature falls to 1062°C any further loss of heat causes the simultaneous precipitation of tridymite and albite. At this temperature the liquid composition is I, and it is in equilibrium with two crystalline phases so that the modified phase rule reduces to $F = 2 + 1 - 3 = 0$ and the equilibrium is isobaric invariant. The temperature at I (1062°C) cannot change until one of the phases is eliminated, and with continued loss of heat the liquid will ultimately be used up to form tridymite plus albite. At that instant the number of phases is reduced to two and further loss of heat will be accompanied by a fall in temperature. It is important to note that in a binary diagram any field labelled as containing two phases describes the stable phase assemblage for that range of temperature and composition, the compositions of the two phases being determined by the extremities of the field at any given temperature. For instance, the point X in Fig. 8 specifies a particular bulk composition i.e. 90% albite and 10% silica at a temperature of 1080°C. The diagram shows that at 1080°C the stable phase assemblage for this composition is albite crystals in equilibrium with a liquid of composition Y. The proportions of liquid and crystals in composition X at 1080°C can be determined by the "lever rule", by which the proportions of albite crystals to liquid are given by the ratio of XY:XZ. This rule may be applied to any point within a divariant field in a binary diagram.

An invariant point such as I is a *eutectic*. The boundary at which a system becomes completely solidified is called the *solidus*.

It has long been traditional when considering phase diagrams involving liquids to think of them in terms of progressive crystallisation with falling temperature, a process which the metamorphic petrologist might term retrograde! In phase diagrams which do not involve liquids, on the other hand, the metamorphic petrologist traditionally thinks in terms of changes produced by increasing temperature (and pressure), i.e. prograde reactions. It can be very enlightening for the igneous petrologist to adopt prograde thinking and consider a phase diagram such as Fig. 8 in terms of increasing temperature. The point S in Fig. 8 describes the composition containing 10% albite and 90% silica at 1000°C, consisting of albite and tridymite crystals in the proportions 10:90. Addition of heat to this assemblage produces an increase in temperature until, at 1062°C, the mixture of crystals begins to melt. The liquid that forms has the composition I, and further addition of heat increases the quantity of liquid of the same composition, without further increase in temperature, until all the albite crystals are consumed. This condition is represented by the point T and further increase in heat produces a rise in temperature through a series of conditions represented by T—U. Any point such as U describes the condition and phase assemblage in that particular bulk composition as outlined in the previous paragraphs. Two important features are highlighted by this prograde consideration of the phase diagram. Firstly, it shows the effect of progressive melting and emphasises the fact that *all mixtures* of albite and tridymite crystals must begin to melt at the same temperature and produce identical liquids until one of the solid phases is consumed. This is the type of process that may be expected to occur in what is essentially a solid Earth. It is much more likely, for instance, than the production within the solid Earth of a liquid, say of composition 10% albite and 90% silica, which would then cool through the stages normally outlined in retrograde discussions of crystallisation. It must be more than a coincidence that in the simple system albite–silica the eutectic liquid at I is analogous to rhyolite and granite composition. Secondly, it is clear that in a heating cycle there will be a lag in the rise in temperature of the system while melting takes place at an invariant point, until one of the phases present is consumed. Thus melting at an invariant point acts as a thermal sink or buffer, delaying the melting of wider composition ranges. In the natural situation the time lag may allow the continuous escape of the invarant liquid from the system leaving behind a much more refractory solid residuum.

The same principles outlined in the above description apply to more complex phase diagrams and these are considered in more detail later. It is worth noting at this point that in a phase diagram which portrays variation of composition, a three component system is the maximum that can be represented in a planar diagram that permits rigorous geometric analysis. Four or more components can be represented in a planar diagram in such a way that the low variance (univariant and invariant) phase assemblages can be completely described but the component variations and phase proportions cannot be systematically analysed.

7. PRESENTATION AND INTERPRETATION OF RESULTS

In the previous section the construction and interpretation of a simple unary and a simple binary phase diagram were used to illustrate the underlying principles of phase equilibria studies. In the present section this will be extended to more complex systems to illustrate the applications of phase diagrams in the interpretation of crystalline rocks. Several features of the underlying philosophy of the discussion should be stated at the outset:

(1) There will be no attempt at an exhaustive coverage of every kind of phase diagram. The aim here is to consider the more widely used diagrams, taking, where possible, examples from the later Chapters. From these the interested reader should have little difficulty in understanding the less common types of diagram that are sometimes used in the literature.

(2) It is not intended to provide a comprehensive coverage of the systematics of phase diagram construction and analysis. Such a coverage is outside the scope of this book and is, in any case, adequately covered by texts on physical chemistry (Ricci, 1951) and, more specifically for geologists, by Ehlers (1972). The present discussion is designed to highlight the low

variance conditions in systems, which are the chief concern of the experimenter, and are the most relevant for the petrologist. Detailed discussion of systematics will therefore be confined to low variance conditions.

(3) Following from the previous section (6), concepts and terms will be defined or elaborated as they are introduced, and can be located through the index, but for an excellent additional source of definitions in common use in experimental petrology the reader may be referred to the Glossary included in "Phase Diagrams for Ceramicists" (Levin, E. M. *et al.*, 1964). The usage in this book is consistent with the definitions in the Glossary.

(4) Where possible, thermodynamic expressions and constructions will be avoided, because, although the concise mathematical treatment of classical thermodynamics combines rigour with elegance, it remains esoteric for many whose main field of study is outside chemical thermodynamics. Part of Bowen's great success in communicating his ideas to non-experimentalists came from his avoidance of specialist terminology. An attempt to follow Bowen's example is unlikely to meet opposition from most petrologists, but for any who might have doubts, it is worth pointing out that the *rigorous* application of classical thermodynamics requires far more knowledge of the state of a system than is normally obtainable. Frankly, that is usually why the empirical approach is necessary.

(5) The order of discussion will be: (i) Principal types of equilibria that involve liquids, and their depiction in phase diagrams, at constant pressure or constant temperature. (Temperature–composition (TX), or Pressure–composition (PX) diagrams). (ii) Equilibria involving only solids, or solids and gases (TX and PX diagrams). (iii) Equilibria with varying temperature *and* pressure (PT diagrams). (iv) Diagrams using other forms of variable: P_{H_2O}; P_{CO_2}; P_{O_2}.

(a) TX DIAGRAMS INVOLVING LIQUIDS

(1) *Congruently melting compounds*

In section 6 the system albite–silica was used to illustrate the construction of a simple binary isobaric phase diagram (Fig. 8). This represents a special case in which the two components form completely separate crystalline solids (cristobalite, tridymite and albite) with no solid solution, nor intermediate compounds. In truth, there is some slight solid solution (notably of alkalis in high temperature silica polymorphs) but its effect on the phase relations is not measurable. The case is also special, because each of the two end-compounds, pure albite or pure silica, melts to produce a liquid with the same composition as the solid. This is known as *congruent melting*, and, in the strict sense is found only in pure compounds, and even then not invariably. An important exception, for instance, is enstatite ($MgSiO_3$) which melts *incongruently* at low pressures, to produce forsterite crystals (Mg_2SiO_4) plus liquid of a more siliceous composition than the original enstatite. Incongruent melting will be examined more fully in a later discussion. Albite, which melts congruently at low pressure, is itself an intermediate compound between nepheline and silica, and the phase diagram for nepheline–silica is shown in Fig. 9. This is a binary system made up of two smaller binaries, nepheline–albite and albite–silica, each containing one isobaric invariant point, which in each case is a eutectic. Starting at an equilibrium state for any given bulk composition, on either side of the albite composition, it is

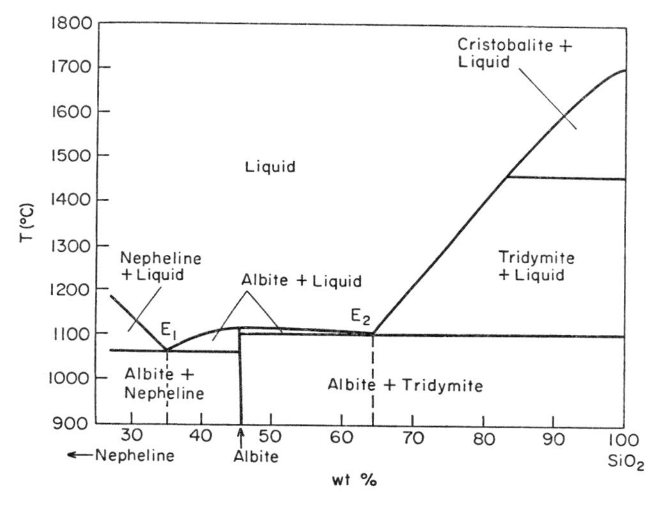

Fig. 9. Part of the system nepheline-silica at one atmosphere pressure. (After Schairer and Bowen, 1956.)

impossible by crystallisation or melting to attain a composition on the other side, i.e. it is not possible to generate an undersaturated product from oversaturated starting material, *nor vice versa.* The system is effectively divided into two separate smaller systems by the albite composition. It will be noticed that there is a temperature maximum, between the two eutectics, at the albite composition. Following Bowen, it has become traditional to describe such maxima as "thermal divides", to indicate that liquids on one side can never generate compositions on the other. But the reader should realise that neither temperature nor heat content have anything directly to do with the separation of the undersaturated and oversaturated regions in Fig. 9. The two regions are separated by a *composition barrier*, at the albite composition, which cannot be breached from either side by crystallisation or melting.

In Fig. 9 it may be seen that as a new component (either nepheline or silica) was added to a liquid of albite composition, the liquidus temperature of albite would be progressively lowered (to an eventual minimum represented by the respective eutectic). The addition of even a small amount of either component to albite crystals would reduce the melting point (the beginning of melting) of the crystal mixture to that of the respective eutectic. These principles are of general application, and can be extended to cover behaviour in more complex systems. For instance, the addition of a third component to albite–silica, provided that no solid solution were involved, would lower liquidus temperatures throughout. In particular, the additional component would lower the temperature (and change the liquid composition) at which tridymite and albite crystallise simultaneously. Expressed in terms of the phase rule (isobaric) the system now consists of 3 components, in the form of 3 phases (albite + tridymite + liquid) and the variance is given by,

$$F = 3 + 1 - 3 = 1 \text{ (univariant).}$$

The addition of the third component means that the temperature and liquid composition are no longer invariant: the phase relations must now be described in a ternary system. Ideally, it would be apt to continue this discussion by considering an actual case of adding an appropriate rock-forming component to albite–silica. Such a component should melt congruently and have no intermediate compounds, or solid solution with either albite or silica. But the plain fact is that there are no relevant ternary systems of such simplicity! As a basis for understanding more complex systems, however, a model system of this nature may be used.

Consider the case of three components, A, B and C, each of which forms a pure, congruently melting solid, and between which there are no intermediate compounds or solid solutions. Taken in pairs they will provide three binary systems, each with a single eutectic, similar to albite–silica. These are shown in Fig. 10(a). If the three components are taken together they form a ternary system (A–B–C) which is bounded by the three separate binaries (A–B, B–C, and A–C). The *liquidus* conditions in the ternary system may be represented in an equilateral triangle in which the apices represent the pure compositions, A, B and C (see Fig. 10(b)). All points within the triangle represent mixtures of A, B and C, in proportions represented by their perpendicular distances from the sides. Liquidus equilibria are described in the triangle by three composition fields, each of which includes all those liquids that first precipitate crystals of A, or B, or C. The fields are described as the primary phase areas for A, B and C, and they may be contoured for liquidus temperatures (as in Fig. 5). The variance in one of these areas in which two phases are in equilibrium (isobaric) is given by:

$$F = 3 + 1 - 2 = 2$$

i.e. such an equilibrium is *divariant*.

In either of the bounding binary systems A–B and B–C, the equilibrium between crystals of B and a liquid is univariant (isobaric), i.e. specifying the temperature of the equilibrium (t, in Fig. 10(a)) specifies the liquid composition;

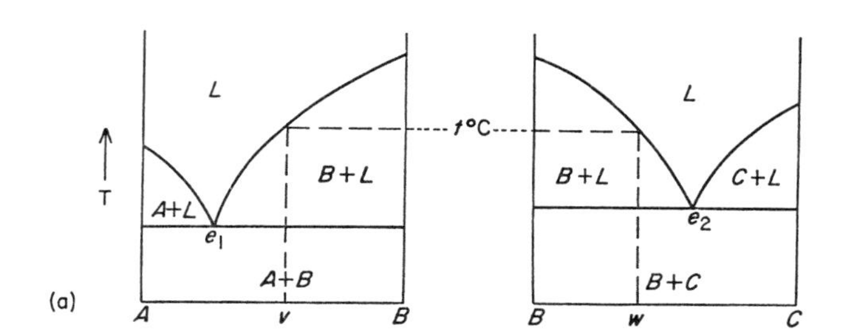

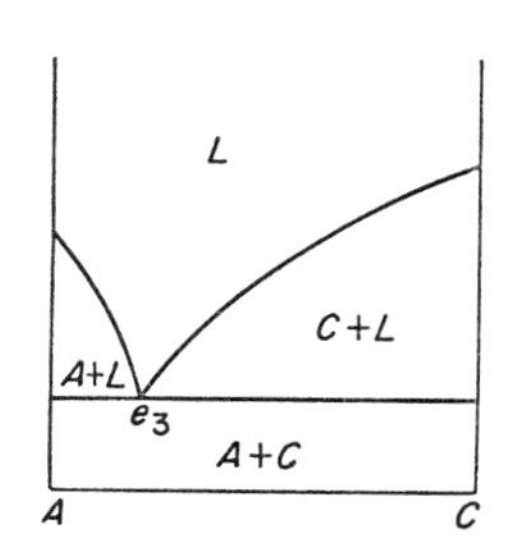

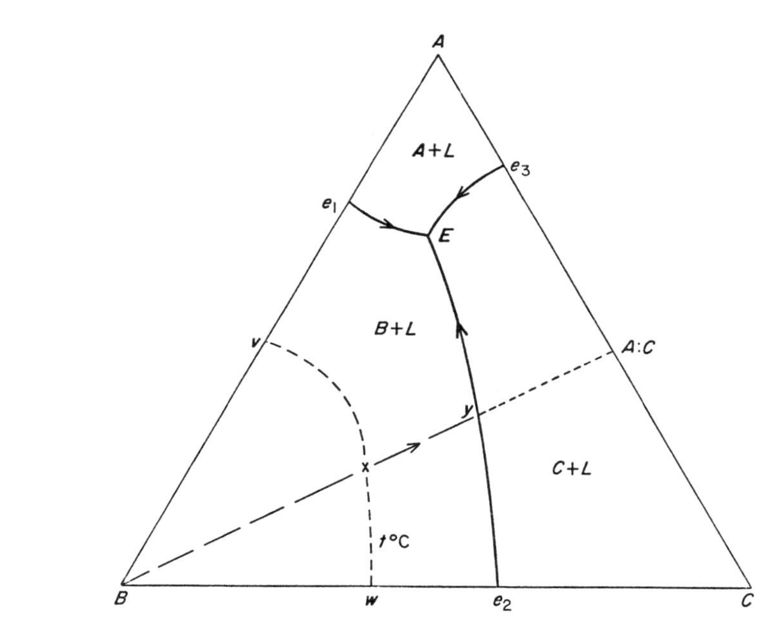

Fig. 10. (a) The three bounding binary systems, *A–B*, *B–C*, and *A–C*, of a model ternary *A–B–C*, showing one point on the liquidus of *B*, at temperature t, in *A–B* and *B–C*. In each binary system there is only one liquid composition (v and w, respectively) that coexists with crystals of *B* at temperature t. (b) Model ternary *A–B–C*, with no solid solutions or intermediate compounds. Composition x is a point on the ternary liquidus, at temperature t (for which a temperature contour is shown). Crystallisation of x produces a series of liquids $x \rightarrow y \rightarrow E$. Arrows indicate falling temperatures.

or, if the composition of the liquid is specified, there is only one possible equilibrium temperature (t). In the ternary system there will be a range of liquids that may coexist with crystals of B at temperature t, i.e. along the temperature contour (t) which runs between the two binaries. Specifying the temperature (t) is not enough, therefore, to completely define the *ternary* equilibrium, it is necessary to know the ratio of $A:C$ in the liquid to fix its position on the contour t. Alternatively, it is necessary to know the liquid composition completely to specify the temperature of the equilibrium. For a ternary liquid two pieces of information are needed to define its composition (e.g. $\%A$ and $\%B$). Either way, two variables must be specified completely to describe the ternary equilibrium, B + liquid, hence it is divariant. It is the addition of a third component (composition variable) to either of the binary systems A–B or B–C, which has added the extra degree of freedom for an equilibrium between crystals of B and a ternary liquid.

Where two primary phase regions meet, in the system A–B–C, e.g. $A + L$ and $B + L$ in Fig. 10(b), the boundary (e_1E) defines the range of liquids that may exist in equilibrium with crystals of A and B. This boundary also defines a range of equilibrium temperatures for the phase assemblage $A + B + L$, and may be marked in the phase diagram with temperature intervals. The variance of this three phase equilibrium (isobaric) is given by $F = 3 + 1 - 3 = 1$; i.e. the condition is univariant. Specifying an equilibrium temperature will fix the position on the boundary e_1E and hence the liquid composition; or, specifying one composition variable in the equilibrium liquid will give its position on the curve e_1E, and hence fixes the equilibrium temperature. It should be noted that the isobaric univariant curve e_1E in the ternary system A–B–C terminates at the isobaric invariant point e_1 in the binary system A–B. Addition of the third component C thus introduces an additional degree of freedom to all equilibria involving A and B.

Temperatures along e_1E fall to a minimum at E, the ternary eutectic, at which four phases, $A + B + C + L$, are in equilibrium, with a variance of $F = 3 + 1 - 4 = 0$. The point E is an isobaric ternary invariant point at which three univariant curves meet. The temperature and liquid composition of the equilibrium E are invariable. All liquids in the ternary system must follow a course of crystallisation which ends at E. A liquid of composition x in Fig. 10(b), will first begin to precipitate crystals of B (at temperature t°C). Further separation of B will change the liquid composition to y, where crystals of B and C separate together, driving the remaining liquid composition along the univariant curve towards E. At E, crystals of B, C and A will separate, with no further change of temperature or liquid composition, until all the liquid is consumed.

Again it is instructive to consider "prograde" reactions. The ternary phase diagram in Fig. 10(b), tells us that a mixture of crystals of A, B and C, in *any proportions*, will begin to melt at the temperature of E, and that the first liquid will have the composition E. Furthermore, the addition of more heat will continue to generate more liquid E (without changing the temperature!) until one of the crystalline phases (A, B or C) is used up.

If a fourth component, D, were to be added to the system it will add an extra degree of freedom to all the equilibria involving A, B and C. The effects are analogous to those caused by adding a third component C to the binary system A–B, but an extra dimension is required to depict the relationships. The usual method is to depict the system as a regular tetrahedron, in which A, B, C and D form the apices. For every point within a regular tetrahedron the sum of the perpendiculars to the faces equals the height, hence the geometric location of the point is a simple expression of its composition in terms of the proportions of A, B, C and D. The method is essentially similar to the use of the equilateral triangle to depict compositional variation in a ternary system. With the aid of Fig. 11, it may be seen that within a regular tetrahedron all the heterogeneous equilibria involving a liquid phase may be depicted as follows:

(1) The primary phase regions are interlocking volumes, each of which encloses all those compositions that have one common solid phase at the liquidus.
(2) Any two adjacent primary phase volumes are separated by a curved divariant surface which defines those liquid compositions that may coexist with two solid phases.
(3) The contact (common edge) between any three primary phase volumes is a curved univariant line defining liquids that may coexist with the three solid phases.
(4) Four volumes can be in mutual contact only at a point, which is the invariant composition of a liquid that may coexist with the four solid phases.

It may be seen from Fig. 11, that the addition of component D to the ternary isobaric invariant point, *abc*, is to make the four phase condition $A + B + C + L$ univariant, since

$$F = 4 + 1 - 4 = 1.$$

A range of quaternary liquids may thus coexist with crystals of $A + B + C$. With increasing contents of component D the temperature of the equilibrium is progressively lowered, until the condition is reached at which crystals of D also separate. This five phase assemblage is a quaternary eutectic (isobaric), $A + B + C + D + L$, of invariant temperature and liquid composition. From this quaternary invariant point, arise four quaternary univariant curves, which terminate in ternary invariant points in the faces of the tetrahedron.

All liquids in the system A—B—C—D will cease crystallising at the quaternary eutectic. All mixtures of the four phases $A + B + C + D$, in *any proportions*, will begin to melt at the temperature of the quaternary eutectic, yielding a liquid of eutectic composition. The amount of liquid will increase with further heating, but it will remain constant in temperature and composition until one of the four crystalline phases disappears.

Addition of a fifth component to the system requires yet another dimension for complete description, and it is no longer possible to do this graphically. It is still possible, however, to display some of the important relationships of low

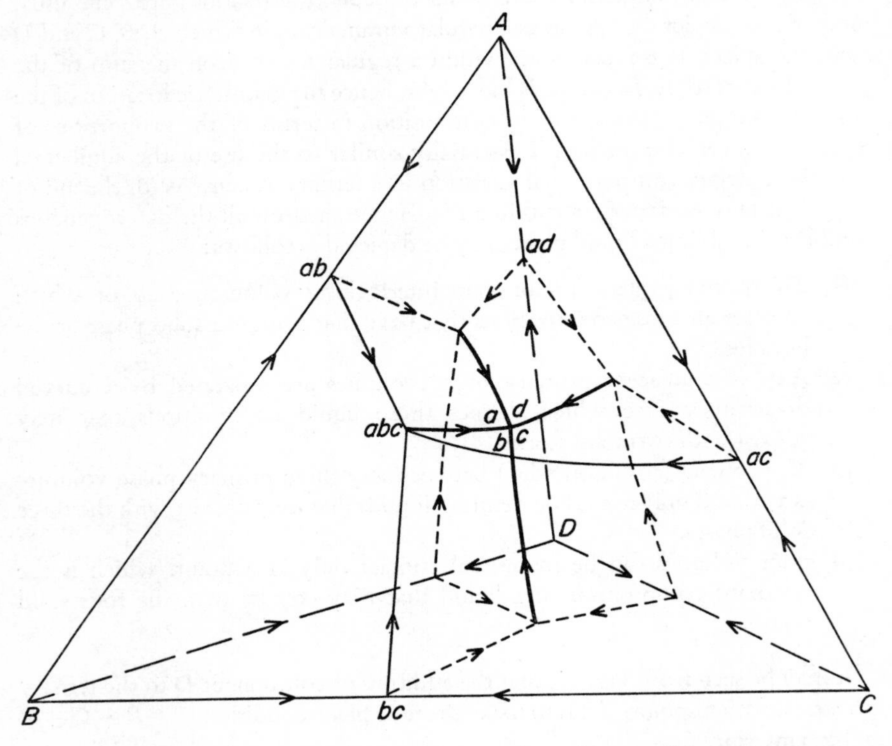

Fig. 11. Simple schematic, isobaric quaternary system, with no solid solutions or intermediate compounds between the pure-component end-members. Quaternary univariant curves shown by heavy lines: arrows indicate falling temperatures. Four primary phase volumes occupy the apices of the tetrahedron and are in contact along the quaternary univariant curves and at the eutectic, *abcd*.

variance equilibria in systems with high numbers of components, by using the phase rule, and extending the principles derived from systems of two, three, and four components.

In any isobaric system of C components the phase rule may be re-cast as:

$$P = C + 1 - F.$$

In an invariant equilibrium, $F = 0$, so that

$$P = C + 1.$$

If one of the phases is liquid, then in an invariant equilibrium the number of *solid phases* equals the number of components. At such an invariant condition the liquid phase is fixed, and so is the temperature. The condition may be represented by a point. The equilibrium can vary only if one of the phases disappears, when a new degree of freedom appears. The condition at such an invariant point can be most simply described by reference to Fig. 12, which is a diagrammatic

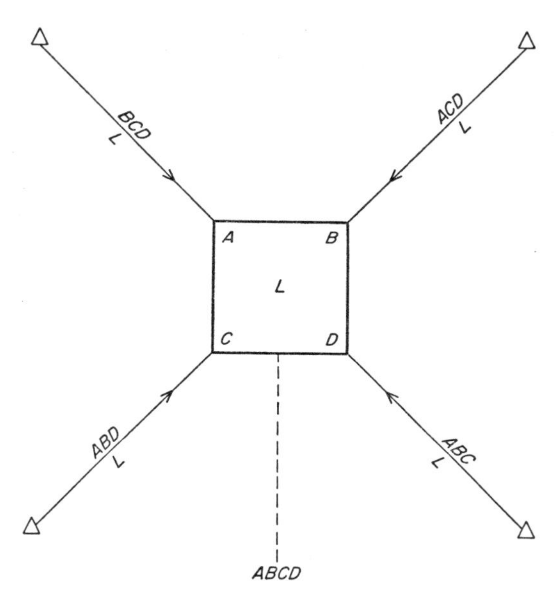

Fig. 12. Basis of construction of an isobaric quaternary "flow-diagram" showing four univariant curves (3 solid phases + liquid (L)) originating in four ternary eutectics and meeting in the quaternary eutectic (A, B, C, D, L). The broken line ABCD indicates the subsolidus equilibrium (isobaric univariant), which is simply the stable coexistence of the four solid phases.

representation of the quaternary invariant point (eutectic) in Fig. 11. At the isobaric invariant equilibrium four solid phases (equals the number of components) coexist with liquid. Temperature and liquid composition cannot change until one of the four solid phases disappears. When this occurs the equilibrium will become univariant, and temperature and liquid composition will vary dependently along one of four univariant paths, which may be represented by four curves originating in the invariant point. The four paths describing equilibria involving a liquid phase are truly isobaric univariant. But invariance will also disappear when the liquid at the invariant point is used up (during cooling), leaving only the four solid phases. This subsolidus equilibrium is different, however, because the four phases are no longer in continuous mutual interaction. In the idealised example chosen for Figs 11 and 12 the four solid phases have no intermediate compounds, nor solid solution, so that immediately liquid disappears they are effectively isolated from each other. The same is largely true of many natural examples, in which the most common isobaric changes below the solidus are polymorphic inversions and exsolutions, within single phases, or reactions between only a few of the solids. Such changes are not dependent on the nature and number of the other crystalline phases present. Consideration of subsolidus equilibria takes us into the realm of metamorphic petrology, the distinctive aspects of which will be considered when our discussion of liquid equilibria is complete.

The principles described above can be applied to any number of components in a system in equilibrium. General conclusions about system behaviour will also apply. For instance, when a solid composed of X crystalline phases begins to melt, the first liquid will form at the temperature of the lowest invariant point in the system. As heat is added, the amount of liquid increases but its composition and temperature remain fixed, until one of the solid phases is used up, whereupon the system becomes univariant (with $X - 1$ solid phases able to coexist with a range of liquid compositions over a range of temperature).

It is the principle embodied in Fig. 12, of an invariant point (of X solid phases + liquid) being the origin of X univariant curves (each with $X - 1$ solids + liquid) which forms the basis of "flow-diagrams", used to describe the crystal $\rightleftharpoons$ liquid relations in an X-component system. Most commonly, "flow-diagrams" have been used to describe quaternary equilibria (e.g. Bailey and Schairer, 1966) but they may be applied to quinary and higher systems, although their construction and interpretation become correspondingly more complex. This is especially true when modes of melting other than the simple congruent type we have so far considered are taken into account. These are described in the section that follows.

(2) *Non-congruently melting compounds*

In the preceding section we considered equilibria involving compounds which melt congruently, i.e. each pure compound melts to a liquid of the same composition. But some important silicates exhibit the phenomenon of melting to form another crystalline phase and a liquid, e.g. enstatite melts yielding crystals of olivine and a liquid more siliceous than enstatite. In this case two solids and a liquid are in equilibrium when enstatite melts. The two solids are, however, of fixed composition. Another mode of melting is found when a solid solution is heated until it melts—it produces a liquid of different composition, and with further heating (and melting) the liquid and solid change composition simultaneously. Because both modes of melting yield liquids of composition different from the starting solid they cannot be described in terms of a single component (the starting solid) as was the case for congruent melting, i.e. non-congruent melting of a single solid phase requires description in a binary system. It is, of course, these two modes of crystal-liquid interaction which Bowen built in to his famous Discontinuous and Continuous Reaction Series.

Before considering other forms of isobaric melting in detail, it would be useful to list the general features of the three main types.

(a) Melting involving only congruent phases. Composition of solids does not vary during melting. The liquid produced always lies within the composition range defined by the solids.

(b) Melting involving at least one incongruent phase. Composition of individual solids does not vary during melting. Liquid lies outside the composition region defined by the solids, if these include the extra phase that forms during incongruent melting (e.g. olivine in the case of enstatite melting).

(c) Melting involving at least one solid solution. Composition of the solid solution changes during melting. Liquid composition always lies within the composition range defined by the solids (providing this is taken to include the possible range of solid solution).

Statements about the liquid in (b) and (c) require further explanation, and this is best done by reference to actual examples.

(i) *Melting of an incongruent compound.* At low pressures, the two most important groups of rock-forming minerals, feldspars and pyroxenes, contain examples of incongruent melting, namely, orthoclase and enstatite. The sodium pyroxene, acmite, also melts incongruently, and, what is more, this behaviour persists beyond 40 Kbar. Historically, enstatite has been treated as the most important case because its incongruent behaviour can profoundly modify the phase relations in the most abundant magma-type—basalt. The melting of enstatite at 1 atm can be described completely only in terms of the binary, forsterite–silica, shown in Fig. 13. From this it can be seen that enstatite begins to melt at 1557°C, forming crystals of forsterite and liquid of composition R. At R, there are three phases coexisting in a two component system, so the condensed phase rule tells us that

$$F = C - P + 1 = 0.$$

Thus, point R is invariant, and further addition of heat will increase the amounts of liquid *and* forsterite at the expense of enstatite, at constant temperature and liquid composition. When enstatite is used up the system becomes univariant, and further addition of heat causes the forsterite to melt, increases the temperature, and changes the liquid composition along the univariant curve RQ. At Q all the forsterite is used up, and the liquid reaches enstatite composition.

The same initial course of melting will be followed by a mixture of enstatite and forsterite (P), but some forsterite will persist to higher temperatures, being finally used up as the liquid reaches the composition of the original mixture of enstatite and forsterite. Equilibrium cooling of this liquid composition follows the reverse sequence: continuous precipitation of olivine, until the liquid composition and temperature reach *R*; followed by reaction between olivine crystals and liquid, to produce enstatite, until the liquid is consumed. For this reason an invariant point of type *R* is generally referred to as a "Reaction Point", because it exhibits the phenomenon of one of the crystalline phases being consumed

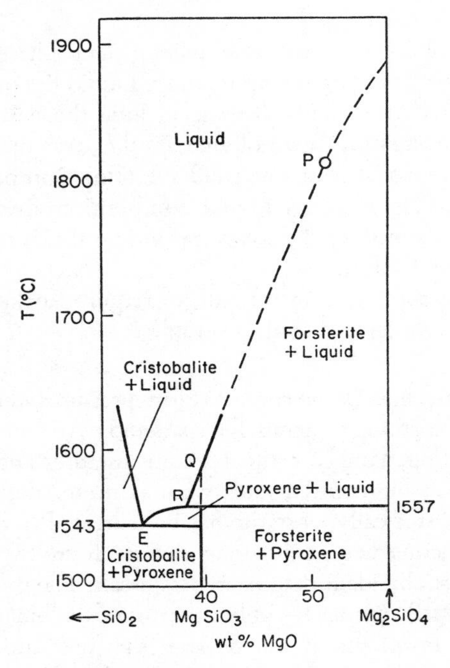

Fig. 13. The system forsterite-silica at 1 atm. (After Bowen and Anderson, 1914.)

during cooling. But it is worth remembering that *during heating*, olivine crystals will be newly formed in this system, indicating that olivine phenocrysts might form during a melting episode!

But surely the most remarkable feature of Fig. 13, is that *R* is the point of initial melting (or final crystallisation) of all mixtures of olivine and enstatite, and

---

Fig. 14. Incongruent melting. If phase *B* melts to form $A + $ liquid (*L*) then in: the binary (a) the invariant liquid for the melting of $A + B$ lies between *B* and *E*; the ternary (b) the invariant liquid for $A + B + C$ lies within the area *BCE*; the quaternary (c) the invariant liquid for $A + B + C + D$ lies within the volume *BCDE*. (No attempt has been made to show the quaternary univariant curves and invariant points in (c) in three dimensions. The reader should attempt schematic ternary diagrams for *ACD*, *BCD*, *ECD* and *ADE*, and construct a "flow diagram", assuming that there are no intermediate compounds or solid solutions.)

(a)

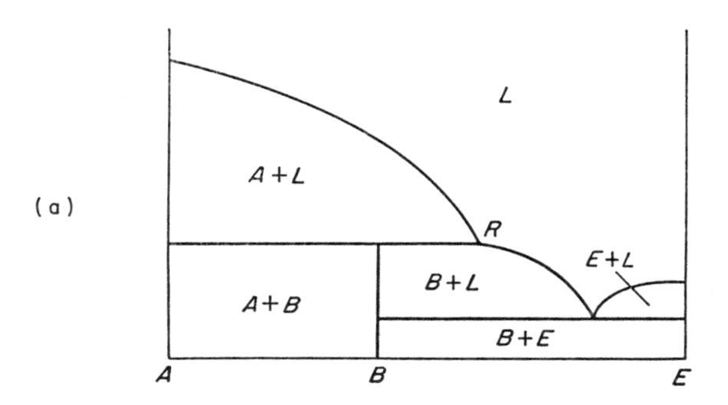

(b)

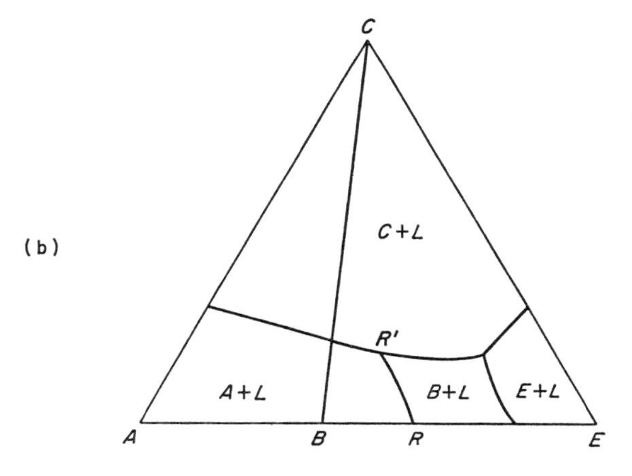

(c)

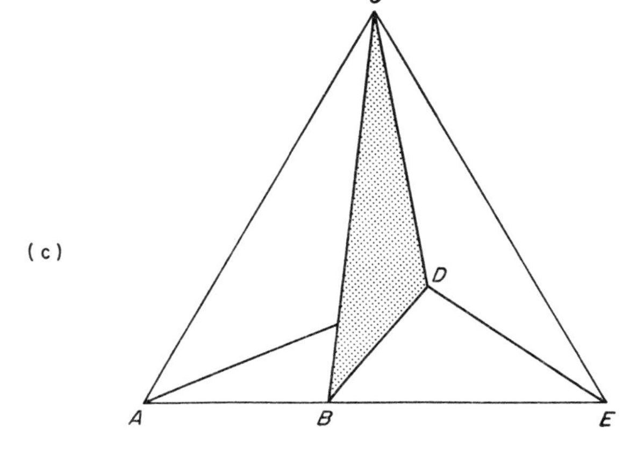

yet the invariant liquid lies outside the composition range defined by the crystals with which it is in equilibrium. This relationship is common to all reaction points, regardless of the number of components in the system. The incongruence of the liquid in reaction point melting, or crystallisation, is completely distinctive. It opens up the possibility of producing liquids that are disparate from their original source material if there is any failure of equilibrium. For instance, if the liquid at *R* is separated from the crystals, it becomes effectively a new starting composition between enstatite and silica, and must ultimately solidify at *E*. In this way, an initial composition undersaturated in silica can, *by equilibrium failure*, yield an oversaturated product. The principle extends to systems of greater complexity, and for a thorough discussion of the applications to basaltic compositions the reader may turn to Yoder and Tilley (1962). It is this principle that forms the basis of Bowen's Discontinuous Reaction Series.

It will be seen from Fig. 13, that to provide a full description of the invariant point for olivine + enstatite, the system must be extended to silica. With the addition of extra components this principle still applies, such that the invariant liquid in equilibrium with any assemblage including olivine and enstatite lies in the system defined by the solid phases at the invariant point *plus* silica. This is exemplified for 2, 3 and 4 components in Fig. 14.

(ii) *Melting of a solid solution*. Most rock forming minerals are solid solutions, indeed, quartz and the aluminium silicates are the only common exceptions. Although a solid solution is a single phase, its composition can only be fully stated in terms of the end members (the compositional variables) which must be two or more. A solid solution is therefore binary or ternary, etc., depending on how many components are needed to describe it, just as in the case of a liquid solution. It follows immediately that the description of the melting of a solid solution will require at least a binary phase diagram. True binary solid solutions are rarely formed in natural mineral assemblages, but many can be considered binary as a first approximation. The plagioclase series was made classic by the early experiments of Bowen (1913) and is shown here as Fig. 15. This diagram is truly binary because it describes synthetic plagioclases (whereas natural plagioclases will contain small amounts of potash feldspar). The diagram shows a departure from previous cases because the composition of the solid phase varies with the composition of the liquid with which it is in equilibrium. In all diagrams up to now the solid phases have had fixed compositions, with only the liquid compositions changeable. The upper curve is the liquidus, describing the compositions of liquids that coexist with crystals, whose compositions are given by the lower curve, the solidus. Any stated temperature will be given by a horizontal line across the diagram, and where such a line intersects the liquidus and solidus it gives the compositions of liquid and crystals in equilibrium at that temperature.

If a composition *x* is heated until it begins to melt the first liquid is much more sodic (y), and with further equilibrium melting there will be continuous reaction between crystals and liquid, such that both crystals and liquid become more calcic (along the liquidus and solidus curves with increasing temperature). Finally, as the liquid reaches composition x, the last crystals (z) are consumed. The reverse will apply if a liquid of composition x is cooled and undergoes equilibrium crystallisation. Either way, equilibrium requires continuous reaction between liquid and crystals, hence, Bowen's Continuous Reaction Series. But such continuous reaction may be well-nigh impossible in many natural situations, because it requires that the crystals are being *continuously re-homogenised*, which must become rapidly more difficult as they grow in size.

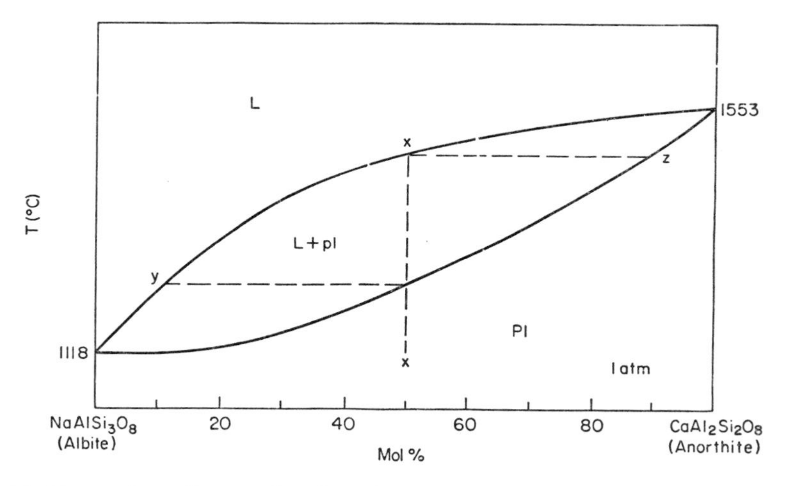

Fig. 15. Binary phase diagram for plagioclase at one atmosphere. (After Bowen, 1913.) See text.

Failure to maintain homogeneity is manifest in the zoning of natural plagioclases, and indeed, all igneous solid solutions. It should be noted that this stringent condition for continuous equilibrium (continuous re-homogenization of the crystals) does not apply where solid solutions are absent—but natural examples of this kind must be exceedingly rare.

One special feature of Fig. 15 must be emphasized. Although it is binary, there are no binary invariant points. In all the phase diagrams considered up to now there have been invariant points at which a wide range of compositions would cease crystallisation, or would begin to melt. In Fig. 15, each plagioclase composition has a different temperature at which it begins to melt, and a different course of equilibrium melting. If to this potential we add the condition that continuous equilibrium may be hard to sustain, the possible variability of behaviour, even in this simple binary, is remarkable. In more complex systems it offers immense scope for variability in the courses of melting or crystallisation.

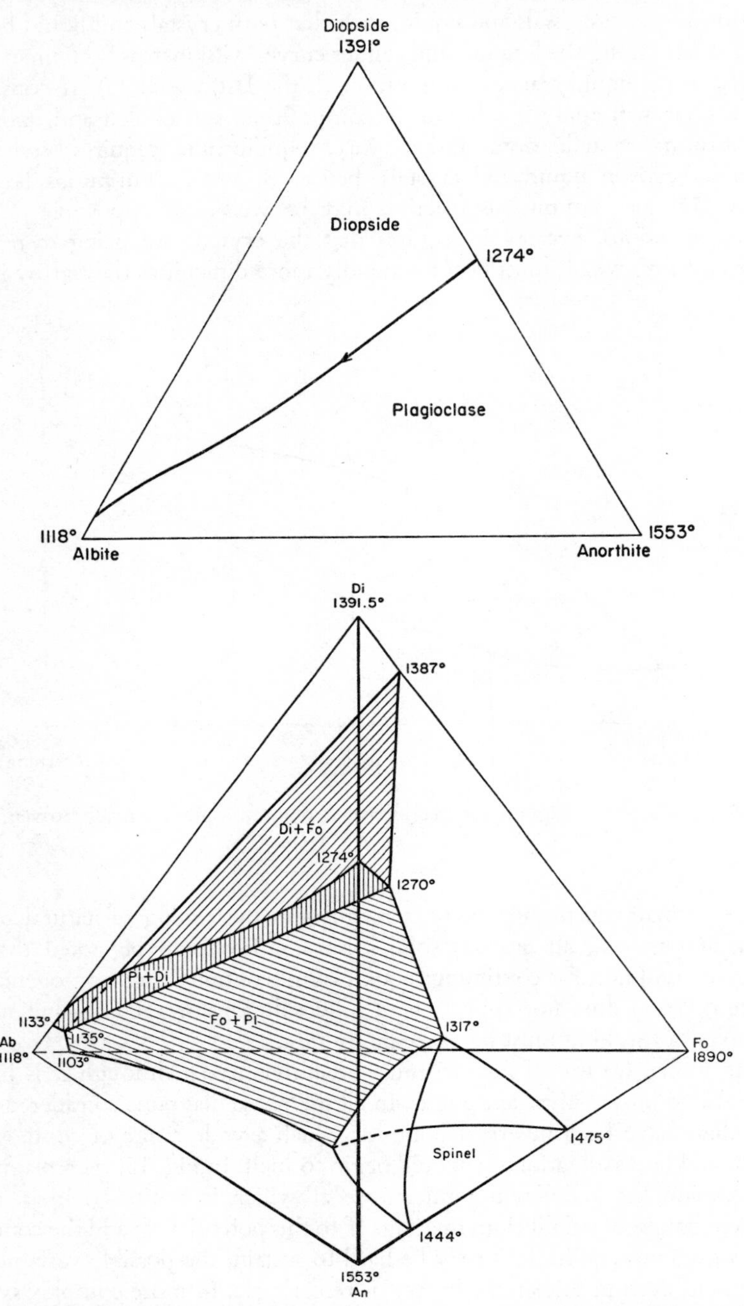

Fig. 16. (a) Ternary diagram, diopside-plagioclase at one atmosphere. (After Bowen, 1928.) (b) "Quaternary" diagram, forsterite-diopside-plagioclase. (After Yoder and Tilley, 1962.) (The equilibria involving spinel are not quaternary.)

This condition of no invariant point, seen in the binary plagioclase system, will extend to all systems involving plagioclase, and indeed applies to all systems in which two or more of the components show extensive solid solution. Figure 16 shows examples of ternary and quaternary diagrams in which the plagioclase series plays this important role. In each case there is no ternary or quaternary invariant point respectively. The increased scope for variability in the course of melting or crystallisation referred to above, becomes increasingly evident in these more complex systems wherein every plagioclase-bearing composition has its own unique course of equilibrium melting or crystallisation. For a detailed description of crystallisation in the ternary the reader may refer to the original discussion by Bowen (1928).

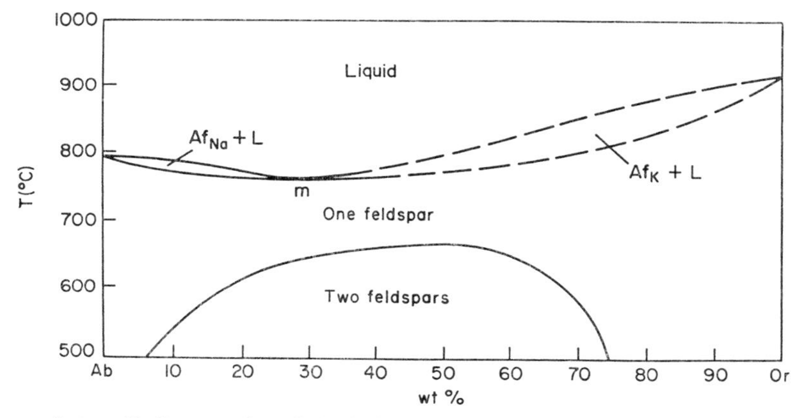

Fig. 17. "Binary" diagram for alkali feldspars, at $P_{H_2O} = P_{total} = 3000$ kg/cm^2. (After Tuttle and Bowen, 1958). The lower boundary (*solvus*) encloses the TX region in which two feldspars are the equilibrium assemblage: their compositions at a given temperature are found by the intercepts on the solvus.

Two other types of binary solid solution diagrams are possible: one in which there is a temperature maximum on the liquidus and solidus; the other, where there is a temperature minimum. The first type is very rare, and there are no examples among rock-forming minerals. High temperature alkali feldspars provide the best known case of a solid solution series with a temperature minimum (Fig. 17). The alkali feldspar system is not wholly binary at low pressures because of the incongruent melting of orthoclase (to leucite + liquid) but with increasing water pressure the stability of leucite is reduced, so that at 3 Kbar the system can be discussed in binary terms. In the strict sense, this system is not binary either, because the liquids contain H_2O and there is a coexisting hydrous vapour (fluid) phase in all assemblages, but these factors may be neglected for present purposes.

The relationships shown in Fig. 17 could be regarded as two simple solid solution loops which meet in a minimum at the composition of an intermediate compound. But although the feldspar of minimum composition has the unique property of melting to a liquid of identical composition it is not strictly an intermediate compound between Ab and Or. Nevertheless, all feldspar compositions to the right of the minimum will behave in a similar fashion to those described in the plagioclase series, and so will all those compositions to the left. It is essential to realise that the minimum in the alkali feldspar solid solution

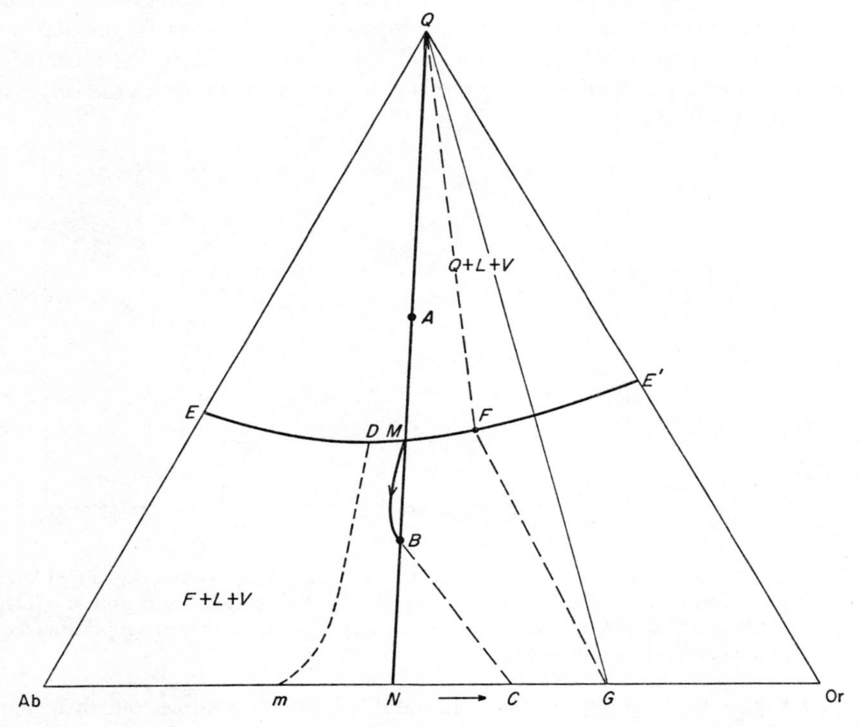

Fig. 18. "Ternary" diagram for the "granite" system at $P_{H_2O} = 1000$ kg/cm². (After Tuttle and Bowen, 1958.) See text for explanation.

loops is not an invariant point. Only one feldspar composition melts or crystallises at the minimum, all other alkali feldspars in the system begin to melt at higher temperatures yielding liquids of different composition from the crystals. Non-equilibrium crystallisation of an alkali feldspar liquid could eventually yield a liquid at the minimum. If, for instance, the alkali feldspar crystals were continuously zoned there would be continuous failure of equilibrium and the liquid would ultimately reach the minimum and precipitate feldspar of that composition.

Addition of extra components results in extension of the behaviour of the high-temperature alkali feldspar solid solutions into more complex systems. The best known example of this is seen in the famous study of the "granite" system by Tuttle and Bowen (1958), from which Fig. 18 is taken. Here it is seen that there is no isobaric invariant point in the "ternary" diagram (strictly, the diagram is a projection from the quaternary system involving H_2O). Instead, there is a temperature minimum (*M*) on the quartz–feldspar boundary (commonly called a cotectic). At the temperature of the minimum, a mixture of quartz and feldspar (of composition *N*) in the presence of excess H_2O, will begin to melt, yielding a liquid of composition *M* until either quartz or feldspar is used up. If feldspar is consumed first (mixture *A*), further heating causes a rise in temperature and the liquid composition shifts progressively from *M* towards *A*. If quartz is consumed first (mixture *B*) the liquid follows a curved path from *M* to *B*, whilst the composition of the feldspar changes from *N* to *C*. Equilibrium crystallisation of liquids *A* and *B* follow the same paths in reverse. For the special range of compositions along the line *Q–N*, therefore, the minimum has the function of an invariant point, *but* for all other compositions in the triangle, melting will begin at a higher temperature and a different liquid composition on the quartz–feldspar cotectic. For instance, all mixtures along the line *Q–G* will begin to melt at a point such as *F*.

The minimum on the quartz–feldspar cotectic is really the "ternary" analogue of the minimum in the alkali feldspar binary diagram, because the liquid of composition *M* has the same alkali ratio (or potential feldspar composition) as the feldspar with which it is in equilibrium. It is the only liquid in the system with this characteristic.

Tuttle and Bowen (1958) also defined a low temperature zone on the feldspar liquidus, which they call the "thermal valley" (mD in Fig. 18). Over much of its length, the liquids in this zone precipitate feldspar of constant composition. It may be visualised as an extension into the ternary of the minimum on the alkali feldspar binary solid solution loops, and indeed, Tuttle and Bowen show the zone originating in the alkali feldspar minimum. But the analogy is not completely apt, because the liquids in the "ternary" low temperature zone do *not* have the same alkali ratios as the feldspars with which they are in equilibrium. The feldspar–liquid relationships along the "thermal valley" have *no* exact counterpart in the binary feldspar system. This is emphasised in the straight section of the valley where feldspar of constant composition precipitates from a liquid of changing alkali ratio. The disparity between alkali ratios in liquid and crystals becomes even greater when the low temperature region on the feldspar liquidus is traced into more complex peralkaline systems (Bailey and Schairer, 1964).

The line of the quartz–feldspar cotectic in the "granite" system (which, for purposes of discussion, may be considered ternary) will extend into a quaternary system as a cotectic surface. In this surface there will be a line of quartz–feldspar minima, one for each increment of the new component. This line of minima does not describe a univariant condition. Each point in the line is a liquid composition that coexists in equilibrium with a specific feldspar composition and quartz, and all three phases must, when added together, fall within a specific and very limited range of bulk composition. If, however, a liquid composition is formed on this line, it will evolve with cooling (by subtraction of feldspar and quartz) along the line until separation of a new phase intervenes, whence the liquid will evolve to the quaternary minimum, at which it is finally consumed. It must be emphasized that this path of liquid evolution is only open to a very restricted range of bulk compositions. Equilibrium melting of similar bulk compositions

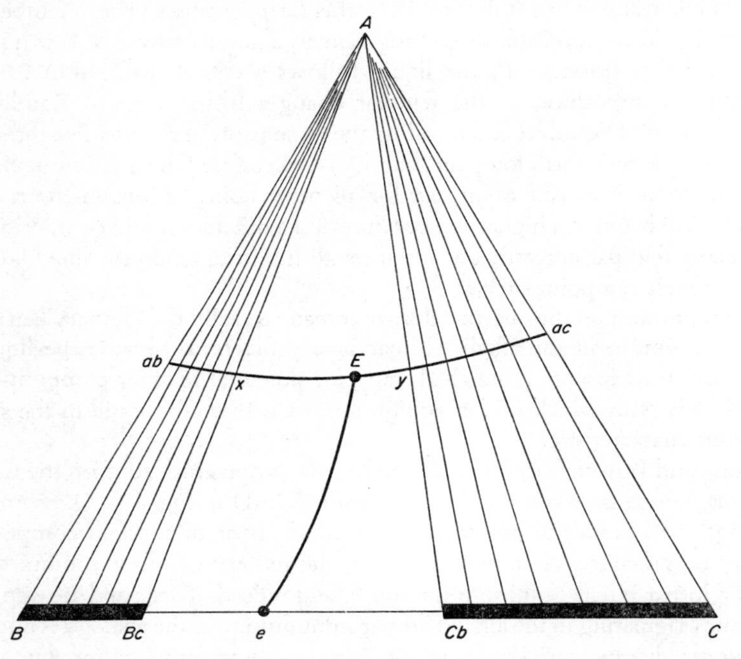

Fig. 19. Schematic ternary diagram, with limited solid solutions (*B–Bc* and *C–Cb*). Melting of mixtures in either of the ruled areas would involve only two solids + liquid (the liquid composition lying between *ab* and *E*, or *ac* and *E*). Melting of a mixture in the unruled area would involve three solids and liquid *E* (i.e. an isobaric invariant equilibrium).

can follow the reverse course. Some related bulk compositions could follow part of the reverse course until either quartz or feldspar were consumed in melting. All other compositions in the quaternary system will follow their own unique courses of crystallisation or melting. If, to this range of possibilities, is added the possibility of failure to maintain equilibrium, the variability of crystal $\rightleftharpoons$ liquid behaviour becomes immense.

In the quaternary there will be no invariant point involving alkali feldspar. The same will be true of higher systems, and the general isobaric case may be summed up by saying that, where a solid solution series forms one boundary of a system, the maximum number of solid phases that may coexist with liquid is one fewer than the components, i.e. $P = C$, (P including the liquid). This means that there will be no isobaric invariant point, only a minimum. Such a minimum is always one phase short of the number necessary for isobaric invariance (namely $P = C + 1$) and there will be no general melting composition for the system. It is important to realise that many natural systems will be of this type.

Many natural solid solutions have only limited ranges, which may lie wholly *within* the system under consideration. In these cases, the lack of an invariant point will still apply to those portions of the system which are effectively

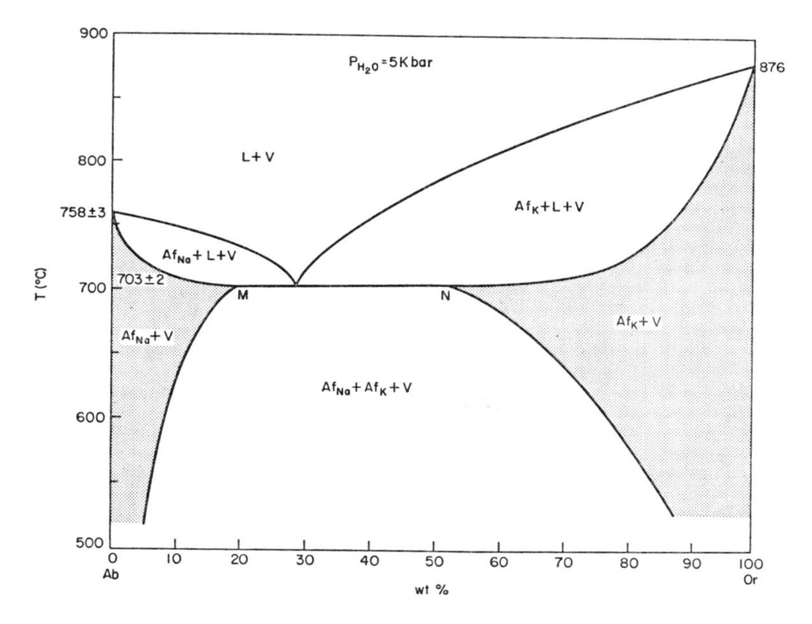

Fig. 20. "Binary" diagram for alkali feldspars at $P_{H_2O} = 5000$ atm. (After Morse, 1970.) The shaded regions show the varying range of feldspar compositions at different temperatures.

bounded by the solid solution, as shown in Fig. 19. The end of the solid solution series within the system then functions as an intermediate compound, which can coexist with the required number of crystalline phases and liquid at an invariant point.

Solid solutions also tend to unmix or exsolve at lower temperatures, in the same way that some liquids may unmix into immiscible fractions. Exsolution in high temperature alkali feldspars is well known in the various forms of perthite, and is expressed in the solvus in Fig. 17. At high pressures of H_2O the solvus is found to intersect the solidus. When this occurs the minimum on the feldspar liquidus is an invariant point because it now describes a liquid in equilibrium with two different feldspar compositions. It may be seen from Fig. 20 that when the solvus intersects the solidus the resulting diagram is like a simple binary in which the two solid phases show limited ranges of composition (shown by shading in Fig. 20). The diagram also shows how the termination of a solid solution within the system divides it into two types of melting regime, one in which solid compositions melt at various temperatures without encountering an invariant point (outside the composition span of the solvus) and another region (within the span of the solvus) where all compositions begin to melt at an invariant temperature, forming a liquid of fixed composition. In an actual melting episode, the *effective* span of the solvus will be determined by the ability of the feldspars exsolved at lower temperatures (and perhaps completely separated) to re-homogenise with increasing temperature.

Figure 19 could be regarded as an example of the "granite" system under conditions where the feldspar liquidus has intersected the solvus, producing two types of melting regime in a similar manner to that described above.

It is worth noting at this juncture, that in the anhydrous system the alkali feldspar solvus may not intersect the solidus at any reasonable pressure: this relationship and its petrologic significance are discussed in Part II, Section B2 (Bailey).

(3) *Projected phase diagrams*

Figure 18 shows the diagram used by Tuttle and Bowen (1958) in their description of the "granite" system. In their account they make it quite clear that triangular diagrams of this type are not ternary, but projections from the quaternary containing H_2O. In essence, all the equilibria shown were in the presence of an additional fluid (composed largely of H_2O) which has been shown as an additional phase (V) in all fields in Fig. 18. Another way of describing this is to say that the H_2O-saturated equilibria have been projected onto the anhydrous base of the quaternary system in which H_2O constitutes the apex. This simple, and apparently reasonable, procedure is not without pitfalls for the unwary, especially if it is applied to rocks without regard to the limiting conditions. A full discussion of the specific case of the "granite" system will be found in Part II, Section B1 (Luth), but the reader will realise that the equilibria (especially temperatures and total liquid compositions) must be different if H_2O is deficient, and that such a condition might be the common one in nature.

Projected phase diagrams are now commonplace, and may contain far greater pitfalls if the point of projection contains components that are important in the solid phases (Bailey and Schairer, 1964).

When the projection technique is combined with component "reduction", e.g. Fe_2O_3 and Al_2O_3 treated as one component, it must be recognised that the resulting phase diagram is, in its essence, hypothetical: other constructions are possible, and may allow significantly different conclusions. The component "reduction" method is sometimes used to compare rock compositions with

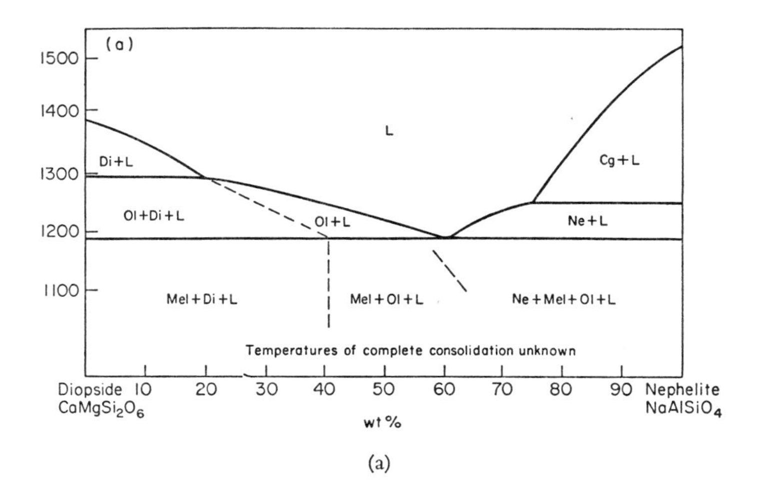

(a)

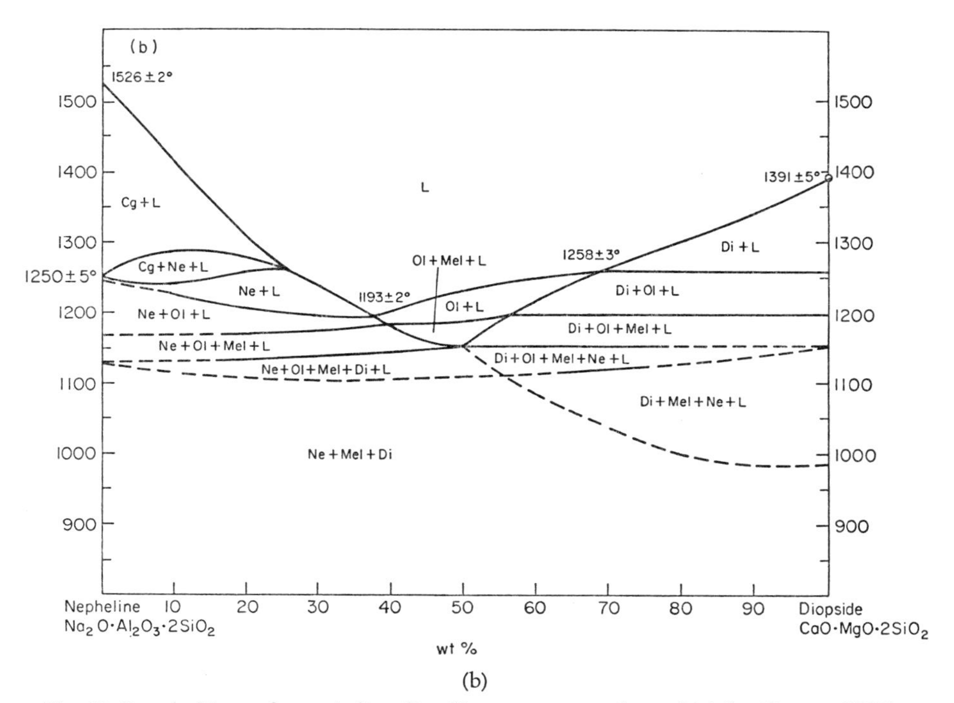

(b)

Fig. 21. Pseudo-binary for nepheline-diopside at one atmosphere. (a) (after Bowen (1922). (b) after Schairer *et al.*, (1962).

synthetic products, and the "reduction" may be sufficiently complex to need a computer programme for routine applications (O'Hara, 1968; Jamieson, 1969). The resulting diagram may be a fair approximation to reality, but the reader should be aware of just how far removed he is now from making his own objective analysis of the data.

In short, projected phase diagrams may contain ambiguities unless the nature and consequences of the projection are clearly and thoroughly understood. Diagrams based on "reduced" components have built-in assumptions about chemical behaviour that at best apply only as approximations, and only then for restricted composition ranges.

(4) *Pseudo-system diagrams*

Sometimes phase-diagrams are used which superficially have, say, a binary form, but are referred to as pseudo-binary. This term is applied whenever the diagram contains phases which cannot be expressed in terms of the two end compositions. An early example of this type was diopside–nepheline described by Bowen (1922) and revised by Schairer and others (1962). Liquids in this range of composition (Fig. 21) crystallise olivine and melilite solid solution, and a complete description of the phase relations can be made only by reference to the quinary system Na_2O-CaO-MgO-Al_2O_3-SiO_2 because all these oxides must be behaving independently. Further discussion on these lines is given in Bailey (1974) and in Part II, Section B2 (Bailey).

Pseudo-binary diagrams have been used more and more in recent years, as a means of summarising high pressure experiments on a series of compositions between two end-member minerals which lie in a multi-component system. Figure 22 shows such a pseudo-binary, diopside–pyrope at 30 Kbar, as given by O'Hara and Yoder (1967). The complete phase relations require description in terms of the quaternary CaO-MgO-Al_2O_3-SiO_2, but the equilibria not involving spinel can be described in ternary terms within the plane enstatite–wollastonite–Al_2O_3. For the full description the reader should consult O'Hara and Yoder (1967). Suffice to say that the system is binary in the two single-phase regions of Cpx solid solution only, and liquid only; the equilibria may also be binary in the two phase region Cpx solid solution plus liquid. Nowhere else can the pseudo-binary be read directly, because all the crystalline phases are solid solutions whose compositions lie outside the diagram—consequently the co-existing liquids lie outside the diagram. What use then, is such a diagram? It gives the phase assemblage (but not necessarily the compositions) for any given bulk composition between diopside and pyrope, at any given temperature. The diagram also shows the minimum melting compositions in the join (but again it does not give the composition of the liquid so formed). A pseudo-binary, therefore, tells only part of the story, but if skilfully combined with determinations of the compositions of the phases (by XRD or electron microprobe) it can be used to define the phase relations in the larger system.

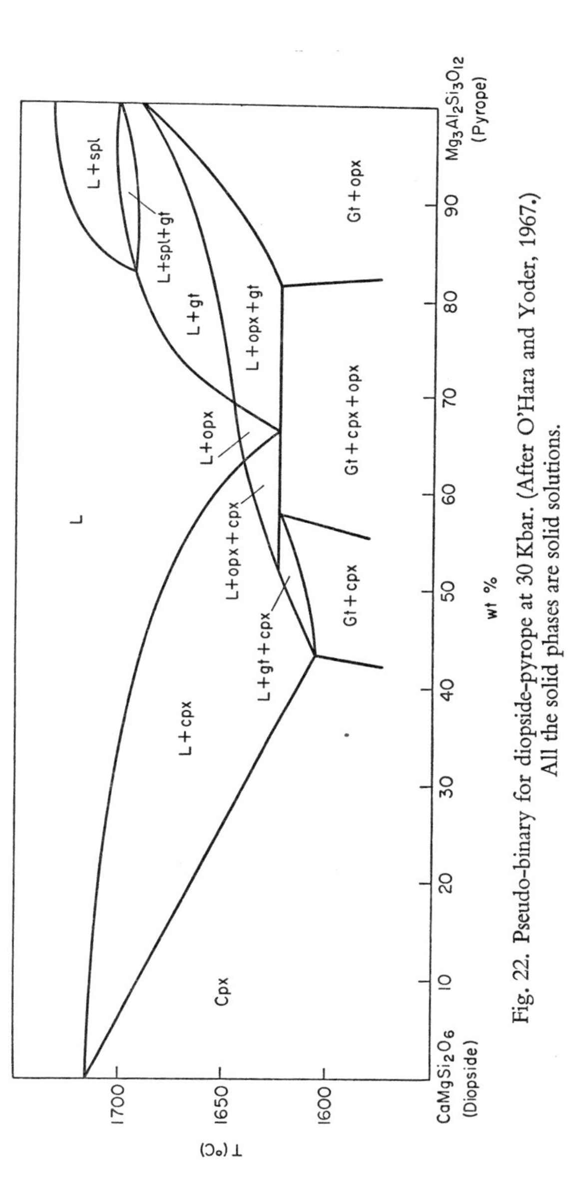

Fig. 22. Pseudo-binary for diopside-pyrope at 30 Kbar. (After O'Hara and Yoder, 1967.) All the solid phases are solid solutions.

Pseudo-ternary diagrams are those in which a phase appears which cannot be described in terms of the components used as the apices of the triangular diagram. Such diagrams are used in a different way from pseudo-binary diagrams. As we have seen previously, it is customary in a ternary diagram to show only liquidus (or primary phase) fields, these being separated by boundaries which define those liquid compositions that precipitate two solid phases simultaneously: within the diagram any three primary phase fields (and their two-phase boundaries) will meet at a point. In the true ternary the compositions of all solid phases lie in the plane of the diagram, and the two-phase boundaries are univariant and the three-phase point is invariant. In a pseudo-ternary (Fig. 23) the composition of one or more of the primary phases may lie outside the diagram; hence, as soon as that phase begins to crystallise the liquid composition moves off the plane of the diagram. Any two-phase boundaries and three-phase points, involving a non-planar phase, will not be univariant and invariant respectively. If, however, the pseudo-ternary is part of a quaternary system, the three-phase point must lie on a univariant curve in the quaternary system. Such a point in the pseudo-ternary is described as a *piercing point*, and its location fixes a temperature and composition on the univariant curve within the quaternary. Experiments to locate the piercing points in carefully selected triangular sections of a quaternary system (e.g. Fig. 23) are the standard method by which a quaternary "flow-diagram" is constructed (Bailey and Schairer, 1966). Such a flow-diagram depicts all the univariant and invariant equilibria in the system. A simple model was given in Fig. 12, and real examples will be found in Part II, Section B2 (Bailey, Figs 8 and 14).

Diagrams to show pseudo-systems of higher order than pseudo-ternary would be of little value, because they would imply the loss of a dimension even before the construction of the pseudo-diagram. A true quaternary, for instance, requires a three-dimensional figure: a pseudo-quaternary would therefore consist of a three-dimensional model of a poly-dimensional system. It is difficult to see how such a model could be used without ambiguity.

(5) *Poly-component systems*

Synthetic systems with more than four components can be treated rigorously only by algebraic methods, not graphically. These methods are available, and readily applicable with machine computing, and the reader is referred to Greenwood (1967 and 1968) for details. So far, little has been done in this way and it is useful to ask why. There are four basic reasons:

(1) It is difficult to communicate poly-dimensional information.
(2) Crystalline rocks, because they have had time to approach equilibria involving several phases, will generally represent low variance states. Low variance states can be represented graphically, and therefore the systematics of poly-dimensional states are largely unnecessary.

(3) Poly-component synthetic systems are usually set up in an attempt to approach closer to rock compositions. More and more experimenters have consequently used rocks as starting materials, on the grounds that the systematics of multi-component synthetic systems begin to look superfluous to the petrological problems at hand.

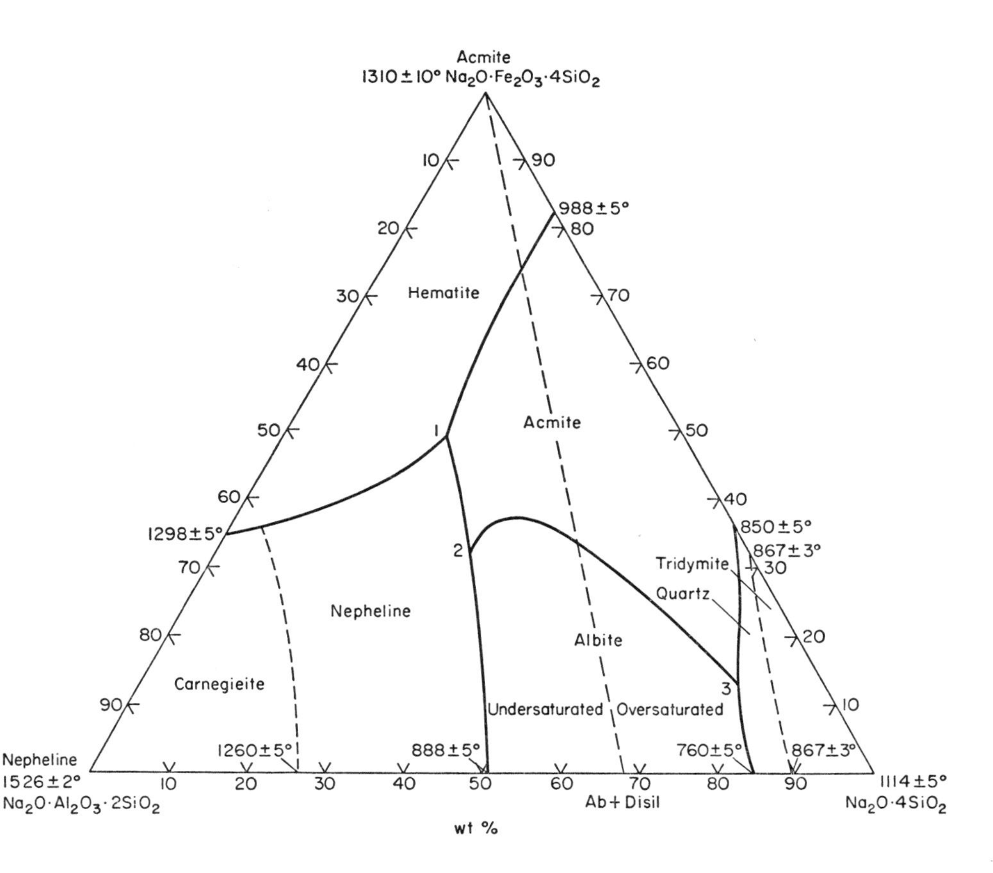

Fig. 23. Pseudo-ternary for the triangular join acmite-nepheline-($Na_2O.4SiO_2$) in the quaternary system Na_2O-Al_2O_3-Fe_2O_3-SiO_2 at 1 atm. (After Bailey and Schairer, 1966.) The 3-phase points (1, 2, 3) are *not* isobaric invariant, they are piercing points where this join intersects univariant lines in the quaternary system.

(4) The electron and ion microprobes are opening up a new approach to the study of low variance equilibria in complex systems, because with these instruments it is possible to analyse the coexisting phases in an equilibrium. Kushiro (1974) provides a clear example of this potential.

(6) *Isothermal sections*

Strictly, these are isothermal sections taken from an isobaric TX diagram. An isothermal section is a device for depicting the variations of phase assemblages with composition in a system at some chosen temperature and pressure. In the case of a binary system, an isothermal "section" would be simply a horizontal line, divided into different lengths by its intersection with phase equilibrium boundaries in the isobaric binary diagram. For isobaric considerations there is little need for such a figure because the information can be read directly from the complete binary TX diagram without difficulty: the figure can, however, be a useful device in describing polybaric, polythermal (PT) variations in a binary

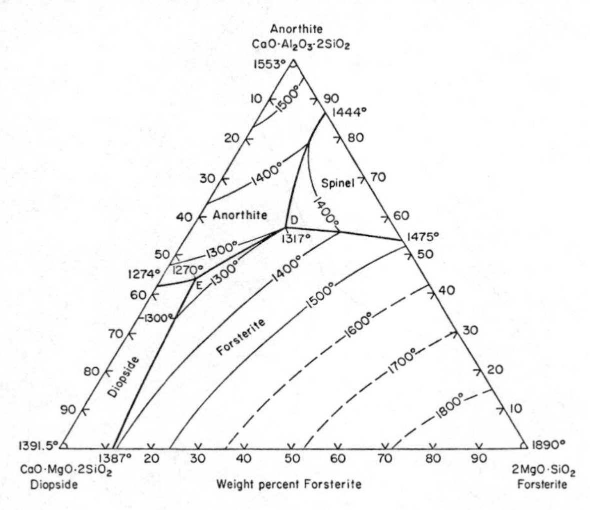

Fig. 24. Liquidus diagram for the system anorthite-diopside-forsterite at 1 atm pressure. (After Osborn and Tait, 1952.)

system (see later, p. 65). The most widespread use of isobaric isothermal sections is to depict phase assemblages in ternary systems, where the effect of temperature variation across the system can only be completely represented on the liquidus surface (by contouring). From a ternary liquidus diagram it is not always easy to visualise the phase assemblages for various compositions at any chosen temperature, and an isothermal diagram then becomes useful. Isothermal diagrams can be made for quaternary systems, but are not capable of rigorous analysis. Similarly isothermal sections can be produced from pseudo-ternary diagrams, but because some of the phases lie outside the plane of the diagram the resulting isothermal diagram must be treated with care.

Ternary isothermal sections are commonly used to depict subsolidus relations, especially where there are rapid changes of phases, phase assemblages, and solid solutions, with temperature. These conditions are not common in silicate systems, but are, for instance, prevalent in sulphide systems, hence the common usage of isothermal sections by experimentalists studying ore mineralisation (see, for instance, Greenwood, Fig. 13, this vol.). In silicate studies, isothermal sections are most commonly used to examine crystal-liquid relations in a ternary TX diagram. Such sections form a convenient basis for analysing the melting relations of different ranges of composition in the system. As an example, take the system anorthite–diopside–forsterite, the liquidus diagram for which is shown in Fig. 24.

Part of this system (the triangular region an–D–fo) is not ternary, because spinel crystallises from compositions within this area, at temperatures above 1317°C. At temperatures below 1317°C, however, any spinel will have reacted out under equilibrium conditions and the whole system is ternary. The lowest temperature liquid in the system is at the ternary (isobaric) eutectic *E* (an + di + fo + *L*) at 1270°C. Composition *E* is the only liquid in the system that can coexist with olivine (fo) clinopyroxene (di) and plagioclase (an): it represents a simplified (haplo-) basaltic liquid (49% di, 43.5% an, and 7.5% fo), i.e. largely composed of plagioclase and pyroxene. Any other liquid in the system must be (a) at a higher temperature, and (b) at most can coexist with only two of the three phases (fo, di, an). It is to depict these possible ranges of phase assemblage that an isothermal section is constructed.

Figure 25 shows three isothermal sections at 1268°C, 1270°C and 1272°C. As the temperature reaches 1270°C, all mixtures of an + di + fo begin to melt, yielding a liquid *E*. The composition of the liquid is constant but the *amount* depends on the quantity of heat put into the system and the proportions of the solid phases. Melting progresses (at constant temperature, 1270°C) as heat is added to overcome the latent heat of fusion of the crystals. Given a continuous heat input, the solid phases melt in the eutectic proportions until one or more of the three is exhausted. For a composition such as *A*, anorthite will be consumed at an early stage of melting, at which the liquid composition and temperature will shift from *E* with further heating. A bulk composition close to *E*, however, will have to be almost completely melted (requiring a big heat input) before the liquid composition and temperature can change. The range of possibilities is best seen using the 1272°C isothermal in Fig. 25.

This diagram is constructed on the 1272°C contour in the original liquidus diagram and essentially the system breaks down into smaller three-sided areas, described as follows:

(1) The small area enclosed by the 1272°C contour (shaded) covers all bulk compositions in the system that are entirely liquid at 1272°C. This is a single-phase area.

(2) Where the contour crosses a primary phase field (e.g. Fo) in the original liquidus diagram it describes the range of liquids (a—b) that are in equilibrium with forsterite at 1272°C. No other liquids in the system can coexist

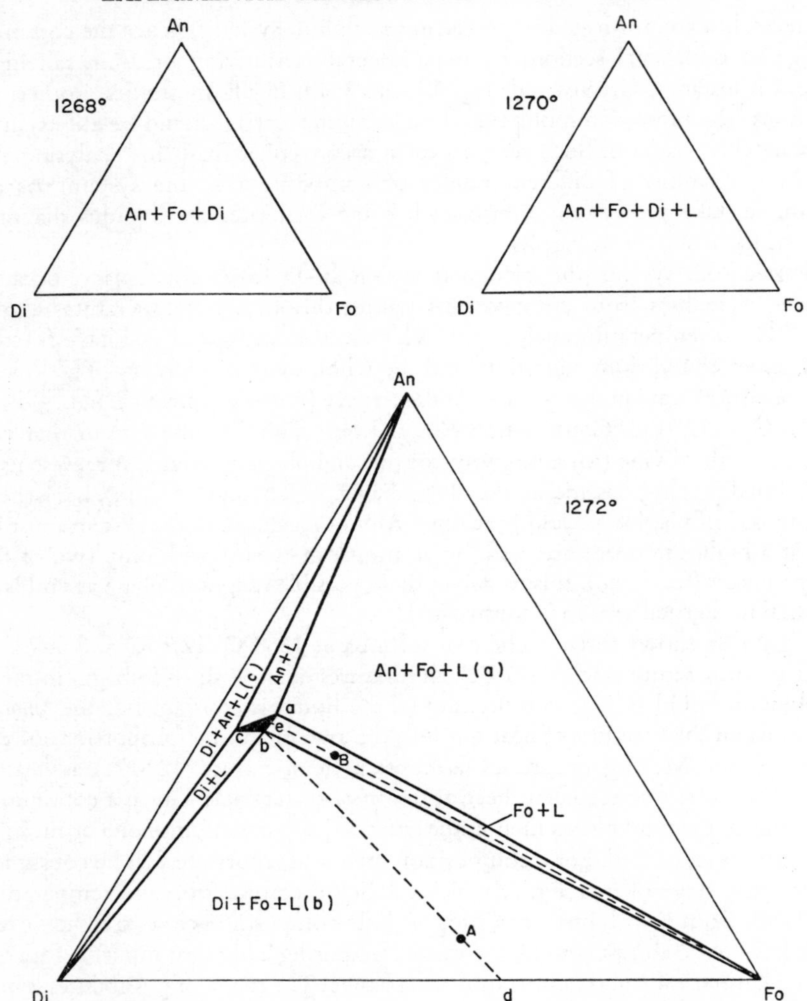

Fig. 25. Isothermal sections of the system anorthite-diopside-forsterite at 1268°C, 1270°C and 1272°C at 1 atm pressure. The 1268°C figure shows the equilibrium sub-solidus assemblage; the 1270°C figure shows the one atmosphere invariant assemblage at the beginning of melting. At each temperature the phase assemblage is the same for all bulk compositions in the system. The third isothermal section is constructed on the 1272°C liquidus contour derived from Fig. 24. Bulk compositions within the 1272°C contour (filled) are completely liquid. Points a, b, c, are where the liquidus contour crosses the three univariant lines leading to the eutectic E (1270°C, see Fig. 24). Drawing tie-lines from each of the univariant liquid compositions, a, b and c, to the compositions of the coexisting solid phases divides the rest of the system into alternate 3-phase and 2-phase regions. The relationships in these regions are described in the text. Bulk composition B is made up of forsterite crystals and liquid, e, at 1272°C. Bulk composition A is made up of diopside and forsterite crystals (mixture d) and the liquid composition b (univariant).

only with forsterite crystals at this temperature. The area a–b–Fo encloses all those bulk compositions that at 1272°C consist only of forsterite + liquid: the proportion of liquid to crystals in such a composition *B* can be obtained by applying the lever rule giving a ratio of BFo:eB. The area a–b–Fo is a 2-phase region in the diagram.

(3) Where the 1272°C contour crosses a univariant boundary in the original liquidus diagram, there is a liquid (e.g. b) that coexists with two solids (diopside + forsterite) at this temperature. The triangle b–Di–Fo encloses

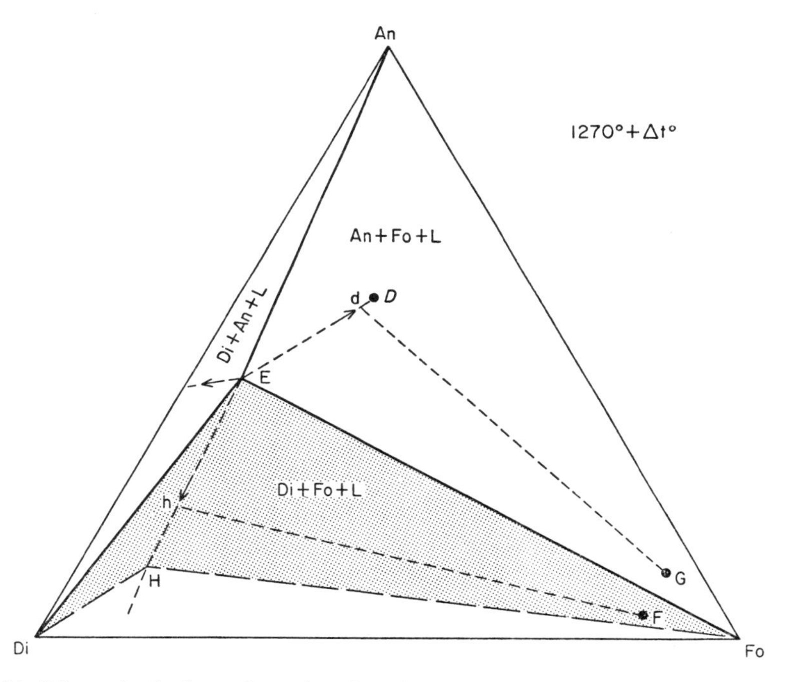

Fig. 26. Schematic isothermal section for the system anorthite-diopside-forsterite at 1270°C plus Δt°. This diagram is designed to illustrate the instant at which the temperature rises above the 1 atm invariant point. At this instant the system is dividing into three 3-phase areas (bounded by heavy lines) which meet in the one atmosphere eutectic *E*. Also shown are the traces of the three univariant curves (broken lines with arrows) which stem from eutectic *E*. *F* and *G* are olivine-rich compositions, $An_5Di_{10}Fo_{85}$ and $An_{10}Di_5Fo_{85}$ respectively. Progressive melting (at temperatures above 1270°C) of composition *G* produces a series of liquids along the path EdG, whereas progressive melting of *F* produces liquids along the path Eh*F*.

The area Di H Fo is a 3-phase triangle at 1360°C: this and the adjacent shaded region are discussed in the text.

all those bulk compositions in the system that will have this 3-phase assemblage at 1272°C. The proportions of the phases in a composition such as *A* may be readily obtained: the ratio of liquid to crystals is dA:bA, and the ratio forsterite:diopside in the crystals is given by dDi:dFo.

Notice that the isothermal construction divides the ternary system into 1-phase, 2-phase and 3-phase regions. When the system was at 1270°C (ternary isobaric invariant point, *E*) it consisted wholly of one 4-phase region. If an isothermal section were drawn at a small increment of temperature above the invariant point (1270°C + Δt) the system splits into three 3-phase regions about *E* (Fig. 26). With increasing temperature the evolutionary path followed by any liquid in a 3-phase assemblage will be along one of the three univariant curves stemming from *E*. The direction taken by any given liquid during progressive melting, therefore, will be governed by which 3-phase triangle encloses the bulk composition. Compositions *F* and *G*, although they are both simplified peridotites, consisting of 85% forsterite, would generate quite different series of liquids during the early stages of partial melting. Composition *G* will generate a series of liquids from *E* to d, becoming enriched in anorthite and depleted in diopside, whereas the liquids generated from *F* over the same temperature interval will have the opposite characteristic. This illustrates the fact that the path followed by a liquid during partial melting may depend critically on small variations in the composition of the source rock, especially when it is largely composed of one mineral, as in peridotite. But it is equally important to remember that the same series of liquids will be generated by widely different source rocks. For instance in Fig. 26 some portion of the series of liquids along *EH* will be generated by partial melting of all compositions in the shaded region. At 1360°C, a liquid such as *H* may therefore be in equilibrium with a crystal residue consisting entirely of diopside (i.e. pyroxenite) or entirely forsterite (i.e. dunite), or any mixture of the two. The isothermal diagram thus nicely illustrates one of the paradoxes of magma generation, in that divergent liquids may be generated by minor differences in the source rock whilst identical liquids can be generated by widely different source rocks. Everything depends on the initial bulk composition and its relationship to the configuration of the phase diagram. The reader should consider whether the standard petrographic classification of rocks (based on percentages of major minerals) is particularly useful in considerations of partial melting.

Where a solid solution is part of a system its composition at any given temperature will depend on the bulk composition. In an isothermal section, therefore, a family of tie-lines between the solid solution and the other phases is needed to describe the variation of the solid solution in different phase assemblages. Two examples from the "granite" system (Tuttle and Bowen, 1958) are given in Fig. 27. These were constructed to illustrate two conditions in the system; one, where the alkali feldspar solvus does not intersect the solidus, the other (at a lower temperature and higher pressure) where the solvus and solidus intersect. The variation in composition of alkali feldspar solid solutions in each phase assemblage are described by the tie-lines. Note that fluid or vapour (*V*) would be an additional phase in all areas if these diagrams were derived from the experimental liquidus diagrams, which were determined in the presence of excess H_2O. If we ignore *V*, however, we can note that "2-phase" areas, involving alkali feldspar solid solution plus liquid, are four-sided, and every liquid along the isothermal liquid edge is in equilibrium with a different feldspar. We may also note that in these isothermals involving solid solutions there are some regions where liquid is *absent* at the isothermal temperature. This should be contrasted with the previous case (with no solid solution) in which all parts of the system began to melt as soon as the eutectic temperature was reached. When a solid solution spans the system, partial melting will initially affect only a narrow band of composition around the minimum. Where solid solutions terminate within the system (as in Fig. 27(b)), initial partial melting will affect all compositions within the triangle *QFE*. A full and lucid discussion of the phase and composition sequences that may develop by different modes of partial melting has been given by Presnall (1969).

(7) *Heating diagrams*

All the preceding discussions of temperature-composition (*TX*) diagrams have concentrated on phase equilibria, and have only occasionally referred to the *amounts* of the phases present in an equilibrium, and the *quantities* of heat (lost or gained) necessary to change the equilibrium. It is instructive to consider phase changes in terms of heat change and one of the simplest graphical means is to plot temperature changes in a system against time, assuming a constant rate of heat input. This is the basis of differential thermal analysis (D.T.A.) in which phase changes are identified by changes in the rate of change of temperature with time. Some model cases, using simple systems as examples, are shown in Fig. 28.

In Fig. 28 the basic model describes melting in various cases given a steady input of heat, but these examples can easily be reversed to consider the case of cooling and crystallisation. The main feature to notice in Fig. 28 is that in each example a change of equilibrium phase assemblage is marked by a step, or a change in slope of the heating curve. Similar forms of heating curve will be found in more complex systems and in each case a horizontal step in the heating curve will correspond with an invariant point and a change in slope will signify a change in variance with increasing degrees of freedom. All the cases in Fig. 28 represent equilibrium melting conditions. Under *perfect fractional* melting conditions, the heating curve would have different slopes between steps because liquid would be removed as fast as it were formed at every temperature in the system (Presnall, 1969). The reader is recommended to try constructing a model heating curve using an example from a ternary system such as composition *A* in Fig. 25.

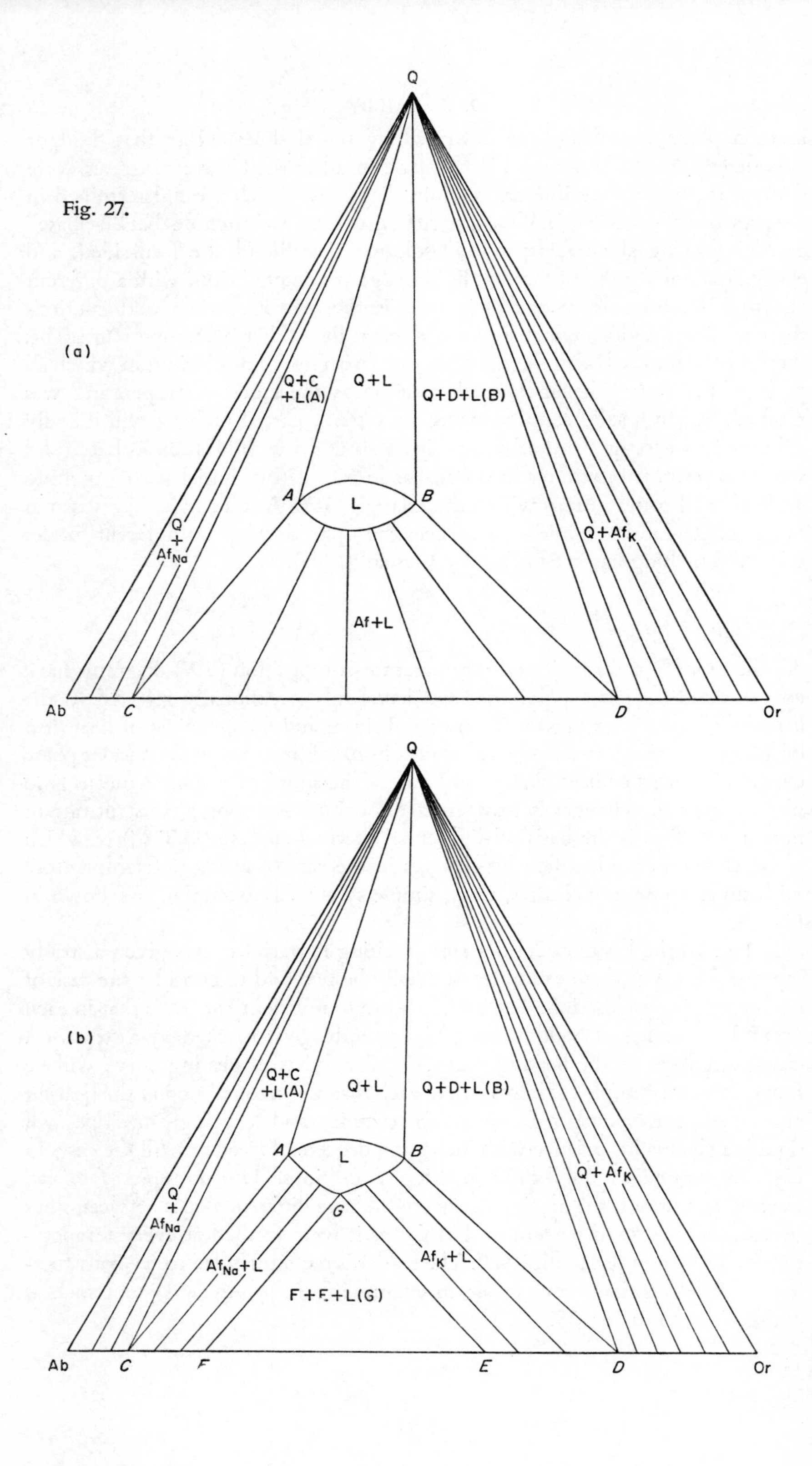
Fig. 27.
(a)
Q
Q+C
+L(A)
Q+L
Q+D+L(B)
A
L
B
Q
+
Af$_{Na}$
Q+Af$_K$
Af+L
Ab
C
D
Or
(b)
Q
Q+C
+L(A)
Q+L
Q+D+L(B)
A
L
B
Q
+
Af$_{Na}$
Q+Af$_K$
G
Af$_{Na}$+L
Af$_K$+L
F + E + L(G)
Ab
C
F
E
D
Or

(b) PX DIAGRAMS INVOLVING LIQUIDS

All the previous examples have been concerned with the effect of varying temperature on a system, whilst pressure has been held constant. It is equally possible to treat the effects of changing pressure at constant temperature, but this is rarely done. The reasons are, in part, traditional, because early experiments were TX studies made at atmospheric pressure. Even today the preliminary studies of many systems are made at one atmosphere, because then the experimenter needs only to control the temperature. Once started in this way the next step is usually to make another TX study at a higher constant pressure for comparison with the one atmosphere results. Also, many of the early systematic studies in experimental petrology were concerned with liquid equilibria, and its applications to magma evolution. The chief influence on magma evolution was considered to be changing temperature, so this too has helped to enshrine the TX study. A PX study would clearly be most apposite to consider the evolution of a liquid rising through the Earth at constant temperature but experimentalists have tended to favour a series of TX diagrams or PT diagrams to explore this eventuality.

In general PX diagrams involving liquids will look like reversals of the TX forms, because the effects of pressure and temperature tend to be opposite. In a PX binary (involving anhydrous silicates), for instance, the completely liquid field will be in the low pressure region, whilst the system will be completely crystalline at high pressures. In this context the TX origins of expressions such as "sub-solidus" are plainly evident.

(c) SOLID $\rightleftharpoons$ SOLID EQUILIBRIA

Up to this point the discussion has centred on equilibria involving a liquid phase. Such cases have been separated deliberately because there are special differences between liquid $\rightleftharpoons$ solids and solids $\rightleftharpoons$ solids equilibria. The distinctions are worth noting before considering specific examples of equilibria between solid phases.

When a liquid phase is present in a system it represents special conditions, because every possible composition, no matter how complex the system, can exist as a liquid. For a solid this is possible only in a one component system, or where a single solid solution spans the whole system. When a liquid is in equilibrium with crystals in any system, its composition must change with either temperature or pressure, except at an invariant point. Continuous change of

---

Fig. 27. (a) Isothermal section of the "granite" system at a temperature (t) above and a pressure (p) below that at which two feldspars appear at the liquidus. (After Tuttle and Bowen, 1958.) (b) Isothermal section of the "granite" system at a temperature (t′) below and a pressure (p′) above that at which two feldspars coexist at the liquidus. (After Tuttle and Bowen, 1958.)

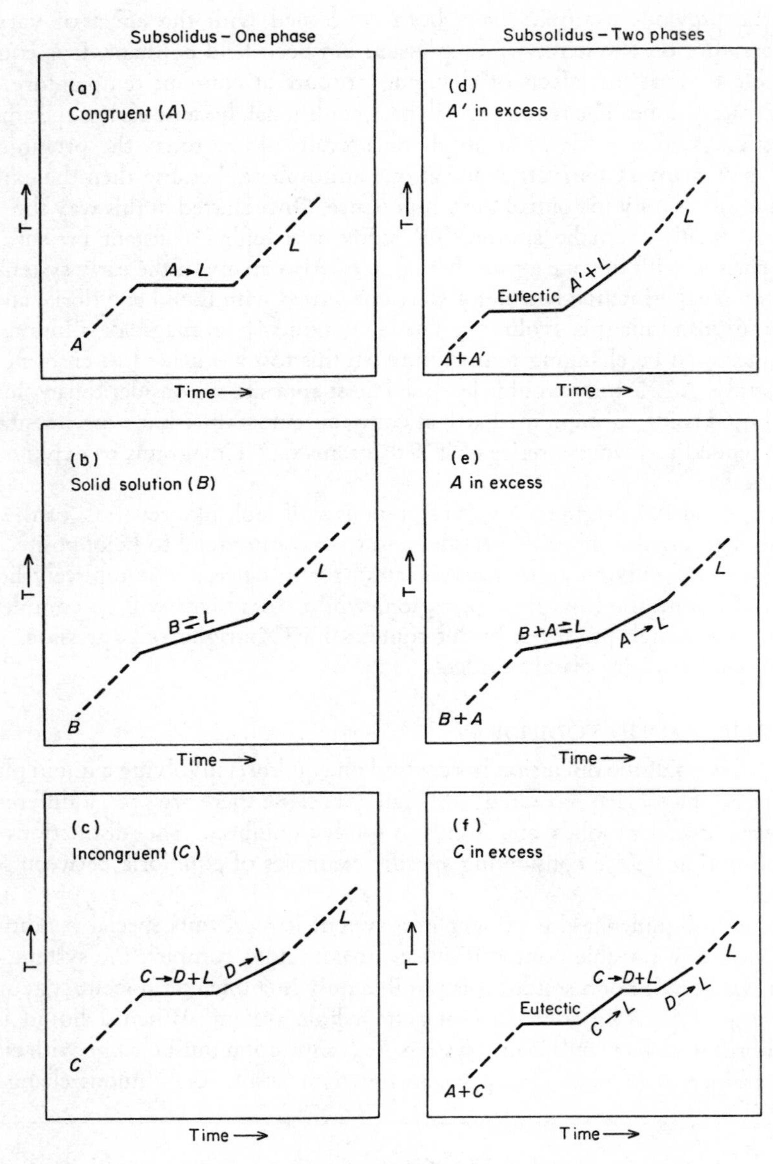
Subsolidus – One phase
(a)
Congruent (A)
A→L
L
A
Time→
T→
(b)
Solid solution (B)
B⇌L
L
B
Time→
(c)
Incongruent (C)
C→D+L
D→L
L
C
Time→
Subsolidus – Two phases
(d)
A′ in excess
Eutectic
A′+L
L
A+A′
Time→
(e)
A in excess
B+A⇌L
A→L
L
B+A
Time→
(f)
C in excess
Eutectic
C→L
C→D+L
D→L
L
A+C
Time→

Fig. 28

pressure or temperature causes continuous change in the phase structure. (A solid solution composition may also be subject to pressure or temperature changes, but the composition range of a solid solution is generally restricted in complex systems, and therefore it is not in continuous interaction with *all* the other phases, in the same way as a liquid.) Consequently, changes in temperature and pressure *below* the solidus do not result in gradual changes in phase composition, and phase proportions, such as occur when liquids are crystallising. Once a composition is crystalline, changes in temperature or pressure produce only step-like changes in the phase structure, when solids interact, or when there is a polymorphic inversion. Figure 29 illustrates these features in a binary isobaric TX diagram. Between the phase change steps, below the solidus, the solids may be thought of as "at" equilibrium (static) rather than "in" equilibrium (dynamic). *Within* these intervals only two phases coexist: they are said to be compatible, or, to constitute a compatibility pair. Any given pair of minerals will also have a range of stability (or compatibility) with pressure, as well as with temperature, as required by the phase rule,

$$F = C - P + 2 = 2 - 2 + 2 = 2.$$

Thus, the equilibrium between two solids (in a binary system) is divariant; or, the two solids may coexist over a range of temperature and pressure. At certain

Fig. 28. Model heating curves for simple systems under isobaric conditions, assuming a constant heating rate and similar thermal capacities when the system is wholly crystalline and wholly liquid (shown by broken lines). The changes in slope (solid lines) between the wholly crystalline and wholly liquid states represent periods of melting where the whole or part of the heat input is used for latent heat of fusion. Each example should be compared with an appropriate phase diagram in preceding figures. (a) Congruent melting of a single phase. This is a one component system throughout, the temperature step representing the isobaric point, crystals → liquid. (b) Melting of a solid solution. Beginning of melting represents the binary condition but is not an invariant point so there is no step in the heating curve, merely a change in slope while the solid solution changes composition until it is eventually exhausted. (c) Incongruent melting of a single phase. This system is binary in the melting interval, when first the phase C melts at the isobaric reaction point to form crystals of D + liquid. When C is used up the equilibrium becomes univariant and the melting curve rises as D melts. When D is exhausted the curve changes to a standard liquid heating slope. (d) Melting in a binary system where both solids are congruent. The step in the curve represents melting at the eutectic. The succeeding slope represents the univariant melting of the phase in excess of the eutectic proportions. (e) Melting in a ternary where two of the components form a continuous solid solution. The first slope of the melting curve represents univariant melting of phases A and B together. The slope changes when one of the two phases is exhausted and represents the divariant melting of the remaining phase. (f) Melting in a binary where one of the phases (in excess) is incongruent. Depending on the bulk composition in relation to the composition of the reaction point, this diagram will take the form of some combination of the melting patterns shown in Figs. (c) and (d). The illustrated curve takes the case where the bulk composition contains phase C in excess of the composition of the reaction point. In this case two invariant points are encountered in the melting interval.

limiting temperatures and pressures, however, a third phase may form. At such an equilibrium $F = C - P + 2 = 1$, i.e. three phases (in a binary system) are in a univariant condition; or, if the pressure is fixed (as in Fig. 29) the three phases can coexist only at one temperature (the condition is isobaric invariant). In Fig. 29 at temperature t_1 the reaction

$$A + B_2 \rightleftharpoons AB$$

takes place; whilst at t_2, the polymorphic change $B_2 \rightleftharpoons B_1$, occurs. The temperatures of both these changes are invariable for the condition of constant pressure depicted in Fig. 29, but the temperatures will be different at other pressures. For each three-phase equilibrium, the locus of pressure and tempera-

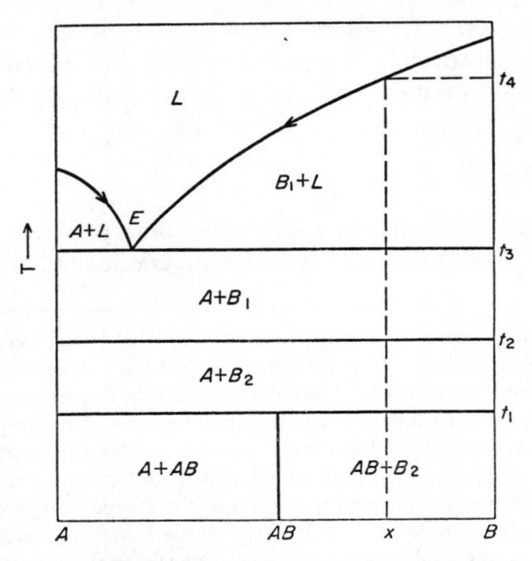

Fig. 29. Model binary to show subsolidus changes. Isobaric, at pressure p.

ture will be a line in a pressure-temperature (PT) diagram, as shown in Fig. 30. Each line describes a univariant condition. The area between the two univariant lines on the low pressure side of *I*, is the range of pressure and temperature in which *A* and B_2 can coexist. A two-phase assemblage can thus be regarded as the general equilibrium state (i.e. spanning a range of temperature and pressure) of a binary system. This principle extends to all systems so that, below the solidus, the general equilibrium state is when the number of crystalline phases equals the number of components. When an additional phase appears it signifies that either a reaction, or a polymorphic change, is in progress. This condition is univariant ($F = 1$); the temperature and pressure of the equilibrium are thus inseparably linked, their locus forming a line in a PT diagram.

As a simple ternary example we can consider Fig. 31, which takes the case that the third component C has neither polymorphs, nor intermediate compounds with A or B. Even for this simple case, three statements are required to describe the different phase assemblages at temperatures $<t_1$, $<t_2$ and $<t_3$. These statements are given most succinctly by the triangular figures, which are known as compatibility triangles. A compatibility triangle will normally be subdivided into smaller triangular regions showing the stable three-phase assemblages for various ranges of bulk composition. The reader will already have noticed that the polymorphic change $B_1 \rightleftharpoons B_2$ must be independent of any

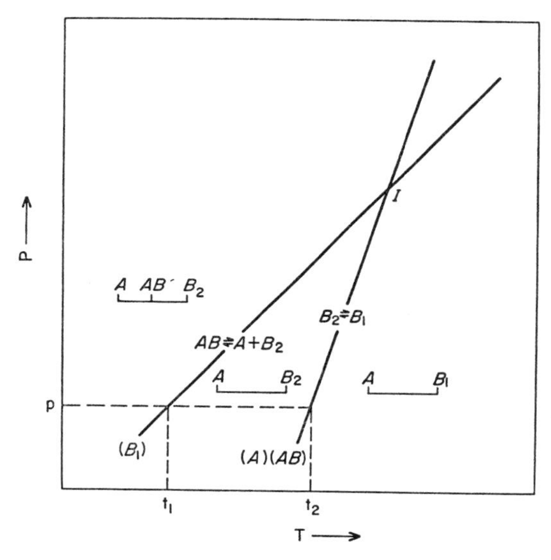

Fig. 30. Schematic PT diagram that might be developed from the points t_1 and t_2 (at pressure p) in Fig. 29. The labels at the ends of the univariant lines (in brackets) indicate any phase that is missing from the univariant reaction. Note that the slopes of the univariant lines are arbitrary, and could not have been predicted from Fig. 29. The bar-symbols in each divariant area are binary compatibility figures showing which pairs of minerals are stable in each area, e.g. in the lowest temperature region the possible pairs are $A + AB$ *or* $AB + B_2$, depending on bulk composition. I is an invariant point (hypothetical) at which $A + AB + B_1 + B_2$ coexist.

other components. In a similar way the equilibrium $A + B_2 \rightleftharpoons AB$ is independent of C (which plays no part in the reaction), so that the subsolidus phase changes in the system A–B–C occur at the same temperatures and pressures as in the binary system A–B. The univariant lines in the PT diagram will be unchanged, but we can put in the compatibility triangles to show all possible three and two phase assemblages in the system, as in Fig. 32. It will be noticed that some simple assemblages can appear in more than one range of temperature and pressure, e.g. $A + C$ occurs over the whole range. The reader is strongly

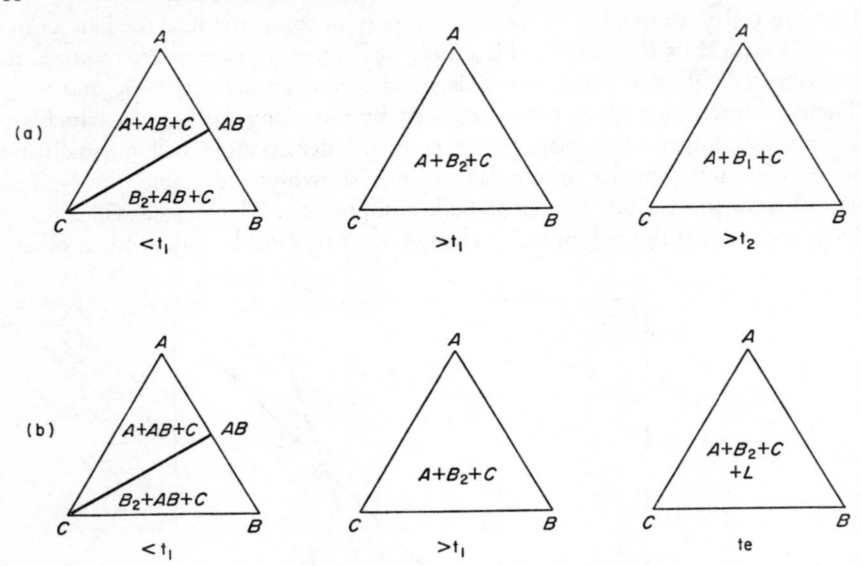

Fig. 31. Compatibility triangles for a ternary in which C is added to A–B (in Fig. 29). Isobaric, at pressure p. Two possibilities are considered because the addition of a third component may reduce the beginning of melting below the temperatures of the binary subsolidus changes. (a) The case where the polymorphic transition temperature (t_2) lies below the ternary beginning of melting (te). (b) The case where melting begins (te) below the polymorphic transition temperature (t_2). The third triangle describes the ternary (isobaric) eutectic at pressure p. (Cp Fig. 25.)

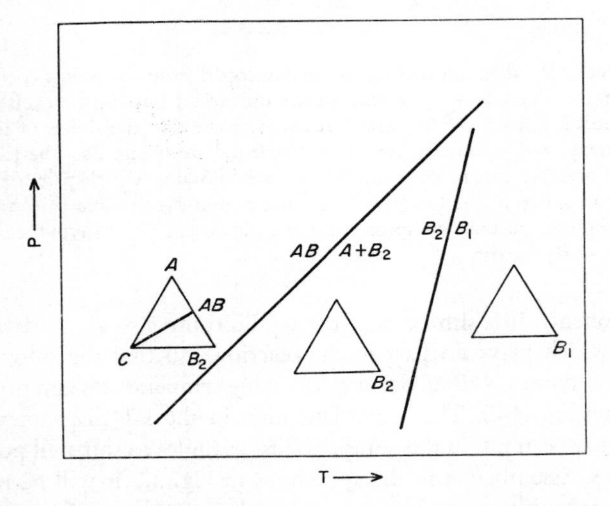

Fig. 32. PT diagram for the case represented by Fig. 31(a). The positions and slopes of the univariant lines are the same as those for the binary in Fig. 30.

urged to write down, for this and subsequent cases, what phases or phase assemblage are necessary to specify each range of conditions in the PT diagram. The results are always illuminating, for instance, the intermediate range in Fig. 32 is not specified by the *presence* of any special phase, only by the absence of AB and B_1!

If a third component, D, forms an intermediate compound with A below the solidus it introduces much more variety because there is a range of possibilities for the overlap, or not, of the stability regions of the compound AD with AB and with the polymorphs B_1 and B_2. If the thermal stability range of the new compound AD is below that of the polymorphic transition, $B_1 \rightleftharpoons B_2$, the distinction between the polymorphs can be ignored for purposes of discussion. Entirely new compatibility triangles are required to describe the new possible assemblages. Their existence, and sequence, depend on the relative stabilities of AB and AD, as shown in Fig. 33, which describes an example where the relative thermal stabilities of AB and AD change with pressure. From the compatibility triangles for the two pressures, p_1 and p_2, it is possible to construct a schematic PT diagram (Fig. 34) for the sub-solidus of the system, A–B–D, (below the polymorphic transition $B_2 \rightleftharpoons B_1$). The more intrepid reader may care to try fitting a polymorphic transition curve to Fig. 34. The construction of Fig. 34 is explained in the caption, because the method is more readily understandable in this way.

An actual example of the type just described is provided by the system forsterite–nepheline–silica (Fo–Ne–SiO_2). This constitutes the base of the "basalt tetrahedron" (forsterite–nepheline–silica–diopside) devised by Yoder and Tilley (1962) to model the behaviour of basaltic compositions. Most of their considerations were devoted to crystal $\rightleftharpoons$ liquid equilibria, but their great intellectual breakthrough came with the concept of different primary basalt liquids, being produced by melting of the different crystalline phase assemblages that would be stable at different pressures in the Earth. Their concept was skilfully illustrated by reference to compatibility triangles in Fo–Ne–SiO_2, by which they showed that the low pressure tie-lines from forsterite and enstatite to albite, were replaced by tie-lines from forsterite and enstatite to jadeite at 33 Kbar: the two relevant compatibility triangles appear at the lowest and highest pressures in Fig. 36. Note here that these triangles are at different pressures but constant temperature. A little consideration reveals the need for other triangles to describe other pressures; firstly because the stability limits of albite and jadeite cannot coincide (except at some possible, but at this stage unknown, invariant point); and secondly, because it is possible that the reactions

$$2\text{Fo} + \text{Ab} \rightleftharpoons \text{Ne} + 4\text{En}, \qquad (1)$$

and

$$\text{Ne} + 2\text{En} \rightleftharpoons \text{Jd} + \text{Fo}, \qquad (2)$$

may occur within the range of conditions under consideration.

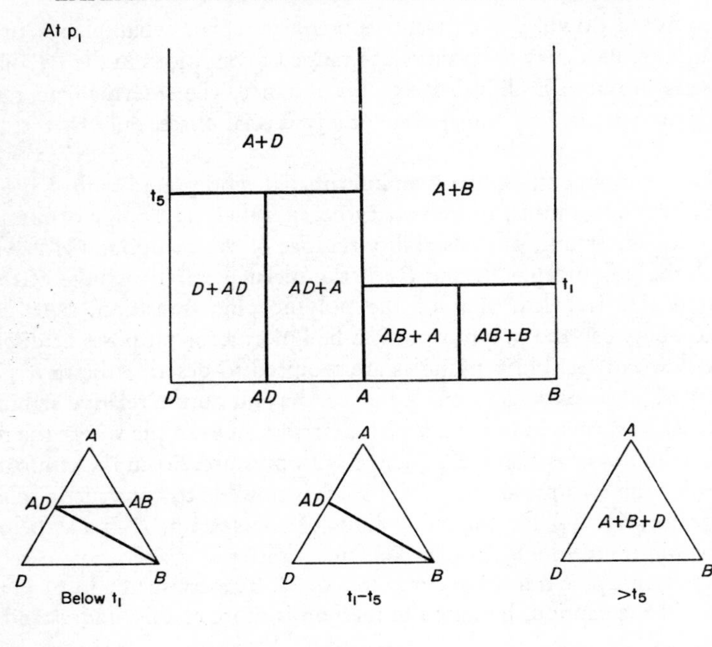
At p_1
A+D
A+B
t_5
t_1
D+AD
AD+A
AB+A
AB+B
D
AD
A
B
A
AD
AB
D
B
Below t_1
A
AD
D
B
t_1–t_5
A
A+B+D
D
B
>t_5

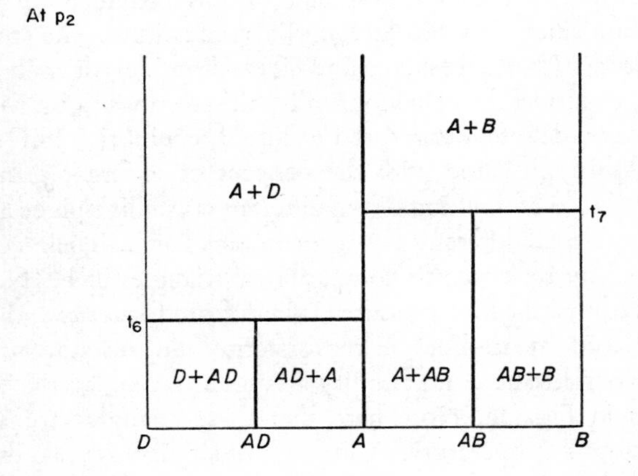
At p_2
A+B
A+D
t_7
t_6
D+AD
AD+A
A+AB
AB+B
D
AD
A
AB
B

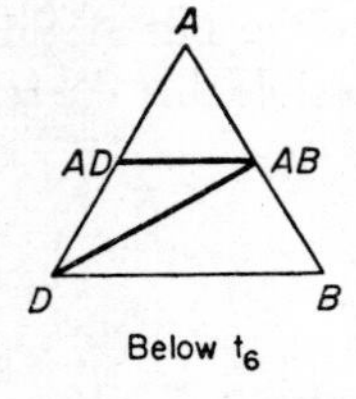
A
AD
AB
D
B
Below t_6

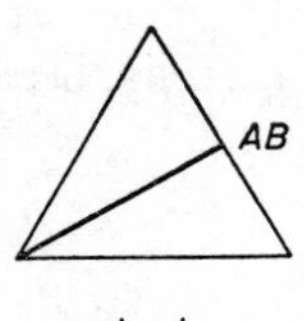
AB
t_6–t_7

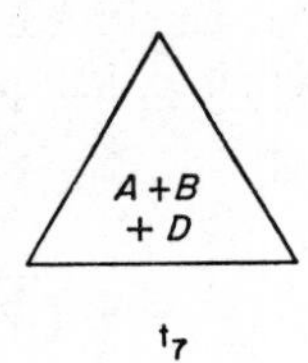
A+B
+ D
t_7

Fig. 33.

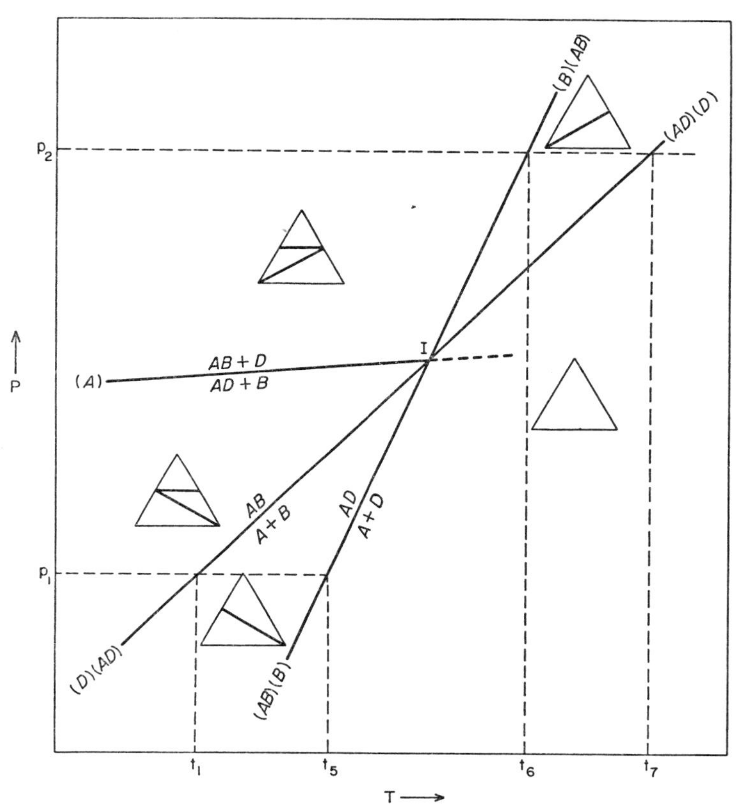

Fig. 34. PT diagram constructed from the information in Fig. 33. Two intercepts on the univariant reaction $AB \rightleftharpoons A + B$ are provided by Fig. 33 (p_1t_1 and p_2t_7), and similarly for the reaction $AD \rightleftharpoons A + D$ (p_1t_5 and p_2t_6). If the reaction curves are taken as linear they must intersect at an invariant point, I, where all five phases are in equilibrium. A fifth univariant reaction curve, $AD + B \rightleftharpoons AB + D$, must also terminate at *I*. No information on univariant reaction, $AD + B \rightleftharpoons AB + D$, is available, so the slope of the line (*A*) is postulated only. Its existence, and relative position are required by the topology of the other univariant lines, and the invariant point *I*($A + B + D + AB + AD$).

Kushiro (1965) showed that with increasing pressure, at 1100°C, the first reaction goes from left to right around 11.5 Kbar, above which the forsterite–albite tie-line is replaced by nepheline–enstatite. At pressures above 22 Kbar jadeite starts to crystallise in appropriate compositions. This latter observation is consistent with the detailed study of nepheline–silica, by Bell and Roseboom (1969), from which the PX diagram (at 1100°C) has been constructed (Fig. 35).

Fig. 33. (a) and (b). TX sections of the binaries A–D and A–B, with compatibility triangles for the ternary A–B–D, at pressures p_1 and p_2 ($p_1 < p_2$). In (b) the temperature t_6 is higher than t_5 in (a).

Bell and Roseboom, incidentally, show that the stability curves for Ab (⇌Jd + Q) and Jd (⇌Ab + Ne) never intersect, so the possible invariant point, referred to above, has no existence. Applying the subsolidus information from this PX binary to the ternary, Fo–Ne–SiO_2, allows the construction of the possible compatibility triangles shown in Fig. 36. The possibilities cannot be fully resolved because the pressure (at 1100°C) at which reaction (2) runs to the right is not known. Kushiro (1965) points out that it must occur between 22 Kbar and 33 Kbar, but the sequence of compatibility triangles depends on whether reaction (2) occurs above or below 28 Kbar (the upper limit of albite at 1100°C): this is why two alternative sequences are given in Fig. 36.

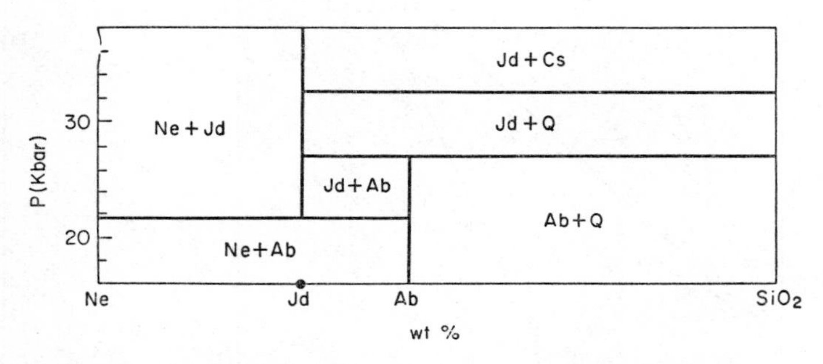

Fig. 35. PX diagram for nepheline-silica at 1100°C (constructed from Bell and Roseboom, 1969).

One general rule is worth knowing when constructing a series of triangles (for either constant T, or constant P). When a new phase (such as jadeite in Fig. 36) appears it will set up new two- and three-phase assemblages with all *adjacent* stable phases. In the general case the new phase will not generate a tie-line that immediately cuts a pre-existing tie-line. If a case were found in which a pre-existing tie-line was so cut, then it would mean that, either the temperature and pressure chosen for the diagram were by chance those of an invariant point, or, the system was not truly ternary (i.e. it is quaternary or higher). An instance from Fo–Ne–SiO_2 is illustrated in Fig. 37.

Similar methods are applied to sub-solidus descriptions of quaternary systems, but here the compatibility figure is a tetrahedron, which encloses a set of smaller tetrahedra, each one representing the four-phase assemblage for a particular range of compositions in a particular PT range. The univariant curves in a PT diagram will describe the loci of conditions, at which various sets of five phases are in equilibrium. Six such univariant curves will meet at an invariant point, which is the unique condition at which six phases are in equilibrium in a quaternary system. Examples of such a PT diagram are given by

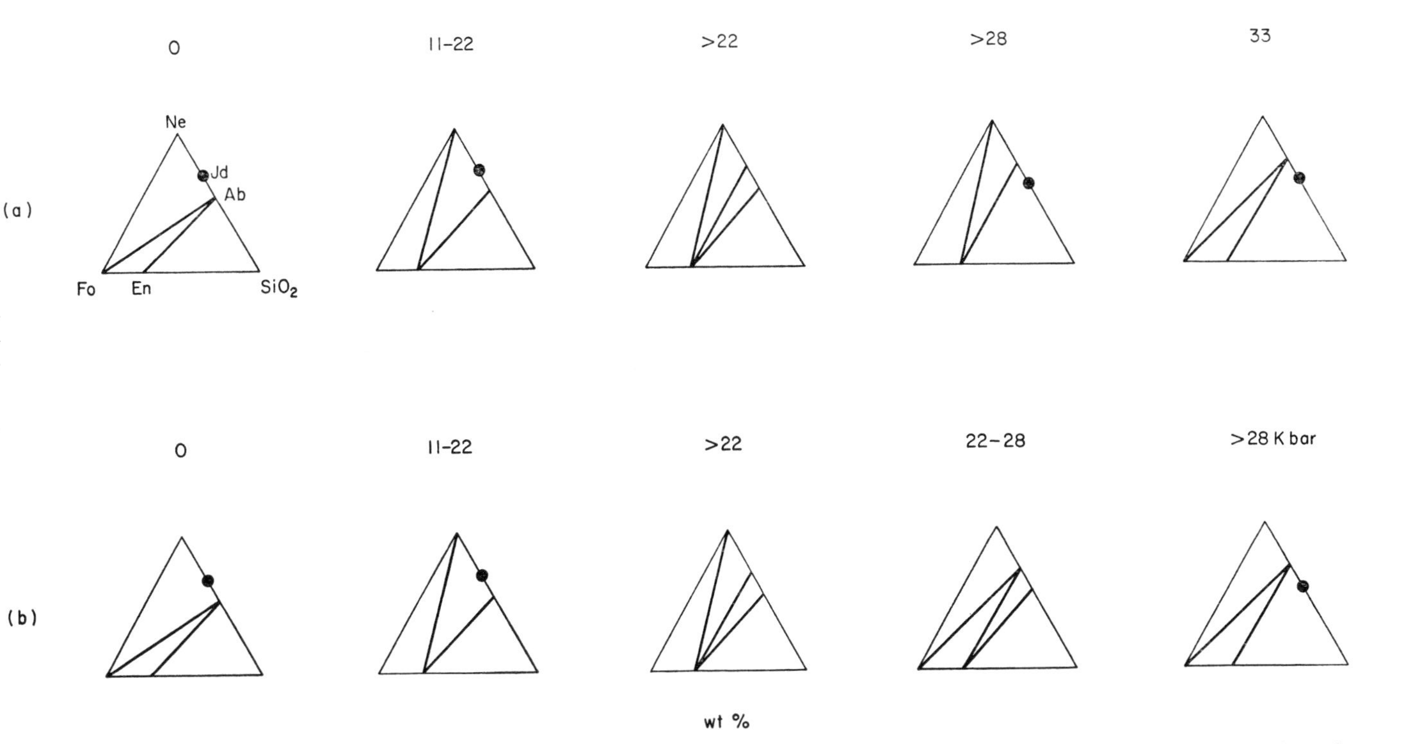

Fig. 36. Compatibility triangles for forsterite-nepheline-silica, for different pressure ranges (all at 1100°C). (a) refers to the condition where the reaction Ne + 2En $\rightleftharpoons$ Jd + Fo occurs at pressures greater than 28 Kbar (the stability limit of Ab). (b) refers to the condition where the reaction occurs between 22–28 Kbar (while Ab is still stable in the presence of quartz).

Luth (this vol., Figs 2 and 17) who also illustrates quinary invariant points (Fig. 17).

For systems higher than quaternary, compatibility diagrams are not possible, and if rigorous description is required it can only be provided in algebraic form. The methods have been described by Greenwood (1967 and 1968). The univariant and invariant conditions can, however, still be accurately described in a PT diagram, which is subject to the conditions previously described. If the number of components is C, then every invariant point must have $C + 2$ phases in equilibrium, and from it must arise $C + 2$ univariant lines, giving the PT loci of every possible set of $C + 1$ phases. In complex systems this leads to complex PT diagrams; some of those that the petrologist

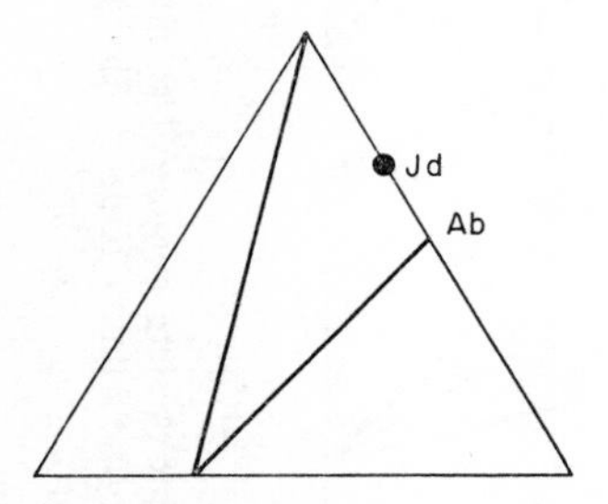

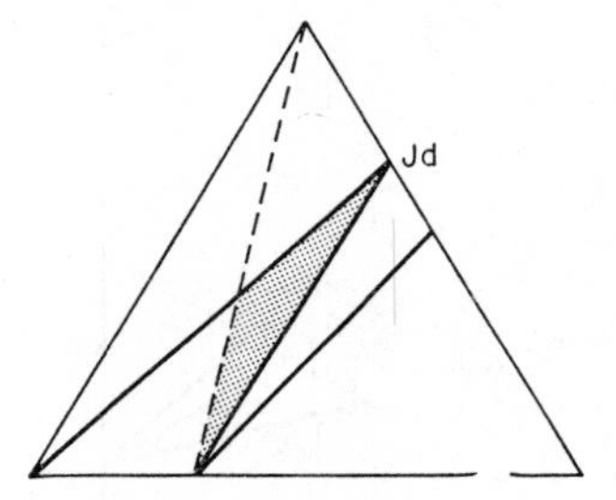

Fig. 37. An example of isothermal compatibility triangles to show that when a new phase becomes stable it will not generally coincide with the collapse of a pre-existing tie-line The first triangle shows the compatible assemblages between 11–22 Kbar in Fo–Ne–Sil (at 1100°C). At 22 Kbar jadeite forms at the expense of nepheline + albite, and new assemblages result. The second triangle depicts the possible case if the En–Ne tie-line were to collapse (replaced by Fo–Jd) as soon as Jd becomes stable: if this were to happen compositions in the shaded area would go from Ne + Ab + En $\leftrightharpoons$ Jd + Fo + En. Five phases occur in this ezuilibrium, which must therefore be ternary invariant. A series of isothermal sections will only fortuitously encounter such a condition.

may find daunting in the literature represent the effort by the experimentalist to treat his data rigorously (see, for instance, Schreyer, Fig. 10, this vol.). If the system is relevant to the group of rocks in which you are interested, and the experimentalist has done his job well, the insight to be gained from the PT diagram will more than repay the effort of understanding it. Considerable circumspection may be needed at first because other lines may be shown in a PT diagram which are not univariant (e.g. liquidus curves) and their intersections are not invariant. A PT diagram is also used frequently to compare results from different systems, when even more care is required! Some of these cases are considered below, but remember, when in doubt always apply the phase rule to any curve or point in a PT diagram. It must be possible for $C + 1$ phases to coexist on the univariant curve and $C + 2$ at the invariant point.

(d) CHARACTERISTICS OF PT DIAGRAMS

In the preceding discussion, PT diagrams were introduced so that the interplay of varying temperature and pressure could be summarised in one figure. Such a diagram can encompass the range of physical conditions that might be expected within a segment of the Earth, and is therefore a powerful way of setting out the low-variance states of any system of geological interest. Consequently, PT diagrams are widely used, but they may convey different sorts of information so some consideration of their construction and properties is needed. The following discussion looks at:

1. The systematic PT diagram describing the low-variance conditions for one particular system; considering first the general case, and then the case of a degenerate system.
2. The PT diagram that combines results from two or more systems.
3. PT diagrams involving liquids.
4. PT diagrams involving gas or fluid (super-critical gas) phases.

(1) *Systematic PT diagram—Schreinemakers analysis*

It was noted above that an invariant equilibrium must be a point in a PT diagram, and in a system of C components there must be $C + 2$ phases at the invariant point. From the point radiate $C + 2$ univariant curves, describing the equilibria of every possible combination of $C + 1$ phases. Between every pair of univariant lines there will be an area which defines the PT region of a divariant equilibrium between different phases.

It will be apparent that the arrangement of the univariant lines, and divariant fields, around an invariant point must conform to certain topological rules if they are to be consistent with the phase rule. The foundations of the topological rules were laid by Schreinemakers (1915–1925) using methods of graphical analysis, and by Morey and Williamson (1918) using algebraic methods. The use of the rules to verify the consistency of the arrangement of lines and areas about an invariant point, in a PT diagram, is generally known as *Schreinemakers analysis*. The method may be used: (a) to construct the complete distribution around an invariant point from a small amount of information (two univariant reaction curves may be enough to predict all the others); (b) to check a postulated construction for consistency; or (c) to construct plausible PT diagrams for geologic systems, using observed parageneses.

The detailed methods for constructing a PT diagram from minimal information, and making rigorous tests of complex PT diagrams, are normally the domain of the experimentalist, and a full discussion would not be appropriate here. In some instances, however, a petrologist can gain valuable insight into a natural system if he knows the rationale of PT constructions, and the simple constraints imposed thereby on any published diagram. For an extensive

TABLE 2

Arrangement of phases A, B, C and D in the binary system, with possible bulk compositions 1, lying between A and B, 2, lying between B and C, and 3, lying between C and D thus:

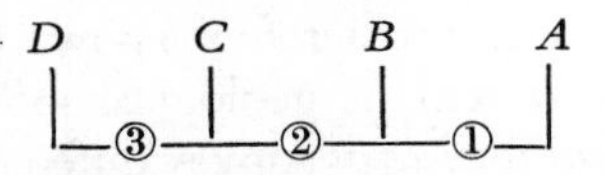

Observed divariant assemblage (critical)	Critical assemblage found in bulk compositions	Possible additional assemblages in this PT range	Compatibility pairs	Phases not stable in this region	PT region in Fig. 40
$A + D$	1, 2 and 3	Nil	D A	B and C	I
$B + D$	2 and 3	$A + B$	D B A	C	II
$C + B$	2	$A + B$, and $C + D$	D C B A	Nil	III
$A + C$	1 and 2	$C + D$	D C A	B	IV

UNIVARIANT REACTIONS

Deduced from critical assemblage	Forming boundaries for critical assemblages	Curve labelled by missing phase
$A + D$	$C = A + D$ $B = A + D$	(B) (C)
$B + D$	$C = B + D$	(A)
$A + C$	$B = A + C$	(D)

The coexistence of $C + B$ (critical assemblage for PT region III) requires that the divariant fields of $A + B$ and $C + D$ must overlap in the divariant field $C + B$ (region III): thus in the above table $A + B$ is listed a possible additional assemblage in regions II and III, while $C + D$ is listed under III and IV.

treatment of the Schreinemakers method the reader may consult Niggli (1954). For any reader requiring a comprehensive coverage, Zen (1966) has brought together all the relevant information from diverse sources (some of which are not easily accessible). His concise, lucid treatment includes diagrams of all possible PT forms for unary, binary and ternary systems, and these are illustrated by some natural examples. In the account which follows, the aim is more modest—it is designed to illustrate the principles underlying all PT constructions, so that the reader can better appreciate published figures.

For the purposes of illustration, we may consider a binary system in which four solid phases (A, B, C, D) may coexist at an invariant point ($C + 2 = 4$, if $C = 2$). The linear compositional relationship between the four binary phases may be described by a line, with the four different phase compositions shown at intervals along it. This is illustrated in the accompanying Table 2, which sets out possible observed assemblages and the deductions that these permit. It is possible to follow the subsequent discussion without reference to the Table, which is provided as additional illustration, and a synoptic statement of the basic information. From the invariant point four univariant lines must radiate, dividing the PT diagram into four sectors, which are divariant areas. Each univariant line describes the equilibria between three phases, and each divariant area is the PT region in which two phases are at equilibrium.

The first rule that can be applied is the Morey–Schreinemakers rule, which says that the angle about the invariant point occupied by a divariant assemblage is never greater than 180°. This condition is illustrated in Fig. 38, for the divariant binary assemblage $A + D$.

It is immediately apparent that at each of the univariant boundaries, $A + D$ must be joined by another phase. Because of the simple nature of the example taken, we can say that one boundary must be where B, and the other where C, enter the equilibria. The following reactions may be deduced:

$$A + D = B \qquad (C),$$

and $$A + D = C \qquad (B).$$

By standard practice, the label for each reaction is given by the phase (written in brackets) which is absent from the equilibrium. On Fig. 38, it will be noted that the choice of which curves should be labelled (C) or (B) is arbitrary. Their existence is required, as boundaries to the field $A + D$, and they must enclose the field at an angle less than 180°, but there is no specific information on their position in the PT diagram.

From the above relationship, however, it is possible to go on to deduce the relative positions of the other two univariant lines (A) and (D). It is a corollary of the Morey–Schreinemakers rule that the univariant curve from which a given phase is totally missing, e.g. (A), must lie on the side of the reaction where

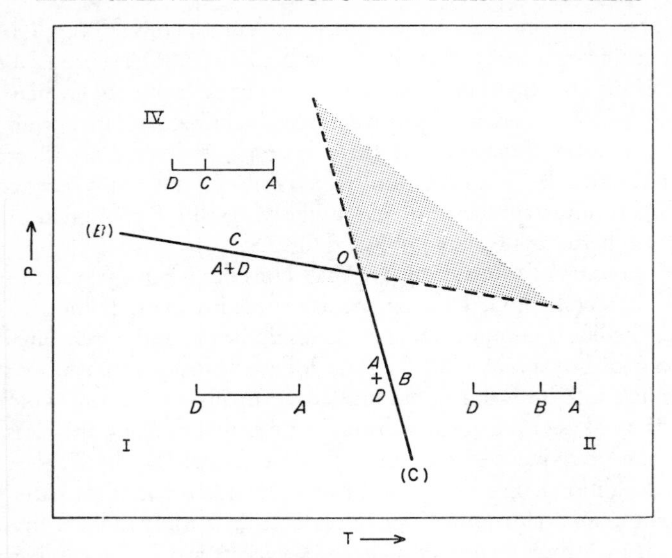

Fig. 38. Schreinemakers analysis of the distribution of univariant lines in a PT diagram about an invariant point (*O*) in the binary system *A*–*D* (with intermediate compounds *B* and *C*). Solid lines indicate the stable curves (broken lines—metastable extensions), with the reactions shown along the curves. The label on the end of a univariant line, in brackets, is the phase missing in that univariant reaction. Compatibility bars (as in Table 2) show all the possible compatibility pairs in the divariant area between two univariant lines. Roman numerals indicate pressure–temperature regions listed in the Table. The shaded region between the metastable curves shows the region in which curves (*D*) and (*A*) must lie.

that phase is missing in all the other univariant curves. For instance, (*A*) must be on the same side as phases *B* and *C* in the above reactions. The relationship in the first equation (*C*) may be expressed in Morey–Schreinemakers terms as follows:

$$A + D = B$$
$$(B)/(C)/(A)(D).$$

Meaning that curves (*A*) and (*D*) will lie on the opposite side of (*C*) from (*B*).

The second equation (*B*) can be expressed:

$$A + D = C$$
$$(C)/(B)/(A)\,(D).$$

Meaning that (*A*) and (*D*) must lie on the opposite side of (*B*) from (*C*).

These two limitations are satisfied only if the univariant curves (*A*) and (*D*) lie in the sector between the metastable extensions of (*B*) and (*C*) through the invariant point (shaded in Fig. 38). The relative positions of (*A*) and (*D*) in this sector are not yet specified, but they may be deduced as follows:

First, the curve (*D*) may be drawn in any arbitrary position within the sector as in Fig. 39. This curve must describe equilibria between the three phases, *A*, *B* and *C*, among which the only possible reaction is,

$$A + C = B, \qquad \text{(D)}.$$

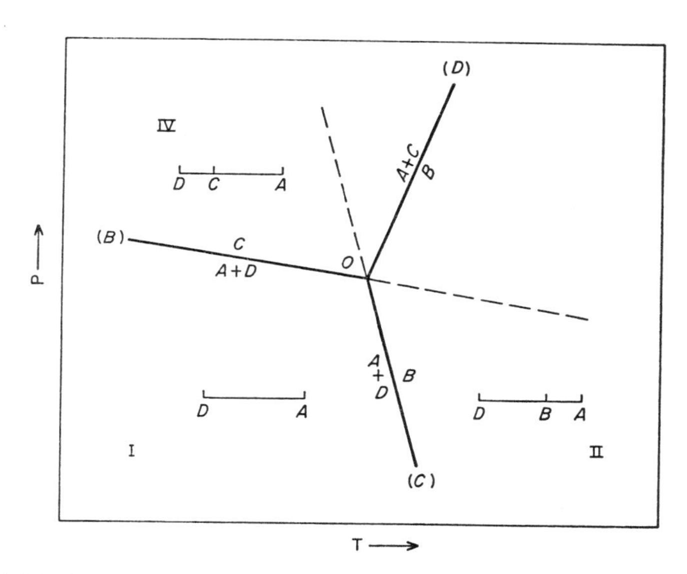

Fig. 39. Positioning curve (*D*). Symbols as in Fig. 38.

The direction of the reaction across this curve (*D*) must be consistent with the reaction directions on the established curves (*B*) and (*C*). The reaction across the curve (*B*)

$$A + D = C \qquad \text{(B)}$$

can generate two stable divariant assemblages depending on bulk composition. If the bulk composition lies somewhere between *D* and *C*, the assemblage above the curve (*B*) will be $D + C$ (phase *A* used up). If the bulk composition lies between *A* and *C*, the assemblage above the curve (*B*) will be $A + C$. These two possibilities are summed up in the compatibility figure above curve (*B*) in Figs 38 and 39. The reaction across the curve (*D*) must, therefore, have $A + C$ on the side compatible with curve (*B*), as shown in Fig. 39. In this arrangement, phase *B* also appears on the correct side of (*D*) with respect to curve (*C*).

Having established the required direction of the reaction across (*D*) we can apply the Morey–Schreinemakers expression to the equation:

$$A + C = B$$
$$(B)/(D)/(A)(C),$$

which states that curve (*A*) must lie on the opposite side of (*D*) from curve (*B*).

The univariant equilibria along (*A*) must involve the three phases *B*, *C* and *D*, among which the only possible reaction is

$$B + D = C \qquad \text{(A)}$$

The only possible direction for this reaction across the curve (*A*) is that which has the divariant assemblage $B + D$ on the same side as curve (*C*) (see Fig. 40, and the compatibility figure between (*C*) and (*A*)).

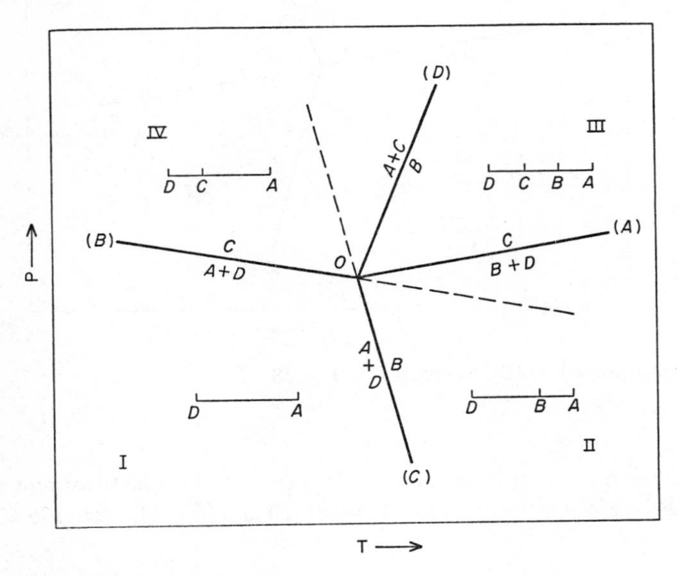

Fig. 40. Positioning curve (*A*). Symbols as in Fig. 38.

This working has produced a PT diagram in which the arrangement of lines and sectors is mutually consistent about the invariant point, in a manner required by the phase rule. It is worth noting the assumptions along the way (taking the Morey–Schreinemakers rule as axiomatic). In order to arrive at the final construction we had to know (or postulate) (a) the nature of the reactions; and (b) the slopes and reaction directions of the first two univariant curves, (*B*) and (*C*). If the positions of (*B*) and (*C*) in Fig. 38, had been reversed we would have derived a consistent set of univariant lines in reversed rotational order about the invariant point. If (*B*) and (*C*) were assigned different slopes, the size and position of the divariant sector, $A + D$, would vary accordingly. As worked out then, Fig. 40 specifies only the consistent relationships between the lines: it tells nothing about their slopes or the actual position of the invariant point, i.e. their true positions in PT space are still unspecified. The power of this method is revealed, however, if we take stock of the information that would be required to make the PT diagram more specific:

(a) *One* determination of the pressure and temperature of *one* reaction (say on curve (*C*)) will anchor the whole arrangement, and fix the rotational order of the curves.
(b) Two determinations on curve (*C*) will give its slope.
(c) Two further determinations on curve (*B*) will then fix its position and slope, and hence, fix the position of the invariant point, and define the sector in which (*A*) and (*D*) must lie.
(d) Only one determination of the pressure and temperature on each of curves (*A*) and (*D*) will then be needed to fix their positions and slopes (because the invariant point is already known).

It is also possible to fix the positions of curves using thermodynamic data on the reactants, such as entropy and volume. Good density values for minerals, and their observed parageneses, can also be combined with the above method to produce a plausible PT diagram for a natural system. The Schreinemakers method, therefore, if coupled with only a small amount of data, is a powerful means of extending our understanding of any system.

Degenerate systems

A univariant reaction involving a small number of components will persist in systems with a greater number of components, in certain ranges of composition. For instance, the polymorphic change from α quartz $\rightleftharpoons$ β quartz, which is a univariant equilibrium involving only one component, will persist into all systems in which free silica can exist as a phase. When a system of C components contains univariant curves which can be described with fewer than C components, it is said to be degenerate. This condition arises when the system contains polymorphs, when three or more phases are colinear in composition, four or more phases are coplanar, and so on. A binary system can be degenerate only by the presence of polymorphs; a ternary system can additionally become degenerate if three phases form a binary reaction (three phases colinear); a quaternary also becomes degenerate if four phases form a ternary reaction (four phases coplanar). When a univariant curve does persist into a higher system, it means that along the curve $C + 1$ phases are present, but not all of them are taking part in the reaction. It means, too, that the same univariant curve will apply equally to a range of systems, as exemplified by a polymorphic transition. It is clear that the condition of degeneracy will be particularly relevant to natural systems, for, although never expressly stated, it is a prime requirement for the setting up of any widely-applicable "petrogenetic grid" (Bowen, 1940) and is basic to the concept that particular minerals, or small-number mineral assemblages, may be used as indicators of metamorphic grade in rocks, which

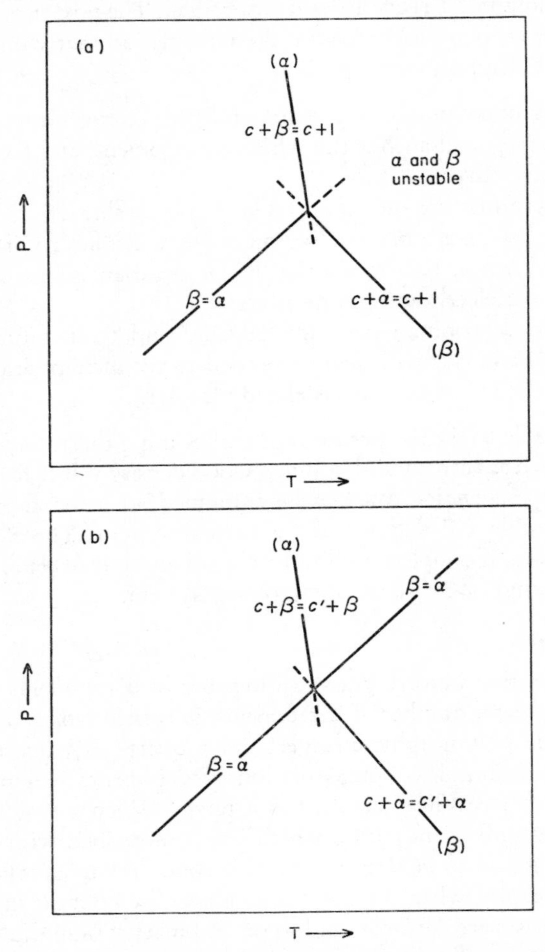

Fig. 41. Degenerate univariant lines and invariant point in a system of C components in which one compound has two polymorphs (α and β) in the range of conditions considered. (a) The polymorphic transition terminates in the invariant point, because β is eliminated in the univariant equilibrium, (α), and α is eliminated in the univariant equilibrium (β). (b) The polymorphic transition curve courses through the invariant point, without change of slope. In this case α or β is stable in all parts of the diagram.

are obviously multi-component systems. In fact, if univariant curves (reaction boundaries) are to be used to define PT ranges of metamorphism, the more phases there are in the reaction the more restricted is the composition range for which the boundary will be applicable. In this sense, degenerate systems are the most important.

(i) *Degeneracy by coincident phases—polymorphism.* When a polymorphic transition curve (univariant) meets a normal univariant curve (with $C+1$ phases), an invariant point occurs, but this is not of the non-degenerate type. From a non-degenerate invariant point, $C+2$ univariant lines emerge, but from the degenerate point involving a polymorphic transition only three or four univariant lines will emerge. The relationships are shown in Fig. 41. If the univariant line with $C+1$ phases is the boundary of the stability region for the polymorphic phase, only three lines are possible. If, however, the polymorphic phase is not destroyed, but merely participates in the reaction, as one of the $C+1$ phases, the polymorphic transition curve courses through the invariant point, giving four lines but with only three different slopes.

The first case (3 univariant lines) arises whenever the composition of the polymorphic phase lies *within* the composition range defined by the other C components. The second case in Fig. 41(b), (four lines), occurs when the polymorphic phase lies on the boundary of the system, i.e. one or other of the polymorphs is present in every univariant and divariant assemblage (sometimes described as the necessary components being "in excess").

(ii) *Degeneracy by colinear phases.* The most important common feature of PT diagrams for degenerate systems is that two or more univariant lines assume the same value of slope at the invariant point. In the above examples (Fig. 41) all possible univariant lines, other than (α) and (β), have the same slope as the polymorphic transition curve, and in the first case they all coincide completely. In a similar manner, a binary reaction curve (three phases) extended into a ternary or higher system will produce the effect that two or more univariant lines will have the same slope at the invariant point. In a ternary system this may result either in five lines meeting at the invariant point, or four only. Five univariant lines is, of course, the same number as in an undegenerate ternary system ($C+2$), but if a binary reaction is involved two of the five lines will have the same slope, but on opposite sides of the invariant point. This case arises when the specified binary lies on the boundary of the system. If the binary lies *within* the system the number of univariant lines reduces to four. In essence, this is the case where the binary three-phase equilibrium is terminated in the PT diagram (just as the polymorphic transition curve was terminated in Fig. 41(a)). In this case two univariant lines in the ternary coincide along the binary reaction curve: an example is sketched in Fig. 42, where it should be noted that the binary univariant curve has two labels, indicating the coincidence of two ternary curves. In all cases of degenerate systems, two or more univariant lines will have the same slope on either side of the invariant point, or will coincide completely, this is known as the "coincidence rule".

An example that illustrates the former case, where a three-phase binary forms a boundary to the ternary system, was used in Fig. 34, which shows two bounding binary systems generating four of the curves at a ternary invariant point. Both binary curves course through the invariant point without change of slope.

Zen (1966) has compiled examples of all possible configurations for binary and ternary degenerate systems. Graphical representation of the compatibility

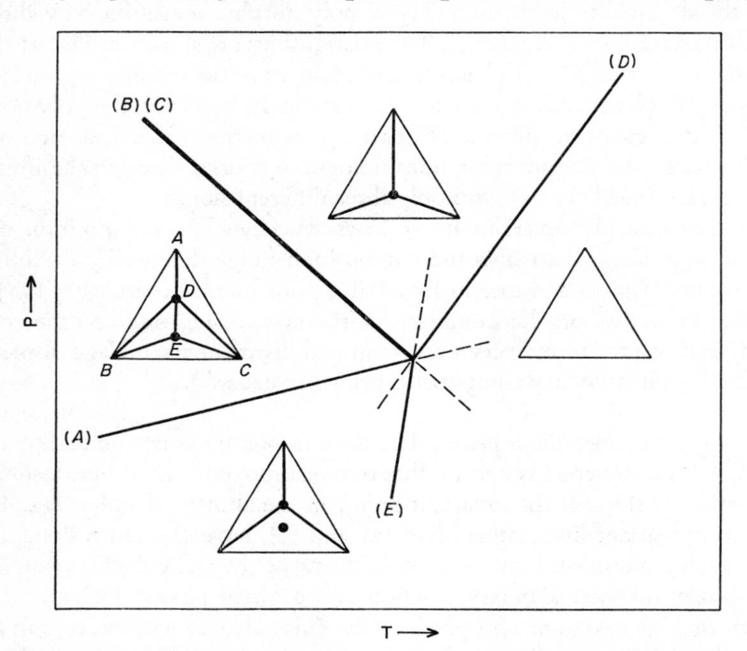

Fig. 42. Degenerate conditions in the ternary *A–B–C* when it *includes* a binary *A–E* (with intermediate compound *D*). The heavy line is the degenerate univariant curve for the binary reaction, $D \rightleftharpoons A + E$, in which phases B and C are indifferent—hence the double label (*B*) (*C*). The reader should deduce the reactions, and the direction of reaction on curves (*A*), (*E*) and (*D*).

figures for systems of components greater than four is impossible (even for four it is difficult). Algebraic methods are then needed for systematic descriptions of the PT diagrams.

In the description of a degenerate univariant reaction, the reactants are described as *singular* phases, whilst the inert phases are described as *indifferent.*

(2) *Combined PT diagrams of different systems*

It is common practice to present PT diagrams which combine data from several systems. This is a powerful means of summarising and comparing a great deal of complex information, and is especially useful in metamorphic petrology, representing the concept of the "Petrogenetic Grid" (Bowen, 1940). Diagrams presenting all the most modern data are given in the subsequent sections (Greenwood, Figs 1 and 2). Greenwood's Fig. 2(b) (reproduced here as Fig. 43) is a good example, showing many univariant curves, but his statements about the limitations of the curves and their applications to rocks should not go unheeded. Some of his general points are repeated here, but there need be no apology for stressing the need for caution in the interpretation and use of combined PT diagrams.

It must always be borne in mind that the univariant curves were determined experimentally in small systems, and such curves may be significantly modified, or disappear completely, in certain composition ranges in a larger system. For instance, many curves from the system $MgO-SiO_2-H_2O$ must be modified by addition of Al_2O_3, and yet further changed by addition of FeO, and so on. Curves defining a dehydration reaction will depend on the H_2O in the system, and will be sensitive to any factor that modifies the activity of water: curves involving iron-bearing phases will be dependent on the redox state of the system. Nor should the possibility of solid solutions in an expanded, or natural, system be forgotten—any solid solutions will increase the degrees of freedom. The effects of moving to a larger system may be: (a) the disappearance of a curve, because one of the phases is not stable in the new composition region (it is eliminated by another reaction); (b) shifts in the curve; or (c) expansion of the curve into a band, because the assemblage becomes divariant in the higher system. Whilst the curves may be used, with good analytical data from natural assemblages, to define *mineral facies*, they can, as emphasised by Greenwood (this vol.) provide only a broad framework for *metamorphic facies*.

In Fig. 43 many of the univariant lines are seen to cross. It is a useful exercise to examine a few of the intersections to decide if, or under what conditions, they might constitute invariant points. If an intersection is an invariant point for a particular system, is it degenerate? Will any of the curves shown as crossing terminate at such an invariant point?

(3) *PT diagrams involving liquids*

When melting occurs in any system it is a singular condition, because all the solid phases (and any vapour or supercritical fluid) must contribute to the liquid phase. Continued melting will eventually eliminate one (or more) of the pre-existing phases, after which all the remaining phases will be in equilibrium with a *continuously changing* liquid phase. This does not alter the principles of constructing a PT diagram, but univariant curves involving liquid (or fluid) are describing equilibria in which the composition of one phase is changing continuously across the diagram. Furthermore the divariant areas between the univariant curves represent broad regions in which the composition of one of the phases (liquid) can be known only by the provision of additional data, either thermodynamic or experimental.

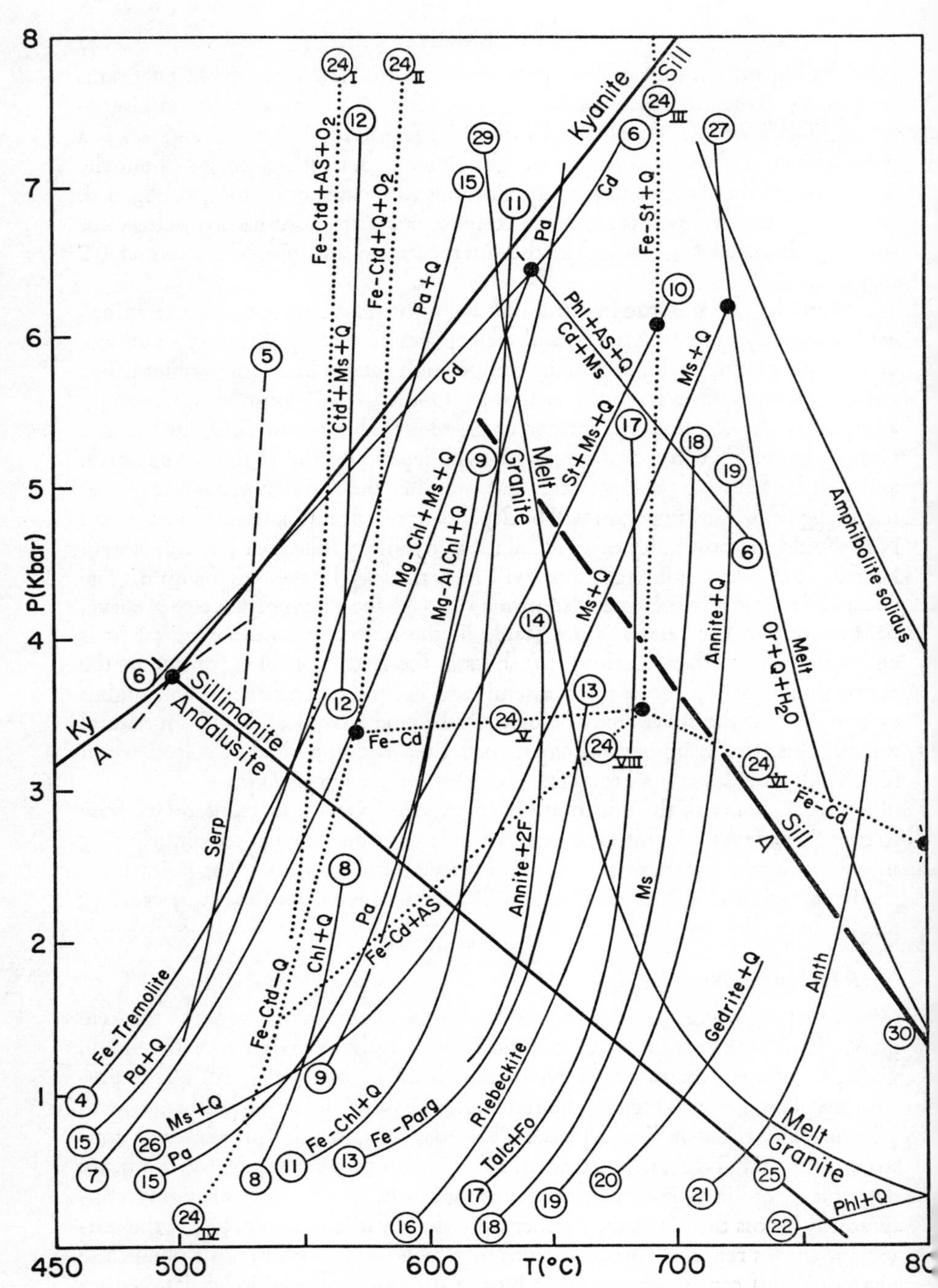

Fig. 43. Univariant equilibria from various systems plotted in one PT diagram. See Greenwood (this vol., Fig. 2(b)) for data sources.

When a PT diagram is used to depict liquid equilibria that are *not* low-variance states, it is well to recognise its ambiguities. A "beginning of melting" curve will represent either a univariant condition, or a series of thermal minima compositions if one of the solids is a major solid solution. In either event it is a low-variance line from which the melting assemblage cannot depart until one of the coexisting phases is eliminated. This may be contrasted with a "liquidus" line, such as is commonly shown in PT diagrams. Such a line represents a multivariate condition. At any point on the liquidus a change of temperature *or* pressure will not necessarily eliminate one of the phases, it may simply induce more of the solid phase to crystallise. In addition, the liquid composition along a PT "liquidus" curve is *constant*, because it refers to *one* specific composition only. It cannot have general application over a range of composition, because every bulk composition has its own unique liquidus (as a glance at any simple binary system will confirm). A further ambiguity can arise when dealing with natural systems, because PT liquidus curves may cross, or partly coincide—for instance, an olivine tholeiite might have an olivine liquidus partly coincident with an olivine nephelinite; in both instances olivine is in equilibrium with liquid at the same temperature and pressure, but the two liquids represent different chemical systems. Similar considerations apply to other PT plots of curves describing equilibria between the liquidus and the solidus. Although these may give useful insight into crystallisation sequences, and relative stabilities of minerals, they are strictly applicable only to the specified bulk composition, and should certainly not be confused with *univariant* equilibria.

If the solidus (beginning of melting) is a univariant curve, rather than a minima line, and the curve has been defined experimentally, it describes an important condition because all phases present will contribute to the initial liquid, which is common to all mixtures of these phases, no matter what the proportions. This condition thus characterises the system in a special way.

A discussion of equilibrium curves for reactions between crystals and liquids is an appropriate place to examine the question of the slope of lines in PT diagrams. Any line in a PT diagram must depict a boundary between two different phase assemblages: the two assemblages will have different molal volumes (densities). Whether the boundary has a positive or negative slope depends on whether an increase in heat content (or rising temperature) favours the low density side or the high. In all cases the assemblage of smaller volume (greater density) must lie on the high pressure side of the boundary, as shown in Fig. 44. For both positive and negative slopes, at constant temperature, the volume will be reduced on crossing the boundary from the low (p_1) to the high pressure side (p_2). Increasing the temperature at any point on either of the two curves produces the opposite effect, in one case the high density assemblage is favoured (negative slope, $1 \rightarrow 2$), in the other the low (positive slope, $2 \rightarrow 3$). This consideration applies to all phase assemblage changes, but is especially important when liquids are present because the volume changes may be great, the slopes may change markedly with changing compressibility of the liquid, and, as seen in the following section, the slopes may be reversed in the presence of a gas or supercritical fluid.

In the absence of a gas or fluid that can react with liquid, the melting of a pure solid, or mixture of solids, commonly produces a liquid of greater molal volume (lower density). Exceptions to this are rare, e.g. ice ⇌ water at low pressure. In general then, those curves in a PT diagram which define conditions of increasing liquefaction (in the absence of a reactive gas or fluid) will have positive slopes, i.e. dT/dP will be positive.

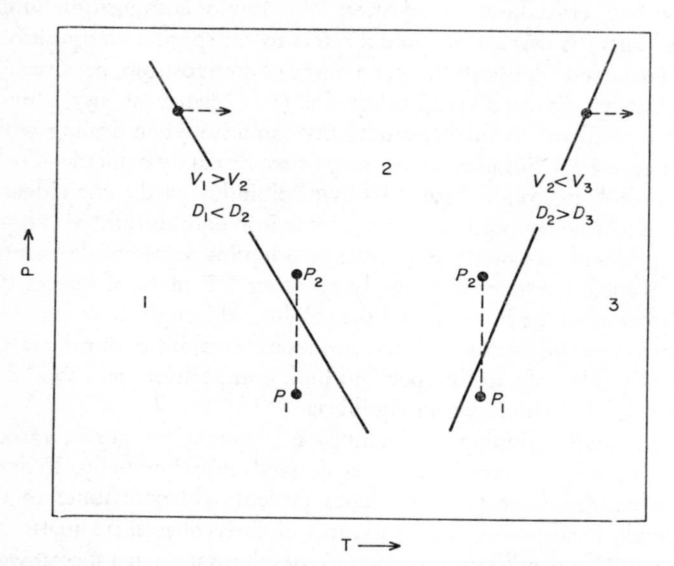

Fig. 44. Slopes of univariant lines, showing the dependence on the response of the reaction to increasing pressure and increasing temperature. V = Molal volume, D = Density ($= 1/V$). Assemblage 2 is more dense than 1 or 3.

(4) *Systems containing gas or supercritical fluid*

Since the development of the Tuttle cold-seal pressure vessel, many experiments have been made in sealed capsules, containing the solid starting material and an excess of water; and many phase diagrams have been published in which P_{H_2O} is plotted against T. The experimental technique has been widely employed, partly because of the need to study systems containing hydrated minerals and hydrated liquids, but also because H_2O facilitates reactions and speeds up the achievement of equilibrium throughout the charge, even when there are no hydrated phases. When P_{H_2O} is given as one axis of a PT diagram it normally means that P_{H_2O} is equal to the total pressure throughout the system. Strictly speaking, this should be expressly stated on the diagram ($P_{H_2O} = P_{total}$) but this is not always done. When $P_{H_2O} = P_{total}$, (or there is "excess H_2O"), it means that a hydrous vapour or supercritical fluid is present in all phase assemblages in the diagram, but this information is frequently omitted because the investigator "takes it as read". In subsolidus reactions among anhydrous phases, the presence or absence of H_2O may make no difference; in such cases the reactions are degenerate in terms of a total system which includes H_2O. In the same way reactions which do not include carbonates or hydrates, would be independent of the presence or absence of CO_2. But for equilibria containing hydrates the presence or absence of a hydrous vapour (fluid) may make profound differences. Before looking at cases, we should consider what is being presented in a phase diagram which uses P_{H_2O} as a variable. Essentially, one of the compositional variables in the system has been singled out and expressed as an *intensive* variable. It is easy to do this with a gaseous component, because its chemical potential or activity can be expressed in terms of its partial pressure, indeed, in ideal states they are directly equivalent. A P_{H_2O} diagram is therefore a form of activity diagram; this, of course, is perfectly valid in terms of the phase rule, which deals with equilibrium states, at which the activity of any component is equal in all phases. Thus a P_{H_2O} diagram is a device for describing equilibria in terms of variation in the activity of one of the components. P_{H_2O} is an easily comprehended variable because everyone is familar with the concept of total pressure. But it is worth bearing in mind that the non-ideal behaviour of H_2O (and other geologically interesting gases), means that the relationship between partial pressure and activity is not linear at high temperatures and pressures. Any tendency for dissociation, or solution of other components, will also modify the activity of H_2O. When $P_{H_2O} = P_{total}$ these matters are of little consequence, but when $P_{H_2O} \neq P_{total}$ there is less risk of ambiguity if H_2O is expressed in terms of its relative concentration in the system. If we wish to describe the behaviour in a system in which P_{H_2O} may vary independently of total pressure, we require three-dimensional co-ordinates; normally three mutually perpendicular axes, one for temperature, one for pressure, and one for P_{H_2O}. The third axis is, in a sense, a special expression of the varying concentration of water in the system and, indeed, diagrams may be constructed from experiments made simply on compositions containing various percentages of water. When $P_{H_2O} = P_{total}$ throughout the system, however, the two axes can be made to coincide: the activity of water is then directly related to the total pressure in all states, and all parts of the system.

The stability field of a hydrate in a PT diagram may be bounded by a curve depicting its breakdown by dehydration, or by its melting curve, or by a combination of these. In the case where the stability field is defined by dehydration, a hydrate will have maximum stability when $P_{H_2O} = P_{total}$. A detailed discussion of this case, and the geological inferences that may be drawn from it, is given by Newton and Fyfe (this vol.). When the hydrate stability field is bounded by the melting curve a similar condition generally applies, but it is worth noting that in certain cases the thermal stability may be slightly enhanced by the presence of another gas, as in the case of phlogopite melting in the presence of CO_2 (Yoder, 1970). Thus, when P_{H_2O} departs seriously from P_{total} there will be major changes in the stability fields of hydrates, and in the univariant curves of reactions involving hydrates, which must be a major factor in the generation of granulites, and in some cases, eclogites. It is worth noting here that in rocks at high pressures it is most unlikely that P_{H_2O} could be less than P_{total} unless H_2O were mixed with another gas or vapour was absent. The alternative would require the existence of pure H_2O in communicating pore spaces that were grain-supported, and it is unlikely that such stress gradients could be maintained by mineral grains for appreciable periods.

Variation in P_{H_2O} is plainly important in subsolidus reactions, as reference to detailed examples in the sections on metamorphic rocks will show, but one of the most remarkable effects is seen in melting curves, where the difference between $P_{H_2O} = 0$ and $P_{H_2O} = P_{total}$ normally means a reversal in slope of the melting curve in any given system.

(vi) *Melting in the presence of* H_2O. For an anhydrous silicate system, the melting curve in a PT diagram normally has a positive slope, because there is an increase in total volume of the system when melting occurs (the broken line in Fig. 45). Addition of an extra component will lower the melting temperature at

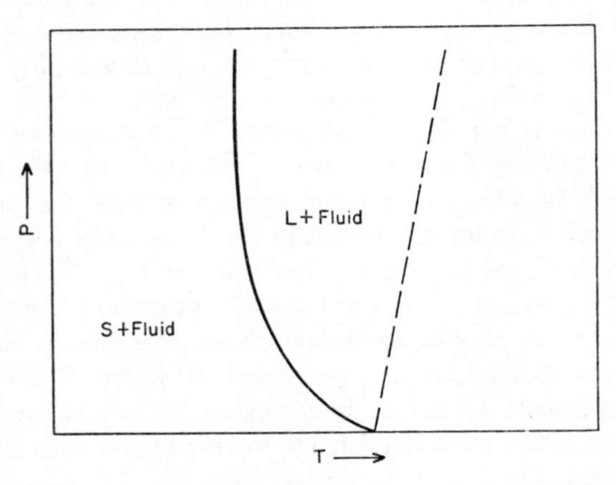

Fig. 45. Melting curve for anhydrous solid (S) when $P_{H_2O} = P_{total}$ (solid line). Broken line is the melting curve when $P_{H_2O} = 0$.

any given pressure—this is the well known phenomenon of "depression of the freezing point"—and addition of H_2O is no exception. In fact, with silicate liquids, the solution of light molecules produces a big depression of the temperature of solidification, because these molecules break the polymeric silicate structures in the liquid. If the available H_2O is insufficient to saturate the liquid, however, the slope of the melting curve is still positive. But when the amount of H_2O exceeds its solubility in the liquid the slope of the melting curve is negative until very high pressures are encountered. The relationships are summarised in Fig. 45.

The change to a negative slope is a consequence of the fact that at the condition of maximum solubility, the reaction at the melting curve is crystals + vapour (fluid) $\rightleftharpoons$ liquid, which at moderate pressures means a reduction of *total* volume on melting. This is because the H_2O occupies a greater volume in the vapour state than when it is dissolved in the silicate liquid. Two opposed effects seek to modify the negative slope of the melting curve. Silicate melts can dissolve greater amounts of water with increasing pressure, which tends to enhance the negative tendency, but at the same time, because of its relatively high compressibility, the molal volume of H_2O decreases rapidly with pressure,

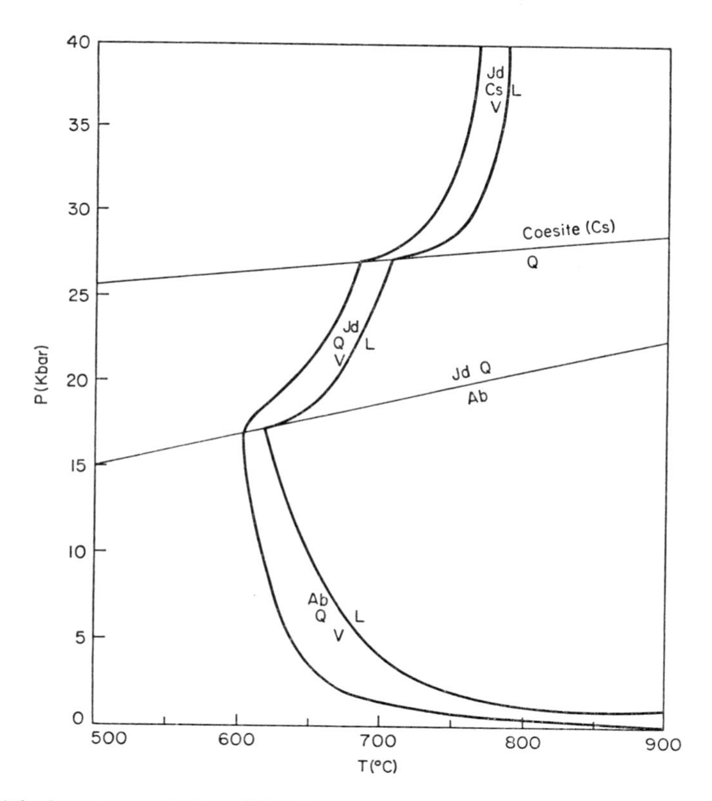

Fig. 46. The lower curve is the solidus for a biotite granite in the presence of excess water. For comparison, the upper curves show the experimentally determined positions of univariant reactions in the SiO_2-excess portion of the system $NaAlSiO_4$-SiO_2-H_2O. (After Boettcher and Wyllie, 1969.)

which diminishes the negative tendency. With rising pressure, the compressibility effect becomes more and more dominant, so that the melting curve bends round towards the vertical, and eventually becomes positive. The inflection to a positive slope may be given a filip in some systems by the inversion of a solid phase to its high density polymorph, because the volume change on melting is increased, e.g. quartz ⇌ coesite in Fig. 46. See also Luth, Figs 1 and 2.

As was noted previously, along an anhydrous melting curve the composition of the first-formed liquid changes continuously with pressure. In the case of the hydrous melting curve the compositional change in the liquid is even greater because of its continuously increasing solution of H_2O.

Between the two extreme curves in Fig. 45 there will be a family of curves describing conditions where P_{H_2O} is less than P_{total}. The forms of these curves will vary according to the way in which P_{H_2O} varies in relation to P_{total}. One possibility is that P_{H_2O} and P_{total} are always in a fixed ratio, which would generate the sort of curves shown in Fig. 47. In a natural system this situation

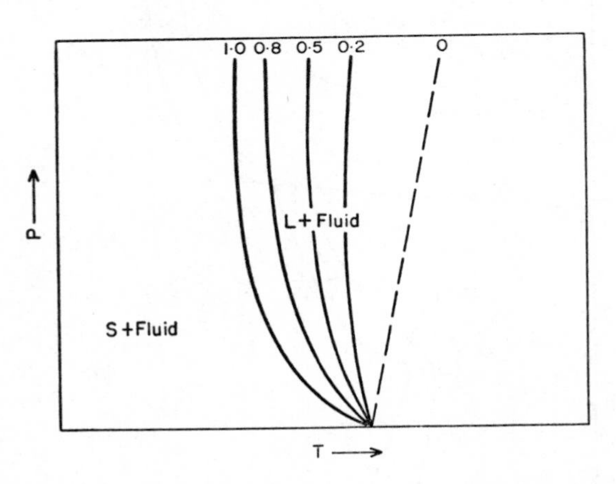

Fig. 47. Melting curves for fixed values of P_{H_2O}/P_{total} (ratios indicated on curves). Ideally, the same curves would represent conditions where H_2O is mixed in fixed proportions with an insoluble vapour ($X^V_{H_2O}$ constant at the ratio on each curve).

would be approximated by rock melting in the presence of a pore fluid, in which H_2O was always mixed with an insoluble gas (such as CO_2) in a proportion fixed by an external source. Effectively, an external environment would be buffering the composition of the pore fluid.

A second possibility is that P_{H_2O} is limited (Fig. 48). Under this condition the low pressure portion of the melting curve will be identical with $P_{H_2O} = P_{total}$, because at low pressures the available H_2O will exceed its solubility in the melt.

When the available H_2O equals its solubility in the liquid, the melting curve departs from the $P_{H_2O} = P_{total}$ curve, because beyond this, P_{H_2O} forms a continuously diminishing proportion of the total pressure. The curve rapidly inflects to a positive slope, similar in form to the anhydrous melting curve, as shown in Fig. 48. This represents an idealized general case where, above a certain

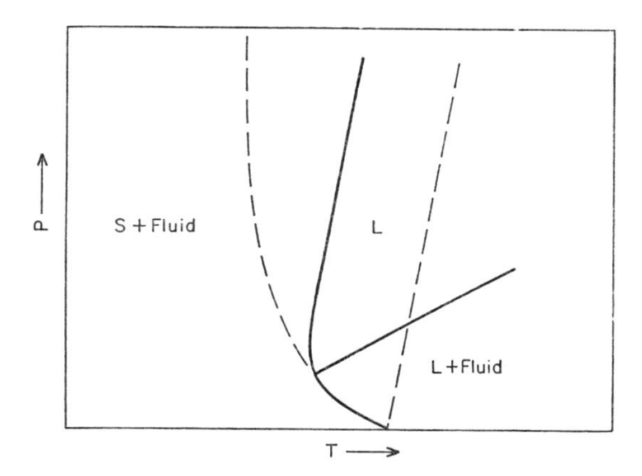

Fig. 48. Melting curve when P_{H_2O} is limited to a maximum value. ("Fluid" = hydrous fluid). This departs from the water-saturated melting curve (broken) when the solubility of H_2O in the melt exceeds its availability. Above this point, in the liquid field (L), any co-existing fluid is not hydrous. A field of liquid + hydrous fluid will be present in the lower pressure, higher temperature region, as shown.

value of P_{total}, the hydrous pore fluid is increasingly diluted with an insoluble vapour, so that P_{H_2O} is held constant. The condition may be approximated by "dry" melting of a mineral assemblage containing a hydrate, which sets the P_{H_2O} of the initial melt below the breakdown curve for the hydrate, but above the water-saturated melting curve; an excellent detailed discussion of the course of closed-system melting of mica schist is given by Burnham (1967). The curve for melting controlled by hydrate breakdown will not be exactly the same as that for a constant P_{H_2O} because the hydrate breakdown curve describes a condition of increasing P_{H_2O} with total pressure.

(e) MIXED GAS SYSTEMS

In a later chapter, Greenwood examines the effects of gases in mineralogic systems, and especially where H_2O and CO_2 are combined, because of their crucial role in carbonate systems. It will suffice here, to outline the general effects of H_2O and CO_2 mixtures, in a coexisting gas (fluid) phase.

In a sub-solidus reaction involving a hydrate (without carbonates), P_{H_2O} will be the vital factor, and the effect of adding CO_2 will be to reduce P_{H_2O} below P_{total}, and thus reduce the stability of the hydrate. In this case CO_2 is behaving as an inert component, in that it does not take part in the reaction, but it profoundly affects the reaction by reducing P_{H_2O}. High P_{CO_2} seems to have been the vital, but cryptic, factor in the granulite metamorphism (low P_{H_2O}) of the Adirondacks, as Greenwood points out.

In the reverse situation, the addition of non-participating H_2O to a carbonate system reduces the stability of a carbonate, as is well known from examples such as the formation of wollastonite from calcite plus quartz:

$$CaCO_3 + SiO_2 \rightleftharpoons CaSiO_3 + CO_2.$$

This reaction is obviously dependent on P_{CO_2}, and the stability of the carbonate will be maximised when $P_{CO_2} = P_{total}$. Such a reaction is dependent on P_{total}, T, and P_{CO_2}, and can be fully described only in a three dimensional (three axis) diagram. The P_{CO_2} can be varied by mixing H_2O, which effectively dilutes the CO_2, in the vapour (fluid).

Equilibria involving both hydrates and carbonates depend on the *interplay* of P_{H_2O} and P_{CO_2}, the added complexities of which are discussed in detail by Greenwood. Whilst a gas phase is present and is a binary mixture of CO_2 and H_2O it is still possible to describe the equilibria in three dimensions; P_{total}, T, and an axis showing the proportions of CO_2 and H_2O. Often the presentation is in the form of isobaric sections, showing T *v* CO_2/H_2O.

It is not possible to describe in a rigorous graphical way, a range of PT conditions where $P_{CO_2} + P_{H_2O} \neq P_{total}$.

(f) VARIATION IN P_{O_2}

When a system contains a multivalent element the equilibria between phases containing that element will be dependent on the activity of oxygen, or, as it is more commonly expressed, oxygen fugacity. The activity, or fugacity, is also loosely expressed as P_{O_2}, as in the case of H_2O and CO_2. Again, the equilibria can be described in a three dimensional diagram with axes for P_{total}, T, and P_{O_2}, but another method is sometimes used which calls for comment. This method shows a given reaction as a curve in a PT diagram, but the curve applies to *changing* P_{O_2} across the diagram! The warrant for this seemingly extraordinary procedure derives from the experimental method of oxygen buffers (Wones and Eugster, 1965). In this method the charge (containing excess H_2O) is sealed in a capsule that is permeable to hydrogen (e.g. Pt or Ag) and surrounded by a reservoir of an equilibrium reaction mixture of metal, metal oxide or silicate (plus water); this "buffer" generates a specific hydrogen fugacity at any given temperature and pressure. The hydrogen fugacity of the external buffer is imposed on the charge, through the permeable capsule walls, and the oxidation state of the charge is thereby fixed. The buffer reaction, and hence the hydrogen fugacity, is a function of temperature and pressure (although the most commonly used buffer reactions are relatively insensitive to total pressure). A PT study of any reaction which is buffered, therefore, will be charting that reaction under a known, but varying, P_{O_2} (P_{H_2}). The variation in P_{O_2} for the standard buffers is given in Fig. 6, and the importance of P_{O_2} is discussed in detail by Greenwood. (See also Appendix II.)

A cautionary note is needed. In attempting to apply the experimental data, derived from buffering, to rocks, the limiting conditions must never be overlooked. One of the most important is that the buffer curves apply only if H_2O is present as a separate phase in the equilibrium reaction. If the rock system was not known to meet this requirement the application of buffer-controlled reactions might be very misleading, as indicated by Whitney (1972). As an example, it has sometimes been claimed that the coexistence of fayalite and magnetite in some igneous rocks indicates that the P_{O_2} during crystallisation was on the QFM (quartz–fayalite–magnetite) buffer. This would only be true if the fayalite and magnetite coexisted with quartz crystals and a hydrous vapour—in all fayalite trachytes known to the author, neither of these two latter conditions is met. Even in rhyolites containing phenocrysts of quartz, fayalite and iron ore, there is usually no unequivocal evidence of a coexisting hydrous vapour phase.

(g) APPLICATIONS OF PT DIAGRAMS TO METAMORPHISM

Metamorphism means change of crystalline form, and each change (if it is to be significant) must mean a change of mineral assemblage. The first appearance of a new mineral in a given range of rock composition is the basis of Barrow's metamorphic grade boundaries, and provides the most easily applied subdivision of metamorphic rocks. Barrow believed that change of grade was a step-wise response to progressive changes in pressure or temperature. Ideally, such changes would represent intercepts on univariant curves in PT space, but because many grade changes involve complex solid solutions, dehydration/decarbonation, or iron-bearing minerals, the "univariant" curves will, in the natural system, be spread into bands. Similar considerations apply to metamorphic facies boundaries, a fuller discussion of this topic being provided by Greenwood. The fact that one rock composition can go through several grades ("univariant" curves) in which new minerals appear whilst the previous grade minerals survive, suggests that the reactions at the grade boundaries are mostly degenerate, i.e. they do not involve all the phases present in the rock. The original staurolite grade gives an example: it is now known that the appearance of staurolite in pelitic rocks is not general, but dependent on special composition. It follows that the common grade boundary reactions in pelites, such as the production of biotite or garnet, must be degenerate in rocks that can form staurolite. It is an interesting implication that if progressive metamorphism results in a progressive increase in the total number of phases present, as in the classic Barrovian zones, then the system reaches non-degenerate univariance only when melting begins.

In any consideration of grade boundaries it is worth differentiating between polyphase reactions and polymorphic changes. Given the appropriate range of bulk composition, a polymorphic change will be unaffected by fluctuations in P_{H_2O}, P_{O_2}, etc., which may cause marked shifts in reaction boundaries. Polymorphic boundaries can, thus, provide the most widely applicable measures of metamorphic grade. They cannot be used in the same way for facies boundaries because the polymorphic phase will not extend through all ranges of composition included in the facies.

Even when $P_{H_2O} = P_{total}$, dehydration reactions occur with increasing T (and ultimately, increasing pressure). Progressive metamorphism is thus marked by a progressive series of dehydration reactions. In the same way, in carbonate rocks, increasing grade (at any given P_{CO_2}) is marked by progressive decarbonation.

If one phase is present in all phase assemblages, e.g. SiO_2 or H_2O, it is customary (and convenient) to set it aside when describing groups of assemblages. This is so usual that there is a danger that the participation of the "excess" phase in some reactions may be overlooked. In a discussion of granulite metamorphism, for instance, De Waard (1965) lists some reactions that lead to the elimination of hydrous phases, under conditions where there is deficiency of H_2O. One of these reactions:

$$\underset{\text{hornblende}}{NaCa_2(Fe,Mg)_4Al_3Si_6O_{22}(OH)_2} + \underset{\text{almandite}}{(Fe,Mg)_3Al_2Si_3O_{12}} + \underset{\text{quartz}}{5SiO_2} \rightleftharpoons$$

$$\underset{\text{orthopyroxene}}{7(Fe,Mg)SiO_3} + \underset{\text{plagioclase}}{NaAlSi_3O_8 + 2CaAl_2Si_2O_8} + H_2O$$

requires a large excess of quartz to eliminate amphibole. A condition of excess quartz will apply to a wide range of rocks, but it should not be overlooked that amphibole may persist in granulite facies rocks that do not fulfill this simple requirement. To some extent, an appreciation of this difficulty is embodied in the maxims of Barrow's metamorphic grades: in the first place a restricted range of composition is laid down, then a new grade is marked by the appearance of the new indicator mineral, *not* by the disappearance of an earlier one. These principles are overlooked in some modern attempts to simplify facies boundaries by relating them to particular mineral stabilities. In the granulite facies, for example, great emphasis is sometimes placed on the *disappearance* of hydrous minerals such as amphiboles, to mark the boundary of the facies. *Persistence* of one mineral may, however, be a vagary of bulk composition, in the complete range encompassed in a facies. This is perhaps another way of saying that although the metamorphic facies principle may allow more precise definition of metamorphic conditions, its application requires far more precise information on rock and mineral chemistry than the simpler system of metamorphic grades.

The upper limit of metamorphism is marked by the onset of melting, which, in the simplest case, is a non-degenerate univariant curve in PT space. Where other univariant curves meet the melting curve some useful invariant conditions will be represented. A simple example would be the beginning of melting of a pelitic rock containing both kyanite and sillimanite: this represents the condition where the polymorphic transition curve intersects the melting curve, as shown in Fig. 49. Such an invariant condition would be represented where migmatites

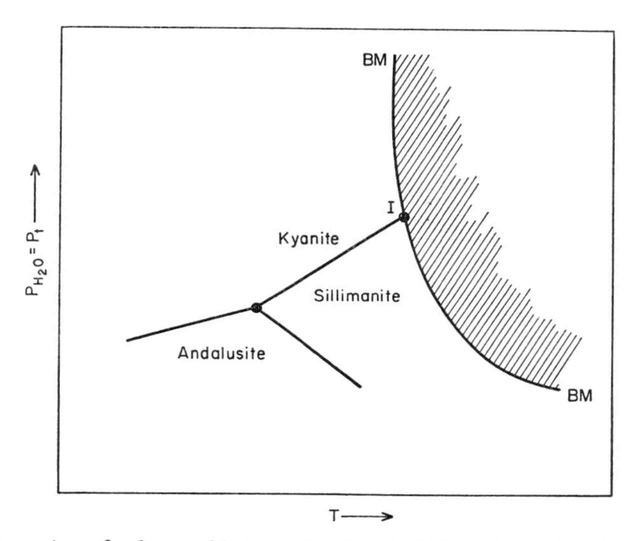

Fig. 49. Intersection of polymorphic transition (kyanite/sillimanite) and hydrous beginning of melting (BM) of sialic rocks (migmatite), at an invariant point I. Note that the position of curve BM will vary with bulk composition of the rocks being melted, and will be especially sensitive to the availability of H_2O.

are found at the kyanite–sillimanite isograd. Using Fig. 43 it is possible to make some estimate of the temperature and pressure, but it will be an *estimate* only, and gives probably a minimum pressure and temperature. This is because the "granite" minimum curve was not determined in the presence of the full complement of pelitic minerals, but also because that minimum melting curve will shift strongly to higher pressures and temperatures if $P_{H_2O} < P_t$. Under the latter conditions the intercept (I) will also shift to higher pressures and temperatures.

REFERENCES

Bailey, D. K. (1974). In "The Alkaline Rocks" (H. Sørensen, ed.). John Wiley and Sons, New York.

Bailey, D. K. and Schairer, J. F. (1964). *Am. J. Sci.*, **262,** 1198–1206.

Bailey, D. K. and Schairer, J. F. (1966). *J. Petrology*, **7,** No. 1114–170.

Bell, P. M. and England, J. L. (1964). *Carn. Inst. Wash. Yr. Bk.*, **63,** 176–178.

Bell, P. M. and Roseboom, Jr. E. H. (1969). *Mineralog. Soc. Am.*, Special Publication No. 2.

Biggar, G. M. and O'Hara, M. J. (1969). *Miner. Mag.*, **37,** 198–205.

Boettcher, A. L. and Wyllie, P. J. (1969). *Am, J. Sci.*, **267,** 875–909.

Bowen, N. L. (1913). *Am. J. Sci.*, **35,** 577–99.

Bowen, N. L. (1922). *Am. J. Sci.*, **3,** 1–34.

Bowen, N. L. (1928). "The Evolution of the Igneous Rocks". Princeton University Press, Princton.

Bowen, N. L. (1940). Petrogenetic Grid. *J. Geol.*, **48,** 225.

Bowen, N. L. and Anderson, O. (1914). *Am. J. Sci.*, **37,** 487–500.

Bowen, N. L. and Tuttle, O. F. (1949). *Geol. Soc. Amer. Bull.*, **60,** 439–460.

Boyd, F. R. and England, J. L. (1960). *J. Geophys. Res.*, **65,** 741.

Boyd, F. R. and England, J. L. (1963). *J. Geophys. Res.*, **68,** No. 1.

Burnham, C. W. (1967). In "Geochemistry of Hydrothermal Ore Deposits", H. L. Barnes, (ed.). Holt, Rinehardt and Winston, New York, 34–76.

De Waard, D. (1965). *Amer. J. Sci.*, **263,** 451–461.

Edgar, A. D. (1973). "Experimental Petrology", Clarendon Press, Oxford.

Ehlers, E. G. (1972). "The Interpretation of Geological Phase Diagrams". Freeman, San Francisco.

Faust, G. T. (1936). *Am. Miner.*, **21,** 735–736, 1936.

Frye, K. (1974). "Modern Mineralogy". Prentice-Hall, New Jersey.

Gibb, F. G. F. (1974). *Miner. Mag.*, **39,** 641–653.

Goranson, R. W. (1931). *Amer. J. Sci.*, **22,** 481.

Greenwood, H. J. (1961). *J. Geophys. Res.*, **66,** 3923–3946.

Greenwood, H. J. (1963). *J. Petrol.*, **4,** 317–351.

Greenwood, H. J. (1967). *Geochim. Cocmochim. Acta*, **31,** 465–490.

Greenwood, H. J. (1968). *XXIII Int. Geol. Congr.*, **6.,** 267–279.

Jamieson, B. J. (1969). *In* "Progress in Experimental Petrology". NERC Report 1, Edinburgh, 152–155.

Kushiro, I. (1965). *Carn. Inst. Yr. Bk.*, **64,** 109–112.

Kushiro, I. (1974). *Earth Planet. Sci. Let.*, **22,** 294–299.

Levin, E. M., Robbins, C. R. and McMurdie, H. F. (eds.) (1964). *Amer. Ceram. Soc.*, Columbus, Ohio.

Morey, G. W. and Williamson, E. D. (1918). *Am. Chem. Soc. Jour.*, **40,** 59–84.

Morse, S. A. (1970). *J. Petrology*, **11,** no. 2, 221–251.

Niggli, P. (1954). "Rocks and Mineral Deposits" (English Ed.). Freeman, San Francisco.

O'Hara, M. J. (1968). *Earth Sci. Rev.*, **4,** 69–133.

O'Hara, M. J. and Yoder, H. S. Jr. (1967). *Scott. J. Geol.*, **3,** 67–117.

Osborn, E. F. and Tait, D. B. (1952). *Am. J. Sci.*, Bowen Vol. II, 413–433.

Presnall, D. C. (1969). *Am. J. Sci.*, **267,** 1178–1194.

Ricci, J. E. (1951). "The Phase Rule and Heterogeneous Equilibrium". Van Nostrand, New York.

Richardson, S. W., Gilbert, M. C. and Bell, P. M. (1969). *Am. J. Sci.*, **267,** 259–272.

Schairer, J. F. (1950). *J. Geol.* **58,** No. 5, 512–517.

Schairer, J. F. (1959). *In* "Physico-chemical Measurements at High Temperatures", ed. Bockris, White and McKenzie. Butterworths, London.
Schairer, J. F. and Bowen, N. L. (1956). *Am. J. Sci.*, **254,** 129–195.
Schairer, J. F., Yagi, K. and Yoder, H. S. Jr. (1962). *Carn. Inst. Yr. Bk.*, **61,** 96–98.
Schairer, J. F. and Yoder, H. S. Jr. (1960). *Am. J. Sci.* **258-A.,** 273–283.
Schairer, J. F. and Yoder, H. S. Jr. (1960). *Carn. Inst. Wash. Yr. Bk.*, **59,** 70–71.
Schreinemakers, F. A. H., (1915–1925). *Koninkl. Akad. Wetenschappen te Amsterdam*. Proc., English Ed., 18–28.
Tuttle, O. F. (1949). *Bull. geol. Soc. Amer.*, **60,** 1727.
Tuttle, O. F. and Bowen, N. L. (1958). *Geol. Soc. Amer. Mem.* **74.**
Whitney, J. A. (1972). *Am. Miner.*, **57,** 1903–1908, 1972.
Wones, D. R. and Eugster, H. P. (1965). *Am. Miner.*, **50,** 1228–1272.
Wyllie, P. J. and Tuttle, O. F. (1960). *J. Petrol.*, **1,** 1–46.
Yoder, H. S. Jr. (1970). *Carn. Inst. Wash. Yr. Bk.* **68,** 236–240.
Yoder, H. S. Jr. (1950). *Trans. Am. geophys. Un.*, **31,** 827–835.
Yoder, H. S. Jr. and Tilley, C. E. (1962). *J. Petrology* **3,** 342–532.
Zen, E-An. (1966). *U.S. Geol. Surv. Bull.*, 1225.

Part II A

Experimental Petrology: Metamorphic Rocks

Section A

Metamorphic Rocks

SUMMARY

In this section experimental metamorphic petrology is reviewed by considering the results and applications in the three major divisions of metamorphic rocks. The first chapter considers the special cases where high pressure is the prime factor in producing distinctive mineralogies. The second chapter reviews experimental results pertaining to the widespread rocks of regional metamorphism, which can be regarded as spanning the range of moderate crustal pressures and temperatures, up to the beginning of melting. The concluding chapter focuses on conditions where temperature is of paramount importance, covering what is generally referred to as thermal metamorphism. The authors have each used different approaches to the subject, consequently there is increased perspective where the ranges of physical conditions overlap.

It is an irony of metamorphic petrology that many of the major facies divisions refer to basic rock compositions, which encompass an enormously complex chemical system. A complete experimental description of the subsolidus of this system is still a distant prospect. In the meantime, therefore, it is fortunate that the boundary conditions of the major facies can be defined in broad terms by the stabilities of individual minerals (as in the first chapter) and by mineral reactions in pelitic and carbonate rocks (as in the second and third chapters).

In the first chapter, on high pressure metamorphism, Newton and Fyfe devote their main attention to the blueschists. Their approach is to look at the experimental stability ranges of the characteristic blueschist minerals, and to relate these to the natural evidence. The effects of changing bulk compositions on mineral stabilities are examined where possible, but the youthfulness of high pressure experimentation in this field is reflected in the paucity of data on complex mineral reactions. Consideration of blueschists leads the authors naturally to a discussion of eclogite stability and, in this context, the special problem where water pressure may be locally much lower than total pressure. Regional low water pressures are then briefly discussed in relation to granulite metamorphism. In the concluding discussion, the authors turn their attention to the possible effects of very high pressures on crystal structures in the mantle.

Experimental conditions in the range representing greenschist and amphibolite metamorphism are the subject of the second chapter. Greenwood's approach is to consider the evolving complexity of experimental systems as they gradually approach natural rock compositions. He focuses attention on the applications to the pelitic and carbonate rocks, and the development of petrogenetic grids based on mineral reactions of low variance. Of special interest is his detailed examination of the effects of different volatiles on mineral reactions; in this he shows how modern quantitative studies of mixed volatile reactions are providing a new cutting edge for petrology.

In the final chapter, Schreyer examines the case where temperature is the crucial factor. His approach is to consider first the stabilities of the major rock-forming minerals, thus defining the boundary conditions of thermal metamorphism. He then proceeds to examine the modifications imposed by additional components, by looking at synthetic systems of increasing complexity, and ending with a consideration of natural systems. Like Greenwood, he concludes from the experimental studies that the strict application of the metamorphic facies principle is unworkable, but may still provide a loose classification for metamorphic rocks. Both authors believe that mineral facies are being accurately defined by the experiments, and that the resulting quantification of the conditions of metamorphism is the prime function of experimental petrology.

1. High Pressure Metamorphism

R. C. Newton and *W. S. Fyfe*

1 INTRODUCTION . . . 101
(a) Problems of time and stress . . . 102
(b) Some thermodynamic principles . . . 103
(c) Fluid pressure at depth . . . 106
(d) Scope of this review . . . 111
2 GLAUCOPHANE–LAWSONITE SCHISTS . . . 112
(a) Geologic setting . . . 112
(b) Introduction to the experimental petrology of blueschists . . . 113
3 ARAGONITE . . . 115
4 AMPHIBOLES . . . 123
5 GARNETS . . . 129
(a) Almandine-rich garnets . . . 130
(b) Spessartine–grossular garnets . . . 133
(c) Hydrogrossular garnets . . . 133
6 MICAS, CHLORITE AND SERPENTINE . . . 134
7 LIME–ALUMINIUM SILICATES . . . 142
(a) Lawsonite . . . 142
(b) Pumpellyite . . . 146
8 SODIC PYROXENES . . . 149
(a) Stability of jadeite and omphacite . . . 149
(b) Stability of analcime relative to jadeite . . . 154
9 ALBITE . . . 156
10 SUMMARY OF GLAUCOPHANE SCHIST CONDITIONS . . . 166
11 OTHER HIGH PRESSURE MINERAL ASSEMBLAGES . . . 169
(a) The ecologite facies . . . 169
(b) Granulites and charnockites . . . 176
12 HIGH PRESSURE PHASES IN THE UPPER MANTLE . . . 179

1. INTRODUCTION

In this chapter we wish to review some current ideas on metamorphism at high pressures. Because of the great interest in the nature of the upper mantle and the interaction of mantle and crust, which forms an integral part of plate tectonic mechanisms, there is good reason to study, as fully as possible, metamorphic events at low crustal levels. Whether or not present experimental data are adequate or reliable, there is no doubt that experimental workers have influenced, if not dominated, recent thinking in this area.

Many writers have constructed diagrams showing the distribution of metamorphic facies in terms of the co-ordinates P (P_{H_2O}) and T; P can be correlated with depth. If we assume that crustal rocks have densities in the range 2.6–3.0, then the lithostatic pressure increases by 260–300 bars per km in depth. To assess possible thermal gradients, we must consider data based on heat flow measurements and data from bore holes. Measurements of heat flow are quite variable and so are estimates of T with depth. Birch (1955) considered such data and from various models concluded that gradients in the range 10–20°C km^{-1} are feasible. High gradients in the range 20–30°C km^{-1} are possible or may be common, and in some hydrothermal areas gradients of 100°C km^{-1} occur. In Fig. 1 we show a facies distribution diagram from Turner (1968). On this diagram we have indicated some possible geothermal gradients. Various geothermal regimes in different environments lead to the concept of facies series (see Turner, 1968). In this chapter we will be mainly concerned with the facies series ranging from glaucophane–lawsonite schists (blueschists) to eclogites and granulites. These are obviously facies which could be encountered at deep crustal levels. Before advancing directly to a discussion of high pressure metamorphism and the experimental work which bears upon it, a few of the important concepts and classic problems of theoretical petrology will be reviewed briefly.

(a) PROBLEMS OF TIME AND STRESS

The temperature and hydrostatic pressure variables of metamorphism are not difficult to reproduce experimentally, but this is not true for some other variables involved in metamorphism. These include stress or departures from hydrostatic conditions, and time. Laboratory times are many orders of magnitude (10^8–10^{10}) shorter than the times involved in a metamorphic cycle. This has encouraged metamorphic petrologists in the belief that metamorphic reactions may approach equilibrium states. There is much obvious evidence (e.g. the finding of all three polymorphs of Al_2SiO_5 on the earth's surface, or even in a single rock sample) to show that this assumption is not generally true. We must thus constantly bear in mind that experiments and rocks may not show equilibrium states; reaction rates are of interest and, at times, may provide valuable limits on tectonic processes (p. 118).

Deformation is a difficult parameter to deal with realistically in the laboratory. Some mechanical properties of rocks are functions of time and mechanism. Some workers have considered that certain mineralogic reactions may have needed rather moderate total pressures but high stress concentrations. These questions are largely unresolved as yet, but a start is being made. Birch (1955) concluded that stress differences of the order of 10^2 to 10^3 bars could persist for prolonged periods.

Recently there have been significant advances in treating the thermodynamics of the stressed state. Coe and Paterson (1969) have measured the effect of directed pressure on the α–β transition in quartz and McLellan (1969) has extended thermodynamic arguments. Perhaps the main conclusion that can be drawn from such studies is that even long-term deformational strain is unlikely to have had a major influence on phase equilibria in the crust. Using the mean pressure in thermodynamic equations will provide a reasonable guess of the effect of elastic deformation.

On the other hand, it is known that stored *plastic* strain energy in artificial and natural materials can reach magnitudes which are thermodynamically significant. For instance, plastic strain damage in quartz crystals acquired during triaxial deformation tests at high temperatures can cause it to invert to coesite at confining pressures far below the stability limit of coesite relative to unstrained quartz (Hobbs, 1968). Naturally acquired plastic deformational energy probably exerts large effects locally on recrystallisation kinetics in rocks undergoing metamorphism, and may eventually be shown to have an influence on mineralogy, although this agency is little understood at present.

(b) SOME THERMODYNAMIC PRINCIPLES

If we compare high pressure and low pressure metamorphic rocks and their mineral constituents certain differences are obvious. High pressure rocks are dense and this results from denser packing of atoms in their constituent minerals. For example, some blueschists contain the pyroxene jadeite in place of the feldspar albite encountered in rocks of the greenschist facies. The large density difference can be correlated with the 6 co-ordination of Al in jadeite and 4 co-ordination in albite. As we move to the upper mantle even more dramatic changes occur, in particular the appearance of structures where Si–O compounds contain silicon in 6 co-ordination (see Ringwood, 1972).

States of low volume are normally states of low entropy, since the two properties are usually correlated. If a phase transition has a large negative volume change it will tend to have a negative entropy change. As the slopes of first order phase transitions are given by the familiar Clausius–Clapeyron equation:

$$\frac{\mathrm{d}P}{\mathrm{d}T} = \frac{\Delta S}{\Delta V}$$

we should anticipate that most such phase boundaries will be positive in slope. Frequently, they tend to have slopes of similar magnitude indicating that entropy and volume changes (ΔS and ΔV) are closely related. If a high-entropy fluid phase is involved in the reaction, slopes tend to be steeper at moderate pressures.

Frequently we wish to describe the variables ($P - T$, etc.) under which a given mineral assemblage formed. Good equilibrium phase diagrams are essential, but normally only allow us to place limitations. It appears that in the future we may rely heavily on isotope fractionation to assign temperatures. Thus Epstein and Taylor (1967) have shown how $0^{18}/0^{16}$ fractionation between rather common minerals can sometimes be used to assign a metamorphic temperature within rather close limits. This type of thermometer has a diffusion-controlled upper limit (Fyfe, 1970). In general, fixing pressure presents greater difficulties.

If certain indicator phases are present, or if certain reactions can be deduced (e.g. kyanite → sillimanite, calcite → aragonite, etc.) some limits on pressure can be set, but these are generally rather broad. At a first glance simple reactions such as these appear to be more useful than complex reactions, but a little reflection shows that multivariant equilibria involving solid solutions may be of greater value. Consider the simple reaction (see Fig. 20).

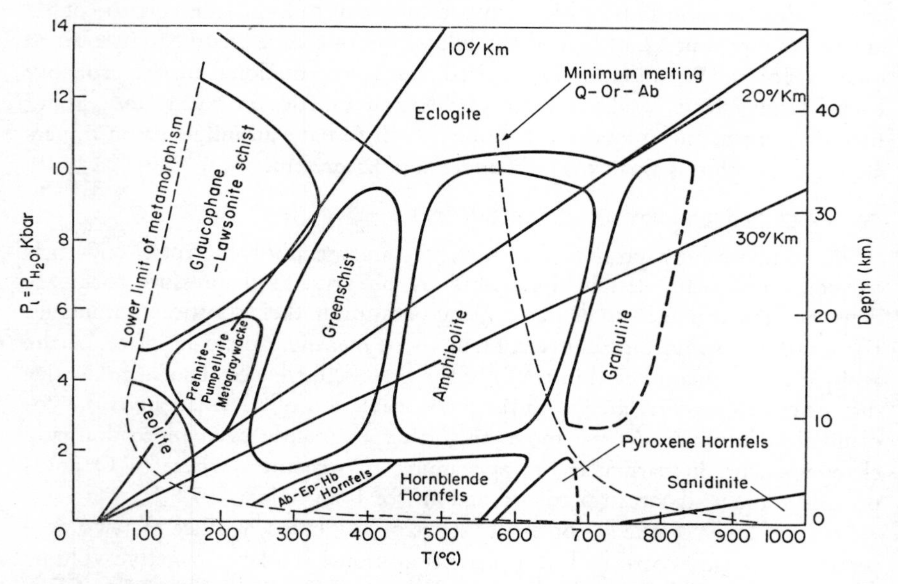

Fig. 1. Generalised pressure–temperature relations of metamorphic facies. (After Turner, 1968.)

$$\underset{\text{albite}}{NaAlSi_3O_8} \rightarrow \underset{\text{jadeite}}{NaAlSi_2O_6} + \underset{\text{quartz}}{SiO_2}$$

The presence of albite indicates lower pressures of crystallisation than the jadeite–quartz assemblage. But if we have solid solutions, e.g.

$$\text{quartz} + \text{omphacite} \rightarrow \text{albite} + \text{pyroxene}$$

$$SiO_2 + \left.\begin{matrix} NaAlSi_2O_6 \\ NaFeSi_2O_6 \\ CaMgSi_2O_6 \end{matrix}\right\} \begin{matrix}\text{solid}\\ \text{solution}\end{matrix} \rightarrow \text{albite} + \left.\begin{matrix} NaFeSi_2O_6 \\ CaMgSi_2O_6 \end{matrix}\right\} \text{s.s.}$$

the phase diagram for such a system will have the form depicted in Fig. 2. In Fig. 3 are shown some experimental data for such a system. It is quite clear that if such systems are present, and if they have been studied experimentally, then a unique pressure can be estimated to match the isotopic temperature. Given these variables, one would then search for a hydrate equilibrium to fix water pressure, an oxygen system to fix oxygen pressure, etc. Thus, the solid solution series

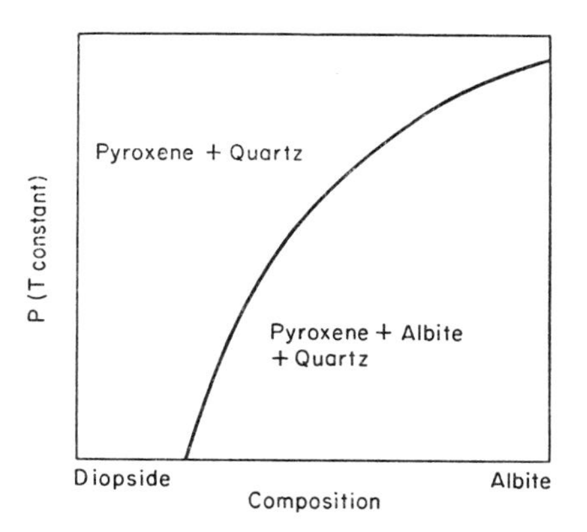

Fig. 2. Form of pressure–composition diagram for the system albite-diopside.

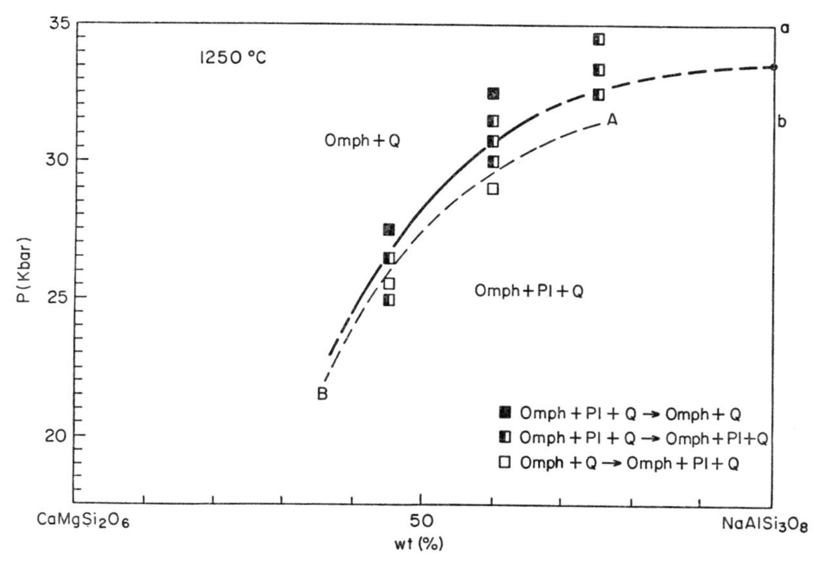

Fig. 3. Phase relations in the system albite–diopside at 1250°C. (After Kushiro, 1969.)

displayed by the majority of minerals provide an excellent opportunity of fixing the variables in a way not otherwise possible.

The number of unique compositional states which may be found in a natural solid-solution series is infinite. Only a limited number can be studied experimentally but frequently interpolations are made by using mixing theory and, often, the simple ideal solution equation. In dealing with any mixing phenomena (or order–disorder phenomenon) some consideration of statistical mixing theory is essential. The origin of the ideal solution equation:

$$\mu_i^{S.S.} = \mu_i^\circ + RT \ln x_i^{S.S.}$$

where x_i is the mol fraction of component i in the solid-solution, μ_i is the chemical potential of i and μ_i° refers to pure i, can be found in the statistical entropy of mixing (see Denbigh, 1955). Such an equation allows us to interpolate solution data and extend known systems to the unique case found in a particular rock. Great care is needed in applying the equation. First it applies only if the heat of mixing is zero. Second, the above equation applies only to random mixing on one atomic site per formula unit. But all these factors can be assessed given a limited amount of experimental data. In the case of Fig. 3, let us estimate the transition pressure for a 50:50 mol % mix (i.e. 45.2 wt % diopside). We shall consider ideal mixing and to a first approximation ignore the anorthite content of the albite. At equilibrium we have:

$$\mu_{Jd}^{ss} + \mu_Q = \mu_{Ab}$$

or

$$\mu_{Jd}^\circ + 2RT \ln x_{Jd}^{ss} + \mu_Q = \mu_{Ab}$$

the factor 2 arising on account of two mixing sites (Ca–Na, Mg–Al) in the pyroxene. We then have

$$\Delta\mu^\circ = -2RT \ln x_{Jd}^{ss}$$

where $\Delta\mu^\circ$ is the free energy change of the Ab $\rightarrow$ Jd + Q reaction at the P and T of equilibrium. The $\Delta\mu^\circ$ may be evaluated approximately as a $\Delta P \Delta V$ where ΔP is the depression of the transition pressure below that for the end-member reaction. Thus, for $x = 0.5$, $\Delta P = 10.4$ Kbar, and equilibrium should occur at about 23.1 Kbar (compared with the experimental value of about 29 Kbar). While the equation describes the general form of the diagram, it clearly is only approximate and one could hardly proceed further without more knowledge of plagioclase compositions. A positive heat of mixing is indicated (see p. 147). Before such equations are used, it is also essential to know from crystallographic data the exact types of sites involved in mixing as the discussion on site preference by Burns (1970) makes clear.

(c) FLUID PRESSURE AT DEPTH

One of the long debated problems of metamorphic petrogenesis concerns the relation of fluid pressures to lithostatic pressure in rocks undergoing progressive metamorphism and deformation. While at this time no final statement can be made, it appears reasonably certain that there may be two common situations.

In one, when fluids are actively evolved, fluid and load pressures may be similar; in the other fluid pressures may be controlled by factors other than, and differ drastically from, the load pressure.

The two situations can be clearly envisaged by reference to a simple experiment. Imagine that we take a container with a close fitting piston. In the container we may place various mixtures in the system $MgO–H_2O$; the hydrate brucite may form depending on the pressure applied by the piston and on the temperature (Fig. 4). If excess water is present (i.e. the mol% H_2O

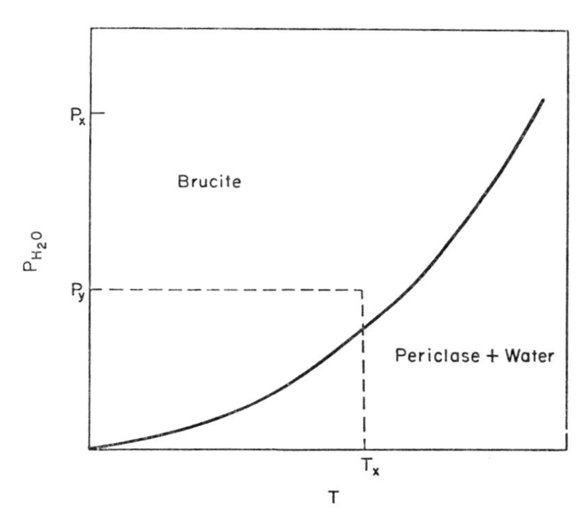

Fig. 4. Form of vapour pressure curve for brucite (for details see text).

exceeds 50%) then at all times $P_{load} = P_{fluid}$. P_{fluid} will nearly equal P_{H_2O}, for the solubility of MgO in the fluid is slight. If the mol% H_2O is less than 50, at low T's and appropriate P's the chamber will contain MgO and $Mg(OH)_2$ and no free vapour phase. We may still define a P_{H_2O} which would be in equilibrium with the system by considering a device whereby we introduce a membrane permeable only to water into the chamber and measuring P_{H_2O}. In this case we would find: $G_{Mg(OH)_2} = G_{MgO} + G_{H_2O}$ when G is the Gibb's free energy. $G_{Mg(OH)_2}$ and G_{MgO} would be determined only by P and T; G_{H_2O} could be given by a function such as $G_{H_2O} = G^{\circ}_{H_2O} + RT \ln P_{H_2O}$ ($G^{\circ}_{H_2O}$ would refer to a standard state. The equation is written as for an ideal gas). At a total pressure P_x and temperature T_x (see Fig. 4) we should find that P_{H_2O} lies at a value such as P_y, above the equilibrium vapour pressure curve. In this case, P_{H_2O} is fixed by the variables P and T and the presence of brucite and periclase. On the boundary curve, $P_{H_2O} = P_{total}$ and is unique for any given T. For the non-permeable system with excess water, P_{H_2O} always equals P_{load}.

It is not difficult to see how such situations could arise in nature. If wet sedimentary rocks are buried and heated, a series of dehydration reactions will occur. If we consider the vast number of reactions that may proceed,

clays → chlorites → micas → anhydrous silicates,
zeolites → epidotes → feldspars,

when almost every reaction involves solid solutions with multivariant vapour pressure curves, then it is probably true that for every temperature increment, dT, there will be a change in water pressure dP_{H_2O} and water will be progressively evolved from the system. The general lack of retrograde metamorphism during unloading shows that most of this water is lost.

When rocks are not deeply buried and highly compacted, one would anticipate a more or less continuous pore system leading to the surface. Comparing the density of liquid water and rocks, we might expect to encounter values of $P_{\text{fluid}} \simeq 1/3\ P_{\text{rock}}$. But as porosity diminishes (and presumably permeability), to obtain finite flow, higher water pressures will be needed, unless water can diffuse easily through crystal structures or along grain boundaries. There is no evidence for this; on the contrary, geological evidence indicates that flow in systems of low permeability is very small, possibly comparable with diffusion coefficients of the less volatile constituents of the rocks.

One might anticipate that the water of dehydration would flow rapidly at depth only if a fluid phase is present and at a pressure equal to load or even slightly greater. This type of behaviour has been seen many times in deep bores (Dickey, Shriram and Pine, 1968; Bredehoeft and Hanshaw, 1968), where at depths below 3–5 km, fluid pressure begins to reach lithostatic pressure. These same high pressures, or even slight excess pressures, may be a necessary requirement for the development of some classes of shallow thrusts; the rocks will be essentially floating on pore fluids.

The second limiting type of environment will be one where large quantities of dry igneous debris are deeply buried without weathering or where igneous rocks are intruded into dry rocks and then metamorphosed. If diffusion does not allow entry of water on a pervasive scale, such rocks could remain dry throughout metamorphism. As we shall see (p. 172) crustal eclogites are probably formed in this situation.

In general, one would expect that wet sedimentation would be more common than dry sedimentation. As metamorphic rocks rarely contain more than 1–5% water, the condition of excess water might be common. Is there any good evidence for this?

Probably, the most convincing evidence arises from the following type of consideration. Let us consider progressive metamorphism as a series of dehydration steps in a system $X - H_2O$ with vapour pressure curves as in Fig. 5. Let us imagine that load pressure and temperature were at point P. If $P_{H_2O} = P_{\text{load}}$, then one phase a will be formed in the rocks. If water is deficient, then we could have assemblages a or $a + b$ or $b + c$ or e $+ d$ or d. In these cases

P_{H_2O} would be fixed by mineralogy. This type of mineralogical chaos is uncommon in progressive metamorphism, though it is seen during partial retrograde metamorphism of dry rocks, as might be anticipated. Frequently, even initially dry rocks may be adequately hydrated if they are enclosed in sediments and have appropriate structures to allow entry of water. Deformation must also play a significant part.

An excellent test of excess H_2O in low grade metamorphism is provided by the zeolite facies reaction:

$$\begin{array}{llll} \text{quartz} & + \text{ analcime} & = \text{albite} & + \text{ water} \\ SiO_2 & + NaAlSi_2O_6 \,.\, H_2O & = NaAlSi_3O_8 & + H_2O. \end{array}$$

This boundary is crossed in many zeolite facies rocks. One of the best described examples is to be found in Coombs (1954). Coombs estimated a minimum

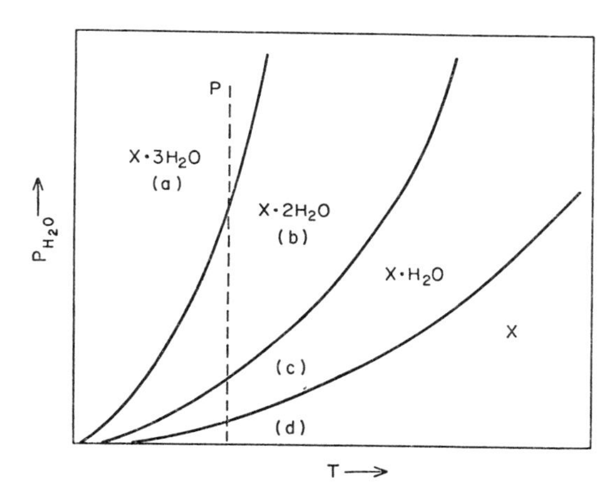

Fig. 5. Normal form of vapour pressure curves for series of hydrates at low pressures.

depth of 5 km for the transition. An upper temperature limit for existence of analcime and quartz has been determined by Campbell and Fyfe (1965) and Thompson (1971). The phase diagram is shown in Fig. 6. It is of unusual form compared to most vapour pressure diagrams, but as we shall see (p. 170), is in fact typical of all such curves for known hydrated minerals except that the regions of negative slopes normally occur at upper mantle and not crustal pressures.

Let us modify this diagram for the conditions that $P_{total} = 3P_{H_2O}$. Consider a point such as x at $P_{H_2O} =$ 1 Kbar and T 195°C. At this point we have:

$$G_{Anal} + G_Q = G_{Ab} + G_{H_2O}.$$

Now let us add 2 Kbar pressure to the solids keeping P_{H_2O} (and G_{H_2O}) constant. The change in free energy of the system (ΔG) must be dependent on the volume changes in the solid phases, as follows:

$$\Delta G = G_{Ab} + \int_{1\ \mathrm{Kbar}}^{3\ \mathrm{Kbar}} V_{Ab}\, dP + G_{H_2O} - G_{Anal} - \int_{1\ \mathrm{Kbar}}^{3\ \mathrm{Kbar}} V_{Anal}\, dP -$$

$$G_Q - \int_{1\ \mathrm{Kbar}}^{3\ \mathrm{Kbar}} V_Q\, dP = \int_{1\ \mathrm{Kbar}}^{3\ \mathrm{Kbar}} \Delta V_{Solids}\, dP.$$

The term ΔV_{Solids} in this case is $-20\ \mathrm{cm}^3$. ΔG will hence be approximately -1 Kcal. Analcime will react with quartz and dehydrate. To retain equilibrium

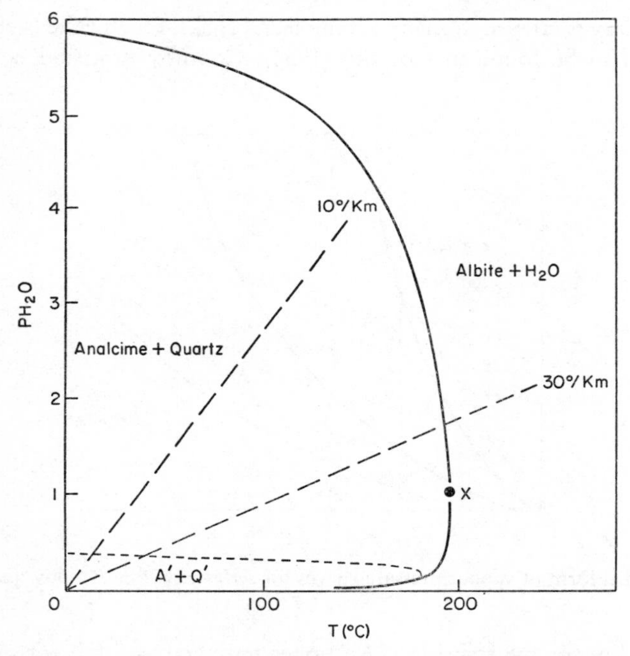

Fig. 6. Phase diagram for the system albite–water at low P and T. The dashed curve shows the approximate analcime field for conditions $P_{total} = 3P_{H_2O}$ (see text).

with P_{H_2O} at 1 Kbar and P_{load} at 3 Kbar we must lower the temperature by an amount determined by the entropy of reaction. As this entropy of reaction is very small in the present case (not greater than 3 cal mol^{-1} deg^{-1} at 190°C) it is clear that analcime is not stable in the presence of quartz even at 25°C. The stability field becomes a loop restricted to very low pressure–temperature regions as shown in Fig. 6. Now if the geothermal gradient was in the range 10–30° km^{-1}, the phase boundary would be crossed at depths of about 1 km. If on the other hand $P_{H_2O} = P_{total}$, crossing would be anticipated at depths exceeding 5 km in accord with observation. The indication is that water pressure must have been high in this thick (10 km+) section of zeolitized volcanic rocks. Glassy volcanic debris could carry large quantities of water to depth.

The model we are considering implies that water will flow through rocks at depth only if a definite fluid phase is present. Even given this condition, passage through massive non-schistose rocks may be difficult. Partially serpentinised ultrabasic rocks demonstrate this impermeability. The serpentine reaction:

$$\text{pyroxene} + \text{olivine} + \text{water} \rightarrow \text{serpentine}$$

involves a large increase in volume of the solid system. Thus when a peridotite is hydrated, it may swell and the accompanying deformation must tend to diminish permeability, even though at shallow depths it could lead to fracture. It is not uncommon to find massive ultrabasic bodies partially serpentinised only for short distances in from joint surfaces (Fyfe, personal observation on Californian ultramafics). Experiments have shown that if water is available the serpentinisation reaction is very rapid (Martin and Fyfe, 1970). In natural cases the reaction must be diffusion controlled.

An analogous diffusion-controlled process has been studied in the laboratory (Martin, 1968). When MgO is hydrated, the resultant $Mg(OH)_2$ occupies a much larger volume. Thus when an $Mg(OH)_2$ layer forms on an MgO crystal, it can block further easy access of water to the reaction interface. From studies of the growth of this interface at temperatures near 300°C, water diffusion coefficients appear to be no greater than 10^{-8} cm sec^{-1}. This diffusion coefficient through a lightly confined brucite layer is approaching the same order of magnitude as some diffusion coefficients of ionic species in solids. Of course, stress build up by the hydration process can induce shear-failure and periodically enhanced permeability.

Over recent years, writers on metamorphic rocks have tended to consider that the condition $P_{H_2O} = P_{total}$ is the general condition, except in carbonate rocks. As we shall see later this condition does not apply for the formation of eclogites or granulites. At levels where partial melting occurs under conditions of very high grade metamorphism, while P_{total} may equal P_{load}, the fluid may be dominantly a silicate melt. These conditions may be appropriate in the generation of some granulites and amphibolites as residues from partial melting.

(d) SCOPE OF THIS REVIEW

There is probably more pertinent experimental work available for minerals characteristic of the blueschist facies than for any other kind of metamorphism. For this reason, the experimental petrology of blueschists at its present level will be reviewed in detail. Experimental petrology of eclogites and granulites will not be analysed so completely because most of the experimental work on these rocks has been done on whole rock compositions, with the focus on broad petrological features rather than on details of mineralogy. Much of the experimental mineralogy which is pertinent to blueschists applies also to eclogites and granulites.

2. GLAUCOPHANE–LAWSONITE SCHISTS

(a) GEOLOGIC SETTING

Few types of metamorphic rocks have attracted more attention in recent years than glaucophane–lawsonite schists, or blueschists. This attention has been fairly evenly divided between field and laboratory. The rocks have a unique mineralogy characterised by dense minerals. They are frequently associated with eclogites, and they have some minerals related to those characteristic of eclogites, which have long been thought to be possible constituents of the upper mantle. They occur in extensive elongate belts characterised by great crustal activity, often on continental-oceanic margins. Hence, such a schist belt contained within a continent could point to a fossil continental margin. Some geophysicists now consider the possibility that crustal materials are dragged deep into the upper mantle by convective forces (Wilson, 1968; Gilluly, 1970). It is interesting to speculate that some dense schists may be explained by such a process.

The scale of the glaucophane schist problem is impressive. Around the Pacific margin large belts or scattered fragments of these rocks are to be found. They have been recorded along much of the West Coast of the U.S.A. and Baja California, in Japan, the Celebes, the Philippines and New Caledonia and New Zealand and Australia. They occur in narrow belts, and are often associated with all the marginal features of mountain-building activity; transcurrent faults, volcanic activity, granitic intrusion, ultrabasic intrusion. Another belt of comparable dimensions runs from the European Alps, through the Greek Islands, Turkey and Northern India. A much older belt is preserved in the Urals. In most of the Pacific area, graywackes are the dominant metamorphosed geosynclinal rocks. They are partly formed from volcanic debris. In California the thickness is known to be great but detailed estimates are lacking, partly because of tectonic complexity. Basic volcanic rocks, such as basalts and spilites, are of secondary importance, as are gabbros and ultrabasic rocks. In the European Alps (Ernst, 1973) there are much greater thicknesses of calcareous rocks. It is in the Alps that the association with crustal eclogites is most clearly demonstrated. In the Zermatt area, Bearth (1967) has described pillow lavas with cores of eclogite and rims of glaucophane schist. This is probably one of the few places where eclogites can clearly be seen to have formed from crustal materials.

Frequently too, glaucophane schists show transitions to rocks of the greenschist facies and amphibolite facies. In other cases there are clear examples of glaucophane schists which have been converted to higher temperature facies (Fry and Fyfe, 1971). Thus in a general way, these rocks appear to carry the imprints of parts of the orogenic process.

In view of the geological associations of glaucophane schists it is not surprising that many of the dense minerals in these rocks are believed to have formed at high pressures (Eskola, 1939). The low temperature nature of the assemblages, which has been revealed by experimental studies and by oxygen isotope thermometry, brings in the special problem of very low geothermal gradients in terrains undergoing blueschist metamorphism. The purpose of this section is to critically evaluate the experimental data by which we hope to arrive at quantitative estimates of the pressure and temperature histories of blueschists. Evidence of this sort will be one of the principal kinds of input in an eventual quantitative description of major earth processes such as mountain building and continental growth.

(b) INTRODUCTION TO THE EXPERIMENTAL PETROLOGY OF BLUESCHISTS

The minerals considered to be most characteristic of the glaucophane schist facies have been listed by Ernst (1963a). They include glaucophane (or crossite), lawsonite, pumpellyite, albite, chlorite, sodic pyroxene, complex garnets, stilpnomelane, phengitic mica, epidote, aragonite, and calcite. Among these, glaucophane, lawsonite, jadeite and aragonite have been considered to be confined to, and diagnostic of, glaucophane schists. Aluminium-poor chlorite, or serpentine minerals, should probably be added to the list of characteristic minerals because of the abundant and striking occurrence of serpentinised ultramafic bodies in most glaucophane schist terrains. The many minerals of high density in blueschists suggest metamorphic conditions involving high pressures and low temperatures. The parent rocks which develop these minerals in metamorphism are eugeosynclinal sediments and volcanics: graywackes, shales, conglomerates, cherts, pillow lavas and tuffs. Volcanic products are conspicuously important in the parent material of blueschists, either as the sedimentary detritus of graywackes as in California, or as the more direct volcanic deposits.

Experimental metamorphic petrology is a young science. Only twenty-seven years have elapsed since the development of high pressure hydrothermal techniques and the first experimental studies of metamorphic minerals (Bowen and Tuttle, 1949). Modern solution calorimetric studies of minerals began at about the same time (Torgeson and Sahama, 1948). A certain amount of experimental work has accumulated concerning most of the diagnostic blueschist minerals. This preliminary work perhaps has served more to emphasise many of the outstanding problems and to suggest additional ones rather than to produce definitive statements about blueschist metamorphism. In this chapter, existing experimental data concerning sodic amphiboles and pyroxenes, albite, garnets, micas, serpentine and chlorites, aragonite, lawsonite and pumpellyite will be reviewed. Several of the more important field problems such as the question of soda-metasomatism, the associated eclogite "tectonic inclusions" and the "overpressure" controversy will be considered as they arise in the discussion of individual minerals.

A meaningful review of experimental petrology of glaucophane schists cannot be separated from simultaneous discussion of the important field aspects; the two approaches to understanding petrology should be thoroughly integrated in order to be most effective. The discussions of the experimental work and of the field problems they bear upon will be presented side-by-side in this article with as much cross-reference as is necessary to make clear the significance of the experiments and to provide ways of criticising their application—and sometimes, of criticising the experiments themselves. Nature can be a good sounding board of the reliability of experimental data. The lack of enough mutually critical collaboration of field petrologists and experimentalists in the past explains much of the uncritical acceptance of preliminary experimental work by the former group and the often overly-enthusiastic and forced application of their diagrams by the latter.

Before starting a discussion of the experimental work, a few criteria for judging the reliability of a mineralogical stability diagram should be outlined. The most important point is the clear-cut demonstration of "reversibility", that is, if the experimenter was able to create one assemblage of crystalline substances ($\pm$ fluid) from the alternative assemblage and then achieve the reverse reaction across their mutual stability boundary in runs as close as possible to the boundary. This is the proof of chemical equilibrium that has been lacking all too often in past experimental studies. Many non-experimental petrologists have seemed not to be aware that runs made on metastable starting materials like glasses or gels can crystallise to metastable assemblages which may persist for long times and give very false indications of stable assemblages. Highly metastable and swiftly reacting starting substances may approach equilibrium only via a series of metastable, though perhaps very persistent, "Ostwald Steps". Fyfe (1960) has analysed several of the ways in which a false indication of equilibrium was obtained in early experimental studies. Field and experimental petrologists should be more aware, perhaps, that equilibrium may be approached in a series of complex metastable steps in nature, and they might be somewhat more wary of far-reaching, quantitative interpretations made on the basis of weakly-recrystallised metamorphic rocks.

The characterisation of the starting material phases is a very important consideration. Synthetic crystalline products might be energetically very unlike natural phases even if they appear superficially similar in X-ray diffraction and optical properties. Minerals with ordered tetrahedral aluminium and silicon are exceedingly hard to synthesise, while their disordered analogs often crystallise readily. Synthetic starting materials, even if crystalline, therefore may yield perturbed equilibrium boundaries. Natural minerals with nearly end-member compositions are desirable but available only for a few systems. Petrologists should be more aware that there may be variation of crystal properties in natural materials as well. When we speak of the breakdown of "albite" we must keep in mind that natural albite exhibits a variety of energy conditions, some dependent on the (Si,Al) ordering state.

Experimental apparatus is an important consideration in criticising experimental work. The ideal high-pressure and high-temperature system is a gas-pressure device, externally or internally heated, with pressures directly measured by a gauge which communicates with the fluid medium. Many experiments have been carried out with flat piston and solid-pressure medium devices in which the pressure conditions are often quite poorly known both as to magnitude and uniformity (despite investigators' claims as to precision of calibration). Some experiments have been carried out inside gas-pressure vessels but with dry, hard, solid reactants sealed in metal capsules: the pressure distribution over a sample of hard jadeite and feldspar grains abrading one another, for example, must be quite non-uniform. Petrologists probably need to be more aware that pressure is applied in nature in most situations by abrasion of interlocking grains. Individual mineral susceptibility to *plastic* strain energy accumulation may be a biasing factor for crystallisation during or immediately after deformation. *Elastic* theory calculations, some of which minimise the perturbing effect of deforming stresses on mineral equilibria (MacDonald, 1957), have little application here.

Confirmation of equilibrium results by independent laboratories using different approaches is a desirable situation which has not occurred very often with important mineral systems. Very desirable also is confirmation of a directly-determined diagram by the method of solution calorimetry in conjunction with measured heat capacities and specific volumes of reactants and products. The calorimetric method has shown that it is capable of yielding precise equilibrium results in a few cases where available direct experimental results were reliable enough to test the method. Needless to say, there are many experimental and interpretational pitfalls associated with heat of solution work also.

The criteria outlined above for deciding whether an experimental system is satisfactorily determined are demanding. It is safe to say that very few systems yet investigated can meet these conditions completely. There is considerable promise, nevertheless, that several important simple mineralogic equilibria pertaining to blueschists will approach the "satisfactory" category in the foreseeable future.

3. ARAGONITE

McKee (1961) and Coleman and Lee (1961) simultaneously announced the discovery that aragonite, $CaCO_3$ of density 2.94 gm/cm^3, was a widely-distributed phase in some of the low-grade metamorphic portions of the Franciscan Formation of the California Coast Ranges. This discovery generated much interest for two reasons. Firstly, conjecture had been developing that some of the other relatively dense minerals such as jadeite pyroxene and lawsonite required elevated pressure conditions for their formation, and secondly, some experimental evidence was available on the stability relations of aragonite and calcite, the less dense and much more common polymorph, so that a means of quantitatively estimating the minimum pressures under which the low-grade metamorphic rocks crystallised seemed suddenly at hand. The experimental work on calcite–aragonite stability relations will be briefly reviewed.

A Gibbs energy difference for the calcite–aragonite inversion was calculated as early as 1893 by F. W. Kohlrausch from electrical conductivity measurements of saturated solutions of calcite and aragonite in pure water at atmospheric pressure over a range of temperature from 0° to 35°C. Later investigators (c.f. Backstrom, 1921; Buchan, 1927) repeated these measurements in various ways, using different kinds of conductivity cells and found a variety of numbers for the ΔG of the reaction. Perhaps the most comprehensive was that of Backstrom (1921) who found a ΔG of —191 cal/mol at 25°C. From Backstrom's number a pressure for the equilibrium inversion of calcite to aragonite at 25°C of 2.83 Kbar may be calculated.

Jamieson (1953) carried out conductivity cell measurements which differed from those of the earlier workers in that they were made at high pressures. He used very pure deionised water and fine-grained Iceland spar and precipitated aragonite. Measurements were made at 29°, 38°, 53° and 77°C. At 29°C, aragonite solutions become less conducting than calcite solutions at pressures above 3.98 Kbar and hence aragonite is stable above that pressure. Jamieson's value for 100°C is 5.2 Kbar. His calcite-aragonite stability boundary is shown in Fig. 7.

A large number of experimental attempts to define the calcite–aragonite transition has been performed. In spite of the fact that reaction kinetics in the $CaCO_3$ system are fast enough to allow experimental reversals of the calcite–aragonite equilibrium under dry conditions above about 450°C and above 100°C hydrothermally in feasible laboratory times, there is not yet good enough agreement on the P-T location of the boundary in the probable blueschist temperature range (100°–300°C) to permit very quantitative interpretations concerning the physical conditions of blueschist metamorphism. The discrepancies probably stem from uncertainty in calibration of solid-pressure experimental devices, unknown effect of strain on the phases, the slow rate of attainment of equilibrium below 200°C and actual stability differences among different samples of calcite and aragonite used for starting material.

In one of the first experimental studies performed in the "simple squeezer" device (Griggs and Kennedy, 1956), MacDonald (1956) succeeded in converting aragonite to calcite and calcite to aragonite at temperatures above 400°C and mapped out a linear calcite–aragonite stability boundary in the P-T plane. His extrapolation to 100°C falls about 1 Kbar below Jamieson's value. Clark (1957) made experiments on dry calcite and aragonite in an internally-heated gas apparatus. He secured a narrow "fix" on the equilibrium at 575° of 13.1 ± 0.2 Kbar and a broader fix at about 450°C. A slope through his points parallel to the MacDonald curve agrees very nicely with Jamieson's indirect determination at lower temperatures.

Simmons and Bell (1963) presented another diagram for the $CaCO_3$ system, based upon runs made in a rotating-piston version of the simple squeezer. The high shearing stress introduced into the sample during the run greatly accelerated reaction rates. Simmons and Bell were able to produce reversals of the equilibrium at room temperature. Their curve lies at 7.3 Kbar at 100°C but nearly

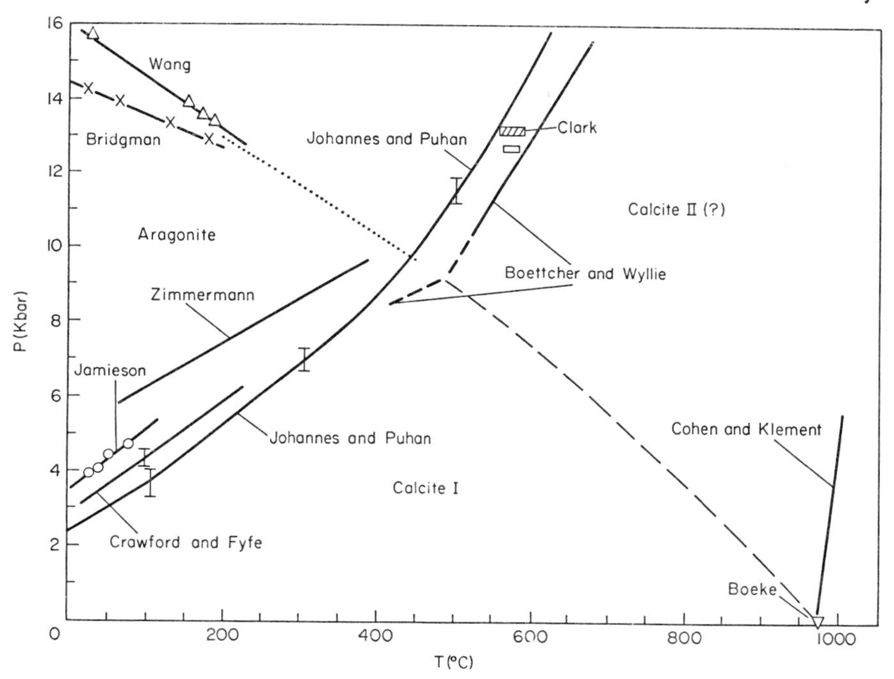

Fig. 7. Experimentally determined phase boundaries in the system $CaCO_3$. Experimental points and uncertainty brackets of some authors shown. Dotted and dashed lines indicate conjectures by Jamieson (1957) and Boettcher and Wyllie (1968a) about a field of stability of calcite II. Data from Boeke (1912), Boettcher and Wyllie (1968a), Bridgman (1939), Clark (1957), Crawford and Fyfe (1964), Cohen and Klement (1973), Jamieson (1953), Johannes and Puhan (1971), Wang (1968) and Zimmermann (1971).

reproduces Clark's result at 575°C, owing to a flatter dP/dT slope than found by other investigators. Serious difficulties of the rotating piston squeezer are a sacrifice in accuracy of pressure determination and the necessity of assuming that shearing stress does not alter the P-T position of the equilibrium.

The most convincing earlier low-temperature determination of calcite–aragonite-relations was the hydrothermal study of Crawford and Fyfe (1964).

They used welded tube-bombs with internal water pressures directly gauge-measured. Starting material was fine-grained synthetic calcite and aragonite and, in a few runs, natural aragonite. Dilute $CaCl_2$ solutions were used to accelerate reactions. Run times were from 6 to 26 days. A fix of 4.35 ± 0.25 Kbar at 100°C was secured. This value favours the MacDonald determination in the low-temperature region and falls nearly 3 Kbar below the Simmons and Bell curve.

Boettcher and Wyllie (1968a) found that the calcite–aragonite boundary began to curve sharply above 400°C and tended to much higher pressures, resulting in an inversion pressure of about 21 Kbar at 800°C. This change in slope of the boundary in the range 400°C–600°C had not been reported explicitly by earlier workers. The measurements of Boettcher and Wyllie were done using liquid fluxes in the system $Ca(OH)_2$-$CaCO_3$-H_2O, which produces very fast growth of carbonate phases on the liquidus. Identification of the stable polymorphs was made with the petrographic microscope. This is a very reliable criterion because the crystal habits of calcite and aragonite differ strikingly in the quenched charges. Boettcher and Wyllie explained the break in slope of the calcite–aragonite curve on the basis of Bridgman's (1939) high pressure metastable calcite I to calcite II inversion which was first discovered by volume measurements and later confirmed by high-pressure X-ray analysis (Jamieson, 1957). Bridgman found a flat negative dP/dT slope for the transition in the range 25°C–150°C which, as Jamieson (1957) pointed out, would by extrapolation intersect the calcite–aragonite line at about 500°C and 10 Kbar, producing a field of stability of calcite II at high temperatures. If a small curvature is allowed, the calcite I–calcite II stability boundary could be made to explain the changes observed at 975°C at 1 atm in calcite crystals studied by Boeke (1912). Jamieson was of the opinion that the intersection of the calcite I–II boundary could not deflect the calcite–aragonite line to any great extent. This opinion was based on the very small volume change of the I–II transition found by Bridgman. The very small volume change has been confirmed by Wang (1968). Goldsmith and Newton (1969) confirmed the change in slope of the calcite–aragonite curve found by Boettcher and Wyllie both by hydrothermal runs and in dry piston-cylinder runs fluxed by Li_2CO_3. The possible phase relations of calcite I and calcite II are shown in Fig. 7.

Some workers close to the blueschist problem seem to prefer the Jamieson–Clark $CaCO_3$ diagram (Coleman, 1967b; Ernst and Seki, 1967), even though Jamieson's determination was an indirect one and Clark's 575°C point is now known to be located on the steep-slope portion of the aragonite–calcite curve. The direct application of the $CaCO_3$ diagram is the most commonly chosen method in attempts to quantitatively estimate the pressures of blueschist metamorphism. Arguments about the inferred P-T history of rocks, such as the metamorphosed portions of the Franciscan Formation, start with a geothermal gradient. Figure 8 shows immediately that the geothermal gradient chosen must be a very low one. A gradient of 20°/km does not even intersect the aragonite field, if burial is the principal source of high pressures. A gradient of 10°/km, one near the lower limits of gradients measured in the field (Spicer, 1942) requires a pressure minimum of 4.5 Kbar to form aragonite stably.

The $CaCO_3$ phase relations proposed by Jamieson and Boettcher and Wyllie are consistent with the idea that Bridgman's "calcite II" is the same as the high temperature, low pressure $CaCO_3$ phase inferred from the thermal analysis work of Boeke. However, Cohen and Klement (1973) have found the thermal event in calcite reported by Boeke and have followed it to 5 Kbar pressure. It seems to

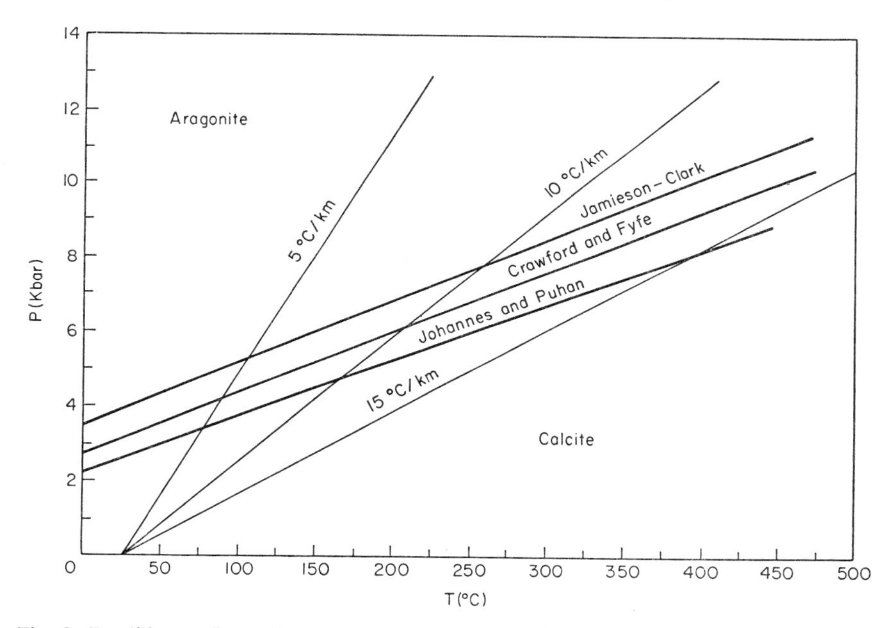

Fig. 8. Possible geothermal pressure–temperature gradients in blueschist metamorphism in relation to the stability fields of calcite and aragonite as determined experimentally in the low temperature region by Jamieson (1953), Clark (1957), Crawford and Fyfe (1964) and Johannes and Puhan (1971).

increase somewhat in temperature with application of pressure, which observation vitiates much of the speculation of previous workers.

Two recent experimental determinations of the calcite–aragonite equilibrium in gas pressure apparatus with reversals of the equilibrium are in considerable disagreement as to the low-temperature location of the boundary (Zimmermann, 1971; Johannes and Puhan, 1971). The latter study gives a value of 3.6 ± 0.5 Kbar at 100°C and thus is the lowest-pressure observation for this curve. Neither study finds evidence for a sharp break of slope above 400°C, although the Johannes and Puhan boundary has considerable curvature in this range.

Figure 7 shows selected experimental results illustrating the present uncertainty of $CaCO_3$ phase relations, especially in the low-temperature region. The ideal criteria for concluding that a transition or reaction has been satisfactorily determined have been outlined in the introduction to this chapter. The calcite–aragonite experimental situation seems to fall quite short of satisfying these criteria at the present time.

Brown, Fyfe and Turner (1962) showed by rate studies of the inversion of dry aragonite to calcite at one atmosphere that temperatures in aragonite-bearing rocks could not have greatly exceeded 300°C. Otherwise the aragonite would have back-reacted to calcite over any conceivable time-scale of unloading of the metamorphic rocks. They concluded that geothermal gradients less than 12°/km must have been prevalent during the metamorphic cycle, in order to intersect the aragonite stability field at temperatures below 300°C. The pressures required would be in the range of 5 to 9 Kbar. The concept of low geothermal gradients needed to preserve the aragonite is supported by the $CaCO_3$ inversion rate studies of Davis and Adams (1964). Their technique employed the monitoring of the growth or decline of calcite and aragonite X-ray diffraction maxima at high pressures in a beryllium-bomb X-ray device. They did not consider the rate effect of the high shearing stresses which must have been present in their apparatus.

The inference from the experimental $CaCO_3$ diagram of pressures of 9 Kbar or more for the genesis of aragonite-bearing rocks constitutes the central problem of blueschist petrogenesis. The stratigraphy of the Franciscan formation is not known well enough to accurately estimate depths of burial, but maximum estimates of the stratigraphic pile that could have accumulated at any site run at 15 km (Bailey, Irwin and Jones. 1964), which is about half the burial necessary, even assuming the low temperature of 300°C at the base of the continental crust. Some authors have preferred great geosynclinal accumulations nevertheless (Ernst, 1965; Essene, Fyfe and Turner, 1967; Newton and Smith, 1967) to account for the pressures needed to enter the aragonite stability field. Others (Coleman and Lee, 1962; Blake, Irwin and Coleman, 1969) have invoked sources of pressure other than greater burial, such as large mountain-building pressures or other "tectonic overpressures" which would have to be supported by the strength of rocks long enough to allow development of high pressure minerals. It would be necessary under this hypothesis that intraformational aqueous solutions could be kept at anomalously high pressures within the "over-pressured" region, because much of the aragonite of the Franciscan terrain is in the form of hydrothermal veins, often cross-cutting foliation planes and apparently late in the metamorphic sequence (Coleman and Lee, 1963).

New concepts of tectonics have been brought to bear on the problem of the high pressures seemingly required to create metamorphic aragonite and other dense blueschist minerals. A growing school of thought interprets the blueschist terrains as fossil "subduction zones", or sites where spreading oceanic crust collides with and is driven underneath continental margins, ultimately to be consumed in the mantle at depths of 400–700 km (Wilson, 1968; Hamilton, 1969; Dickinson, 1970; Ernst, 1970). Such an analysis seems to account plausibly for the chaotic nature of some blueschist terrains, notably the Franciscan formation of California, and could explain the apparently very low geothermal gradients required by the evidence from experimental work on aragonite. A main feature of the subduction concept is that depth of burial is not necessarily correlated with stratigraphic thickness.

Recently a few contributions of field petrology have indicated certain problems of interpretation which result from the direct use of the experimental $CaCO_3$ relations together with the assumption that deep burial is the main source of high pressures. Blake, Irwin and Coleman (1967) found calcite to be the dominant $CaCO_3$ polymorph in largely weakly metamorphosed regions of coastal northern California and southern Oregon, with aragonite, as well as lawsonite and pumpellyite, coming in as one traverses eastward, apparently up-section toward a great boundary thrust. The concept of progressively greater pressure as a function of burial in the section to produce aragonite and lawsonite does not seem to be borne out on a regional scale in northern California. A very clear example of such "upside-down" metamorphism has been reported from New Caledonia by Brothers (1968). The concepts of deep geosynclinal burial and subduction along a continental margin do not seem to throw much light on this peculiar stratigraphic association. Blake, Erwin and Coleman (1967) attach considerable importance to the occurrence of abundant lawsonite and aragonite in close proximity to the boundary thrust, intimating that an "overpressure" of tectonic origin may have been generated by the thrusting. Preliminary experimental deformation of natural graywackes has not yet eliminated the uncertainty in the minds of some petrologists about whether rocks undergoing glaucophanitic metamorphism could support such overpressure long enough for high pressure minerals to be generated (Brace, Ernst and Kallberg, 1970; Robertson, 1972). In other Franciscan areas of California, such as the Cazadero area, the occurrence of aragonite is mysteriously patchy, with greater abundance of aragonite-bearing veins in very local areas of somewhat greater recrystallisation (i.e., the "type III glaucophane schists" of Coleman and Lee (1963)). Ghent (1965) noted the occurrence of aragonite in rocks of otherwise typically greenschist grade, containing albite, chlorite, actinolite and epidote at Black Butte, California. In other localities, such as the Kanto Mountains, Japan and the western Pennine Alps, calcite is present instead of aragonite even though jadeite and quartz, thought to be a still higher pressure assemblage, coexist. Vance (1968) described the occurrence of metamorphic aragonite on a regional scale as veins and cavity fillings in greenstones and replacing fusulinid limestones in the San Juan Islands, N.W. Washington. These rocks contain prehnite and pumpellyite and lack typical blueschist minerals other than aragonite. This fact and lack of stratigraphic evidence for deep burial or evidence for structural control of pressure on a *regional* scale led Vance to question the direct application of the calcite–aragonite P-T diagram to this aragonite occurrence. Thompson *et al.* (1968) have described aragonite veins associated with serpentinite dredged from the Mid-Atlantic Ridge. The serpentinites are sometimes associated with weakly metamorphosed greenstones containing albite, epidote, chlorite and pumpellyite. The carbonate mineralogy of the greenstones has not been studied yet.

Barnes and O'Neil (1969) have found widespread development of aragonite and other carbonate minerals forming in open fissures in some ultramafic bodies in the Franciscan terrain which appear to be undergoing present-day serpentinisation at low temperatures. Coleman (1967a), on the other hand, applied the conventional interpretation of high pressure genesis to the occurrence of aragonite veins in one of these same ultramafic bodies (New Idria), and concluded that the pressure of serpentinisation was at least 9 Kbar. This might be a correct interpretation of some of the aragonite veins in the New Idria body, but it seems apparent that we have as yet no clear criteria for distinguishing between aragonite metastably formed and stably formed. The fact that aragonite crystal orientation is often systematic with respect to the fabric of coexisting minerals (Coleman and Lee, 1962) is not a guarantee of stable formation. Some of the N.W. Washington samples show considerable aragonite fabric also.

The well-known phenomenon of deformational strain-energy accumulation in calcite (Paterson, 1959; Gross, 1965) has not received much attention from metamorphic petrologists, even though it has been shown that a mild mechanical deformation could supply strain energy sufficient to lower the pressure requirements for aragonite stability relative to the strained calcite by 3 Kbar or more (Gross, 1965). The several modes of twin- and translation-gliding in calcite (Turner *et al.*, 1956) account for the ability of calcite to soak up plastic strain energy. The lack of attention given by petrologists to this possible source of energetic bias of the calcite–aragonite equilibrium is probably due to a prejudice in favour of consideration only of stable parageneses. Aragonite growth after strained calcite would be a metastable phenomenon at low pressures. Special kinetic factors would have to operate for this mechanism to account for apparent production of aragonite at reduced pressures, as perhaps in the N.W. Washington example. It would have to be shown that aragonite could nucleate and grow at the expense of the deformed calcite faster than "strain recovery" by thermal annihilation of crystal dislocations (Thomas and Renshaw, 1965) and that nucleation of fresh unstrained calcite could be inhibited somehow in metamorphism. Newton, Goldsmith and Smith (1969) experimentally showed that these conditions could be met under certain circumstances. They were able to grow aragonite at the expense of calcite treated by prolonged mortar grinding at pressures several kilobars below the aragonite stability limit, in a variety of aqueous solutions and at temperatures up to 300°C. Experiments on the strained calcite with pure water and calcium and ammonium chloride solutions at pressures of one atmosphere to 2.4 kilobars and temperatures in the range 100–200°C yielded aragonite growth but also nucleation and growth of a fresh unstrained calcite. Solutions of magnesium and calcium chloride with Mg/Ca — 2:1 resulted in fast aragonite growth and quantitative destruction of the highly ground calcite at 120°C and 1–2 Kbar in a few hours. Calcite regeneration is severly inhibited by solutions of high Mg/Ca ratio (Bischoff and Fyfe, 1968). Northwood and Lewis (1970) showed by X-ray analysis of severely ground calcite that deformational energy can be a significant parameter in stability considerations relative to aragonite.

The foregoing evidence indicates that several features concerning metamorphic aragonite remain to be clarified. Several low-pressure occurrences of metastable aragonite growth are known even in cases where elevated temperatures were active, as in the fumarolic aragonite–sulphur deposits of Girgenti, Italy (Palache, Berman and Frondel, 1951, p. 190). Fyfe and Bischoff (1965, p. 12) state that "From repeated observation of primary precipitation, . . . it is easier to precipitate aragonite from solutions near the boiling point than at lower temperatures." Aragonite is known as a vein and cavity filling associated with zeolite in altered lavas (Deer, Howie and Zussman, 1962, Vol. V, p. 312) and it may be that solutions resulting from alteration of volcanic rocks may favour nucleation of metastable aragonite. Connate sea water, with Mg/Ca = 5:1 would be effective. Aragonite, once deposited and sealed off from further invasion by solutions may persist for times as long as 10^9 years at temperatures below 200°C (Brown, Fyfe and Turner, 1962). Even when the aragonite seems to be replacing calcite *in situ* as in some of the N.W. Washington rocks, consideration of the strain energy state of the calcite may render the quantitative interpretation of metamorphic pressure ambiguous. The interpretation is an important one, as much of the present concept of very low thermal gradients and very high pressures in glaucophane schist genesis leans heavily on the calcite–aragonite equilibrium. Much future experimental work, together with field work, will be needed to establish definitely the circumstances under which aragonite forms stably in nature and how these circumstances may be recognised.

4. AMPHIBOLES

Bluish amphibole of the glaucophane, $Na_2Mg_3Al_2Si_8O_{22}(OH)_2$—riebeckite, $Na_2Fe_3^{2+}Fe_2^{3+}Si_8O_{22}(OH)_2$ series is the most characteristic component of blueschists and the mineral for which the metamorphic facies was named (Eskola, 1939). A common variety is crossite with a typical composition of about $Gl_{65}Ri_{35}$. Magnesioriebeckite, $Na_2Mg_3Fe^{3+}Si_8O_{22}(OH)_2$ is found in some rocks assigned to the glaucophane schist facies (Ernst, 1964) but aluminous sodic amphiboles are the rule. Occasionally nearly pure riebeckite appears in metacherts and meta-ironstones.

A greenish actinolite, $Ca_2(Mg,Fe^{2+})_5Si_8O_{22}(OH)_2$ is common in rocks of the blueschist facies, especially in rocks considered transitional between blueschists and greenschists (Ernst, 1964). Intergrown glaucophane and actinolite are known in blueschists, as in the mafic "tectonic blocks" associated with many outcrops of the Franciscan Formation in California (Coleman and Lee, 1963), and may reflect partial readjustment to lower-temperature conditions of an originally actinolite-bearing rock or of increased oxygen pressure in the blueschist environment (Iwasaki, 1963). Ernst (1968, p. 31–32) indicates the existence of a glaucophane–actinolite miscibility gap and systematic partitioning of Mg and Fe between the phases, based on a few preliminary electron microprobe measurements of coexisting blueschist amphiboles. The coexisting amphiboles from Sanbagawa terrane, Shikoku and Honshu, Japan, show less pronounced major element fractionation than some coexisting amphiboles from the Franciscan formation, in which actinolite shows a marked preference for the magnesium atom. Ernst believes that this difference in fractionation reflects somewhat higher temperatures of origin of the Japanese blueschists, in accordance with comparison of mineralogy of the two terranes (Ernst and Seki, 1967). Herein lies a possibly important field for future experimental work.

Experimental work on the stabilities of the sodic amphiboles, glaucophane and magnesioriebeckite, reveals surprisingly high thermal stability in view of the fact that they are restricted to low-temperature metamorphic facies in nature. Figure 9 shows the thermal decomposition curves for glaucophane and magnesioriebeckite above 1 Kbar P_{H_2O} deduced experimentally by Ernst (1961, 1960). Also shown is the stability of glaucophane + 2 quartz. According to Ernst, the assemblage glaucophane + quartz is stable to the melting point above 0.75 Kbar P_{H_2O} and co-exists with liquid over a wide temperature interval. More recent experimental work by Carman (1972) casts some doubt on this conclusion. A problem, acknowledged by Ernst, exists in his glaucophane dehydration curve below 1.5 Kbar, in that an entropy value for glaucophane calculated from the known thermodynamic properties for the breakdown products forsterite, albite, enstatite and vapour seems impossibly low, if the method of Fyfe, Turner and Verhoogen (1958) for estimating entropies of silicates is valid. This could be a result of the fact that equilibrium in many silicate systems is extremely hard to obtain at water pressures below 1 Kbar, even at quite elevated temperatures. But the problem is also complicated by the fact that the additivity rules for estimating entropies of silicates are less reliable for sheet and band structures.

The thermal stability limit of glaucophane may perhaps be somewhat lower than in the low pressure region shown in Fig. 9, but available experimental data indicate that glaucophane by itself is quite a refractory phase. Ernst (1963a) suggested that the probable low temperature reactions which limit the stability of glaucophane in nature are expressed by equations like:

$$5 \text{ glaucophane} + 3 \text{ lawsonite} = 10 \text{ albite} + \text{chlorite} + \text{actinolite.}$$

This ideal iron-free reaction has been deduced from comparison of blueschist and greenschist assemblages by other workers as well and may be a crucial one in definining the blueschist facies for several reasons. (1) The composition 5 glaucophane + 3 lawsonite makes a close approach to an iron-free basalt–water system, with 53% SiO_2. (2) The left-and-right-hand assemblages are diagnostic of blueschist and greenschist facies rocks, respectively. Thus there is the suggestion from nature that the reaction may be confined to a narrow P–T band; that is, it may be nearly univariant. (3) The reaction does not involve dehydration. Therefore, its use, when experimentally determined, is not subject to ambiguities regarding departure of fluid pressure from total pressure and in all

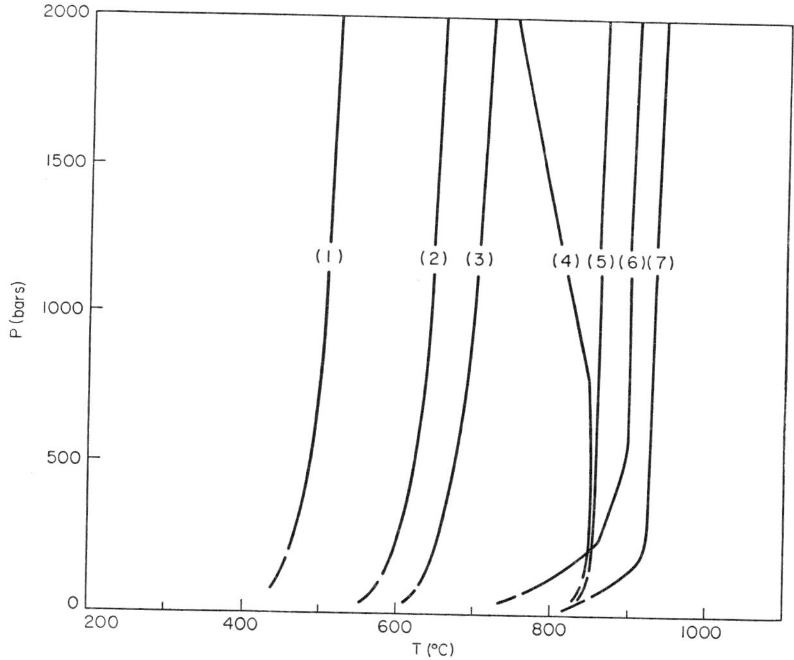

Fig. 9. Upper-temperature stability limits of various sodic amphiboles Curve (1) is for riebeckite, hematite–magnetite buffer (Ernst, 1966). Curve (2) is for riebeckite, Ni–NiO buffer (Ernst, 1966). Curve (3) is for riebeckite, quartz–fayalite–magnetite buffer (Ernst, 1966). Curve (4) is for glaucophane + 2 quartz (Ernst, 1961). Melting occurs above 750 bars. Curve (5) is for glaucophane (Ernst, 1961). Curves (6) and (7) are for magnesioriebeckite, quartz–fayalite–magnetite and hematite–magnetite buffers, respectively (Ernst, 1961).

probability the equilibrium boundary or band will be relatively temperature insensitive, which is desirable for a reaction which is to be used as a geobarometer. Many other reactions which run at low temperature and elevated water pressures could be written to define the conditions under which glaucophane and crossite form in nature. These will provide an interesting field for future experimental petrology.

A pronounced change of unit-cell dimensions of glaucophane synthesised under different P-T conditions was noted by Ernst (1963b). Because of a potentially important application to interpretation of blueschists, his diagram is reproduced with permission (Fig. 10). The rectangle sizes indicate Ernst's estimates of the pressure and temperature uncertainties of the syntheses. Numbers beside the plotted points indicate the size of the glaucophane unit cell

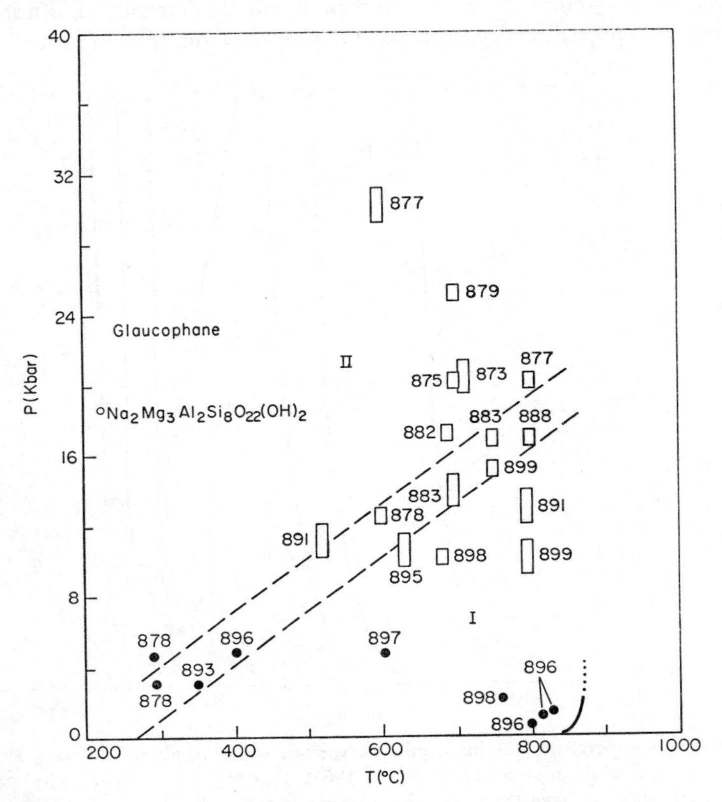

Fig. 10. Experimental data of Ernst (1963) for the transformation of glaucophane I to glaucophane II. Unit cell volumes of synthetic glaucophanes are shown. Dashed lines enclose approximate interval of the transformation. High temperature limit of glaucophane stability shown. (From *Amer. Mineral.* with permission.)

volume. Glaucophane of relatively small unit cell, designated "glaucophane II" by Ernst, is shown to be found on the high pressure and low temperature side of a P-T band whose width is about 3 Kbar. The crucial runs made by Ernst are those at 3 Kbar and 300°C and 350°C. The starting material for the 300°C run was glaucophane I. The run time was 1096 hours and the product was a glaucophane II of unit cell volume 878 A°3. The run at 350°C started with glaucophane II and ended with glaucophane I of 893 $A^{\circ 3}$ after 1097 hours. These two runs appear to verify the observed effect; the polymorphism was "reversed".

Ernst ascribed the 2% volume difference between the two glaucophane polymorphs to order–disorder relations among Al and Mg in the M_2 and M_3 cation positions in the amphibole structure. All natural glaucophane samples examined by Ernst belong to the glaucophane II category. It has been verified by Papike and Clarke (cited by Ernst, 1968) that natural well-crystallised glaucophane has a marked concentration of Al in the M_2 site.

No corresponding pressure-induced unit cell change could be found for magnesioriebeckite or riebeckite. Synthetic crossite of composition Gl_{50} Ri_{50} showed a transition with a much smaller volume change, apparently located at considerably higher pressures, but the transition was diffuse and was not studied in detail.

The mere fact that natural glaucophanes seem to be of the cation-ordered variety does not preclude their having been formed as glaucophane I, for, if the ordering process is so rapid that it can be accomplished in laboratory times at 300°C, there seems little chance of expecting the disorder to be "frozen-in". The prospects of using the polymorphism directly as a geobarometer or geothermometer are, therefore, not encouraging. However, major element partitioning between glaucophane and coexisting phases, especially with respect to Fe^{2+} and Fe^{3+}, might be quite different for the two polymorphs, and such tell-tale partitioning patterns could conceivably be "frozen-in". Also, consideration of the molar volume, as well as of an entropy of disordering, will seriously affect calculations of the stability relations of glaucophane to other minerals.

Maresch (1973) reports that he was unable to synthesise "glaucophane I" at low pressures and he suggests that the stable synthetic amphibole in the low pressure experiments is related to magnesiorichterite ($Na_2Mg_6Si_8O_{22}(OH)_2$). He deduces that the stability field for all true glaucophane is at high pressures, and moderate to low temperatures.

A certain amount of experimental work relating to the stability of actinolite is available. Boyd (1954) determined the stability limits of tremolite, $Ca_2Mg_5Si_8O_{22}(OH)_2$, to 2 Kbar P_{H_2O} with reversals of the breakdown reaction to enstatite + diopside + quartz + fluid. Tremolite is another example of a relatively low-grade mineral which has high thermal stability: 830°C at 1 Kbar and 865°C at 2 Kbar. Addition of iron to the amphibole replacing magnesium greatly lowers the thermal stability, however. Figure 11 shows a comparison of the breakdown curve of tremolite with that of ferro-tremolite, $Ca_2Fe_5^{2+}Si_8O_{22}(OH)_2$ with oxygen fugacity defined by the magnetite–fayalite–quartz buffer (Ernst, 1966). Further increase in f_{O_2} drastically reduces the field of stability of ferrotremolite, because magnetite must appear at the expense of ferrous iron silicates at the higher oxygen fugacities. Ernst (1966) indicates that ferrotremolite stability may nowhere intersect the Ni–NiO buffer. It is apparent that variations in oxygen fugacity will greatly affect the paragenesis of iron-rich amphiboles.

It must be remembered that relatively small amounts of crystal substituents replacing iron may have large effects on stability with respect to oxidation (Ganguly, 1968).

Several perplexing problems of the natural occurrence of glaucophane arise from field observations. The amount of glaucophane in partially recrystallised graywackes may show a dependence on the amount of shearing of the rock (McKee, 1962). Glaucophane itself does not require shear stress to form, as shown by the experiments of Ernst (1960), but it is possible that shearing

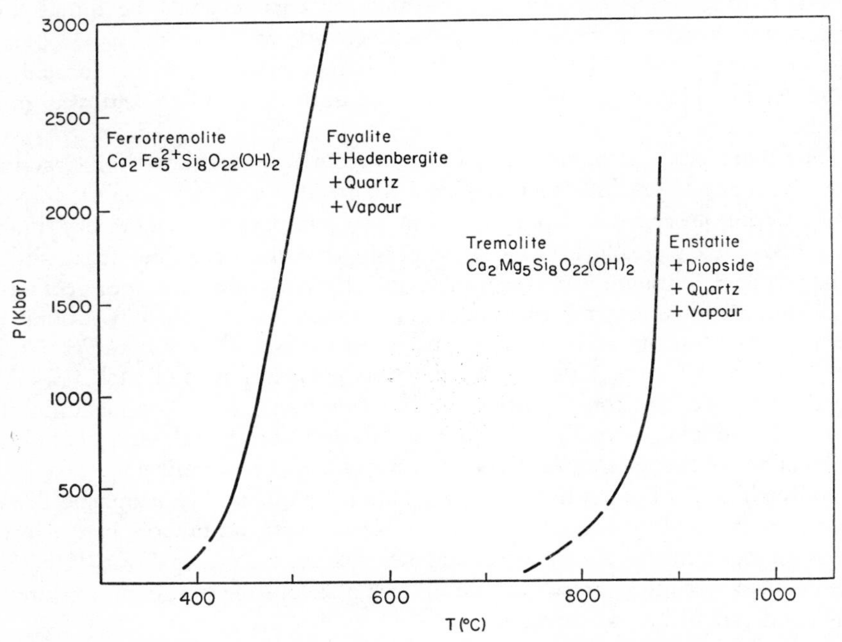

Fig. 11. Thermal decomposition curves of tremolite and ferrotremolite, as deduced experimentally by Boyd (1954) and Ernst (1966) respectively.

facilitates recrystallisation by mechanical activation and by increasing permeability to recrystallising fluids. An argument could be made on this basis that shear stress is after all essential to the development of glaucophane schists in that it provides an opportunity for approach to equilibrium in low temperature circumstances under which metastable volcanic and sedimentary assemblages could otherwise persist indefinitely. Another problem is the debate concerning the glaucophane aureoles or halos that Taliaferro (1943) insisted were common adjacent to serpentinite intrusions. He argued for introduction of soda into the country rock during serpentinisation. Several authors have taken exception to this view. The most convincing arguments are comparisons of the compositions of glaucophane-rich metasediments and basalts with the compositions of those rocks lacking glaucophane. No large systematic differences have been found to account for the presence of glaucophane, although there is a suggestion that some of the analyses of metabasalts tend to show slightly more Na_2O than typical basalts (see analyses of Ernst, 1963). Coleman and Lee (1963) state that some of the glaucophanised basalts may have originally had compositions that would qualify them as spilites. Thus, the soda-metasomatism controversy of glaucophane schists may be akin to the controversy of whether introduction of soda from sources such as sea water or connate brine content of submarine volcanics and eugeosynclinal sediments could account for spilitic basalts (Turner and Verhoogen, 1951, p. 212). Fyfe, Turner and Verhoogen (1958) take the view that the intense shearing commonly observed at the contacts of serpentine bodies could have had a role in generation of glaucophane aureoles by virtue of increased mechanical activation and permeability. The question of glaucophane fabric is one which conventional experimental petrology does not bear upon but which deformation-recrystallisation studies may eventually elucidate. Phyllosilicates and amphiboles often define a fabric in Franciscan rocks by preferred orientation which is parallel to original bedding planes (Bailey, Irwin and Jones, 1964). This suggests that elevated pressures were generated by deep sedimentary burial rather than by deformation mechanisms. These and other special problems of glaucophane genesis may be approached through combined field and experimental petrological studies.

5. GARNETS

Garnet is a conspicuous member of the phases which characterise glaucophane schists. It is found in metasediments, cross-cutting veins and replacement pods, and is most spectacularly developed in the coarsely-crystalline gneissic or schistose blocks which are erratically distributed in many glaucophane schist terranes as stream boulders or among landslide debris but which are never found in unambiguous stratigraphical position. Similar coarse-grained schists have been found as pods in the abundant serpentinised ultramafic intrusives of the Californian Franciscan Formation. This may be the source of all such blocks. Eclogites, abundantly garnetiferous, are also found as "tectonic blocks". There is a continuous gradation of mineralogy from that characteristic of eclogites (garnet–omphacite–rutile $\pm$ quartz) to that characteristic of mafic glaucophane schists (glaucophane–chlorite–epidote–sphene). Most of the eclogite-like tectonic blocks show some incipient development of chlorite, epidote and amphibole. While alteration has been termed retrogressive, it may merely reflect readjustment of an original eclogite to the facies conditions of glaucophanitic metamorphism.

Garnets of glaucophane schist terrains fall naturally into three categories: almandine-rich garnets found in eclogites and coarse mafic blueschists, almandine–spessartine–grossular garnets of meta-sediments and replacement veins, and the hydrogarnets found in aureoles of lime minerals, at the margins of some serpentinite bodies. These groups will be discussed as separate entities though it is recognised that the groups present some common problems of petrogenesis.

(a) ALMANDINE-RICH GARNETS

Garnets of this type are found in the more coarsely crystalline blueschists and eclogites associated with the Franciscan terrain. Pabst (1931) gave several analyses of such garnets from California and New Caledonia. He reported a pyrope content of 10–20% in some specimens, which is significantly higher than in typical garnets of amphibolites. The highest pyrope contents are undoubtedly relict, and probably do not represent the typical products of blueschist-grade metamorphism. Pabst's analyses showed high grossular content (8–30%) and relatively low spessartine. More recently published analyses (Lee *et al.*, 1963; Coleman *et al.*, 1965) of garnets from the coarse tectonic blocks show similar compositions, though generally smaller pyrope contents. All of the eclogite garnets of the latter reference are more or less chloritised.

Experimental work on the stability of almandine immediately raises some difficult problems of interpretation of almandine-rich garnets in low-grade metamorphism. According to Hsu (1968) almandine can co-exist with water only above about 550°C at water pressures above 1 Kbar. At lower temperatures, Fe-chlorite, magnetite, quartz and fluid are stable. Hsu's experiments were done in cold-seal apparatus with various oxygen-fugacity solid buffers. The position of the dehydration reaction did not vary greatly over the broad range of oxygen fugacity between the iron-wüstite and fayalite–magnetite + quartz buffers. At higher temperatures and lower oxygen fugacities, the stability limit of almandine was found to be in general agreement with the unbuffered runs of Yoder (1955), with Fe-cordierite + fayalite + hercynite as the breakdown products. The field of almandine was found to diminish drastically with increasing fo_2: on the quartz + magnetite–fayalite buffer, the upper almandine limit is down to 650°C at 1 Kbar and the whole almandine field is only 80° broad. On the Ni–NiO buffer, Hsu found no field of stability for almandine. Occurrence of almandine-rich garnet in glaucophane schists is thus difficult to understand in terms of the high water pressures and relatively high oxygen fugacities which are believed to prevail in blueschist metamorphism.

Two possible mechanisms for stabilising almandine with respect to hydrous phases at low temperatures are the effect of garnet substituents and impure aqueous crystallising fluids. The stabilising effect of solid solution of the spessartite, pyrope and grossular molecules may be evaluated from available experimental work. Garnets in low-temperature metamorphic rocks take a larger proportion of manganese than coexisting chlorite, and this tends to enlarge somewhat the stability field of garnet relative to chlorite. However, Hsu (1968) found that pure spessartine was stable only down to about 400°C, independently of oxygen fugacity. Below 400°C, spessartine is replaced by manganese chlorite and quartz. If Mn-chlorite forms a continuous stable solid solution series with Fe-chlorite, the field of iron–manganese garnets would not be much greater than for the end members. One problem that Hsu recognised in his spessartine runs was the presence of a considerable amount of hydration, probably metastable, in spessartine synthesised below 600°C. The hydro-spessartites might exhibit somewhat different stability relations from anhydrous spessartine.

The addition of pyrope molecule does not contribute much to the stabilisation of iron-rich garnet with respect to hydration, according to the results of Hsu and Burnham (1969). They performed a large number of experiments on the almandine–pyrope join at 2 Kbar water pressure using the fayalite–quartz + magnetite buffer. They found that no garnet in this system could exist below 570°C in the presence of pure water. The iron-rich assemblage below this temperature is chlorite + magnetite + quartz + vapour. Higher water pressures do not extend the stability field of almandine–pyrope garnet. Yoder and Chinner (1960) found that amphibole becomes a very stable phase in the system almandine–pyrope at 10 Kbar water pressure. Only a narrow field of garnet solid solutions exist at temperatures between 750°C and 1050°C and compositions more Fe-rich than $Alm_{60}Py_{40}$.

It would appear that substitution of relatively small amounts of Mn and Mg into almandine does not offer a solution to the problem of iron-rich garnets in blueschist terrains. Although no explicit experiments have been carried out, it seems probable that the grossular component would not have a greater stabilisation effect on almandine than do spessartine and pyrope.

The problem of almandine stability with respect to oxidation would be diminished somewhat by garnet substituents and high pressures. Ganguly (1968) showed that substitution of manganese into almandine has a relatively great effect in promoting stability under high P_{O_2}. It is difficult to calculate the exact magnitude of the effect in lack of appropriate thermodynamic data. Ganguly found experimentally that almandine and water have a mutual field of stability relative to magnetite-bearing assemblages on the Ni–NiO buffer at 10.5 Kbar, and Ganguly and Newton (1968) found that almandine appeared as an apparently stable breakdown product of chloritoid above 10 Kbar on the hematite–magnetite buffer.

Crystallising fluids in which the water fugacity is reduced by admixture of CO_2 or other gaseous species or by electrolyte solutes would favour appearance of anhydrous phases such as garnets relative to hyd ous phases. The nature of a "typical" blueschist mineralising fluid has been much debated. Kerrick and Cotton (1971) found field evidence that the ratio of CO_2 fugacity to H_2O fugacity may have governed the appearance of jadeitic pyroxene in some metagraywackes of the Franciscan Formation. Ernst (1972), on the other hand, estimated from experimental data in the system $CaCO_3$-SiO_2-TiO_2 that CO_2 could not have exceeded 4 mol% in the Franciscan and Sanbagawa (Japan) terrains, based on the presence of sphene and the absence of wollastonite. Gresens (1969) argued that hypersaline reducing solutions play an important role in blueschist metamorphism. Connate sea water suggests itself as a possible crystallising agent.

The significant pyrope content reported for some garnets from blueschist terrains seem to support the widespread opinion that elevated pressures were important in the metamorphism. The demonstration by Boyd and England (1959) that pyrope is stable only at quite high pressures fulfilled expectations of many petrologists who suspected this on the basis of natural occurrences. Later reconnaissance work of Schreyer (1968) confirmed the long extrapolation of the very high temperature reversal experiments of pyrope stability of Boyd and England and showed that pyrope has a considerable range of P–T stability in the presence of water. Below about 750°C, however, pyrope is replaced by assemblages containing hydrous minerals. Because of this fact the stability of pyrope–almandine solid solutions, which will be stable at lower pressures than is pure pyrope, cannot be investigated by high pressure hydrothermal methods.

Yoder and Chinner (1960) therefore attempted to get some idea of the stability of pyrope–almandine garnets in the absence of water by calculations based on experimental data for the end members. Inasmuch as there is a linear dependence of the garnet unit cell constant on composition on the join almandine–pyrope, and because the volume change of the garnet low-pressure breakdown products has a similar relationship, it was assumed that the partial free energy change of the breakdown reaction is, as a first approximation, linear with pressure. This assumption is, as the author pointed out, rather crude in that the low pressure, high-temperature breakdown products for pyrope and almandine are dissimilar (enstatite + sapphirine + sillimanite for pyrope, forsterite + spinel + cordierite for almandine) and the breakdown assemblages of intermediate garnets must vary considerably in the pressure interval between the stability lines of the end members. Figure 12 shows Yoder and Chinner's calculated stability curves for intermediate pyrope–almandine solid solutions mapped into the P-T plane. In spite of the uncertainties of the calculation, a significant point regarding blueschist garnets emerges: garnet on the almandine–pyrope join could probably contain as much as 80 mol% of pyrope. Therefore, some factors other than pressure limit the amount of pyrope actually observed. The most important factor is certainly high water fugacity. The highest amounts of pyrope observed in garnets from blueschist terraines are probably commonly inherited from a former higher grade metamorphism with incomplete readjustment to blueschist facies conditions.

The balance of field and experimental evidence points in the direction that almandine–pyrope garnets are not stable in the glaucophane schist facies except in a situation where, for some reason, water pressure is abnormally low. An example of this situation might be the pillow-lava eclogites described from the low-grade Western Alps terrain by Bearth (1966).

(b) SPESSARTINE-GROSSULAR GARNETS

Intensive analysis of the minerals of glaucophane schists and associated rocks in recent years has revealed another type of garnet which may be more characteristic of glaucophane schist facies metamorphism. Analyses of small garnet euhedra from the in-place, regionally-distributed low-grade metagraywackes and siliceous sediments and from veins formed during metamorphism often reveal high spessartine contents. Coleman and Lee (1962) give an analysis of a garnet in a vein containing aragonite, stilpnomelane and glaucophane which shows nearly 30 mol % spessartine with a large grossular and very minor pyrope contents. Lee *et al.* (1963) give several analyses of spessartine-rich garnets from regionally-distributed low-grade glaucophane schists ("Type III" schist of Coleman and Lee, 1963), with spessartine contents ranging from 9 to 61 mol % and grossular contents as large as 30%. The highest pyrope content in garnets from Type III rocks measured by Lee *et al.* is 5%. Seki (1964) and Hashimoto (1968) also report finds of spessartine–grossular garnets of low magnesium content from glaucophane schist areas.

Stability relations on the join spessartine–grossular have not been thoroughly investigated. Ito and Frondel (1968) succeeded in crystallising a series of homogeneous garnets from gel starting material on the spessartine–grossular join at 550°C–700°C and 2–3 Kbar water pressure and concluded that a stable continuous solid solution series exists between the two end members. Attempts to form garnet below 550°C failed in runs of two days duration. It is apparent that investigation of the stability of garnets on this join at temperatures approaching the blueschist range will be hampered by very slow experimental reaction rates.

(c) HYDROGROSSULAR GARNETS

Hydrous lime garnets of the hydrogrossular series, with hypothetical end-members grossular ($Ca_3Al_2Si_3O_{12}$) and $Ca_3Al_2(OH)_{12}$ are the most distinctive minerals of the metasomatic "rodingite" aureoles at the margins of some serpentinite bodies in blueschist terrains (Coleman, 1967b). Other minerals of rodingites are idocrase, prehnite and chlorite. Calcium and magnesium apparently are lost from the surrounding rocks during serpentinisation. Metasomatism and serpentinisation are believed by Coleman to have proceeded during the main phase of blueschist metamorphism.

The silica content of hydrogarnets was used by Coleman as a geothermometer based on the hydrothermal syntheses of Yoder (1950a) and Carlson (1956) from glass of grossular composition in the pressure range 0.2–2 Kbar and temperature range 400°C–800°C. These experimenters found that silica-deficient, water-rich garnets would form, along with variable amounts of free silica, from Ca_2Al_2-Si_2O_{12} glass, with SiO_2 content of the garnet increasing with temperature to values characteristic of nearly anhydrous grossular at 800°C. The same effect was observed by Pistorius and Kennedy (1960) on syntheses in the simple squeezer from oxide mixes. They found that normal grossular (based on d-spacings) would form only above 750°C, with silica content of garnet at temperatures below 750°C dependent on temperature and nearly independent of pressure. Yoder (1950a) and Pistorius and Kennedy (1960) believed that anhydrous grossular was unstable in the presence of water at elevated pressures at temperatures below 750°C–800°C. Roy and Roy (1957) found evidence, however, that grossular is stable in a hydrous environment at temperatures at least as low as 500°C, on the basis of very long runs. These conflicting results indicate that more work needs to be done on the stability of hydrogarnets before reliable use can be made of them as temperature indicators. The breakdown temperatures of hydrogarnets determined by Carlson (1956) provide only upper temperature limits for rodingite development.

6. MICAS, CHLORITE AND SERPENTINE

Blueschists and closely related rocks characteristically are not as rich in micas as rocks of higher metamorphic grade. Biotite is rare and apparently confined to mafic rocks of the highest glaucophanitic grade (Banno, 1964). Stilpnomelane and chlorite are more usual ferromagnesian sheet structures. The most characteristic blueschist mica is a faintly pleochroic sericite containing relatively large amounts of Mg and Fe and which is somewhat richer in SiO_2 than pure muscovite. A solid solution series between celadonite, $KR^{2+}R^{3+}Si_4O_{10}(OH)_2$, and muscovite, $KAl_3Si_3O_{10}(OH)_2$, where R^{2+} can be Mg^{2+} or Fe^{2+} and R^{3+} can be Al^{3+} or Fe^{3+}, was deduced by Schaller (1950). The intermediate members are called *phengite* and are confined in occurrence to low-grade metamorphic terrains except where lower-grade retrogressive metamorphism has been superposed on higher-grade rocks, as in the Adula district, Switzerland (Müller, 1958).

Ernst (1963c) gives analyses of several phengites from low-grade schists containing associated albite, chlorite, quartz $\pm$ lawsonite, glaucophane and jadeitic pyroxenes. The micas studied by Ernst are all of the $2M_1$ polytype, although several 3T phengites have been described by Heinrich and Levinson (1955). The analyses suggest a solvus on the muscovite–celadonite join, with a phengite solid solution limit close to $Mu_{55}Ce_{45}$. Ernst found an important property of the solid solution series: the interlayer distance $C \sin \beta$ becomes minimal for a composition of about $Mu_{67}Ce_{23}$. The unit cell volume is also smallest for this composition (928A°^{3}). As a consequence, certain postulated thermal breakdown reactions for phengite will be accompanied by significant volume increases. One reaction given by Ernst is:

$$\underbrace{8KR^{3+}_{1\cdot5}R^{2+}_{0\cdot5}Si_{3.5}R^{3+}_{0\cdot5}O_{10}(OH)_2}_{\text{phengite}} + \underbrace{R^{2+}_5R^{3+}Si_3R^{3+}O_{10}(OH)_8}_{\text{chlorite}}$$

$$= \underbrace{5KR^{3+}_2Si_3R^{3+}O_{10}(OH)_2}_{\text{muscovite}} + \underbrace{3KR^{2+}_3Si_3R^{3+}O_{10}(OH)_2}_{\text{biotite}} + 7SiO_2 + 4H_2O$$

which, if $P_{H_2O} = P_{total}$ during the reaction, will have a ΔV of $\sim$4%. Hence, elevated water pressure and reduced temperature should favour phengite stability. Ernst's analyses seem to show that solid solution of the celadonite components decreases with metamorphic grade. Such restriction of crystalline solution with rising temperature is uncommon in mineral systems but is in keeping with the volumetric argument. He suggested that the phengite–chlorite reaction is a reasonable model for the classical biotite isograd. For phengite stability by itself Ernst deduced a reaction:

$$6 \text{ phengite} = 3 \text{ muscovite} + \text{biotite} + 2 \text{ K-spar} + 3 \text{ quartz} + 2\, H_2O.$$

The first serious attempt to study phengite stability was that of Crowley and Roy (1964). They used co-precipitated hydrous gels as the starting material for cold-seal bomb runs in the range 0.2–4 Kbar. They found an apparent abrupt solid solution limit for MgAl celadonite in muscovite at $Mu_{50}Ce_{50}$. For compositions more magnesian and for temperatures below about 700°C, the stable assemblage was found to be "two micas" plus sanidine. The nature of the micas exsolving from phengite was not investigated in detail. Presumably, one would be nearly muscovite in composition and one phlogopitic. The thermal decomposition temperature for phengite found by them was about 700°C, nearly independent of pressure. The most Mg-rich phengite (Ce_{50}) was found to have the highest thermal stability. They encountered the 1M and $2M_1$ polytypes in the higher temperature range except for compositions rich in Mg, which seemed to favour the 3T polytype. The idea of increased solid solution limits of the celadonite components at reduced temperatures and elevated pressures seemed not to be upheld by the work of Crowley and Roy.

Velde (1965a) reinvestigated the phengite problem using dehydrated gels and conventional hydrothermal methods. He prepared equilibrium diagrams for the muscovite–celadonite join using celadonite compositions (MgAl), ($MgFe^{3+}$), ($MgAl_{.5}Fe^{3+}_{.5}$), ($Fe^{2+}_{.5}Mg_{.5}Al$), and ($Al_{.5}Mg^{2+}_{.5}Fe^{3+}_{.5}Fe^{2+}_{.5}$). The temperature range of the runs was 300°C–700°C and water pressures were 2.0 and 4.5 Kbars, with one synthesis run on the Mu–(MgAl)Ce join at 10 Kbar and 350°C in an internally heated vessel. The iron-bearing charges were thought to be approximately buffered by the alloy of the bombs (Haynes Stellite), although Velde did observe a colour zonation in some of the quenched charges containing iron.

Velde's results contrast markedly with those of Crowley and Roy. Figure 13 shows the results for the Mu–(Mg,Al)Ce join, which is representative of most of the joins studied. The negative dependence of solid solubility on temperature which was expected on the basis of the compositions and unit-cell volumes of natural samples was verified. Velde was able to "reverse" most of the solid solution boundaries in the range 400°C–450°C; that is, he observed the same solid solution values starting both with the breakdown products of phengite and with phengite over-saturated in Mg and Fe. The greatest solid solution found at 300°C, which is probably an upper temperature limit for application to blueschists, was about Ce_{35} for the $(MgFe^{3+})$ series. The least solid solution at 300°C was Ce_{17} for the $(MgAl_{.5}Fe^{3+}_{.5})$ series. All of the systems except the $(Mg_{.5}Fe^{2+}_{.5}Al_{.5}Fe^{3+}_{.5})$ compositions showed large increases of solid solution with increase of pressure. In most cases the Ce saturation limits were increased by a factor of about 1.5 in going from 2 Kbar to 4.5 Kbar. The one synthesis run at 10 Kbar in the (MgAl) join indicates that below about 300°C a nearly complete solid solution series exists for this range of compositions and pressures. A discrete "glauconite" field of celadonite-rich solid solutions was found only for the

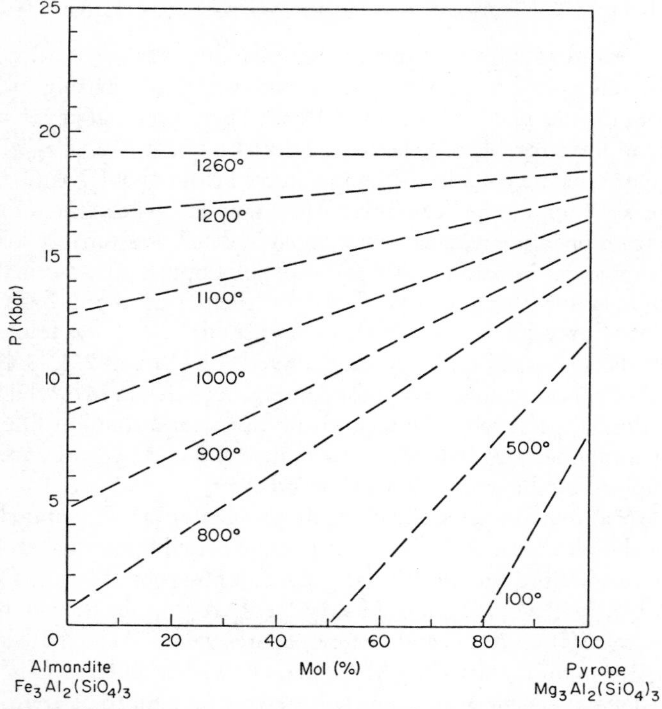

Fig. 12. Isothermal stability curves for pyrope–almandine garnets in the dry system calculated by Yoder and Chinner (1960). Garnet of a given composition is stable at a given temperature above the isotherm. (From Ann. Rept. of the Dir. of the Geoph. Lab., Carn. Inst. Wash. Yr. Bk. 59, with permission.)

$(MgFe^{3+})$ series. All runs yielded the $2M_I$ polytype except the 10 Kbar run, an 8-hour run which crystallised 1M mica.

Figure 14 shows Velde's thermal stability limit for a "representative" phengite of composition $Mu_{70}(MgAl)_{30}$ as a function of pressure, compared with Evan's (1965) curve for the breakdown of muscovite to sanidine plus corundum plus water and Kerrick's (1968) estimate of the stability limit of pyrophyllite. The phengite breakdown curve provides estimates of upper temperature limits

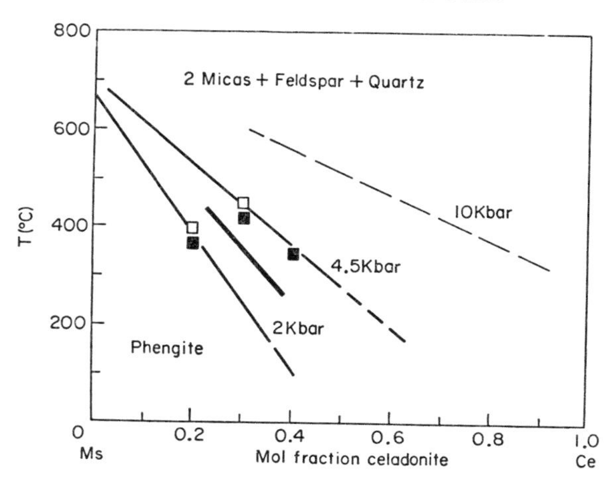

Fig. 13. Stability field of (MgAl) phengite determined experimentally by Velde (1965a) at 2, 4.5 and 10 Kbar. "Ms" denotes muscovite, $KAl_3Si_3O_{10}(OH)_2$. "Ce" denotes celadonite, $KMgAlSi_4O_{10}(OH)_2$. Reversal data of Velde are shown. The 10 Kbar line is based on one synthesis run only. Also shown is a decomposition line for a natural phengite at 2 Kbar (bold line).

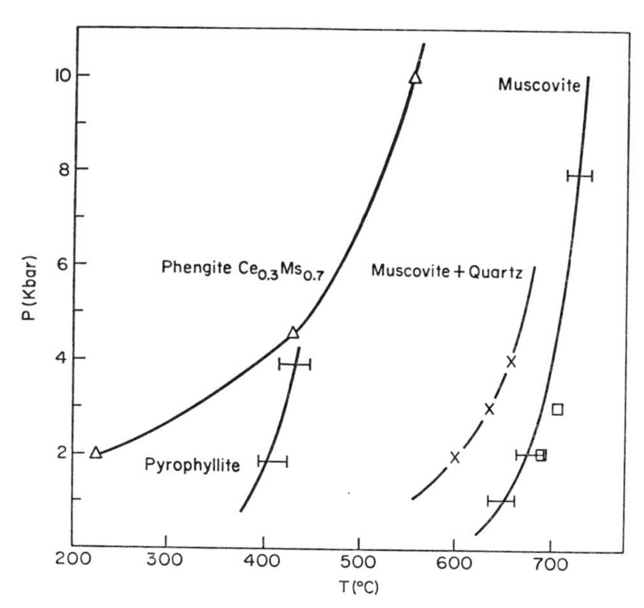

Fig. 14. Experimental mica stability curves under conditions of $P_{H_2O} = P_{total}$. Data of Velde (1965a) for (MgAl) celadonite, Kerrick (1968) for pyrophyllite, muscovite + quartz (Evans, 1965) and muscovite (Evans, 1965; squares; and Velde, 1965: brackets).

for phengite-bearing assemblages and hence may be quite useful. If phengite can in silica-rich situations decompose to chlorite, pyrophyllite and K-spar, as Velde proposed:

$$10\,\underset{\text{phengite}}{KR^{2+}_{0.5}Al_{1.5}Si_{3.5}Al_{0.5}(OH)_2} + \underset{\text{quartz}}{26SiO_2} = \underset{\text{chlorite}}{R^{2+}_5Al_2Si_4O_{10}(OH)_8}$$

$$+ \underset{\text{pyrophyllite}}{4Al_2Si_4O_{10}(OH)_2} + \underset{\text{K-spar}}{10KAlSi_3O_8} + 2H_2O$$

then the pyrophyllite stability curve is an upper temperature bound for phengite stability, as seems consistent with Fig. 14 below 5 Kbar.

The possibility that the compositions of natural phengites might provide a geobarometer for blueschists in conjunction with independent criteria for metamorphic temperatures, such as those derived from oxygen isotope evidence, is suggested by the experimental work. The pressure scale would have to be carefully calibrated by many more experiments, to judge from composition dependence of the pressure variation of the solid solution expressed particularly in the slight decrease with pressure of the $(Al_{.5}Fe_{.5}Fe^{3+}_{.5}Mg_{.5})$ series. Such runs would have to be effectively oxygen-buffered to be assured of reliability.

Equilibrium relations among the mica polytypes have not been completely clarified yet, although Velde (1965) presented experimental evidence that $2M_I$ is the only stable polymorph of $KAl_3Si_3O_{10}(OH)_2$ composition. A systematic difference in field occurrence was found by Coleman and Lee (1963). Their "Type II" or less recrystallised metabasalts at Ward Creek, California, contain IM white mica and the "Type III" or more recrystallised metabasalts contain the $2M_1$ polymorph. It is possible that the change in polytype is composition or temperature induced. It is also possible that the IM type is a metastable relic of diagenesis. Surviving metastable solid solutions may explain wide compositional variations like those observed in phengite from the low-grade greenschists of Eastern Otago, New Zealand (Brown, 1967).

The magnesian phyllosilicates are distinctive in blueschist mineralogy. Their roles are not well understood, but their appearance in characteristic environments indicates that experimental petrology of the chlorites and serpentine minerals can yield significant information on blueschist petrogenesis. Chlorite is a common and characteristic mineral of glaucophane schist facies metapelites and metabasalts. Serpentinised mafic and ultamafic bodies are widespread in the Franciscan Formation, as a glance at Koenig's (1963) Santa Rosa sheet of the geological map of California will show. The role of serpentine formation in blueschist metamorphism has been evaluated as incidental by some workers and as actually causal by others, and is currently a subject of debate.

A compositional series from $Mg_{12}Si_8O_{20}(OH)_{16}$ (serpentine) to $Mg_8Al_8Si_4$-$O_{20}(OH)_{16}$ (amesite) expresses the ideal compositions of iron free serpentine and chlorites. The compositional variation is accomplished by substitution of Al-Al for Mg-Si. Most chlorites have small to major quantities of ferrous and, to a lesser extent, ferric iron. Natural chlorites have an interlayer spacing of about 14.3 A°, within the compositional range from $Mg_{11}Al_2Si_7O_{20}(OH)_{16}$ (penninite) to amesite, with Si values of 4.5 to 7 accounting for most specimens (Deer, Howie and Zussman, 1962, p. 164). Serpentine minerals with structures based on a fundamental 7.3 A° basal spacing, are confined to Al-poor compositions and are generally very low in iron also.

Magnesian chlorites have high thermal stability, even in the presence of quartz, as revealed by the experiments of Fawcett and Yoder (1966). The most stable chlorite investigated by them was somewhat more aluminous than clinochlore, $Mg_{10}Al_4Si_6O_{20}(OH)_{16}$. At 5 Kbar water pressure chlorite and quartz persisted to 625°C. The iron-analogues have relatively high stablity with quartz also, with dehydration temperatures 65°C–160°C lower than for the magnesium chlorites (Turnock, 1960). The laboratory evidence is in agreement with field observation of a wide range of chlorite-quartz compatibility in the lower grades of metamorphism.

Albite commonly occurs with chlorite and quartz in apparently stable paragenesis in rocks of blueschist grade. The relationship of albite and chlorite to glaucophane is not clear. Miyashiro and Seki (1958) and others have called attention to possible glaucophane-producing reactions such as:

$$\underset{\text{antigorite molecule of chlorite}}{Mg_3Si_2O_5(OH)_4} + \underset{\text{albite}}{2NaAlSi_3O_8} = \underset{\text{glaucophane}}{Na_2Mg_3Al_2Si_8O_{22}(OH)_2} + H_2O.$$

It is possible that a complex reaction takes place in which chlorite loses some magnesium and silicon and gains aluminum in the reaction with albite to produce glaucophane. Albee (1962) stated that chlorites generally become more aluminous as metamorphic grade advances. The possibility seems to exist that the amount of aluminum in chlorite coexisting with albite may be an indication of the physical conditions of blueschist metamorphism that could be approached experimentally. There is, of course, the possibility that glaucophane may be generated by reaction of albite with metastable Al-poor phyllosilicates produced by low temperature alteration of volcanic ferromagnesian silicates.

The areal association of serpentinised mafic and ultramafic bodies is a conspicuous feature of some glaucophane schist terrains. There is the strong suggestion that the processes of alteration of large masses of gabbro and peridotite and their emplacement at continental margins are an integral part of the blueschist story. The exact nature of the blueschist–serpentine interaction awaits clarification.

An older notion for the origin of serpentinite bodies is that they were peridotite of the upper mantle which were incorporated somehow in the crust by deep dislocation processes of mountain building. The process of serpentinisation gave increased mobility and buoyancy to the bodies, which then rose by solid state flow to higher levels of geosynclinal metamorphism. This hypothesis is attractive in explaining the lack of high-temperature thermal aureoles of the exotic eclogite and amphibolite masses of the Franciscan terrains, which may have been sheared loose from deeper levels in the upward transport of the serpentine, the intensely sheared margins of many bodies, and the often marginal nature of the serpentinisation. It has been pointed out (Osborn, 1969) that the buoyancy hypothesis suffers from consideration of the densities involved. Many partly serpentinised bodies are too heavy to have risen through the crust in the assumed manner.

A hypothesis for serpentinite emplacement which arises from new concepts of tectonics is that the bodies are pieces of the oceanic crust and upper mantle which have been incorporated onto continental margins at the site of collision with oceanic plates. Some of the serpentinite may have formed in the sub-oceanic environment and some at the supposed continental margin site of blueschist metamorphism as an oceanic plate moved toward and under the continent. A layer of sheared-off serpentine may have "greased" the suture of the two plates. This suture now appears in field relations as a major boundary thrust separating blueschists from rocks of other facies or from unmetamorphosed rocks. This concept would explain the pronounced association of serpentinised peridotite bodies along the eastern boundary fault of the central portion of the Franciscan outcrop in California (Ernst, 1970). Blake, Irwin and Coleman (1967) thought that "tectonic overpressures" generated along such a suture could cause localisation of high pressure minerals near the thrust plane in such a way as to explain the "upside-down" metamorphic gradients.

Taliaferro (1943) recognised augmented development of glaucophane schists in metasedimentary rocks immediately adjacent to some serpentinite bodies in the Franciscan Formation. Taliaferro thought that a metasomatic exchange occurred during serpentinisation in which the peridotite released its small initial soda content to the surrounding rocks, resulting in local concentration of glaucophane. Exportation of lime by such a process is now a recognised fact (Coleman, 1967b). Ernst (1963a, 1965) denied that clearcut evidence for local serpentine–glaucophane association exists and cited rock analyses to show that glaucophane-rich rocks have formed for the most part isochemically from original sediments and lavas. Gresens (1969) assigned a most important role to serpentine formation. He thought that the process would yield highly reducing and saline solutions which would favour recrystallisation of country rocks to assemblages containing dense minerals such as aragonite and jadeite, and that some of these assemblages might well be metastable, so that the direct application of the experimental P-T stability diagrams would not be valid. It is certainly true that the initial assemblages of sedimentary detritus, for the most part of high temperature volcanic origin, are highly unstable in the presence of water at elevated pressures. One cannot assume that the first metamorphic assemblages found will be those of ultimate lowest free energy: the approach to final equilibrium may well be via a series of metastable "Ostwald steps". Incipient or incomplete recrystallisation is one of the most commonly observed characteristics of the Franciscan metamorphic rocks. The few coarsely crystalline glaucophane schists which are found lack aragonite and jadeite (McKee, 1962). These are probably exotic blocks metamorphosed at deeper levels. Another explanation for their lack of jadeite and aragonite is that they are products of a higher temperature facies than the rocks which surround them. A weakness of the Gresens hypothesis is that there is as yet no experimental or other evidence to support the contention that highly reducing and highly saline solutions such as might result from the serpentinisation process would indeed favour metastable growth of some of the dense metamorphic minerals at pressures below their actual stability limits.

Natural serpentine occurs in three major polymorphs, called chrysotile, lizardite and antigorite, differing from one another in alternative stacking arrangements of the structural layers. Chrysotile and lizardite are characteristic of serpentinites in blueschist terrains, while antigorite, the densest of the three, is common in alpine serpentinites of higher-grade metamorphic regions (Page, 1967a; Thayer, 1967). Stability relations among the $Mg_3Si_2O_5(OH)_4$ phases have not been defined experimentally. The serpentine-producing reaction of most interest to petrologists is the hydration of olivine:

$$\underset{\text{forsterite}}{2Mg_2SiO_4} + 3H_2O = \underset{\text{serpentine}}{Mg_3Si_2O_5(OH)_4} + \underset{\text{brucite}}{Mg(OH)_2}.$$

The iron in olivine is largely oxidised to magnetite in the process, but some enters the brucite produced (Page, 1967b). Several investigators have studied this reaction experimentally in the iron-free system. In one of the early investigations of mineral-water systems at elevated water pressures, Bowen and Tuttle (1949) synthesised serpentine from oxide mixes at temperatures below 500°C and water pressures from 2000–40,000 p.s.i. The olivine hydration reaction was not reversed. The data of Bowen and Tuttle serve to place upper temperature limits of about 450°C, nearly independent of pressure, on the formation of serpentine in bulk compositions on the join olivine–water. Pistorius (1963) and Kitahara, Takanouchi and Kennedy (1966) reinvestigated the same reaction at higher pressures in solid-pressure-medium devices and claimed reversal of the equilibrium. Their results, when extrapolated to lower pressures, agree moderately well with the earlier data of Bowen and Tuttle. Chrysotile is the polymorph which forms fastest, and it is likely that all experimental work has involved this phase.

Two more recent studies indicate that the equilibrium breakdown temperatures of serpentine in the system Mg_2SiO_4–H_2O found by previous workers are considerably overestimated. Scarfe and Wyllie (1966) showed in hydrothermal runs that synthetic serpentine and brucite mixtures will slowly break down in the presence of forsterite seeds at temperatures as low as 330°C. They did not achieve the reverse reaction of forsterite hydration. Raleigh and Paterson (1965) in deformation experiments on a natural brucite-bearing serpentinite from a low-grade metamorphic terrain found a sudden loss of strength and ductility of the rock at 300°C at a confining pressure of about 3 Kbar. They attributed this effect to an incipient dehydration reaction to olivine. Occasional newly-formed olivine grains were detected petrographically in the artificially strained samples. The water liberated in the reaction was believed to be the weakening agent. Raleigh and Paterson propose that such partial dehydration and loss of strength is an important mechanism in the rise of serpentinite into crustal levels.

These results suggest an upper limit for the stability of chrysotile on the join $MgSiO_4$-H_2O at pressures of a few Kbar. The presence of iron would bring the effective temperature of serpentine formation downward. Thus it appears that serpentine dehydration to olivine can take place in the temperature range usually ascribed to blueschist metamorphism. If the metasediments and meta-basalts of blueschist terrains are underlain by serpentinised oceanic crust and upper mantle, the partial dehydration of subducted serpentine may be a factor limiting the ultimate temperatures and pressures which most blueschist-grade rocks experience. If, as seems likely, there is some serpentine formation during blueschist metamorphism, the hydration reaction of olivine will be non-equilibrium and strongly exothermic. The solutions coming into contact with olivine will be reduced and concentrated. Whether or not these thermal and solution events would be instrumental in local blueschist recrystallisation may be decided by further detailed field studies and possibly by some very pointed experimental work.

7. LIME–ALUMINUM SILICATES

Several different hydrous lime–aluminum silicate minerals are found in low grade metamorphic rocks: lawsonite, pumpellyite, prehnite, epidote, and the lime zeolites. Many of these are formed from the hydrous alteration of the volcanic plagioclase in the eugeosynclinal suite of rocks. Interpretation of the roles of these minerals is hampered by the complexity of the phase relations among them. To further complicate the situation, it appears that, in at least some natural occurrences, the number of coexisting lime silicates is too large to be considered an equilibrium assemblage (Whetten, 1965). The amount of experimental work on stability relations of the low-grade lime-aluminum silicates is still small.

Only lawsonite and pumpellyite are characteristic of the glaucophane schist facies, though discussion of the experimental phase relations of these two minerals necessarily brings into consideration many of the other lime silicates whose P-T fields of stability border upon, and help to define, the physical conditions of blueschist metamorphism.

(a) LAWSONITE

Lawsonite, $CaAl_2Si_2O_7(OH)_2H_2O$, is one of the most definitive of the glaucophane schist suite of minerals. It is found in nearly every kind of metasediment and metavolcanic rock in the blueschist suite, from incipiently altered plagioclase in the first stages of breakdown to the thoroughly recrystallised "Type IV" schists. The high specific gravity of lawsonite (3.02 gm/cm^3) and its high water content have suggested to many workers that its crystallisation is favoured by high water pressures (Fyfe, Turner and Verhoogen, 1958, p. 174). For this reason a number of investigators have attacked the problem of lawsonite stability.

Pistorius, Kennedy and Sourirajan (1962) made a reconnaisance of some high pressure phase relations in the system CaO-Al_2O_3-SiO_2-H_2O using the simple squeezer (externally heated flat anvil device). They used natural lime zeolites and other metastable starting materials. They found that lawsonite could be synthesised only in the pressure range above 20 Kbar. Below these pressures zoisite-bearing assemblages were produced. The pressures needed to produce lawsonite decreased somewhat with increasing temperature, but on the basis of the first returns lawsonite seemed to be a mineral that could not be produced stably in the crust.

Newton and Kennedy (1963) reinvestigated the stability of lawsonite in a piston-cylinder device. They produced reversals of the equilibria:

$$\underset{\text{lawsonite}}{4CaAl_2Si_2O_7(OH)_2 \cdot H_2O} = \underset{\text{zoisite}}{2Ca_2Al_3Si_3O_{12}(OH)} + \underset{\text{kyanite}}{Al_2SiO_5} + \underset{\text{quartz}}{SiO_2} + 7H_2O$$

and

$$\underset{\text{zoisite}}{2Ca_2Al_3Si_3O_{12}(OH)} + \underset{\text{kyanite}}{Al_2SiO_5} + \underset{\text{quartz}}{SiO_2} = \underset{\text{anorthite}}{4CaAl_2Si_2O_8} + H_2O.$$

Starting materials were pure natural minerals mixed in about equal amounts of products and reactants. The runs were done under conditions of water pressure equal to total pressure. By making a long extrapolation of their experimental results on the two reactions to an intersection at about 5 Kbar and 400°C, they deduced a stable portion for the reaction lawsonite = anorthite and vapour at lower temperatures and pressures (Fig. 15). Their estimated slope for the equilibrium is 45.7 bars/°C. A straight-line projection of this slope would intersect the zero-pressure axis at about 280°C. The discrepancy between this result and that of Pistorius, Kennedy and Sourirajan is a consequence of the use of highly metastable starting materials in the earlier investigation. A large amount of pressure is needed to produce the stable phase lawsonite at a finite rate at low temperatures. At higher temperatures this condition is ameliorated somewhat, causing an apparent negative slope for the boundary of the synthesis field of lawsonite.

Crawford and Fyfe (1965) made a thorough experimental investigation of lawsonite stability relations. They produced a number of reversals of the lawsonite–anorthite stability boundary using cold-seal hydrothermal bombs in which pressures were hydrostatic and well-measured. They used pure natural minerals as starting materials. Anorthite and lawsonite were sealed in equal molar proportions with excess water in silver capsules. Run times were long—up to 76 days. Major changes in amounts of products and reactants were detected optically and with X-rays. Many of the runs produced a good indication of the direction of the reaction. At pressures near 7 Kbar, a few runs yielded small amounts of zoisite, which would agree with the findings of Newton and Kennedy (1963). The ability of Crawford and Fyfe to reverse the equilibrium in a metastable range (with respect to zoisite) illustrates the power of "seeding" in promoting a desired reaction. The result of Crawford and Fyfe is shown in

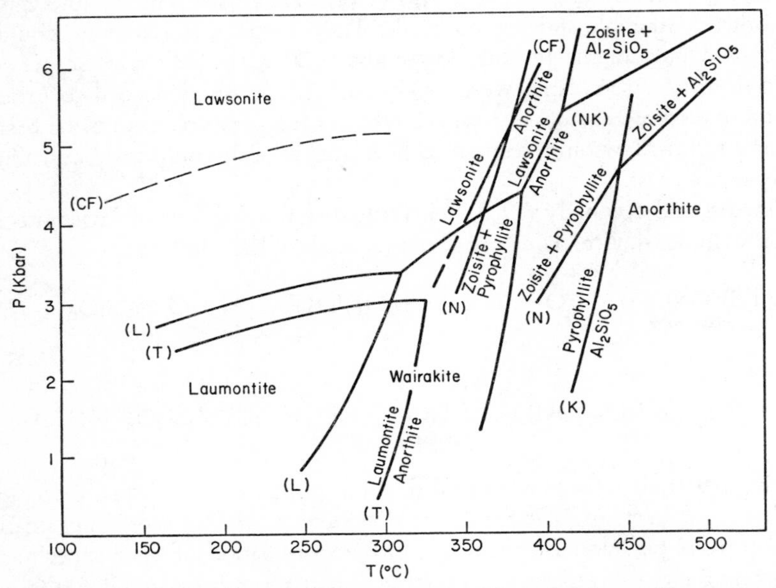

Fig. 15. Experimentally-determined stability curves of lime-aluminum silicate minerals in quartz-excess systems under conditions of $P_{H_2O} = P_{total}$. CF = Crawford and Fyfe (1965), K = Kerrick (1968), L = Liou (1970), N = Nitsch (1971, 1972), NK = Newton and Kennedy (1963), T = Thompson (1970).

Fig. 15. Their measured slope for the reaction is $41^{bars}/°C$. Crawford and Fyfe showed that their experimental determination was in good agreement with the dehydration boundary calculated from the heats of solution of lawsonite and anorthite measured by Barany (1962) and the entropy data of lawsonite measured by King and Weller (1961).

The most recent hydrothermal study of lawsonite stability is that of Nitsch (1972). He found by reversal experiments under well-controlled conditions that the equilibrium thermal decomposition assemblage of lawsonite is zoisite + aluminium silicate + pyrophyllite + vapour, even in the lower pressure range. The possibility that pyrophyllite is a stable phase in this system at the temperatures of lawsonite breakdown was anticipated by Newton and Kennedy (1963).

They did not anticipate the large change in slope of the stability boundary of the anorthite + H_2O field when pyrophyllite becomes stable which was found by Nitsch (1971). This effect appears to prevent a stable equilibrium boundary between lawsonite and anorthite + H_2O. The results of Nitsch (1971, 1972) are shown in Fig. 15. The high temperature lawsonite stability boundaries of Crawford and Fyfe and Nitsch appear to be inconsistent at pressures below their intersection at about 5 Kbar.

The later studies of lawsonite stability limits are in reasonable agreement and indicate that lawsonite *by itself* does not require great pressures for stablisation. The data do provide rather rigorous upper temperature limits for rocks containing lawsonite, because a hydrous mineral is not likely to survive for long above its dehydration temperature in nature. This is one of the few instances where a general statement stemming from experimental work can be made about equilibrium occurrence in natural materials. If the pressures of lawsonite genesis are about 4 Kbar, the upper temperature limit is about 350°C. The temperatures of lawsonite-bearing rocks cannot get much above 400°C under any conceivable circumstances, except possibly in the case of rare lawsonite eclogites (Watson and Morton, 1968) where the pressures of crystallisation may have been subcrustal. Other minerals co-occurring with lawsonite can only further limit its thermal stability range.

Many low-grade rocks transitional to the greenschist facies contain other lime–aluminum silicates, such as epidote, $Ca_2(Al,Fe)_3Si_3O_{12}(OH)$; prehnite, $Ca_2Al_2Si_3O_{10}(OH)_2$; laumontite, $CaAl_2Si_4O_{12}\cdot 4H_2O$; wairakite, $CaAl_2Si_4O_{12}\cdot 2H_2O$ and thomsonite, $CaAl_2Si_2O_8\cdot 2\cdot 4H_2O$. Relations of the lime zeolites in low-grade metamorphism are discussed by Coombs *et al.* (1959). Some data are available on the phase relations of lawsonite with the zeolites. Crawford and Fyfe (1965) calculated a P–T stability boundary separating the assemblages laumontite and lawsonite + quartz + water using published thermodynamic data for the minerals. Thermochemical data are available for leonhardite, $CaAl_2Si_4O_{12}\cdot 3.5H_2O$, rather than laumontite, but Crawford and Fyfe argued that the two zeolites are very closely related, with probably only a small free-energy difference between them. Their calculated laumontite–lawsonite boundary is shown in Fig. 15.

Liou (1970) has obtained reversible equilibrium data for the reactions of laumontite to wairakite + water and of wairakite to anorthite + water. His experiments were done in conventional cold seal apparatus with pure natural phases as starting material. Run times were as long as 6 weeks. His curves are shown in Fig. 15. He obtained data placing an upper pressure limit on the breakdown of laumontite to lawsonite, quartz and water. The boundary shown in Fig. 15 is nearly parallel to, and falls about 1.5 Kbar below the boundary calculated by Crawford and Fyfe, and is in agreement with an earlier reconnaissance determination by Nitsch (1968). Construction of these phase relations of the lime–aluminum silicates in the P–T plane shows that laumontite should undergo a metastable reaction to anorthite, quartz and water at about 3.3 Kbar and 300°C. This is in agreement with the deductions of Coombs *et al.* (1959) and the calculations of Crawford and Fyfe (1965).

Thompson (1970) provided valuable confirmation of the equilibrium relations between laumontite, lawsonite + quartz and anorthite under conditions of $P_{H_2O} = P_{total}$ that had been outlined by earlier workers. He detected the direction of reaction in mixtures of natural laumontite, lawsonite and quartz and laumontite, anorthite and quartz by weighing the quartz, introduced as large ground plates, before and after runs. A weight gain was interpreted as laumontite breakdown and a weight loss as laumontite growth. The laumontite field of stability so determined is quite similar to that found by Liou (1970). The intersection of the stability boundaries of laumontite with respect to lawsonite and anorthite falls almost exactly on an extrapolation of the lawsonite–anorthite phase boundary determined by Crawford and Fyfe and therefore supports their version of the equilibrium thermal decomposition of lawsonite.

The phase relations between laumontite and lawsonite seem consistent with field relations. Bailey, Irwin and Jones (1964) report the occurrence of veins of "sugary" laumontite over large areas of the Franciscan formation which lack lawsonite. The probability that the hydrous mineral laumontite could not persist for long outside its dehydration limits makes it appear that pressures must have been below 3 Kbar and temperatures below 275°C ever since the laumontite veins were formed.

(b) PUMPELLYITE

Pumpellyite is a characteristic mineral of metavolcanics of blueschist grade and of rocks transitional to blueschists. The ideal formula is $Ca_4MgAl_5O(Si_2O_7)_2(SiO_4)_2(OH)_3 \cdot 2H_2O$, where Mg is commonly partly replaced by Fe^{2+} and Al by Fe^{3+} (Galli and Alberti, 1969). Field relations suggest that a number of relatively simple chemical reactions may govern the appearance in SiO_2-excess systems of pumpellyite or of alternative chemically equivalent assemblages such as prehnite, $Ca_2Al_2Si_3O_{10}(OH)_2$, + chlorite (Seki, 1965), or grossular + zoisite + chlorite. The former alternative assemblage is probably stable at lower pressures than is pumpellyite, from density considerations, and the latter is a higher grade assemblage. For these reasons the stability of pumpellyite has come under investigation recently. Hinrichsen and Schürmann (1969) experimentally investigated the equilibria:

$$\underset{\text{pumpellyite}}{4Ca_4MgAl_5O(Si_2O_7)_2(SiO_4)_2(OH)_3 \cdot 2H_2O} + \underset{\text{quartz}}{2SiO_2}$$

$$\rightleftharpoons \underset{\text{prehnite}}{8Ca_2Al_2Si_3O_{10}(OH)_2}$$

$$+ \underset{\text{amesite}}{Mg_4Al_4Si_2O_{10}(OH)_8} + 2H_2O$$

and:

$$20Ca_4MgAl_5O(Si_2O_7)_2(SiO_4)_2(OH)_3 \cdot 2H_2O \text{ (pumpellyite)}$$

$$\rightleftharpoons 16Ca_2Al_3Si_3O_{12}(OH) \text{ (zoisite)} + 16Ca_3Al_2Si_3O_{12} \text{ (grossular)}$$

$$+ 5Mg_4Al_4Si_2O_{10}(OH)_8 \text{ (amesite)}$$

$$+ 14SiO_2 \text{ (quartz)} + 42H_2O.$$

They claimed reversals of these equilibria from synthetic crystalline materials at water pressures of 5 to 7 Kbar. Runs as long as several months were necessary to produce reactions. Their results, shown in Fig. 16 show a field for pumpellyite stability which closes out at low pressures: a projected invariant point at about 4.7 Kbar forms the lower limit for iron-free pumpellyite. The presence of free vapour on the low-temperature side of the reaction of prehnite + chlorite to pumpellyite + quartz is most unusual: only the zeolites laumontite and wairakite (Liou, 1970) and analcime (Manghnani, 1970) have previously been found to have a similar feature in their dehydration curves. In the case of zeolite breakdown reasons for this peculiar behaviour can be found in the large molar volumes of zeolites, so that the vapour-containing breakdown products may be denser than the zeolite, and in the loosely-bound nature of the zeolitic water, which has nearly as high entropy as if it were a free vapour phase. The reasons for the temperature-induced hydration of prehnite + chlorite to pumpellyite in the diagram of Hinrichsen and Schürmann (1969) are not apparent.

Liou (1971) put forth an experimentally determined equilibrium diagram for the reaction of prehnite to zoisite, grossular, quartz and water, based on reversals in mixtures of synthetic materials. He found that prehnite is stable up to about 400°C under conditions of $P_{H_2O} = P_{total}$, nearly independently of pressure in the range of 3 to 5 Kbar. He pointed out that this curve should go through an invariant point in the system MgO-CaO-Al_2O_3-SiO_2-H_2O involving the phases prehnite, chlorite, pumpellyite, grossular, zoisite, quartz and vapour, if one exists. If the two experimental curves of Hinrichsen and Schürmann are produced to an intersection, which would not be much below 5 Kbar, according to their data, the prehnite breakdown on its own composition must be a consequence of that intersection. This version of the phase relations of prehnite and pumpellyite is labelled "A" in Fig. 16. The experimental results on prehnite stability of Liou are considerably removed from the indicated invariant point. Liou presented a different version of the phase relations arguing from geometrical considerations and field evidence. His suggested invariant point relations are shown in Fig. 16 at "B". According to this scheme there may not be any lower pressure threshold for the appearance of Mg–pumpellyite in low-grade metamorphics. Liou's curve for the reaction of pumpellyite + quartz to prehnite + chlorite + H_2O has the H_2O liberated on the relatively high temperature side of the boundary, which seems more reasonable than the Hinrichsen and Schürmann version. Liou claimed that the postulated disposition of his curves is consistent with the order observed by Seki (1969) in sequences

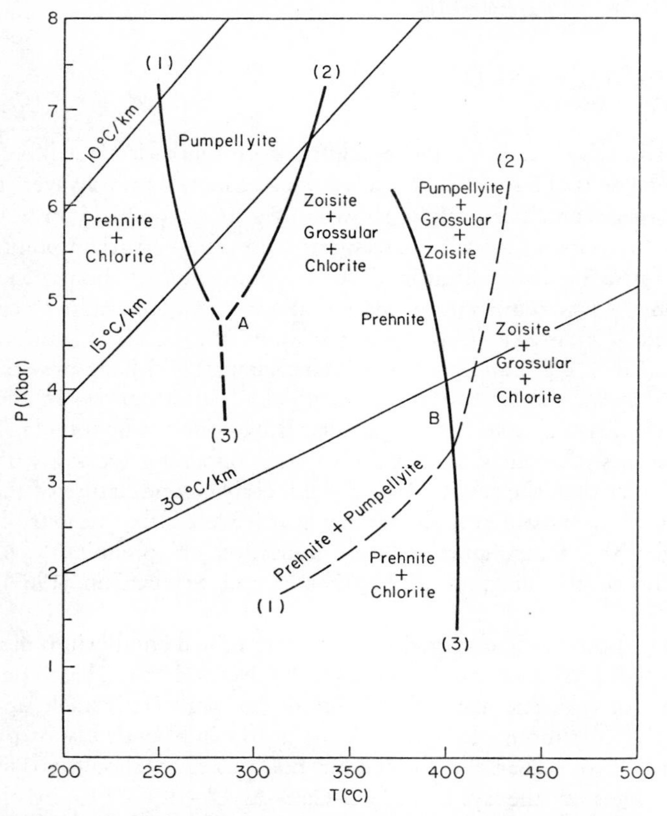

Fig. 16. Experimentally deduced phase relations of prehnite and pumpellyite in the system CaO-MgO-Al_2O_3-SiO_2-H_2O in a quartz-excess system under conditions of $P_{H_2O} = P_{total}$, according to Hinrichsen and Schürmann (1969), labelled "A", and Liou (1971), labelled "B". The dashed lines are inferred boundaries. Similarly numbered boundaries are corresponding. Geothermal gradients shown are used to illustrate the differing interpretations which results from the two schemes (see text).

of increasing burial in regions of normal geothermal gradient, viz, prehnite + chlorite → prehnite + pumpellyite → pumpellyite + epidote. If this is true, the dP/dT slope of the pumpellyite field boundary must be considerably less steep than suggested by Liou (see Fig. 16). The burial sequence observed by

Seki could be satisfied by the Hinrichsen and Schürmann stability relations if the geothermal gradients were considerably smaller than 30°C/km. Both experimental studies indicate that pumpellyite is confined to a rather low temperature range.

As Zen (1961) and Crawford and Fyfe (1965) have pointed out, admixture of CO_2 in mineralising fluids could have large effects in limiting the stability fields of the lime–aluminum silicates. Lawsonite coexisting with carbonate and quartz is common in California blueschists. If an appreciable CO_2 partial pressure was present during metamorphism the pressure requirements for this assemblage might be considerably lower than the experimental lawsonite–laumontite boundary would suggest. This might also be true if the pore solutions were brines rather than pure water.

8. SODIC PYROXENES

Pyroxenes characteristic of blueschist grade rocks are soda-rich varieties. Coleman and Clark (1968) give a summary of pyroxenes characteristic of various blueschist chemical environments from California rocks. Metabasalts feature green soda–pyroxenes with roughly equal amounts of the jadeite ($NaAlSi_2O_6$) and diopside ($CaMgSi_2O_6$) molecules and 10–25% acmite ($NaFeSi_2O_6$), with smaller amounts of $CaFeSi_2O_6$, $CaAl_2SiO_6$ and $MgAl_2SiO_6$ molecules. A few metabasalts contain pyroxene quite rich in iron (Essene and Fyfe, 1967). Soda–pyroxenes of this intermediate composition are called omphacite. Omphacite is the characteristic pyroxene of veins in metabasalts and metasediments and in the common eclogite boulders of indeterminate origin in Franciscan Formation terrains. Metacherts tend to contain very iron-rich pyroxenes, approaching the acmite end-member. Soda–pyroxenes of $NaAlSi_2O_6$ content great enough to be classified as jadeite ($\geq$80%) occur in some low grade metagraywackes distributed over large areas of the California Coast Ranges. The jadeite commonly replaces detrital albite and its amount increases as albite decreases (Bloxam, 1956), although albite and jadeite commonly coexist. Pyroxenes other than soda-rich types are found sometimes in rocks in blueschist terrains, but in most cases petrographic study reveals that they are unstable relics of volcanic rocks which are often partly replaced by omphacite in incipient blueschist metamorphism.

(a) STABILITY OF JADEITE AND OMPHACITE

Much interest is attached to the pyroxenes having a high percentage of the jadeite molecule because experimental work has demonstrated that elevated pressures are required to stabilise jadeite relative to the isochemical assemblage albite plus nepheline. Since the discovery of nearly pure jadeite stream boulders in areas of the Franciscan outcrop in California (Yoder and Chesterman, 1951), numerous occurrences of jadeite pyroxene have been recorded as alteration products of graywackes and basalts in the Franciscan Formation (Bloxam, 1956; McKee, 1962), in Celebes (DeRoever, 1955), in Japan (Seki, 1960) and in other places of blueschist metamorphism. Especially interesting are clear-cut cases where jadeitic pyroxene and quartz occur in a compatible relation together since experimental and thermochemical evidence indicate that very high pressures are needed to stabilise jadeite and quartz relative to albite. This aspect of the jadeite problem is most conveniently referred to the section on albite stability.

The coarse-grained nearly pure jadeite found as masses and veins in Franciscan

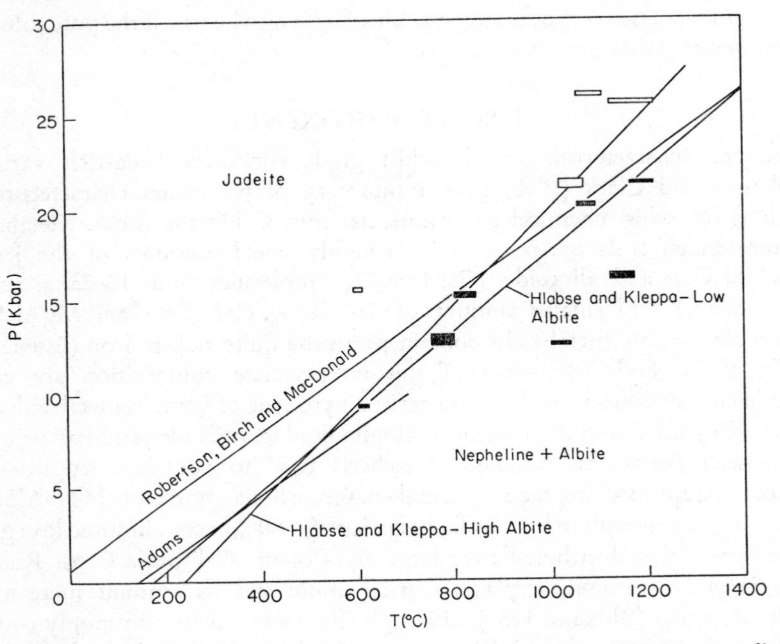

Fig. 17. Experimental and calculated field of stability of jadeite, $NaAlSi_2O_6$, according to Adams (1953), Hlabse and Kleppa (1968), and Robertson, Birch and MacDonald (1957). Reversal run data of last authors are shown, indicating experimental uncertainties of runs.

serpentinites (Coleman, 1961) and as stream boulders probably derived from serpentinite masses reflects a silica-deficient environment, as do such silica deficient minerals as hydrogrossular in close association with Franciscan serpentinites. In this connection, experimental work bearing on the stability of jadeite in the absence of quartz is useful in interpretation.

An early attempt to determine the pressure–temperature stability of jadeite was the calculation of Adams (1953) based on the HF solution calorimetry of Kracek, Neuvonen and Burley (1951). Figure 17 shows the points calculated by

Adams. Considerable uncertainty in the P-T position of the calculated equilibrium:

$$\underset{\text{albite}}{NaAlSi_3O_8} + \underset{\text{nepheline}}{NaAlSiO_4} = \underset{\text{jadeite}}{2NaAlSi_2O_6}$$

results from the large scatter in the heat of solution data. Some of the variation in the measured heats of solution seems to be systematic: a specimen of albite from Varutrask, Sweden, consistently gave a higher heat of solution than the albite from Amelia Courthouse, Virginia. There was also an apparent contamination problem with the jadeite heats of solution resulting from grinding of the hard jadeite in agate and mullite mortars. Adams used the average values of the heats of solution of Kracek, Neuvonen and Burley for the minerals.

The first synthesis of jadeite was accomplished in September 1948 by L. Coes (cited by Robertson, Birch and MacDonald, 1957) in a piston-cylinder apparatus. Early unsuccessful attempts to make jadeite are summarised by Yoder (1950b). Robertson, Birch and MacDonald (1957) made a thorough-going experimental study of jadeite stability in a very high pressure gas-medium apparatus. They used pure jadeite from Burma, Amelia albite and synthetic nepheline as starting material and were able to "reverse" the equilibrium in dry experiments in runs of a few hours duration in the range 600°C–1100°C. Pressure and temperature uncertainties during the runs were large and the authors used a slope of 20 bars/°C projecting from the range of their experimental data to lower temperatures and pressures. The slope value was calculated from the volume and entropy data for the phases. Figure 17 shows their experimental jadeite stability curve. They concluded that their curve was not in serious disagreement with Adams' calculated curve despite the discrepancy of 2–4 Kbar between curves in the low-temperature range.

Hlabse and Kleppa (1968) redetermined the enthalpy of the jadeite = nepheline + albite reaction in a new calorimetric device (Yokokawa and Kleppa, 1964) which employs a PbO-B_2O_3 melt as solvent. Excessive grinding of the samples and consequent contamination is eliminated by the powerful solvent action of the melt. Their calculated curve, based on enthalpies of solution of Amelia albite, synthetic nepheline and a very pure natural jadeite from California and on the heat-content data of Kelley *et al.* (1953) for the minerals agreed well with Adams' (1953) calculated curve in the low temperature region on the HF calorimetry data but did not satisfy the high temperature experimental data of Robertson, Birch and MacDonald. Hlabse and Kleppa noted that, since albite undergoes a low–high inversion somewhere in the range of 500°C–700°C (MacKenzie, 1957) by virtue of (Si,Al) disordering in the tetrahedral framework of the crystal structure, the entropy of transition of jadeite to albite and nepheline should increase by an amount approximately equal to the configurational entropy increase of albite disordering. The ΔS of the transition as determined from heat-content measurements on low albite, synthetic nepheline and jadeite (Kelley *et al.*, 1953) should be increased by about 3.5 to 4.5 cal/°C per mol of albite, depending on assumptions concerning the ordering state of Amelia albite. The enthalpy change of the low–high albite transition, 2.4 to 3.6 kcal/mol, has been determined by oxide melt solution calorimetry by Holm and Kleppa (1968). An order–disorder problem may also be considered in the case of nepheline but was ignored by Hlabse and Kleppa because the heat content and heat of solution measurements were made on high-temperature synthetic nepheline, which was undoubtedly disordered. The calculated stability curve of jadeite subject to these corrections agrees with the high temperature experimental work of Birch, Robertson and MacDonald and is shown in Fig. 17. It is clear from Fig. 17 that while jadeite alone does not require high pressure for stabilisation in the low temperature range, increased pressure favours its development at low temperatures.

A wide range of solid solution exists between the jadeite, acmite and augite components in blueschist pyroxenes (Essene and Fyfe, 1967). Soda-rich varieties predominate, but varieties of omphacite having nearly equal proportions of the three end members are common in metabasalts and veins. Because of the potential application of solid solution limits of minerals to geothermometry and geobarometry, it is important to enquire what sort of miscibility gaps or solvi, if any, exist in the pyroxene system, especially at low temperatures and elevated pressures.

By plotting many pyroxene analyses on a triangular diagram of apices jadeite, acmite and augite, Dobretsov (1962) concluded that a miscibility gap exists on the jadeite–augite join from about 80% to about 55% jadeite. He cited evidence of two coexisting pyroxenes of approximately these compositions in blueschist-grade rocks of the west Sayan and northern Balkash areas, U.S.S.R. He inferred a band of immiscibility across the traingular diagram extending to the jadeite–acmite join, where a gap of approximately Jd_{65} to Jd_{20} was presumed to exist. Coleman and Clark (1968) accepted the existence of such an immiscibility band. Their summary of pyroxene relations for the blueschist grade is shown in Fig. 18. Of especial interest is the structural dissimilarity of metabasalt omphacites near the composition $Jd_{50}Di_{50}$, which have space group P2 as was demonstrated by Clark and Papike (1966) by single crystal X-ray diffraction studies. In these pyroxenes a well-ordered cation arrangement exists with regular alternation of Na-rich and Ca-rich sites and of Al-rich and Mg-rich sites in chains along the C-axis. This observation may support the existence of a solvus gap on the jadeite–diopside join because it implies restriction of cation site occupancy as opposed to free substitution.

The compilation of analyses of Franciscan blueschist rocks of Essene and Fyfe (1967) showed that the pyroxene solid solution picture is more complex than some authors acknowledge. They produced many new electron microprobe analyses of jadeite pyroxenes from metagraywackes and omphacite from meta-gabbros, hydrothermal veins and eclogites. They noted large compositional variation of pyroxenes from the same locality; indeed, zonation of the jadeite content by as much as 25% was common in individual crystals. Many of their analyses fall in Dobretsov's inferred gap on the jadeite–acmite join, but an area near the jadeite–augite join in the position of Dobretsov's solvus seems to be conspicuously vacant. More recently, W. L. Brown (1973) has compiled all the available analyses for omphacite associated with low-grade terrains and concludes that the data indicate a solvus on the jadeite–diopside join which extends from about $Di_{90}Jd_{10}$ to $Di_{25}Jd_{75}$, and which disappears rather rapidly with admixture of acmite molecules. The centre of Brown's gap is interrupted by the appearance of the P2 pyroxenes, and Brown emphasises the possible

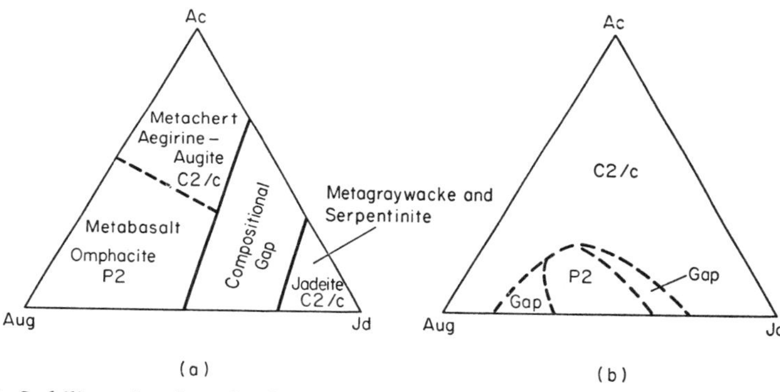

Fig. 18. Stability triangles of jadeite–acmite–augite (diopside + hedenbergite) pyroxenes at pressure and temperature conditions of the glaucophane schist facies, according to (a) Coleman and Clark (1968) and (b) W. L. Brown (1973). Areas in which pyroxenes of the space groups P2 and C2/c occur are shown.

relationship between onset of cation ordering and pyroxene immiscibility. His scheme of pyroxene stability relations for the glaucophane schist facies is shown in Fig. 18.

Experimental work to determine the extent of pyroxene component miscibility at different temperatures and pressures is still in a preliminary stage. Bell and Davis (1965) presented experimental evidence that there is a wide 2-pyroxene solvus on the join diopside–jadeite at 30 Kbar pressure. Their runs were done on glasses in the piston-cylinder apparatus and their criterion for the two-pyroxene field was slight splitting of some of the X-ray reflections of the quenched charges, notably the (220) peak. Additional evidence for a high pressure solvus was found in a positive volume of mixing of quenched charges of intermediate Jd–Di pyroxene. Their diagram suggests that the solvus may extend from Jd_{90} to Di_{90} at temperatures below about 1300°C. They applied a regular solution calculation to the solvus to show that increasing pressure would increase the solvus critical temperatures, and so, presumably, further decrease the solid solution ranges. They ascribed the observed wide solid solution ranges of eclogite pyroxenes to influence of other components. They concluded that the diopside molecule would have a negligible effect on jadeite stability. Specifically, they calculated that a pyroxene of composition $Di_{20}Jd_{80}$ would have very nearly the same temperature–pressure stability as pure jadeite. Bell and Kalb (1969) found a wider range of solid solution for diopside in jadeite in the range 20–25 Kbar and 1150°C–1225°C. They investigated the stability of omphacite solid solutions relative to the low-pressure alternative assemblage plagioclase + nepheline + omphacite (richer in jadeite). They claimed reversal of the solvus at 1225°C based on runs on synthetic crystalline materials of about equal proportion on both assemblages. They found that 30 mol % of diopside produced no observable stabilisation of jadeite to lower pressures, but that 50 mol % resulted in a reduction of about 5 Kbar in the pressure needed to stabilise the omphacite at 1225°C. Ganguly (1973) has calculated from the experimental omphacite stability boundary of Bell and Kalb that the jadeite–diopside join at high pressures and temperatures is nearly ideal and has therefore questioned the existence of a solvus under the experimental conditions of the Bell and Davis study.

The preliminary indication of the experimental and theoretical work seems to be that diopside and jadeite do not have a large thermodynamic affinity, so that jadeite–diopside solid solutions are not greatly stabilised to lower pressures relative to pure jadeite, whether or not a high temperature solvus exists. Another preliminary indication is that very high pressures may cause decreasing solid solution possibility. It must be remembered that solid solution investigations in dry systems, even at very high temperatures, are hampered by slow diffusion rates and that reversals of solid solution equilibria are very hard to demonstrate. The problem of producing ordered pyroxenes is also a tough one—the necessity of working at high temperatures in order to get good reaction rates may prevent direct synthesis investigations involving the ordered P2 polymorph.

(b) STABILITY OF ANALCIME RELATIVE TO JADEITE

Another reaction which may restrict the formation of jadeite-rich pyroxene in nature is the hydration reaction to form analcime, which has an ideal composition of jadeite plus one water, but which can have variable water and silica contents and some calcium substitution for sodium (Saha, 1959). Analcime is common in veins cutting glaucophane schists in some parts of the Franciscan Formation (Coleman and Lee, 1963). The pressure–temperature stability limits of analcime are difficult to determine experimentally because of the low temperature character of the dehydration reactions, especially the reaction to jadeite.

The reaction of analcime at elevated temperatures to albite, nepheline and water was investigated by Greenwood (1961), who produced reversal runs demonstrating the P-T location of the equilibrium. The problem of (Al, Si) order–disorder relations of albite and nepheline (as well as in analcime) were not considered by Greenwood. The analcime which Greenwood synthesised was somewhat deficient in silica relative to the ideal analcime formula. Greenwood was able to show that the P-T field of analcime is considerably restricted for conditions of water pressure much less than total pressure. Greenwood's curve for $P_{H_2O} = P_{total}$ is shown in Fig. 19. Analcime dehydration relations were somewhat revised and extended by Peters, Luth and Tuttle (1966). They found the most stable analcime at the solidus in the system $NaAlSiO_4$-SiO_2-H_2O to have an SiO_2/Na_2O molecular ratio near 3.4, in contrast to the ideal value of 4.0. The relatively high thermal stability of analcime, in conjunction with certain

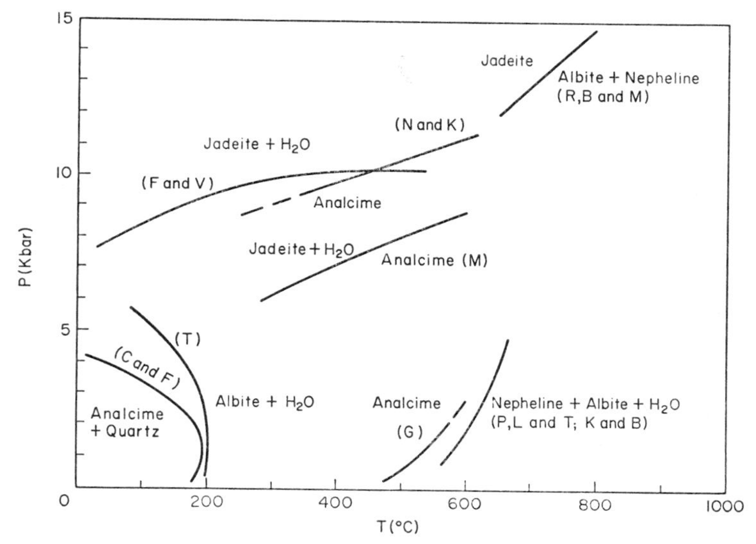

Fig. 19. Experimental curves for analcime stability. (F and V) denotes Fyfe and Valpy (1959). (M) denotes Manghnani (1970). (N and K) denotes Newton and Kennedy (1968). (G) denotes Greenwood (1961). (P, L and T) denotes Peters, Luth and Tuttle. (K and B) denotes Kim and Burley (1971). Also shown is the jadeite–albite + nepheline curve of Robertson, Birch and MacDonald (1957) and the stability relations of analcime + quartz according to Campbell and Fyfe (1965) and Thompson (1971). Melting relations of analcime are omitted for simplicity.

field evidence, suggests that analcime can be a primary mineral of certain igneous rocks. The curve of Peters, Luth and Tuttle is shown in Fig. 19. The experimental equilibrium data of Kim and Burley (1971) are in good agreement with those of Peters, Luth and Tuttle. The data of Kim and Burley suggest that the composition of analcime in SiO_2-sufficient systems should tend toward $NaAlSi_2O_6 \cdot H_2O$ with lowering temperatures.

Griggs and Kennedy (1956) presented a breakdown curve for analcime to jadeite plus water determined in a "simple squeezer" device using analcime as starting material. Their curve lies at extremely high pressures (above 20 Kbar), in the temperature range of interest to blueschist petrology. Fyfe and Valpy

(1959) showed that these pressures are unreasonably high. Their deductions were made on the basis of some preliminary work on the synthesis fields of analcime and analcime plus quartz and the thermochemical data for the phases given by Kelley *et al.* (1953). Their calculated curve for the lower pressure limit of stability of jadeite in the presence of water is shown in Fig. 19. The discrepancy between this curve and that of Griggs and Kennedy undoubtedly is due to the fact that analcime breakdown is a very slow process at low temperatures and that as a consequence upper pressure limits of analcime must be overstepped by large amounts in order to produce breakdown at finite rates. Campbell and Fyfe (1965) have pointed out that the thermochemical data of Kelley *et al.* (1953) for analcime must be considerably in error and that the preliminary synthesis work on the basis of which the Fyfe and Valpy curve is estimated is much higher in temperature than the true stability limit of analcime plus quartz.

A more complete experimental boundary for the dehydration of analcime to jadeite was presented by Newton and Kennedy (1968), who claimed reversals of the equilibrium in piston-cylinder apparatus. Their results confirmed the suspicion of Fyfe and Valpy that analcime breakdown might involve liberation of water on the low-temperature side of the breakdown boundary. This is a rare situation in the dehydration equilibria of silicate minerals, the other known examples being the experimentally-determined breakdown curves of the zeolites laumontite and wairakite (Liou, 1971), and possibly, Mg-cordierite (Newton, 1972), which has a porous zeolite-like structure which can contain loosely-bound water. More recently Manghnani (1970) has produced an experimental diagram for the analcime–jadeite reaction based on piston-cylinder reversal runs on natural minerals. His boundary generally confirms the results of Newton and Kennedy, though it lies at somewhat lower pressures than the previous estimates. Manghnani's and Newton and Kennedy's curves are shown in Fig. 19.

If water pressure during metamorphism is less than the total pressure, either due to the rigidity of rocks or to diluting substances such as CO_2, the field of analcime becomes restricted and the field of jadeite is enlarged. Solutes like NaCl which diminish the activity of water also favour jadeite. Those possibilities are used by Coleman and Clark (1968) to explain the close association of analcime and jadeite in some serpentinite bodies. Their quantitative reasoning is based on the diagram of Griggs and Kennedy (1956) and is therefore not likely to be very realistic. The uncertainty regarding the constitution of mineralising fluids during blueschist metamorphism is nevertheless a large unknown quantity which will have to be considered at length in interpretations of the pressures necessary to produce jadeite rather than analcime in metamorphism under aqueous conditions.

9. ALBITE

Albite is the dominant feldspar in blueschists and consideration of its stability limits plays a key role in genetic interpretations. Other feldspars are, relatively speaking, much less important. Calcic plagioclase is absent in low-grade metamorphism, its place being taken by hydrous lime silicates such as zoisite, laumontite, prehnite and lawsonite. Microcline and orthoclase are among the surviving relics of sedimentary detritus in weakly recrystallised graywackes, but they are uncommon (Bailey, Irwin and Jones, 1964). Adularia in hydrothermal veins has been found at several localities in the Franciscan Formation (Bailey, Irwin and Jones, 1964).

Albite is present as detrital relics in recrystallised graywackes, as a recrystallised matrix mineral in metasediments, and as a vein filler, sometimes associated with jadeite, analcime, lawsonite, or omphacite. Since 1955 it has been recognised that many blueschist facies metagraywackes show detrital albite incipiently or completely replaced by jadeite pyroxene. Examples are known from Celebes (DeRoever, 1955), Japan (Seki, 1960), California (Bloxam, 1956; McKee, 1962) and the western Pennine section of the Alps (Bearth, 1966). The macroscopic and microscopic sedimentary textures of the jadeitised metagraywackes are often well preserved and metamorphism may not be recognisable in the hand specimen except by the criterion of "heft": jadeite-rich metagraywackes are likely to have a specific gravity in excess of 2.70 (Bailey, Irwin and Jones, 1964). Although albite commonly coexists with the jadeite in quartz-rich rocks (DeRoever, 1955; Seki, 1960), possibly as metastable relics of detrital plagioclase, the amount of albite often decreases as the pyroxene increases, which fact has suggested to many workers that the relationship of albite to jadeite is simply:

$$\underset{\text{albite}}{NaAlSi_3O_8} = \underset{\text{jadeite}}{NaAlSi_2O_6} + \underset{\text{quartz}}{SiO_2}.$$

The right-hand side of the equation has a much smaller molar volume than the left-hand side; hence, the breakdown reaction of albite will be favoured by high pressures. The possibility of another quantitative geobarometer suggests itself.

Certain pieces of information from the field evidence indicate that the breakdown of albite in metagraywackes may not be as simple as this. The typical pyroxenes are not pure jadeite. An analysis of the jadeite from the metagraywacke of Angel Island, California, studied by Bloxam (1960), shows one of the highest published $NaAlSi_2O_6$ contents yet reported for pyroxenes associated with quartz (Coleman and Clark, 1968), viz., $Jd_{87}Di_8Ac_5$. The jadeite pyroxene of the Celebes metagraywacke of DeRoever (1955) has about 20 mol % non-jadeite substituents as has the pyroxene reported by Seki (1960) from the Kanto Mountains, Japan. These are estimates from optical data based on the chart correlating composition with optics of Iwasaki (1963). Microprobe examination of the Pacheco Pass, California, jadeitised graywacke, which is distributed on a regional scale (McKee, 1962), reveals widely varying amounts of mafic pyroxene substituents, varying from 10 to 25 mol % (Newton and Smith, 1967). The albite in the Pacheco Pass rocks is very pure but often partially jadeitised. The common co-existence of albite and jadeite implies either that the albite is a metastable relic or that the presence of the pyroxene substituents causes the breakdown reaction of albite to depart widely from univariancy, in which case the pressure requirements for breakdown of albite would be reduced. The albite in veins would have to be formed as a retrograde product at lower pressure after the jadeite-producing main phase of metamorphism (Coleman and Lee, 1963). Finally, the peculiar field distribution of jadeite is a problem. Nearly unmetamorphosed graywacke containing detrital volcanic plagioclase is often found in close field association with completely jadeitised graywacke, for example at Ward Creek, California (Coleman and Lee, 1963) and Panoche Pass, California (Ernst, 1965). Grains of unaltered and completedly jadeitised plagioclase can sometimes be found in a single thin section (Bloxam, 1960). McKee (1962) found a mappable field boundary at Pacheco Pass between graywackes containing dominantly jadeite and rocks containing abundant albite, but the jadeite apparently is concentrated towards the stratigraphic top of a relatively coherent exposed section, which is not the relationship expected of burial-induced mineral breakdown. Blake and Cotton (1969) found evidence in the distribution of jadeite in the Pacheco Pass and neighbouring areas of the Diablo antiform south of San Francisco, California, that a jadeite-rich horizon marks the sole of the fundamental thrust fault bounding the Franciscan Formation at its top. The production of jadeite might have been governed by some thrust-related agency such as tectonic overpressure (Blake, Irwin and Coleman, 1969). Ernst (1971) on the basis of a more complete sampling study, found no compelling evidence of a relation between jadeite in metagraywacke and proximity to thrust faults bounding the Franciscan terrain in this area. McKee (1962) noticed a correlation between shearing and amount of glaucophane and jadeite in the rocks.

The obvious experimental starting point is the univariant reaction of albite to jadeite plus quartz. A P–T location for the reaction was calculated by Adams (1953), from the heats of solution of albite and jadeite measured by Kracek, Neuvonen and Burley (1953) in an HF calorimeter. Adams used their average values of heats of solution, which included rather large and unexplained experimental scatter, and preliminary entropy data for the minerals which have been improved upon since then. Adam's points are shown in Fig. 20, corrected for more modern entropies (Kelley *et al.*, 1953). Birch and LeComte (1960) carried out the first direct experimental determination of the equilibrium using an internally-heated gas-pressure vessel. They used pure natural minerals as starting materials. Many runs were "reversals" of the equilibrium. Temperatures and pressures external to the sealed sample capsule were well controlled and well measured. Their determination still stands as a solid piece of early experimental petrology upon which later investigators have not improved very much. The use of dry natural starting materials necessitated experiments at temperatures above 700°C in order to get recognisable reactions. This in turn demanded a long extrapolation from the temperature region of the experimental points to that of application to blueschists. The high temperature entropy data suggests that the dP/dT slope of the equilibrium should change from 21 bars/°C to 17 bars/°C over the temperature range 0°C to 800°C. Birch and LeComte rather arbitrarily adopted a constant slope of 20 bars/°C. Their curve, which has been often cited in literature on blueschists and eclogites, is shown in Fig. 20, along with their experimental reversal data.

Newton and Smith (1967) produced additional data in the range 500°C–600°C. Their runs made use of mixtures of pure natural minerals (including Amelia albite) with water as a flux. Equilibrium was demonstrated by detection of the

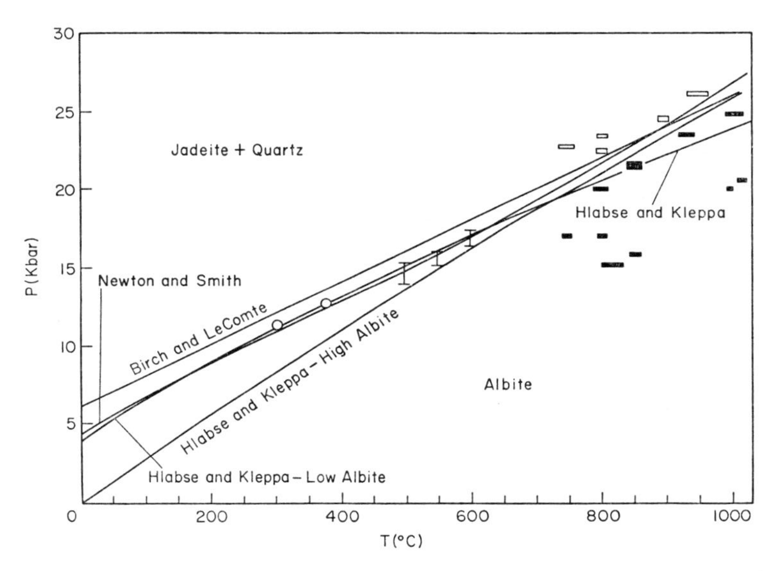

Fig. 20. Experimental and calculated curves for albite breakdown. Reversal data of Birch and LeComte (1960) are plotted. Sizes of rectangles indicate their pressure and temperature uncertainties of runs. Brackets are data of Newton and Smith (1967). The curve of Newton and Smith involves an equilibrium (Si; Al) disordering transition of albite assumed to occur in the range 500°C–700°C. Circles are calculated points of Adams (1953) based on heat of solution measurements of Kracek, Neuvonen and Burley (1951), corrected for more recent entropy data of albite and jadeite. The curves of Hlabse and Kleppa are their calculations (1968) from heat of solution measurements.

growth of one assemblage at the expense of the other by differential growth or decrease of X-ray diffraction maxima of the charges. The experiments were done in the solid medium piston-cylinder apparatus, for which apparatus the pressures are not as well measured as in the high-pressure gas apparatus, though quite hydrostatic because of the presence of water in the sample capsule. Above 650°C, melting begins in the system albite–water at pressures near the jadeite + quartz inversion, and reactions become very slow below 500°C. The data of Newton and Smith lie about 1 Kbar below the extrapolation of Birch and LeComte in this temperature range. Boettcher and Wyllie (1968b) extended the measurements of albite $\rightleftharpoons$ jadeite plus quartz to somewhat higher temperatures by selecting experimental compositions in the system Na_2O-Al_2O_3-SiO_2-H_2O such that either albite or jadeite plus quartz coexisted with water-rich liquid over a range of temperatures. These hydrous silicate fluids are very reactive and clear reversibility data were obtained in runs of a few hours. The runs were done in piston-cylinder gear. The data seemed to reinforce the determination of Newton and Smith, subject to the uncertainty of the pressure calibration in solid-pressure apparatus. Subsequently, piston-cylinder determinations of the albite breakdown reaction under hydrothermal conditions at 600°C have been carried out in six laboratories, with good agreement on a value of 16.3 ± 0.4 Kbar (Johannes *et al.*, 1971). This may be compared to the extrapolated Birch and LeComte value of 18 Kbar.

Any attempt to construct a realistic equilibrium P-T curve for the high-pressure breakdown of albite must take account of the transition with increasing temperature from low albite, in which the silicon and aluminum atoms are nearly completely ordered (Ribbe and Gibbs, 1969) to high albite, with nearly complete (Si, Al) disordering. Experimental petrology which bears on the nature of the low–high albite transition is important in this connection. A continuous variation in the ordering state of albite synthesised hydrothermally from glass was found by MacKenzie (1957) for syntheses covering the range 500°C–1000°C. His criterion of ordering state was the separation in Bragg angle of the $(1\bar{3}1)$ and (131) X-ray diffraction maxima (Smith and Yoder, 1956). He was able to show that the disordering is at least partially reversible by producing increased disorder at 800°C in an intermediate-state albite originally prepared at 600°C. He believed that the transition low albite to high albite was of higher order, with a succession of equilibrium intermediate states as a function of temperature. McConnell and McKie (1960), by applying kinetic theory to the data of MacKenzie, concluded that most of the disordering takes place over the temperature range 575°C–625°C. Holm and Kleppa (1968) measured the heats of solution of Amelia albite and heat-treated Amelia albite (high albite). They found that a step-wise disordering transformation takes place. A 3 day treatment of low albite at 1040°C produces a change associated with 2·4 kcal/mol enthalpy increase, and longer heat treatments, sufficient to prepare high albite, further increase the enthalpy by 1 kcal/mol. From the enthalpy change and the entropy data for the albite polymorphs they concluded that the interpretation of McKie and McConnell was correct, viz., that albite undergoes a "smeared transition" in the neighbourhood of 600°C. Martin (1969) was able to synthesize hydrothermally albites whose ordering states, based on X-ray diffraction criteria, were technically in the low-albite range. He used gel starting material containing excess sodium disilicate, which is more soluble than aluminosilicate and which seems to be capable of promoting more complete solution and recrystallisation of aluminium and silicon atoms. Martin's lowest feldspars were produced at the lowest temperatures (300°C–400°C) and highest pressures (10 Kbar) of his study. Martin made the suggestion that the character of the crystallising fluid could have as great an effect on the character of a feldspar formed during metamorphism as the physical conditions.

Newton and Smith (1967) constructed an equilibrium breakdown curve for albite which allowed for the disordering of albite. They assumed that the disordering transition takes place between 500°C and 700°C. Various estimates of the configurational entropy increase of disordering range from 2.7 to 4.4 cal/°C/mol, depending on the structural scale of disordering. Newton and Smith adopted an average value of 3.5 entropy units, which would increase the slope of the albite–jadeite + quartz reaction by nearly 50%. The experimental points of Birch and LeComte were accepted, although it is not certain if they are equilibrium determinations with respect to albite order–disorder relations. Their reactions of jadeite and quartz certainly produced high albite (stable in the range of their runs), which is much easier to synthesise than low albite, but it is doubtful whether their starting material low albite would have disordered at temperatures below 1000°C in the time durations of their runs. More probably their points for breakdown of albite to jadeite and quartz represent the metastable breakdown of low albite. The curve in this range should have a considerably steeper slope than 20 bars/°C and, as a consequence, the breakdown curve will move to somewhat lower pressures than the Birch and LeComte extrapolation in the low-temperature range. Newton and Smith found a breakdown pressure of 4.2 Kbar at 0°C compared with the 6 Kbar of Birch and LeComte.

Hlabse and Kleppa (1968) determined a breakdown curve for albite from their measurements of the heats of solution of Amelia albite, pure jadeite from Santa Rita Peak, California, and quartz in an oxide-melt calorimeter. Their curve for low albite breakdown almost exactly coincides with the curve of Newton and Smith (1967). Their curve for the stability limit of high albite, assuming an entropy of disordering of 3.5 entropy units per mol, is shown in Fig. 20. In the application of their results to petrology, Hlabse and Kleppa make an intriguing point. The detrital plagioclase of the graywackes associated with blueschist terrains is mostly high-temperature volcanic plagioclase. In addition, the plagioclase structure may have been further disturbed by migration of calcium ions through the structure to segregate into lawsonite or another hydrous lime silicate. It is possible, therefore, that the breakdown of albite in nature is governed by the stability limit of high albite, instead of low albite. The former curve falls several Kbar lower in pressure than the latter, and this hypothesis might help to ameliorate the extreme nature of the physical conditions (low geothermal gradient, high pressure) apparently needed to break down albite in nature. Of course, the hypothesis involves a metastability aspect, which will be discussed below.

Some petrologists invoke diopside and acmite substituents in jadeite as a means of reducing the pressures required for stabilising jadeite in the presence of quartz. Coleman and Clark (1968) argue that the presence of albite-bearing veins in jadeite–quartz metagraywacke indicates pronounced divariancy of the breakdown reaction. The jadeite commonly has about 20 mol % of non-jadeite substituents and it is possible that the impure jadeite, quartz and albite all form in stable association well below the breakdown pressure of pure low albite. A small amount of theoretical and experimental work has been done to determine the effect of acmite and diopside solid solution on jadeite stability in the presence of quartz. Essene and Fyfe (1967) applied ideal solution theory and concluded that at 200°C, 20 mol % of acmite substitution would lower the pressure requirements for the stabilisation of jadeite and quartz by only 400 bars. Twenty percent of diopside substitution would lower the pressure by about 800 bars because twice as many atoms are mixing in the solid solution as in the case of acmite substitution. However, Ganguly (1973) has argued that the linked nature of the substitution CaMg $\rightleftharpoons$ NaAl in jadeite pyroxene will restrict the entropy of mixing such that $CaMgSi_2O_6$ substitution into jadeite will not produce as much stabilisation of the pyroxene relative to albite as Essene and Fyfe suppose for the case of ideal solid solution.

These arguments suggest that much of the jadeitic pyroxene of the metagraywackes has very nearly the same pressure stability as pure jadeite in the presence of quartz. A similar conclusion was reached by Newton and Smith (1967) in high pressure experiments at 600°C on the join $NaAlSi_3O_8$–$NaFeSi_3O_8$. They used glass starting material and water as a flux. No hydrous phases form in this system at 600°C but melting relations prevent experiments at higher temperatures at high pressures in the wet system. Their diagram, shown in Fig. 21, indicates that the pressure lowering for pyroxene + quartz stability relative to albite, quartz and a pyroxene richer in acmite, is only about 400 bars for pyroxene of composition $Jd_{80}Ac_{20}$. They achieved an indication of a reversal by apparently increasing the size of albite seeds at the expense of synthetic pyroxene of composition $Jd_{86.5}Ac_{13.5}$ and quartz at a pressure only 800 bars below the stability limit of pure jadeite plus quartz at 600°C. It must be kept in mind that their experiments were made in a piston-cylinder apparatus, in which there is considerable uncertainty in the pressure measurement, and also without oxygen fugacity buffers. No magnetite was noted in the charges, however, and a pyroxene synthesis from the $Jd_{86.5}Ac_{13.5}$ glass crystallised with a magnetite–hematite buffer had d-spacings identical to those of an unbuffered run. An indication of metastability in the syntheses was the presence of up to 5 mol % of $NaFeSi_3O_8$ molecule in the albite, as determined by microprobe analyses. Natural metagraywacke albites from Pacheco Pass showed no such large iron contents. Popp and Gilbert (1972) found that the stability of acmite–jadeite solutions at the acmite-rich end of the diagram is also what would be predicted by nearly ideal pyroxene mixing. This was convincingly demonstrated by growth or decline of very small amounts of albite in pyroxene–quartz mixtures which were treated hydrothermally at 4 Kbar and 400°C–600°C. At 600°C and 4 Kbar the maximum amount of the $NaAlSi_2O_6$ molecule in acmite is between 4 and 5 mol %.

The preliminary indication of the experimental work is that jadeite–acmite

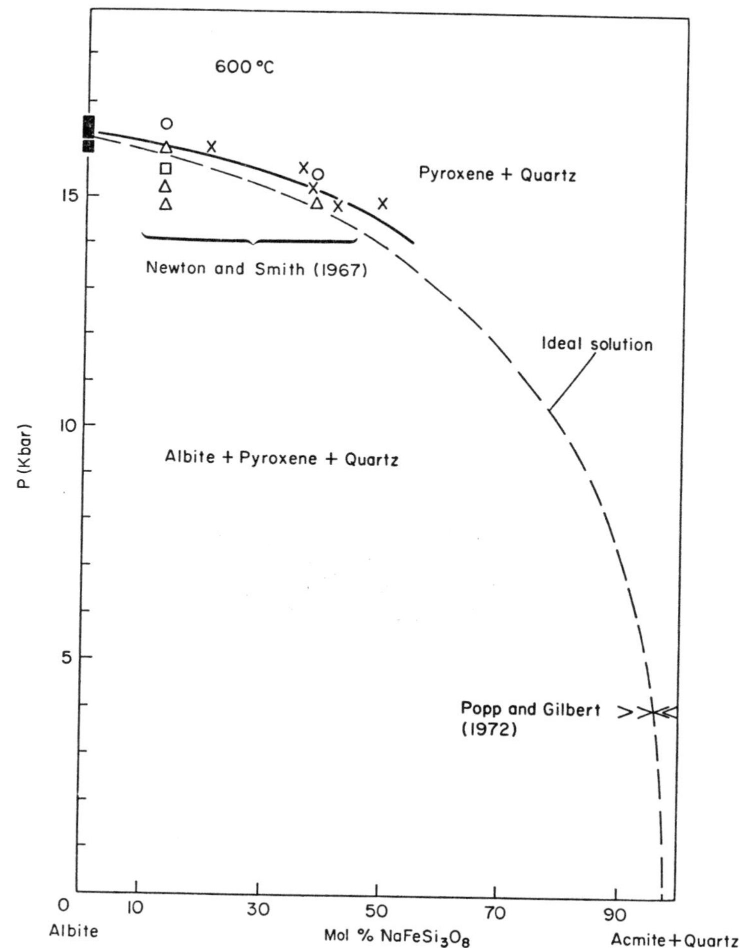

Fig. 21. Experimental join $NaAlSi_3O_8$-$NaFeSi_3O_8$ at 600°C. Solid rectangle is stability limit of pure albite relative to jadeite and quartz, as determined by Johannes *et al.* (1971). The size of the symbol indicates pressure uncertainty of determination. Circles and triangles denote products of hydrothermal syntheses from glass of Newton and Smith (1967). A square denotes a reversal run of Newton and Smith, in which albite seeds grew at the expense of synthetic pyroxene and quartz of the indicated bulk composition. X's denote compositions of pyroxenes synthesised by Newton and Smith, as determined by d-spacings. Arrowheads denote products of hydrothermal runs on mixtures of synthetic pyroxene, quartz and albite of Popp and Gilbert (1972).

pyroxenes are nearly ideal solutions for all compositions and that a jadeite pyroxene of about 20 mol % acmite has nearly the same pressure stability as pure jadeite + quartz.

Kushiro (1969) performed reversal runs at 1250°C on synthetic crystalline mixtures of pyroxene + plagioclase + quartz with bulk compositions on the diopside–albite join. For pyroxene of composition $Di_{40}Jd_{60}$ he showed that plagioclase forms from pyroxene + quartz below 29 Kbar and disappears in favour of pyroxene above 32.5 Kbar. For more albite-rich compositions, he was unable to get clear-cut reversals in runs of 6 hours. His data show, at least for high temperature plagioclase and pyroxene, that the jadeite plus quartz stability field is increased by no more than 2.5 and probably only about 0.5 Kbar at 1250°C by addition of 20 mol % of diopside. Again the indication that small amounts of pyroxene substituents do not greatly stabilise jadeitic pyroxene relative to plagioclase seems to be upheld. The results are shown in Fig. 3.

The stability of albite may be limited in another fundamental way at lower pressures by a hydration reaction generating analcime:

$$\underset{\text{albite}}{NaAlSi_3O_8} + H_2O = \underset{\text{analcime}}{NaAlSi_2O_6} + \underset{\text{quartz}}{SiO_2}.$$

The occurrence of analcime and quartz in veins cutting blueschists has been noted by Coleman and Lee (1963, p. 288) who recognised the possibility that the assemblage might be used as an indicator of physical conditions during metamorphism. Campbell and Fyfe (1965) summarised earlier attempts to deduce the conditions of analcime stability with quartz relative to albite and water. They emphasised that analcime can form metastably from highly reactive starting materials at temperatures far above its true stability limit in the system $NaAlSi_3O_8$–H_2O, and that the synthesis field for analcime (with possibly some SiO_2 excess over the theoretical composition) and quartz extends to about 275°C at water pressures in the range 0.5–2 Kbar, this temperature being only an upper limit for a possible field of true stability. Campbell and Fyfe (1965) performed some weight change experiments on single crystals of albite suspended in water presaturated with analcime and quartz in the temperature range 160°C–245°C and at pressures along the vapour–liquid equilibrium of H_2O in the temperature range. The crystals gained weight consistently above 200°C but showed erratic gain or loss at lower temperature, with losses predominating. They thought that 190°C might represent an equilibrium point for the reaction at 12 bars water pressure, but were careful to point out that this could be rigorously defended only as an upper limit to analcime + quartz stability. They noted also that the weight gains probably only represent growth of high albite, which would be metastable. They calculated a preliminary P–T stability boundary for analcime plus quartz from their available thermodynamic data, shown in Fig. 19. Thompson (1971) performed more detailed experiments on the stability of analcime + quartz relative to albite by the crystal weight change method. He found that quartz single crystals exposed to mixtures of natural analcime and low albite at temperatures of 120°C to 240°C and water pressures of 2 to 6 Kbar showed systematic weight losses or gains which gave indications of the stability limits of analcime + quartz. Weight changes of natural low albite single crystals were consistent with the inferences from the quartz runs. The stability boundary of analcime + quartz so determined is shown in Fig. 19. The general agreement with the deductions of Campbell and Fyfe (1965) is evident.

Campbell and Fyfe (1965) advanced an argument similar to that of Hlabse and Kleppa (1968) that alteration of disordered detrital volcanic albite in graywacke metamorphism can result in an apparently expanded field for the formation of analcime plus quartz. They considered that analcime and quartz might under some circumstances form as a metastable intermediary between high albite and low albite. The existence of natural analcime with silica percentages nearly as high as albite (Saha, 1959) suggest metastable development and persistence. A practical upper pressure limit for this process might be at the limit of analcime stability with respect to jadeite, in that a hydrous mineral may not be expected to persist long in nature outside its dehydration limit, even at low temperatures. On the other hand, Crawford and Fyfe point out that the action of natural brine solutions such as connate sea water in recrystallisation would severely reduce the stability region of analcime.

In summary, experimental work on feldspar in blueschist metamorphism should be critically appraised and cautiously applied. Petrologists sometimes may still be too hasty to draw unwarranted field conclusions from very preliminary experimental data, although some authors are beginning to take a more realistic, if somewhat disenchanted, view of interpretation based on experimental petrology (see Coleman, 1967a, p. 495). Many perplexing features of albite and its relation to jadeite and other minerals remain to be elucidated. Fyfe and Zardini (1967) observed that in a conglomerate from the eastern part of Pacheco Pass, California, volcanic pebbles had jadeitized plagioclase phenocrysts but in metagranitic fragments albite was present. They stated that (p. 821) "... there seems to be no rational explanation for this fact ...". The present authors are indebted to Professor J. V. Smith for an interesting interpretation, namely, that the volcanic plagioclase may have been disordered high-temperature feldspar, whereas, the granitic albite was probably more ordered and, hence, more stable. Thus detrital albite might break down at lower pressures in graywackes than the experimental work implies. Ernst (1971) points to the fact that most albite found in blueschist-grade metagraywackes is of a low structural state, inferring that such must have been the case during metamorphism. This does not necessarily follow—one must consider also that the surviving albite may owe its preservation to better initial ordering than other albite which was transformed to jadeite during metamorphism, and that the surviving albite has had a great many million years to order its silicon and aluminum atoms since metamorphism.

Preliminary experimental work seems to indicate that non-jadeite pyroxene substituents will not reduce greatly the pressure requirements for jadeite stability. However, in assessing the role of the mafic components in jadeite production, one must remember that very subtle metastable reactions are thermodynamically possible as the incompatible association of detrital graywacke minerals approaches equilibrium, perhaps in a sequence of "Ostwald steps". One example might be in oxidation of chlorite minerals in the groundmass to produce ferric iron for jadeite–acmite solid solutions. The net reaction of chlorite plus oxygen plus albite to jadeite plus quartz might have thermodynamic permission to proceed at much lower pressures than the simple reaction albite → jadeite + quartz. A complex oxidation reaction in graywackes involving albite, phengite, chlorite, lawsonite and iron ore was inferred by Kerrick and Cotton (1971) to have produced jadeitic pyroxene in a section of metagraywackes along the western flank of the Diablo Range in California. Variations in water fugacity and oxygen fugacity could have governed the appearance of jadeite at the expense of albite, though these components are not involved in the simple reaction of albite to jadeite + quartz. Thus, even minor amounts of pyroxene substituents in jadeite might, in a more extended analysis, represent fairly large reduction of the pressures necessary for albite breakdown. The fact that the jadeite so formed would be itself metastable with respect to more impure jadeite plus low albite is not necessarily a contradiction. After all, the jadeite plus quartz that we observe at the surface is metastable and must have survived prolonged exposure to elevated temperatures outside its field of stability during unloading. Much more experimental and observational work must be done before the picture of albite breakdown will be clear enough to permit confident quantitative interpretation of the temperatures and pressures involved.

10. SUMMARY OF GLAUCOPHANE SCHIST CONDITIONS

Experimental petrology has begun to formulate a consistent picture of the physical conditions of glaucophane schist metamorphism. We may conclude from the experimental data that conditions of low temperature and elevated pressure prevailed. To produce these conditions by burial in the near-surface regions of the earth, associated geothermal gradients must have been low. The limited amount of data available and the uncertainties concerning the proper application of these data discourage highly quantitative interpretation at this point. Nevertheless, it is instructive to attempt to analyse the conditions of glaucophanitic metamorphism by means of a "petrogenetic grid" of selected experimental results.

Figure 22 shows some of the experimental stability curves for blueschist and related minerals which allow the most unambiguous interpretations. The outlined P–T area in Fig. 22 is consistent with the observed mineralogy of blueschists; that is, it contains P–T fields of the assemblages zeolites + albite +

calcite, lawsonite + albite + calcite, lawsonite + albite + aragonite, and lawsonite + jadeite + aragonite in SiO_2-excess rocks. The successive assemblages named could be a progressive metamorphic sequence with increasing burial pressure in a suite of shales and graywackes with sufficient water content. Such progressive sequences have been broadly mapped in individual terrains, such as the coastal Northern California area (Blake, Irwin and Coleman, 1967). This apparent consistency of field and experimental observations has encouraged petrologists in the expectation that a nearly quantitative application of the stability diagrams may be made (Ernst, 1971). On this basis, Fig. 22 suggests that

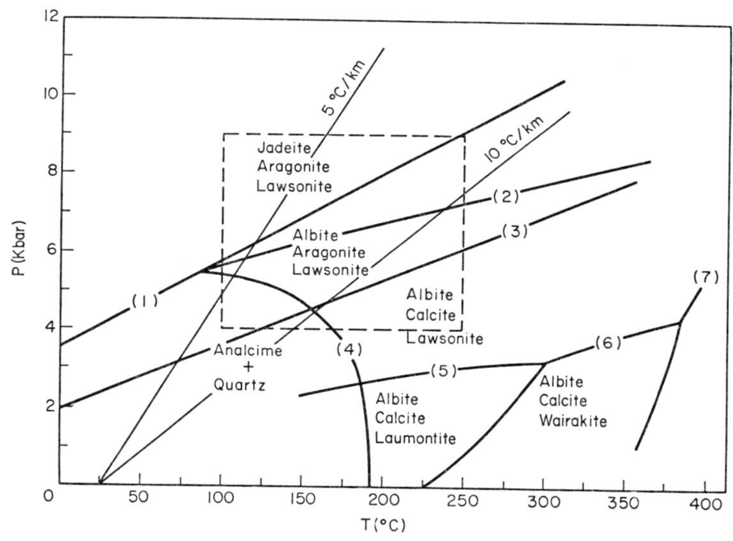

Fig. 22. Selected experimental results and application to glaucophane schist metamorphism. Fields of compatibility of some diagnostic assemblages shown for quartz-excess systems under conditions of $P_{H_2O} = P_{total}$. Curve (1) is albite $\rightleftharpoons$ jadeite + quartz for an albite of nearly perfect (Al, Si) ordering, from Newton and Smith (1967), Hlabse and Kleppa (1968) and Johannes *et al.* (1971). Curve (2) is analcime $\rightleftharpoons$ jadeite + H_2O from Manghnani (1970). Curve (3) is calcite $\rightleftharpoons$ aragonite from Johannes and Puhan (1971). Curve (4) is analcime + quartz $\rightleftharpoons$ albite + H_2O from Thompson (1971). Curves (5) and (6) are laumontite $\rightleftharpoons$ lawsonite + quartz + H_2O and wairakite $\rightleftharpoons$ lawsonite + quartz + H_2O, respectively, from Liou (1970). Curve (7) is dehydration limit of lawsonite after Newton and Kennedy (1963), Crawford and Fyfe (1965) and Nitsch (1972). Area formed by dashed lines suggests P–T conditions for blueschist metamorphism which are compatible with the experimental curves and oxygen isotope evidence of Taylor and Coleman (1968), Clayton O'Neil and Mayeda (1972) and O'Neil, Clayton and Mayeda (1969).

geothermal gradients necessary to produce jadeitic pyroxene must have been less than 10°C/km and that pressures of 7 Kbar or more were reached. The O^{18}/O^{16} distributions in blueschist minerals from the Cazadero, California area measured by Taylor and Coleman (1968) lead to temperature estimates in the range 100°C–200°C using the newer experimentally-determined calcite–water (O'Neil, Clayton and Mayeda, 1969) and quartz–water (Clayton, O'Neil and Mayeda, 1972) fractionation data. The associated temperatures could not have been much higher than 200°C. These temperatures and geothermal gradients are very low compared to those that must have operated in other kinds of metamorphism, and the pressures seem extreme in terms of familiar geosynclinal models of crustal metamorphic cycles. Application of plate-tectonic concepts of continental-margin underthrusting by spreading oceanic crust and upper mantle provides an appealing alternative to the classical ideas (Wilson, 1968; Hamilton, 1969; Ernst, 1970). The chaotic or "melange" character of much of the Franciscan Formation (Hsu, 1968) would be quite consistent with such tectonism.

Some petrologists have cautioned against overly straight-forward and simplistic use of the experimental diagrams in quantitative interpretation of the physical conditions of blueschist metamorphism. They have stressed that not enough is known about several important factors. These include quite complex reactions to produce the diagnostic minerals about which we have very little experimental knowledge at present; the nature of and determinative effect of the mineralising pore fluids; the effect of large plastic deformation on the minerals developed; and possible factors of kinetics and metastability, which in some cases may be just as important as chemical equilibrium in determining the observed assemblages.

The probable effects of most of the factors mentioned, if they are influential, would be to lower the pressures needed to generate many of the blueschist minerals. For instance, if albite gives way to jadeitic pyroxene by complex monotropic reaction with associated ferromagnesian minerals, it could do so at pressures below the ultimate stability limit of low albite by itself. A jadeitic pyroxene so created might survive in the presence of quartz indefinitely because of the high kinetic barrier against a heterogeneous reaction to produce highly ordered low albite. Similarly, detrital volcanic plagioclase has thermodynamic permission to react with water to produce lawsonite, jadeite and quartz at pressures well below the absolute stability limit of low albite. Pore fluids highly concentrated in solutes would diminish the field of the zeolites, as several authors have suggested (Campbell and Fyfe, 1965; Coleman and Clark, 1968) and thus lower the pressures necessary to create lawsonite. Reduced water activity would favour dense anhydrous minerals such as omphacite and garnet relative to amphibole and chlorite. Very impure pore fluids, operating in conjunction with certain kinetic factors, have been invoked to produce many blueschist minerals metastably at low pressures (Gresens, 1969). Accumulated plastic strain energy can exert a bias toward production of aragonite rather than calcite at reduced pressures in the low-temperature range and this tendency could be augmented by pore fluid solutes inhibiting calcite regeneration (Newton, Goldsmith and Smith, 1969).

For reasons similar to the foregoing it might be reasonable to regard the experimental evidence from the simple synthetic systems as yielding upper estimates for the pressures necessary to produce the glaucophane schist facies which may be realized in cases where the most complete approach to chemical equilibrium in the rocks can be demonstrated. The incomplete and often incipient recrystallisation of initially non-equilibrium assemblages of detrital minerals should warn the petrologist not to ignore possible kinetic and metastability considerations in his analysis.

11. OTHER HIGH PRESSURE MINERAL ASSEMBLAGES

In the preceding section the experimental petrology of blueschists was considered in detail, and we concluded that the indicated conditions of formation were low temperatures and high pressures. Some of the stability relations, and the deductions, are also relevant to eclogites, some varieties of which are associated with blueschists. A short discussion of eclogite stability, therefore, follows naturally. This in turn brings in the possibility of a new set of conditions, when $P_{total} \gg P_{H_2O}$. These conditions forge a link between eclogites and granulites, which are discussed in the sequel.

(a) THE ECLOGITE FACIES

Eclogites are found in various situations which suggest that they are stable over a wide P–T range. These include the eclogites associated with diamond-bearing kimberlites, as inclusions in extrusive basalts, as lenses in migmatites and in association with glaucophane schist, greenschist and amphibolite facies of alpine orogenic zones. A temperature range perhaps from 300°C–1200°C and pressures from a few to tens of kilobars might be implied by the variety of occurrences. This would also be in accord with the mineralogical differences of eclogites demonstrated by widely different ratios of elements in co-existing garnets and omphacites in eclogites from the different environments (Coleman *et al.*, 1965). This wide stability field is shown in all facies diagrams (Fig. 1).

Details of the chemistry and mineralogy of the eclogites may be found in Church (1968). We note here that, chemically, eclogites have a rather wider range than the major groups of basalt; some are more analogous to cumulate igneous rocks, or have certain features compatible with metasomatic activity during their formation, e.g. very low K_2O and very low SiO_2. Certain other eclogites, found in association with glaucophane schists, have been derived from spilitic parents.

The dominant minerals in eclogite are garnet (py-alm-gross, with py normally $> 30\%$) and omphacite (di-ac-jd-hy, with di $+$ jd $\gg$ ac $+$ hy). They frequently also contain minor amounts of many other minerals such as quartz, epidote, zoisite; amphiboles such as hornblende-actinolite-glaucophane; kyanite, rutile, sphene, muscovite. Eclogites do not contain plagioclase. The plagioclase components have been taken up in garnet and pyroxene. The many phases to be found in eclogites may appear to be in equilibrium or may obviously be retrograde hydration products. Eclogites appear to be subject to late hydration; for example, garnet may be pseudomorphed by chlorite and pyroxene by amphibole.

Two types of experiment have provided information on the stablity relationships of eclogite. Firstly, experiments on basaltic compositions with excess water (Yoder and Tilley, 1962) have shown that eclogite would not be stable

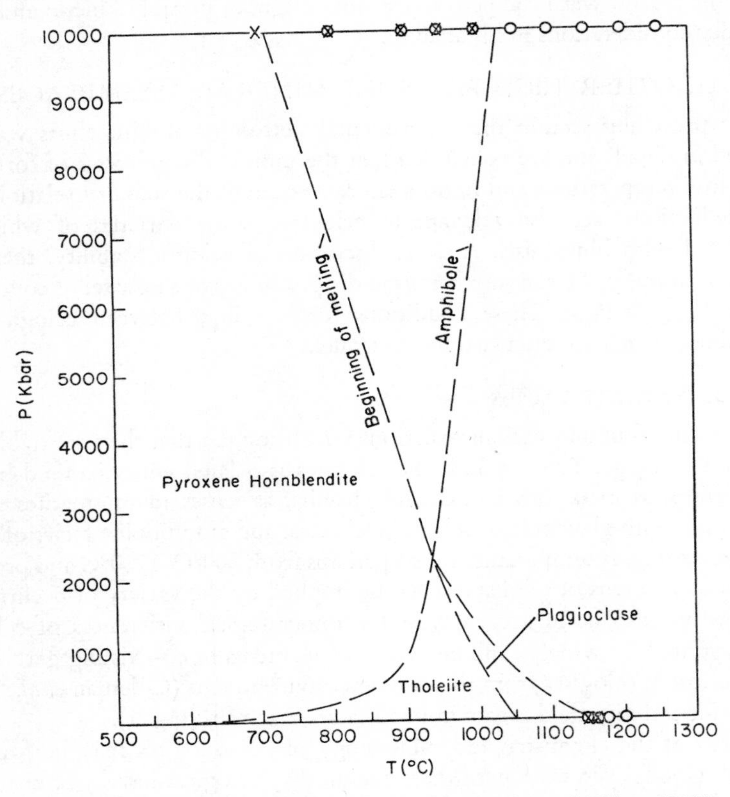

Fig. 23. Stability of hornblende in basaltic rocks. (After Yoder and Tilley, 1962.)

within the crust, i.e. at pressures less than 10 Kbar, the stable products of the hydrous metamorphism of basalts being amphibolites (and greenschists).

Results from the second type of experiment, those conducted in "dry" basalt systems, are summarised in Fig. 24, which is largely derived from the work of Green and Ringwood (1966). The low temperature portion of this diagram is extrapolated from much higher temperatures and could be seriously in error, though one would anticipate boundary curves similar in slope to those found in the jadeite and pyrope systems. Yoder and Tilley (1962), for example, show a very different type of slope. The boundary proposed by Ito and Kennedy (1971) has a slope similar to that of Green and Ringwood (1966), but is displaced towards higher pressures, in fact to near the albite → jadeite + quartz reaction.

There is clearly considerable uncertainty in the position of this reaction at low-temperature, and geological data are not sufficiently precise to permit a soundly-based choice. However, we accept the suggestion of Green and Ringwood (1966) and Green (1967) that eclogites can form from dry basaltic

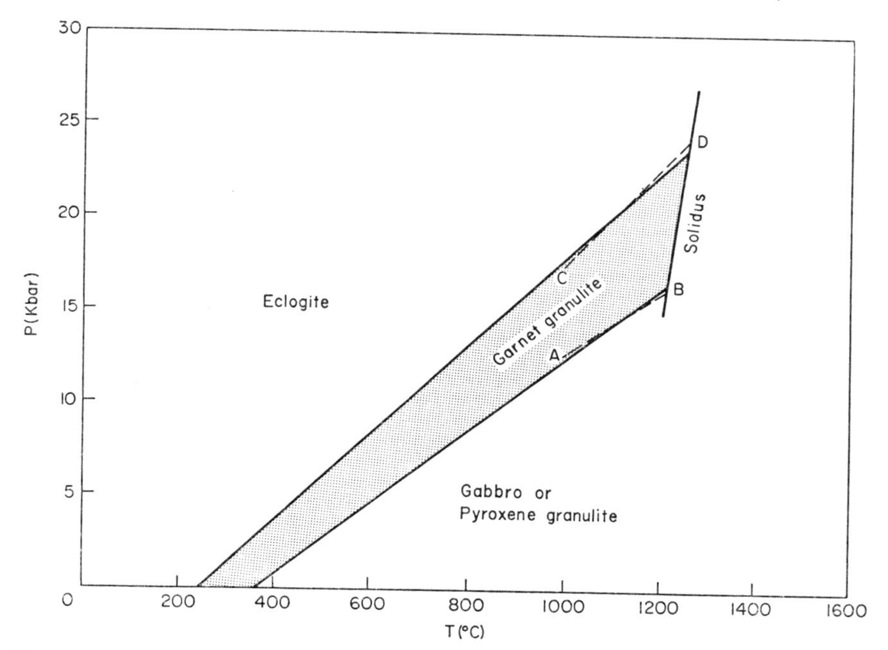

Fig. 24. Relationships between eclogite and gabbro or basalt. (After Green and Ringwood, 1966.)

rocks within the crust, via an intermediate assemblage of gt + px + pl (garnet granulite). The exact P-T conditions at which the various reactions occur depend in large measure on the initial composition of the basaltic parents, especially their degree of silica saturation.

Experiments seem to indicate that both eclogite and amphibolite can form within the crust, from dry and wet basic rocks respectively. Yet in Turner's (1968) facies diagram, eclogites are shown as being stable at crustal depths under high water vapour pressures. This had earlier been suggested by Essene and Fyfe (1967), who were impressed by the association of eclogites–glaucophane schists and serpentinites. The latter members indicate ample water. Also, in such rocks it is not uncommon to find garnet veins and omphacite veins so that it seems that parts of the eclogite mineralogy at least are stable in the presence of copious water. Further, the chemical characteristics of some eclogites are suggestive of metasomatism, again implying activity of a fluid phase. Clearly some further investigation is required as to the relationship between amphibolite, eclogite and water vapour pressures.

The slope of the amphibolite–eclogite boundary will be determined by the entropy and volume change. It is clear from density relations, that at moderate pressures:

$$V_{\text{Eclogite}} + V_{H_2O} < V_{\text{Amphibolite}}.$$

These figures mean that at some high water pressure amphibolites will be transformed to eclogites. Again, because of the usual correlation between

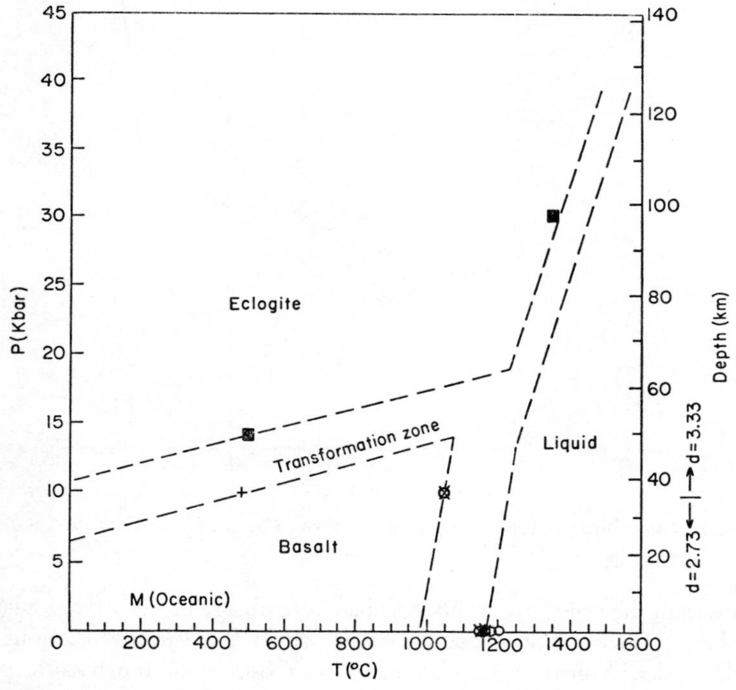

Fig. 25. Phase diagram for basalt. (After Yoder and Tilley, 1962.)

entropy and volume, and since ΔV is large and the amount of water is small, Essene and Fyfe suggested that ΔS and ΔV could have the same sign. This could lead to a phase diagram such as that in Fig. 26 in accord with Turner's facies diagram.

We know little of the entropy relations of amphiboles, but data for tremolite suggest that pyroxenes have relatively higher entropies than amphiboles. We do not yet understand why. If this is also true for omphacite and hornblende entropies, then Essene and Fyfe's guess could be wrong; ΔS could be positive and ΔV negative, and the boundary slope could be negative.

There are few experimental data in a lower temperature region to resolve the dilemma, there being, for example, no direct study of the wet reaction, amphibolite → eclogite + water. But, as is frequently the case, thermodynamic methods can help. If we examine Yoder and Tilley's data (1962) on the stability of hornblendes in the wet basalt system and Green and Ringwood's data (1966), it appears that at some temperature an estimate of the pressure of equilibrium of the above reactions can be made.

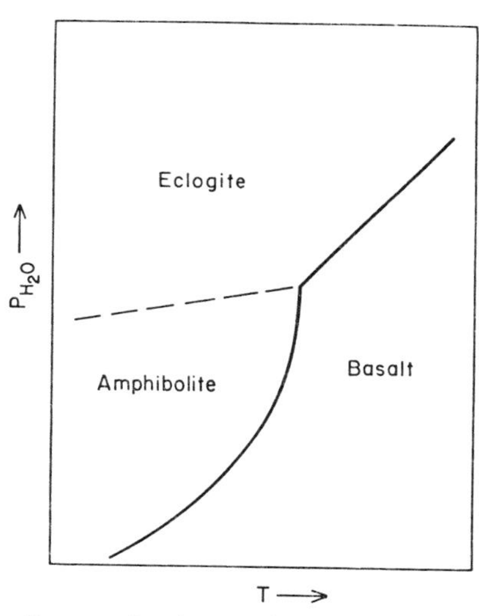

Fig. 26. A possible phase diagram for the wet basalt system. Thermodynamic arguments show this to be most unlikely.

Thus at 900°C the reaction

$$\text{amphibolite} \rightarrow \text{basalt} + H_2O \tag{1}$$

is in equilibrium at P_{H_2O} 1 Kbar while the reaction

$$\text{basalt} \rightarrow \text{eclogite} \tag{2}$$

is in equilibrium at about 15 Kbar assuming nearly univariant reactions. At 1 Kbar the free energy of reaction (2) can be found from the relation

$$\left(\frac{\partial \Delta G_2}{\partial P}\right)_T = \Delta V_2$$

where ΔV_2 is the volume change of the basalt–eclogite transition. At 900°C and 1 Kbar the $\Delta_1 G$ of (1) is zero. Thus, by addition the ΔG of reaction

$$\text{amphibolite} \rightarrow \text{eclogite} + H_2O \tag{3}$$

is simply that of reaction (2) at 1 Kbar and this is given by

$$\Delta G_3 = \int_{1\,\text{Kbar}}^{15\,\text{Kbar}} \Delta V_2 \, dP.$$

Given data on the volume of water, amphibolite and eclogite, we can now find the pressure of equilibrium between a plagioclase-bearing amphibolite and eclogite at 900°C. This is estimated to be about 35 Kbar (water pressure). Fry and Fyfe (1969) have carried out the calculations and, approximate though they may be, there is little doubt that the phase diagram in the wet system must be as shown in Fig. 27. The steep line representing the calculated amphibolite–eclogite boundary produces a maximum eclogite field for at equilibrium the plagioclase amphibolite will transform before the eclogite transition to a pyroxene–garnet amphibolite of even greater stability. Lambert and Wyllie (1968) and Essene *et al.* (1970) have presented some data indicating that the form of Fig. 27 may be reasonable though at low temperatures the pressure may be lower than indicated. In the discussion above we have ignored partial melting which will occur over much of the high temperature portion of Fig. 27. This will not influence arguments about the form of the low temperature part of the diagram.

It thus seems rather certain that eclogites cannot form in the crust if $P_{H_2O} \simeq P_{total}$. We must have dry rocks of appropriate composition and conditions of $P_{total} \gg P_{H_2O}$. Deeply buried impermeable basaltic or spilitic rocks could become eclogites. The conditions under which the reaction occurs will be controlled by reaction kinetics of the essentially solid state reaction. Judging from the metamorphic facies which do and do not contain basalts converted to eclogite, we might guess that minimal temperatures for significant rates of reaction are in the range of 300°C–500°C (Fry and Fyfe, 1971).

It is unlikely that any rock could be perfectly dry. In this case early reactions will generate hydrated minerals, such as epidote, amphibole and mica, and these minerals will then accompany the eclogite minerals when they form and may be in complete equilibrium with them.

We thus encounter the first group of metamorphic rocks where fluid pressure and load pressure differ. The restriction of this facies to basaltic compositions now becomes a result of their primary dryness. Let us not forget that a dry granite under the same conditions might show few obvious phase changes from its primary mineralogy of feldspar–quartz–hornblende–mica. Plagioclase may persist in granites well above the basalt–eclogite transition. It is clear that we must recognise wet and dry metamorphism of the same primary rock type under the same conditions of total pressure and temperature.

What of possible metasomatism and why are some eclogites almost totally devoid of any hydrated minerals? We could propose a model for this type of situation and pillow lavas may provide an excellent example (Bearth, 1967). Imagine that basaltic lava is poured out on the ocean floor forming pillows. The margins of pillows may be quenched to glass while the cores, cooling a little more slowly, crystallise with a normal igneous mineralogy. Rims and spaces between pillows, will tend to be mixed with wet marine muds. Let us now imagine that this material is depressed and buried. There will be opportunity for ion exchange with NaCl rich fluids. The most reactive sites will be in the marine muds and the more glassy pillow rims. Here phases will nucleate and tend to concentrate species required for their growth while porosity and permeability lasts. Thus the clay minerals will become sites to fix potassium in muscovite and even to attract silica in some cases. The same will be true for pillow rims. The less reactive phases will tend to be preserved and leached as they will not provide the best sites for facile nucleation. Thus the reactivity of the system may lead to localised metasomatism at a very early stage in the diagenesis and early metamorphic history. To form the final eclogite, portions of the rock must become sealed and impermeable before the basaltic mineralogy is totally destroyed.

An obvious question to be answered is whether or not the basalt–eclogite transition is a solid-state transition or occurs via a fluid with a low partial pressure of water. This question has been discussed by Fry and Fyfe (1971) who suggested that brines derived from trapped sea water might be possible solvents. Recently an electron microscope study by Champness, Fyfe and Lorimer (1974) has revealed voids, possible fluid inclusions, in the omphacites of a low temperature eclogite from California. It seems probable that a gas phase was present but the exact nature of the fluid composition is unknown.

As regards the status of the eclogite facies the following remarks from Fry and Fyfe (1971) perhaps summarise the position.

"There is little doubt that eclogites can form from dry rocks within the P–T fields where wet rocks of otherwise identical composition would be glaucophane–lawsonite, greenschist or amphibolite rocks. We already recognise the necessity to distinguish silica deficient or excess assemblages. Clearly we must do the same with water, possibly recognising degrees of deficiency.

As for the eclogite facies, it may still be too soon to restrict this facies to the upper mantle. If facies are to be treated as mutually exclusive P–T fields, each defined by stability of a different mineral assemblage in rocks with excess water, then clearly the eclogite facies must be consigned to mantle conditions. In this sense it does not exist as a metamorphic facies.

However, within certain types of metamorphic terrain eclogites have widespread distribution, though each outcrop may be small. Eclogites are easily recognised and (*pace* Green and Ringwood, 1967) represent, these authors believe, elevated pressures. They may be sufficiently indicative of certain plate tectonic phenoma, and resulting types of metamorphic terrains, to justify retaining the eclogite facies as a high pressure facies defined by dry rocks."

(b) GRANULITES AND CHARNOCKITES

Rocks of the granulite facies present some analogous problems to the eclogite facies discussed above, but there are many differences. Granulites are not limited to basic compositions, as are the eclogites. In common with eclogites, they contain dominantly anhydrous minerals, but plagioclase is still a stable phase. Unlike eclogites, they appear to be largely restricted to the Pre-Cambrian and often the very old Pre-Cambrian of the stable shield areas. They are commonly associated with migmatites in regions of partial fusion. The above, when

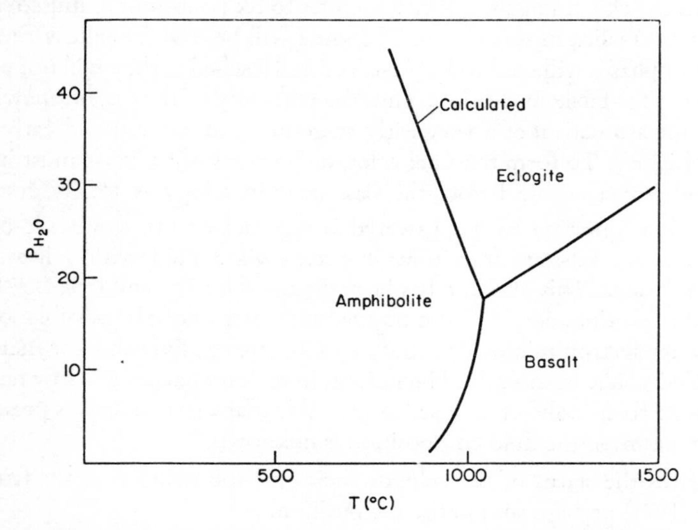

Fig. 27. Calculated phase diagram for the wet basalt system, melting ignored.

considered in relation to our discussion concerning eclogites, gives the impression of a high temperature and medium pressure metamorphism, perhaps at times being coincident with eclogite conditions but most frequently at lower pressures tending to the conditions of the pyroxene hornfels facies of contact metamorphism. A brief discussion of the complex subject of granulite metamorphism, to establish the comparison with the eclogites, is all that can be attempted here, but an excellent summary of data concerning the granulite facies will be found in Turner (1968).

The granulite facies is characterised by the following types of mineral assemblages.

In pelitic rocks:

Quartz-orthoclase-plagioclase-almandine-biotite-sillimanite (kyanite)-cordierite

In calcareous rocks:

calcite-dolomite-forsterite-diopside-phlogopite-scapolite.

In charnockitic rocks (granitic to gabbroic in composition)

quartz-orthoclase-plagioclase-biotite-hornblende-almandine-hypersthene-diopside.

The facies is distinguished from amphibolites by the incoming of pyroxenes and lack of muscovite, and from the hornfels facies by the lack of phases such as wollastonite, andalusite, grossularite.

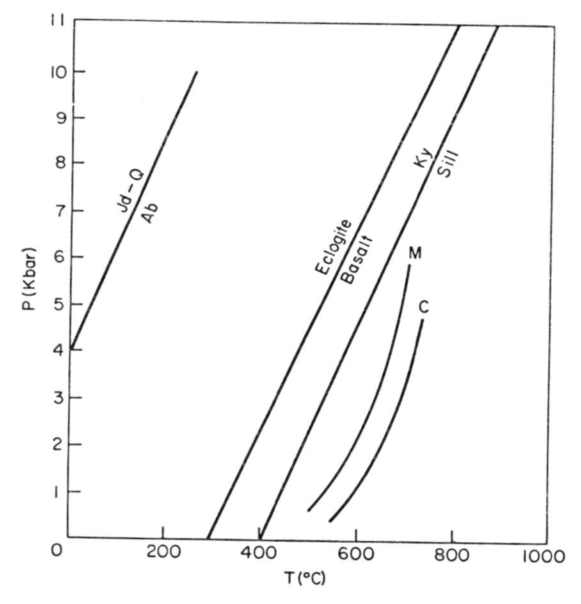

Fig. 28. Some phase boundaries relevant to the conditions of formation of granulites. Curves represent albite—jadeite + quartz reaction, eclogite—basalt, kyanite—sillimanite, upper limit of muscovite (M) and upper limit of calcite (C).

Figure 28 shows the basalt–eclogite transition, the kyanite–sillimanite transition, the albite–jadeite–quartz reaction, the calcite–quartz–wollastonite curve ($P_{CO_2} = P_{total}$), and the upper limit of muscovite in quartz rich systems ($P_{H_2O} = P_{total}$). With exception of the basalt–eclogite transition, these curves are reasonably established (see Turner, 1968).

First, it is clear that granulite metamorphism involves dry rocks; if it did not, basaltic members would be represented by hornblende rich assemblages rather than by plagioclase–hypersthene–garnet–diopside assemblages. Second, it is clear that many are formed below eclogite pressures, particularly if the rocks are basaltic or if they contain sillimanite as opposed to kyanite. Third, it is clear that some granitic granulites could be formed under eclogite pressures if they contain kyanite and have a low Fe-Mg content. Plagioclase breakdown will then occur above the eclogite transition and near the albite–jadeite + quartz boundary (see Fig. 28).

As most granulites contain an assemblage of diopsidic pyroxene, and one or two feldspars each with the albite component, it seems that detailed studies of the conditions of formation of granulitic pyroxenes may provide the data to place pressures of formation on these rocks (see discussion on p. 103) and hence to unravel the story of thermal gradients in the early crust. Such data are not yet available. Another promising approach involves cordierite–garnet compositional equilibria (Hensen and Green, 1971; Currie, 1971). Application of such data to some Archaean granulite assemblages indicates crystallisation at temperatures of about 800°C and pressures in the range 4–6 Kbar (Leonardos and Fyfe, 1973).

It is a striking fact that granitic rocks with an assemblage quartz–jadeite–orthoclase–biotite–hornblende–grossularite–kyanite have not yet been found for they would represent the ultimate high pressure metamorphism of relatively dry granodioritic rocks. It seems that the necessary crustal thicknesses and thermal gradients are never encountered before the onset of large scale fusion.

The primary problem with granulite facies rocks again involves the question of water-pressure and total pressure. If water was in excess, with crustal thicknesses up to over 30 km, from the data of Yoder and Tilley (1962, see Fig. 23) we would expect basic assemblages dominated by hornblende. This does not seem to be the case. Plagioclase–diopside–hypersthene–quartz assemblages (see Fyfe, Turner and Verhoogen, 1958) would not be anticipated in wet crustal rocks. Thus at least with these rocks the situation looks similar to that of eclogites, except that the total pressure may be lower.

It is possible that the chemistry and mineralogy of granulites may be explained by assuming that these rocks are the residue of partial melting (Brown and Fyfe, 1973; Fyfe, 1973; Heier, 1973). The partial melting reactions would be of the type:

biotite schist → granodioritic melt + pyroxene amphibolite
hornblende schist → granodioritic melt + pyroxene granulite

Under such conditions, while fluid pressure may be similar to load pressure, the fluid is a relatively dry silicate melt. Heier (1973) has also stressed that partial pressures of carbon dioxide may be high in this environment (see also Greenwood, this volume).

Some rocks of the granulite facies may thus represent the typical metamorphic residue of a fusion process at the base of the more acid crust. The observation that they are commonly members of very ancient mobile belts would then be explained if ancient thermal gradients were steeper and the acid crust thinner than at the present time. The distribution of granulites, eclogites and glaucophane schists perhaps reflects changes in heat production and tectonic processes as the Earth has evolved.

12. HIGH PRESSURE PHASES IN THE UPPER MANTLE

At the present time we tend to restrict our ideas on metamorphism to the crust. But the ideas developed from such studies must carry over to problems in other parts of the solid Earth. They will have particular application if regions exist where crustal materials are buried by convective forces and where mantle convection occurs, a process which must be accompanied by phase changes. Over the past few years considerable advances in experimental techniques have allowed study of reactions in excess of 200 Kbar (600 km depth) and in a general way, we now have a good general picture of the types of phases which could be present at great depths within the earth. These data must eventually have important bearing on our understanding of the dynamics of the crust–mantle system.

One of the striking features of these very high pressure studies is that at 100 Kbar, practically no common crustal minerals are stable; all will transform into denser materials. Ringwood (1972) has summarised such data. Some of the most significant reactions are as follows. Where structure and not a specific mineral is implied these are identified with an asterisk.

quartz → coesite → stishovite (6 co-ordinated)
olivine → spinel*, density increase 10%
(in silicate spinels*, Si is in 4 co-ordination)
spinel* → ilmenite + oxide (6 co-ordinated)
spinel* → AO (NaCl) + B_2O_3 (corundum)

pyroxene → ilmenite*	18%
pyroxene → spinel*–rutile*	18%
pyroxene → garnet	10%
eclogite → garnetite	
garnet → perovskite*	15%
garnet → ilmenite*	10%
orthoclase → hollandite*	50%

kyanite → oxides (including stishovite).

Up to the present time studies have involved anhydrous materials. What of hydrates? In Fig. 6, the stability field of the low density zeolite analcime was shown. If we consider the pressure range of the upper mantle, it is apparent that dehydration curves of most mineral hydrates (amphiboles, micas, serpentines) will be of the same form. The necessary condition for dehydration, even at high water pressures,

$$V_{\text{hydrate}} > V_{\text{anhydrous}} + V_{H_2O}$$

applies to reactions such as

amphibole → pyroxene + water
mica → dense K feldspar + corundum + H_2O.

Of all known minerals, phlogopite should show maximum mantle stability because the required ΔV relation only occurs near the stishovite transition region where orthoclase transforms to a much denser phase. The stability of hydrates in the mantle has been summarised by Wyllie (1973). This implies that if water is present in the deep mantle, it will be present in phases not at present familiar to us. Hydroxy silicates such as the hydro-garnets may be possible mantle candidates (Fyfe, 1970). With them, small quantities of water could be stabilised due to the very large entropy of mixing of the $(OH)_4^{4-}$ group with the $(SiO_4)^{4-}$ group.

These are problems of the future, which only high pressure experimentation will solve.

REFERENCES

Adams, L. H. (1953). *Amer. J. Sci.*, **251,** 299–308.
Albee, A. L. (1962). *Amer. Mineral.*, **47,** 851–870.
Backstrom, H. L. J. (1921). *Z. Physik Chem.*, **97,** 179–228.
Bailey, E. H., Irwin W. P. and Jones, D. L. (1964). *Calif. Div. Mines and Geol. Bull.*, **183,** 177.
Banno, S. (1964). *J. Fac. Sci. Univ. Tokyo. Sect. II*, **15,** 203–319.
Banno, S. (1970). *Phys. Earth Planet. Interiors*, **3,** 405–421.
Barany, R. (1962). *U.S. Bur. Mines Rept. Inv.*, **5900,** 17.
Barnes, I. and O'Neil, J. R. (1969). *Geol. Soc. Amer. Bull.*, **80,** 1,947–1,960.
Bateman, P. C. and Eaton, J. P. (1967). *Science*, **158,** 1,407–1,417.
Bearth, P. (1966). *Schweiz. Mineral. Petrogr. Mitt.*, **46,** 13–23.
Bearth, P. (1967). "Die Ophiolithe der Zone von Zermatt-Saas Fee", Kummerly and Frey Ag, Verlag, Bern.
Bell, P. M. and Davis, B. T. C. (1965). *Carn. Inst. Wash. Yr. Bk.*, **64,** 120–123.
Bell, P. M. and Kalb, J. (1969). *Carn. Inst. Wash. Yr. Bk.*, **67,** 97–98.
Birch, F. (1955). *Geol. Soc. Am. Spec. Pap.*, **36,** 101–118.
Birch, F. and LeComte, P. (1960). *Am. J. Sci.* **258,** 209–217.
Bischoff, J. L. and Fyfe, W. S. (1968). *Am. J. Sci.* **266,** 65–79.
Blake, M. C., Irwin, W. P. and Coleman, R. G. (1967). *U.S. Geol. Surv. Prof. Paper*, **575-C,** 1–9.
Blake, M. W., Irwin, W. P. and Coleman, R. G. (1969). *Tectonophysics*, **8,** 237–246.
Blake, M. C. and Cotton, W. R. (1969). *Geol. Soc. Amer. Abstr.*, **2,** Programs, I, pt. 2, 6–7.
Bloxham, T. W. (1956). *Amer. Miner.*, **41,** 488–496.
Bloxham, T. W. (1960). *Amer. J. Sci.*, **258,** 555–573.
Boeke, H. E. (1912). *Neues. Jahrb. Mineral. Geol.*, **1,** 91–212.
Boettcher, A. L. and Wyllie, P. J. (1968a). *J. Geol.*, **76,** 314–330.
Boettcher, A. L. and Wyllie, P. J. (1968b). *Geochim. et Cosmochim. Acta*, **32,** 999–1,012.
Bowen, N. L. and Tuttle, O. F. (1949). *Geol. Soc. Amer. Bull.*, **60,** 439–459.
Boyd, F. R. (1954). *Ann. Rept. of the Dir. of the Geoph. Lab., Carn. Inst. of Wash. Yearbook*, **53,** 108–111.
Boyd, F. R. and England, J. L. (1959). *Ann. Rept. of the Dir. of the Geoph. Lab., Carn. Inst. of Wash. Yearbook*, **58,** 82–89.
Brace, W. F., Ernst, W. G. and Kallberg, R. W. (1970). *Geol. Soc. Amer. Bull.*, **81,** 1,325–1,328.
Bredehoeft, J. D. and Hanshaw, B. B. (1968). *Geol. Soc. Am. Bull.*, **79,** 1,097–1,122.
Bridgman, P. W. (1939). *Am. J. Sci.*, **237,** 7–18.
Brothers, R. N. (1970). *Contr. Mineral. Petrol.*, **25,** 185–202.
Brown, E. H. (1967). The greenschist facies in part of eastern Otago, New Zealand. *Contr. Mineral. Petrol.*, **14,** 259–292.
Brown, G. C. and Fyfe, W. S. (1972). *Proc. 24th I.G.C.* Section 2, 27–34.
Brown, W. L. (1973). Symmetry and possible phase relations of clinopyroxenes in eclogites (manuscript).
Brown, W. H., Fyfe, W. S. and Turner, F. J. (1962). *J. Petrol.*, **3,** 566–582.
Buchan, J. L. (1927). *Trans. Faraday Soc.*, **23,** 668–671.
Burns, R. G. (1970). "Mineralogical Applications of Crystal Field Theory." Cambridge University Press.
Campbell, A. S. and Fyfe, W. S. (1965). *Am. J. Sci.*, **263,** 807–816.
Carlson, E. T. (1956). *Nat. Bur. Stds. J. Res.*, **56,** 327–336.

Carman, J. H. (1969). "The Study of the System $NaAlSiO_4$-Mg_2iO_4-SiO_2-H_2O from 200 to 5000 Bars and 800°C to 1,100°C and its Petrologic Applications." Ph.D. thesis, The Pennsylvania State University.

Carmichael, I. S. E. (1967). *Contr. Mineral. Petrol.*, **14,** 36–64.

Champness, P. E., Fyfe, W. S. and Lorimer, G. W. (1974). *Contr. Mineral. and Petrol.*, **43,** 91–98.

Church, W. R. (1968). Eclogites *in* "Basalts: II". Hess, H. H. and Poldervaart, A., (Eds.), Interscience, New York.

Clark, J. R. and Papike, J. J. (1966). *Science*, **154,** 1,003–1,004.

Clark, S. P. Jr. (1957). *Amer. Mineral.*, **42,** 564–566.

Clayton, R. N., O'Neil, J. R. and Mayeda, T. D. (1972). *J. Geoph. Res.*, **77,** 3,057–3,067.

Coe, R. S. and Paterson, M. S. (1969). *J. Geoph. Res.*, **74,** 4,921–4,948.

Cohen, L. H. and Klement, W. (1973). *J. Geol.*, **81,** 724–727.

Coleman, R. G. (1961). *J. Petrol.*, **2,** 209–247.

Coleman, R. G. (1967a). *Tectonophysics*, **4,** 479–498.

Coleman, R. G. (1967b). *U.S. Geol. Surv. Bull.*, **1247,** 49.

Coleman, R. G. and Clark, J. R. (1968). *Am. J. Sci.*, **266,** 42–59.

Coleman, R. G. and Lee, D. E. (1961). *Proc. Geol. Soc. Amer.*, Cordilleran Sect. Mtng. in San Diego, California Mar. 27–29, 53.

Coleman, R. G. and Lee, D. E. (1962). *Am. J. Sci.*, **260,** 577–595.

Coleman, R. G. and Clark, J. R. (1968). *J. Petrol.*, **4,** 260–301.

Coleman, R. G., Lee, D. E., Beatty, L. B. and Brannock, W. W. (1965). *Geol. Soc. America Bull.*, **76,** 483–508.

Coombs, D. S. (1954). *Roy. Soc. New Zealand Trans.*, **82,** 65–109.

Coombs, D. S., Ellis, A. J., Fyfe, W. S. and Taylor, A. M. (1959). *Geochim. et Cosmochim. Acta*, **17,** 53–107.

Crawford, W. A. and Fyfe, W. S. (1964). *Science*, **144,** 1,569–1,570.

Crawford, W. A. and Fyfe, W. S. (1965). *Am. J. Sci.*, **263,** 262–270.

Crowley, M. S. and Roy, R. (1964). *Am. Mineral.*, **49,** 348–362.

Currie, K. L. (1971). *Contr. Mineral. Petrol.*, **33,** 215–226.

Davis, B. L. and Adams, L. H. (1965). *J. Geoph. Res.*, **70,** 433–441.

Deer, W. A., Howie, R. A. and Zussman, J. (1962). "Rock Forming Minerals. Vol. 3, Sheet Silicates", Longmans, London.

Deer, W. A., Howie, R. A. and Zussman, J. (1962). "Rock Forming Minerals. Vol. 5, Non-silicates", Longmans, London.

Denbigh, K. G. (1957). "The Principles of Chemical Equilibrium." Cambridge University Press.

DeRoever, W. P. (1955). *Am. J. Sci.*, **253,** 283–298.

Dickey, P. A., Shriram, C. R. and Paine, W. R. (1968). *Science*, **160,** 609–615.

Dickinson, W. R. (1970). *Rev. Geophys.*, **8,** 813–860.

Dobretsov, N. L. (1962). *Akad. Nauk. SSSR Doklady*, **146,** 676–679.

Epstein, S. and Taylor, H. P. (1967). Variation of O^{18}/O^{16} in Minerals and Rocks, *in* "Researches in Geochemistry **2**" (Abelson ed.), 29–62, Wiley, New York.

Ernst, W. G. (1960). *Geochim. et Cosmochim. Acta.*, **19,** 10–40.

Ernst, W. G. (1961). *Am. J. Sci.*, **259,** 735–765.

Ernst, W. G. (1962). *J. Geol.*, **70,** 689–736.

Ernst, W. G. (1963a). *J. Petrol.*, **4,** 1–30.

Ernst, W. G. (1963b). *Am. Mineral.*, **48,** 241–260.

Ernst, W. G. (1963c). *Am. Mineral.*, **48,** 1,357–1,373.

Ernst, W. G. (1964). *Geochim. et Cosmochim. Acta*, **28,** 1,631–1,668.

Ernst, W. G. (1965). *Geol. Soc. Amer. Bull.*, **76,** 879–914.

Ernst, W. G. (1966). *Am. J. Sci.*, **264,** 37–65.

Ernst, W. G. (1968). "Amphiboles." Springer-Verlag, New York.
Ernst, W. G. (1970). *J. Geophys. Res.*, **75,** 886–902.
Ernst, W. G. (1971). *J. Petrol.*, **12,** 413–437.
Ernst, W. G. (1972). *Geochim et. Cosmochim. Acta*, **36,** 497–504.
Ernst, W. G. (1973). *Geol. Soc. Am. Bull.*, **84,** 2,053–2,078.
Ernst, W. G. and Seki, Y. (1967). *Tectonophysics*, **4,** 463–478.
Eskola, P. (1939). *in* "Die Entstehung der Gesteine", T. F. W. Barth, C. W. Correns and P. Eskola, Springer-Berlin, 367–368.
Essene, E. J. and Fyfe, W. S. (1967). *Contr. Mineral. Petrol.*, **15,** 1–23.
Essene, E. J., Fyfe, W. S. and Turner, F. J. (1965). *Beiträge Z. Mineral. u. Petrog.*, **II,** 695–704.
Essene, E. J., Hensen, B. J. and Green, D. H. (1970). Physics Earth Planet. Interiors, **3,** 378–384.
Evans, B. W. (1965). *Am. J. Sci.*, **263,** 647–667.
Fawcett, J. J. and Yoder, H. S. (1966). *Am. Mineral.*, **51,** 353–380.
Forbes, R. B. (1965). *J. Geophys. Res.*, **70,** 1,515–1,521.
Fry, N. and Fyfe, W. S. (1969). *Contr. Mineral. Petrol.*, **24,** 1–6.
Fry, N. and Fyfe, W. S. (1971). Verh. Geol. B-A.
Fyfe, W. F. (1960). *J. Geol.*, **68,** 553–566.
Fyfe, W. S. (1969). *in* "Mechanism of Igneous Intrusion", (Rast, ed.) Geol. Soc. Liverpool.
Fyfe, W. S. (1970). *Phys. Earth Planet. Interiors*, **3,** 196–200.
Fyfe, W. S. (1973). The granulite facies, partial melting and the Archaean crust. *Phil. Trans. R. Soc. Lond. A.*, **273,** 457–461.
Fyfe, W. S. and Bischoff, J. L. (1965). *Soc. Econ. Paleon. Mineral. Spec. Publ.*, **13,** 3–13.
Fyfe, W. S. and Goodwin, L. H. (1962). *Am. J. Sci.*, **260,** 289–293.
Fyfe, W. S. and Valpy, G. W. (1959). *Am. J. Sci.*, **257,** 316–320.
Fyfe, W. S. and Zardini, R. (1967). *Am. J. Sci.*, **265,** 819–830.
Fyfe, W. S., Turner, F. J. and Verhoogen, J. (1958). *Geol. Soc. Amer. Mem.*, **73,** 259.
Galli, E. and Alberti, A. (1969). *Acta Cryst.*, **B25,** 2,276–2,281.
Ganguly, J. (1973). *Earth Plan. Sci. Lett.*, **19,** 145–153.
Ganguly, J. and Newton, R. C. (1968). *J. Petrol.*, **9,** 444–466.
Garlick, G. D. and Epstein, S. (1967). *Geochim. et Cosmochim. Acta.*, **31,** 181–214.
Ghent, E. D. (1965). *Am. J. Sci.*, **263,** 385–400.
Gilluly, J. (1971). *Geol. Soc. Am. Bull.*, **82,** 2,383–2,396.
Gilluly, J., Reed, J. C. and Cady, W. M. (1970). *Geol. Soc. Am. Bull.*, **81,** 353–376.
Goldschmidt, V. M. (1922). *Naturwiss.*, **10,** 918–920.
Goldsmith, J. R. and Newton, R. C. (1969). *Am. J. Sci.*, Schairer Vol. **267A,** 160–190.
Green, D. H. (1968). *in* "Basalts: I". (Hess, H. H. and Poldervaart, A. (eds)). Interscience, New York.
Green, D. H. and Ringwood, A. E. (1966). *Dept. Geophysics and Geochemistry, Australia National University. Pub.*, **444,** 1–103.
Greenwood, H. J. (1961). *J. of Geoph. Res.*, **66,** 3,923–3,946.
Gresens, R. L. (1969). *Contr. Mineral. Petrol.*, **24,** 93–113.
Griggs, D. T. (1972) *in* "The Nature of the Solid Earth", 361–384. E. C. Robertson (ed.), McGraw-Hill, New York.
Griggs, D. T. and Blacic, J. D. (1965). *Science*, **147,** 292–295.
Griggs, D. T. and Kennedy, G. C. (1956). *Am. J. Sci.*, **253,** 722–735.
Gross, K. S. (1965). *Phil. Mag.*, **12** (8th ser.), 801–813.
Hamilton, W. (1969). *Geol. Soc. Amer. Bull.*, **80,** 2,409–2,430.
Harris, P. G., Kennedy, W. G. and Scarfe, C. M. (1969). Volcanism Versus Plutonism: The Effect of Chemical Composition, *in* "Mechanism of Igneous Intrusion". (Rast ed.), Geological Society Liverpool.
Hashimoto, M. (1968). *J. Geol. Soc. Jap.*, **74,** 343–345.

Heier, K. S. (1973). *Phil. Trans. R. Soc. Lond. A.* **273,** 429–442.
Heinrich, E. W. and Levinson, A. A. (1955). *Amer. Mineral.*, **40,** 983–995.
Hensen, B. J. and Green, D. H. (1970). *Physics Earth Planet. Interiors,* **3,** 431–440.
Hinrichsen, T. and Schurmann, K. (1969). *Neues. Jahr. Mineral. Monatsch.*, **10,** 441–445.
Hlabse, T. and Kleppa, O. J. (1968). *Amer. Mineral.*, **53,** 1,281–1,292.
Hobbs, B. E. (1968). *Tectonophysics,* **6,** 353–401.
Holm, J. L. and Kleppa, O. J. (1968). *Amer. mineral.*, **53,** 123–133.
Hsü, K. J. (1968). *Geol. Soc. Amer. Bull.*, **79,** 1,063–1,074.
Hsu, L. C. (1968). *J. Petrol.*, **9,** 40–83.
Hsu, L. C. and Burnham, C. W. (1969). *Geol. Soc. Amer. Bull.*, **80,** 2,393–2,408.
Ito, J. and Frondel, C. (1968). *Amer. Mineral.*, **53,** 1,036–1,038.
Ito, K. and Kennedy, G. C. (1970). *Min. Soc. Am. Spec. Paper,* **3,** 77.
Iwasaki, M. (1963). *J. Fac. Sci., Univ. Tokyo, Sect. II,* **15,** 1–90.
Jamieson, J. C. (1953). *J. Chem. Phys.*, **21,** 1,385–1,390.
Jamieson, J. C. (1957). *J. Geol.*, **65,** 334–344.
Johannes, W. and Puhan, D. (1971). *Contr. Mineral. Petrol.*, **31,** 28–38.
Johannes, W., Bell, P. M., Boettcher, A. L., Chipman, D. W., Hays, J. F., Mao, H. K., Newton, R. C. and Seifert, F. (1971). *Contr. Mineral. Petrol.*, **32,** 24–38.
Kelley, K. K., Todd, S. S., Orr, R. L., King, E. G. and Bonnickson, K. R. (1953). *U.S. Bur. of Mines Rept. Inv.*, **4955.**
Kerrick, D. M. (1968). *Am. J. Sci.*, **266,** 204–214.
Kerrick, D. M. and Cotton, W. R. (1971). *Am. J. Sci.*, **271,** 350–369.
Kim, K. T. and Burley, B. J. (1971). *Can. J. Earth Sci.*, **8,** 311–337.
King, E. G. and Weller, W. W. (1961). *U.S. Bur. Mines Rept. Inv.*, **5855,** 1–8.
Kitahara, S., Takenouchi, S. and Kennedy, G. C. (1966). *Am. J. Sci.*, **264,** 223–233.
Koening, J. B. (1963). Santa Rose sheet, geologic map of California. O. P. Jenkins edition, Califoronia. Div. Mines and Geol.
Kracek, F. C., Neuvonen, K. J. and Burley, G. (1953). *J. Wash. Acad. Sci.*, **41,** 373–383.
Kushiro, I. (1969). Stability of omphacite in the presence of excess silica. *Ann. Rept. Div. Geoph. Lab., Carn. Inst. Wash. Yr. Bk.*, **67,** 98–100.
Lambert, I. B. and Wyllie, P. J. (1968). *Nature,* **219,** 1,240–1,241.
Lee, D. E., Coleman, R. G. and Erd, R. C. (1963). *J. Petrol.*, **4,** 460–492.
Leonardos, O. H. and Fyfe, W. S. (1974). *Contr. Mineral. Petrol.*, **46,** 201–214.
Liou, J. G. (1970) P-T stabilities of laumontite, wairakite, lawsonite, and related minerals in the system $CaAl_2Si_2O_8$-SiO_2-H_2O. *Jour. of Petrol.*, **12,** 379–411.
MacDonald, G. J. F. (1956). *Am. Mineral.*, **41,** 744–756.
MacDonald, G. J. F. *J. Sci.*, **255,** 266–281.
MacKenzie, W. S. (1957). *Am. J. Sci.*, **255,** 481–516.
Manghnani, M. (1970). *Phys. Earth Plan. Inst.*, **3,** 456–461.
Maresch, W. V. (1973). *Earth Plan. Sci. Lett.*, **20,** 385–390.
Martin, B. (1968). "Some Theoretical and Experimental Observations on Rates of Hydration in the System MgO-SiO_2-H_2O". Ph.D. thesis. Manchester University.
Martin, B. and Fyfe, W. S. (1970). *Chem. Geol.*, **6,** 185–202.
Martin, R. F. (1969). *Contr. Mineral. Petrol.*, **23,** 323–339.
McConnell, J. D. C. and McKie, D. (1960). *Mineral. Mag.*, **32,** 436–454.
McKenzie, D. P. (1972). Plate tectonics, *in* "The Nature of the Solid Earth", 323–360. E. C. Robertson (ed.), McGraw-Hill, New York.
McKee, B. (1961). *Proc. Geol. Soc. Amer. Cordilleran Sect. Mtng. San Diego, Calif., Mar.* 27–29, 1961, 53.
McKee, B. (1962). *Am. J. Sci.*, **260,** 569–610.
McLellan, A. G. (1969). *Proc. Roy. Soc. London A,* **314,** 443–455.
Miyashiro, A., Shido, F. and Ewing, M. (1971). *Phil. Trans. Roy. Soc. Lond. A.* **268,** 589–603.

Miyashiro, A. and Seki, Y. (1958). *Jap. Geol. Geog.*, **29,** 199–227.
Müller, R. O. (1958). *Schweiz. Miner. Petrog. Mitt.*, **38,** 404–473.
Newton, M. S. and Kennedy, G. C. (1968). *Am. J. Sci.*, **266,** 728–735.
Newton, R. C. (1972). *J. Geol.*, **80,** 398–420.
Newton, R. C. and Kennedy, G. C. (1963). *J. Geoph. Res.*, **68,** 2,967–2,983.
Newton, R. C. and Smith, J. V. (1967). *J. Geol.*, **75,** 268–286.
Newton, R. C., Goldsmith, J. R. and Smith, J. V. (1969). *Contr. Mineral. Petrol.*, **22,** 335–348.
Nitsch, K. H. (1968). *Naturwissenschaften*, **55,** 388.
Nitsch, K. H. (1971). *Fortschr. Mineral.*, **49,** Beich. 1, 34–36.
Nitsch, K. H. (1972). *Contr. Mineral. Petrol.*, **34,** 116–134.
Northwood, D. O. and Lewis, D. (1970). *Can. Mineral.*, **10,** 216–224.
O'Neil, J. R., Clayton, R. N. and Mayeda, T. K. (1969). *J. Chem. Phys.*, **51,** 5,547–5,558.
Osborn, E. F. (1969). *Geochim. et Cosmochim. Acta*, **33,** 307–324.
Pabst, A. (1931). *Amer. Mineral.*, **16,** 327.
Page, N. J. (1967a). *Amer. Mineral.*, **52,** 545–548.
Page, N. J. (1967b). *Contr. Mineral. Petr.*, **14,** 321–342.
Palache, C., Berman, H. and Frondel, C. (1951). "Dana's System of Mineralogy," J. Wiley and Sons, Inc. New York.
Paterson, M. S. (1959). *Phil. Mag.*, **4,** (8th Ser.). 451–466.
Peters, T., Luth, W. C. and Tuttle, O. F. (1966). *Amer. Mineral.*, **51,** 736–753.
Pistorius, C. W. F. T. (1963). *Neues Jahrb. Mineral. Monat*, **II,** 283–293.
Pistorius, C. W. F. T. and Kennedy, G. C. (1960). *Am. J. Sci.*, **258,** 247–257.
Pistorius, C. W. F. T., Kennedy, G. C. and Sourirajan, S. (1962). *Am. J. Sci.*, **260,** 44–56.
Popp, R. K. and Gilbert, M. C. (1972). *Amer. Mineral.*, **57,** 1,210–1,231.
Raleigh, C. B. and Paterson, M. S. (1965). *J. Geoph. Res.*, **70,** 3,965–3,986.
Ribbe, P. H. and Gibbs, G. V. (1969). *Amer. Mineral.*, **54,** 85–94.
Ringwood, A. E. (1969). "Phase Transformations in the Mantle." Dept. Geophysics and Geochemistry, Australian National University. Pub. No. 666.
Ringwood, A. E. (1972). Mineralogy of the Deep Mantle: Current Status and Future Developments, *in* "The Nature of the Solid Earth", 67–92. E. C. Robertson (ed.), McGraw-Hill, New York.
Robertson, E. C. (1972). Strength of Metamorphosed Graywacke and Other Rocks, *in* "Nature of the Solid Earth", 631–659. E. C. Robertson, (ed.), McGraw-Hill, New York.
Robertson, E. C., Birch, F. and MacDonald, G. J. F. (1957). *Am. J. Sci.*, **255,** 115–137.
Roy, D. M. and Roy, R. (1960). Fourth Int. Symp. on the Chem. of Cement, Wash. D.C., 1960, Paper III–59, 307–314.
Rutland, R. W. R. (1965). Tectonic Overpressures *in* "Controls of Metamorphism", W. S. Pitcher and G. W. Flinn, (eds.), John Wiley and Sons, New York, 119–136.
Saha, P. (1959). *Amer. Mineral.*, **44,** 300–313.
Scarfe, C. M. and Wyllie, P. J. (1967). *Nature*, **215,** 945–946.
Schaller, W. T. (1950). *Min. Mag.*, **29,** 406–415.
Schreyer, W. (1968). *Ann. Rept.* of the Dir. of the Geoph. Lab., *Carn. Inst. Wash. Yr. Bk.*, **66,** 380–392.
Segnit, R. E. and Kennedy, G. C. (1961). *Am. J. Sci.*, **259,** 280–287.
Seki, Y. (1960). *Am. J. Sci.*, **258,** 705–715.
Seki, Y. (1964). *J. Geo. Soc. Japan*, **70,** 348–349.
Seki, Y. (1965). *Saitama Univ. Sci. Rept. Ser. B*, **5,** 29–43.
Seki, Y. (1969). *J. Geol. Soc. Japan*, **75,** 255–266.
Simmons, G. and Bell, P. M. (1963). *Science*, **139,** 1,197–1,198.
Smith, J. R. and Yoder, H. S. (1956). *Amer. Mineral.*, **41,** 632–647.
Spicer, H. C. (1942). *Geol. Soc. of Amer. Spec. Pap.*, **36,** 279.

Taliaferro, N. L. (1943). *Amer. Assoc. Petrol. Geol. Bull.*, **27,** 109–219.

Taylor, H. P. and Coleman, R. G. (1968). *Geol. Soc. Amer. Bull.*, **79,** 1,727–1,756.

Thayer, T. P. (1966). *Amer. Mineral.*, **51,** 685–710.

Thomas, J. M. and Renshaw, G. D. (1965). *Trans. Faraday Soc.*, **61,** 791–796.

Thompson, A. B. (1970). *Am. J. Sci.*, **269,** 267–275.

Thompson, A. B. (1971). *Am. J. Sci.*, **271,** 79–92.

Thompson, G., Bowen, V. T., Melson, W. G. and R. Cifelli (1968). *J. Sed. Petr.*, **38,** 1,305–1,312.

Torgeson, D. R. and Sahama, Th. G. (1948). *J. Am. Chem. Soc.*, **70,** 2,156–2,160.

Turner, F. J. (1968). "Metamorphic Petrology". McGraw-Hill, New York.

Turner, F. J. and Verhoogen, J. (1951). "Igneous and Metamorphic Petrology". McGraw-Hill, New York.

Turner, F. J., Griggs, D. T., Clark, R. H. and Dixon, R. H. (1956). *Bull. Geol. Soc. Amer.*, **67,** 1,259–1,294.

Turnock, A. C. (1960). *Ann. Rept. of the Dir. of the Geoph. Lab., Carn. Inst. Wash. Yr. Bk.*, **59,** 98–103.

Tuttle, O. F. and Bowen, N. L. (1958). *Geol. Soc. Amer. Mem.*, **74.**

Vance, J. A. (1968). *Am. J. Sci.*, **266,** 299–315.

Velde, B. (1964). *Ann. Rept. of the Dir. of the Geoph. Lab., Carn. Inst. Wash. Yr. Bk.*, **63,** 141–142.

Velde, B. (1965a). *Am. J. Sci.*, **263,** 886–913.

Velde, B. (1965b). *Amer. Mineral.*, **50,** 436–449.

Wang, C. (1968). *J. Geoph. Res.*, **73,** 3,937–3,944.

Watson, K. D. and Morton, D. M. (1968). *Amer. Mineral.*, **54,** 267–285.

Whetten, J. T. (1965). *Amer. Mineral.*, **50,** 752–754.

Wilson, J. T. (1968). *Geotimes*, **13,** 10–16.

Wyllie, P. J. (1971). "The Dynamic Earth", John Wiley and Sons, New York.

Wyllie, P. J. (1973). *Tectonophysics*, **17,** 189–209.

Yoder, H. S. (1950a). *J. Geol.*, **58,** 221–253.

Yoder, H. S. (1950b). *Am. J. Sci.*, **248,** 225–248; 312–334.

Yoder, H. S. and Chesterman, C. W. (1951). *Calif. Div. Mines. Spec. Rept.*, **10-C.**

Yoder, H. S. and Chinner, G. A. (1960). Grossularite-Pyrope-Water system at 10,000 bars. Almandite-Pyrope-Water system at 10,000 bars. *Ann. Rept. of the Dir. of the Geoph. Lab., Carn. Inst. Wash. Yr. Bk.*, **59,** 78–84.

Yokokawa, T. and Kleppa, O. J. (1964). *J. Phys. Chem.*, **68,** 3,246–3,248.

Zen, E. A. (1961). *Am. J. Sci.*, **259,** 401–409.

Zimmermann, H. O. (1971). *Nature*, **231,** 203–204.

2. Metamorphism at Moderate Temperatures and Pressures

H. J. Greenwood

1 INTRODUCTION 187
(a) Range of conditions considered 187
(b) Organisation and presentation of material 188
(c) Rock types and chemical systems 188
2 PRINCIPLES 189
(a) Metamorphic facies and mineral facies 189
(b) Application of experimental results to real rocks 196
3 THE PHASES OF METAMORPHIC ROCKS 198
(a) The fluid phase in metamorphic systems 198
(b) The fluid phase in synthetic systems 198
(c) Buffering and gas equilibria 203
4 OXIDE AND SULPHIDE EQUILIBRIA IN NATURAL AND SYNTHETIC SYSTEMS . . 207
(a) Oxide equilibria 207
(b) Sulphide equilibria 212
5 EQUILIBRIA IN PELITIC ROCKS AND THEIR SYNTHETIC ANALOGUES . . . 215
(a) The system SiO_2-Al_2O_3-MgO-FeO-K_2O-H_2O and its subsystems . . 216
(b) The influence of other components 232
6 EQUILIBRIA IN SYSTEMS CONTAINING CARBONATES 233
(a) Carbonates 234
(b) Carbonates in the presence of H_2O 237
(c) Decarbonation reactions in the absence of H_2O 237
(d) Dehydration reactions in the absence of CO_2 239
(e) Decarbonation and dehydration reactions in H_2O–CO_2 mixtures . . 241
(f) The system CaO-MgO-SiO_2-H_2O-CO_2 244
(g) The system CaO-Al_2O_3-K_2O-SiO_2-H_2O-CO_2 247
(h) Metasomatic exchange between pelitic and carbonate rocks . . . 250
7 SUMMARY 251

1. INTRODUCTION

A large fraction of all the metamorphic rocks we see either reached their maximum temperature in the moderate T-P range or passed through this range *en route* to the conditions that last affected the mineralogy.

(a) RANGE OF CONDITIONS CONSIDERED

The expression "moderate temperatures and pressures" will doubtless seem immoderate to some readers, mainly because the boundaries are arbitrary. The range of temperature is some 450°C, from the upper stability limit of zeolites in the presence of quartz at about 300°C, to the second sillimanite isogradic reaction at about 750°C, where muscovite and quartz react to give potassium feldspar, sillimanite, and H_2O. Precise temperature limits are not possible without specifying more about the compositions of the minerals, the pore fluids, and the relation of total pressure to the equilibrium pressures of the reacting volatiles such as H_2O, CO_2, H_2, O_2, and so forth. The range of pressure considered is also arbitrary, running from approximately 1 Kbar to approximately 8 Kbar, or a depth in the earth of about 24 km. There is unavoidable overlap with Part II, Section A(2), by Schreyer, dealing with low pressures and high temperatures and with Part II, Section A(1), by Newton and Fyfe, dealing with high pressures.

(b) ORGANISATION AND PRESENTATION OF MATERIAL

The petrologist would naturally prefer to have the information on experimental petrology arranged according to petrologic occurrence. Such a format might be arranged in terms of rock types and the user could simply turn to the section dealing with the type of rock that interests him. From the chemical standpoint such a presentation is unwieldy, resulting in duplication, as many of the simpler sub-systems are important in a variety of different rock-types.

Alternatively, efficient presentation might be achieved by treating the whole problem as one of combinations of chemical systems, starting with all the unary systems, followed by all the binary systems, progressing thus through all ten or fifteen common constituents of metamorphic rocks. This approach is as clumsy as the first for if we consider but ten components it would require description of 1023 separate systems!

In view of the foregoing, the chemically logical progression has been broken into groups most resembling real rocks, and within these groups information is arranged in terms of increasing chemical complexity. Examples of petrologic application appear along the way for illustration. It is clearly impossible to discuss here every metamorphic problem or every experimental study, and so considerable selection has been made. It is hoped that the principles displayed in the examples will allow the reader to make his own applications.

(c) ROCK TYPES AND CHEMICAL SYSTEMS

Pelitic rocks are considered here to include all rocks derived by metamorphism from fine grained clastic sediments having a significant proportion of clay minerals. Thus, for our purposes, pelitic rocks include many that would be classified as wackes or sandstones in their original sedimentary condition. The major minerals found in these rocks and their metamorphosed equivalents can be represented in terms of K_2O-MgO-FeO-Al_2O_3-SiO_2-H_2O. This is by no means complete, but it provides a framework for consideration of additional constituents which further complicate the mineralogy. These extra constituents include Fe_2O_3, carbon, MnO, CaO, and Na_2O. With the exception of carbon, these constituents enter into solid solution in the major minerals and in general increase the degrees of freedom associated with the mineral assemblage.

Carbonate-bearing rocks are considered separately because, even though they contain many of the same constituents as the pelites, they are ordinarily separable in the field due to their distinctive mineralogy. The mineralogy of most metamorphosed carbonate-bearing rocks can be approximately described in terms of CaO-MgO-Al_2O_3-SiO_2-H_2O-CO_2 with modifications due to the presence of carbon, FeO, Fe_2O_3, MnO, K_2O, Na_2O, and (Na,K)Cl.

Intermediate to mafic rocks of igneous parentage are among the most difficult to deal with both experimentally and in nature. This is due in large part to the wide range of solid solution exhibited by most of the important minerals in these rocks. Consequently there are no entirely satisfactory graphical projections of the phase relations, and few simple chemical analogues for experimental study. The common minerals can be represented, to a fair approximation, in terms of CaO-MgO-FeO-Al_2O_3-SiO_2-H_2O, with very important changes resulting from the addition of Fe_2O_3, Na_2O, K_2O, and CO_2.

Sulphide and oxide minerals, frequently dismissed as "opaques" in petrographic descriptions, are of vital importance in considering the total phase assemblages in metamorphic rocks. They are especially useful because, besides being a common occurrence, they buffer the activities of sulphur and oxygen gas species during the metamorphic process and their presence permits us to estimate some of these activities from the final assemblage. In addition, many of these minerals do not enter directly into solid solution with the main silicate minerals. This enables one to treat them as chemically isolated sub-systems, eliminating from consideration some of the chemical complexity of the natural rocks. The main oxide and sulphide minerals fall essentially into the system S-O-Fe, with minor amounts of phases that include essential Cu, Pb, Zn, Mn, and Al.

2. PRINCIPLES

(a) METAMORPHIC FACIES AND MINERAL FACIES

The terms "metamorphic facies" and "mineral facies" appear in much of the petrologic and petrographic literature that post-dates Eskola's 1920 contribution (Eskola, 1920) in which he proposed the term "mineral facies" after having previously (Eskola, 1915) proposed the term "metamorphic facies". The original definitions and all the popular paraphrases of the definitions include the following essential ingredients. Equilibrium was attained throughout the specimens studied. A particular bulk composition is characterised in a particular facies, either mineral or metamorphic, by a particular set of minerals. It is implied that in different facies a particular bulk composition should be characterised by different sets of minerals. These concepts are consistent with the Gibbs phase rule, which must apply to rock systems that have come to equilibrium. One of the consequences of the phase rule is that all of the intensive-variable-space is divided up into regions having different degrees of freedom. In the familiar P–T space, the regions are divariant (two degrees of freedom), univariant (one degree of freedom, corresponding to the coexistence of both the products and reactants of a mineral equilibrium), and invariant (zero degrees of freedom, corresponding to an invariant point, where there co-exist $C + 2$ phases in equilibrium, at the intersection of three or more univariant equilibria).

The characteristics of mineral facies have been concisely stated by Thompson (1957). "It is not any one assemblage but the set or 'ensemble' of assemblages that defines the (mineral) facies". (Thompson, op cit., p. 855.) Thus every pigeonhole in the petrogenetic grid (Bowen, 1940) constitutes a mineral facies, and every move from one mineral facies to another is attended by a chemical–mineralogical reaction that changes one set of stably co-existing minerals into another. There are thus, if we consider all possible rocks, an immense number of mineral facies, each separated and distinguished from its nearest neighbours by reactions and accompanying discontinuous changes (Thompson, 1957) in the relevant phase diagram. In order to decide to which mineral facies a group of rocks belongs it is naturally necessary to find a rock of a composition suitable to demonstrate which of the critical assemblages is stable.

The term "metamorphic facies" has come to be applied over the years in a rather loose way, while carrying with it the essential notions of chemical equilibrium and a specific reproducible dependence of mineralogy of a rock on its bulk composition within a definite facies. However, as there are only a few commonly used metamorphic facies names, each must contain a very large number of mineral facies, and the boundaries between adjacent metamorphic facies must also be boundaries between adjacent mineral facies. The principal areas where unanimity is lacking centre on the choice of the most appropriate univariant boundaries to choose as facies boundaries, and whether or not to further subdivide the metamorphic facies into subfacies on the basis of other univariant equilibria. It is not proposed here to provide a rigorous determination of the most appropriate limiting equilibria to use, but only to indicate where in the conventionally-used metamorphic facies certain mineral facies fall, and to emphasise that the concepts of metamorphic and mineral facies constitute natural subdivisions of the mineralogic variations of metamorphic rocks. Because of this naturalness and perfect consistency with thermodynamic theory and petrographic experience the general notions of metamorphic facies and mineral facies are retained.

Figure 1 shows the metamorphic facies and is a simplification of Fig. 2. Figure 1 has been constructed on the following basis. First, there seems to be general agreement about the relative positions of the major metamorphic facies and there seems little profit in modification here. Second, Fig. 2 shows that there are regions in P–T space that are much more densely packed with univariant equilibria than others. These regions in Fig. 2 containing many reactions have been emphasised in Fig. 1 as shaded zones and taken as facies boundaries. These coincide, probably not fortuitously, with the positions usually adopted for the facies boundaries.

Perhaps this approach requires a word in its defence, along with a word in support of retaining the essence of the facies classification. Earlier paragraphs have emphasised the immense number of mineral facies and the view that the boundaries between the metamorphic facies should also be boundaries between mineral facies, and should be sharp, univariant and commonly detectable in rocks. While the criterion of univariance is appropriate to the boundaries of mineral facies it is not appropriate for boundaries of metamorphic facies. First, if the metamorphic facies concept is to be useful its boundaries must be detectable in a wide variety of rock types and bulk compositions. Otherwise one could rarely decide to what facies a rock belongs. This requires the existence of a number of degenerate sub-systems exhibiting reactions in many complicated larger systems and which are unaffected by the addition of extra components. The polymorphs of Al_2SiO_5 and SiO_2 are among the few that qualify, and many rocks contain no Al_2SiO_5 phases while the SiO_2 polymorphs are not preserved in metamorphic rocks. Therefore it seems sensible to compromise and use a region of many reactions as a "boundary-zone" recognisable in many bulk compositions and admittedly somewhat diffuse. Secondly, if we were to arbitrarily select one truly univariant boundary as a facies boundary it is certain that the very arbitrariness would be repellent to many, and that any worker could equally well choose another, confusing the subject further. Thirdly, there exist more variables in metamorphism than can be shown in a simple P–T diagram, and each of these tends to displace the curves shown on such a diagram. Of these one might mention in addition to P_{total} and T; P_{H_2O}, P_{CO_2}, P_{H_2}, P_{O_2}, P_{CH_4}, P_{S_2} open systems, metasomatism, and solid solution of additional elements not explicitly included in the reactions on the P–T plot. The combined effect of these factors is to change simple univariant boundaries into divariant or polyvariant regions thus bringing us back to the need to use "boundary-zones" to separate metamorphic facies. Finally, in support of retaining metamorphic facies it should be remembered that frequently a worker has no opportunity or immediate facilities for detailed study of the mineral assemblages but is still able to convey a good impression of metamorphic grade by stating the rocks to be, for example, "greenschists", or "garnet–staurolite–cordierite amphibolites", and so forth. A broad classification like the metamorphic facies concept is not suited to precise delimitation of physical conditions and should not be used in this way. It should be used, rather, to convey the general impression of metamorphic grade and gradients, while detailed inference on conditions must fall back on the

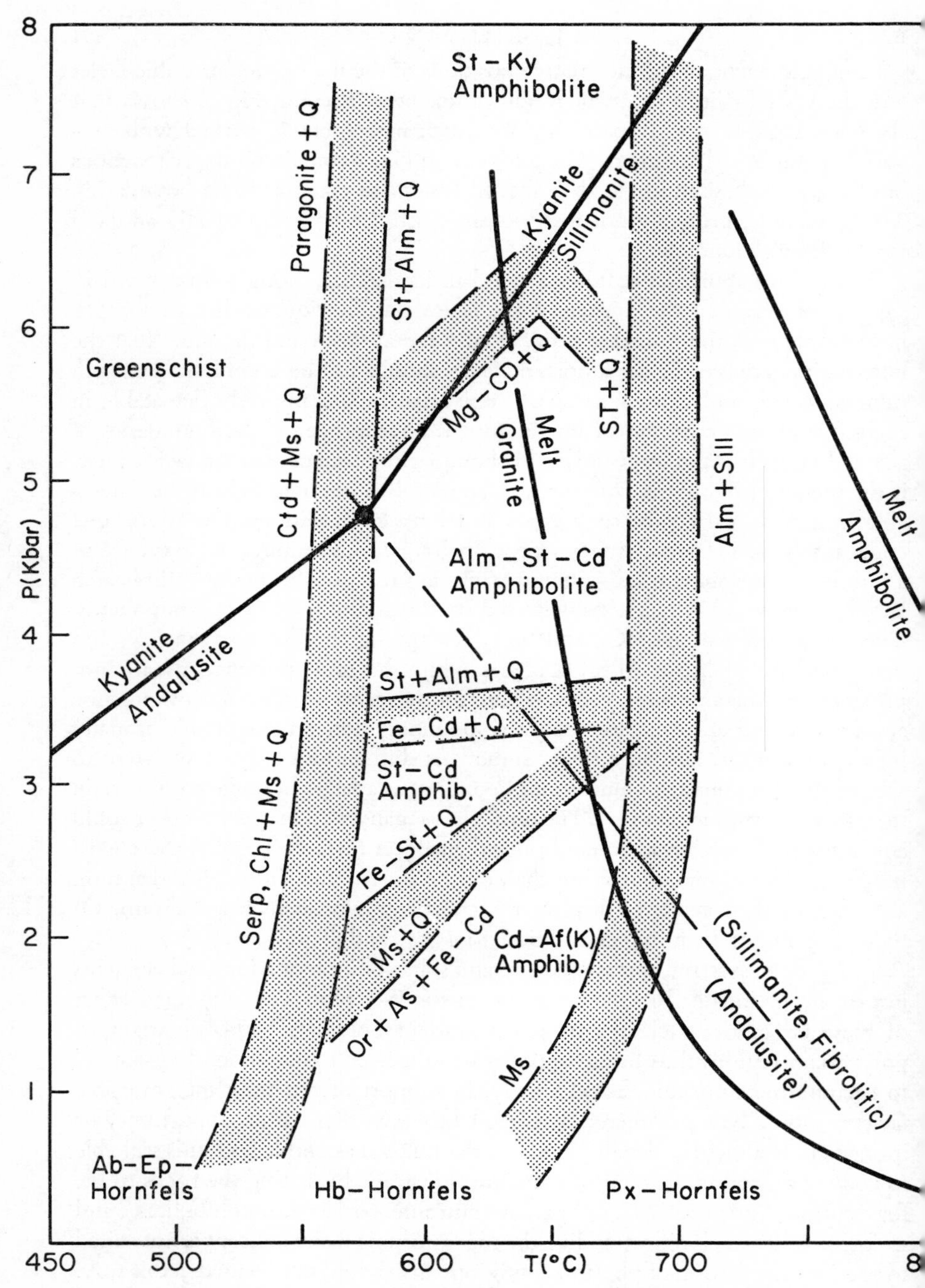

Fig. 1. Metamorphic facies in pressure—temperature projection. Derived from Fig. 2(b) by taking regions containing many curves and re-drawing them as shaded bands, and taking these diffuse bands as facies boundaries. Note that in this diagram the position of the stability curve for andalasite = sillimanite (and hence the Al_2SiO_5 triple point) is based on petrographic evidence. This is discussed on p. 217.

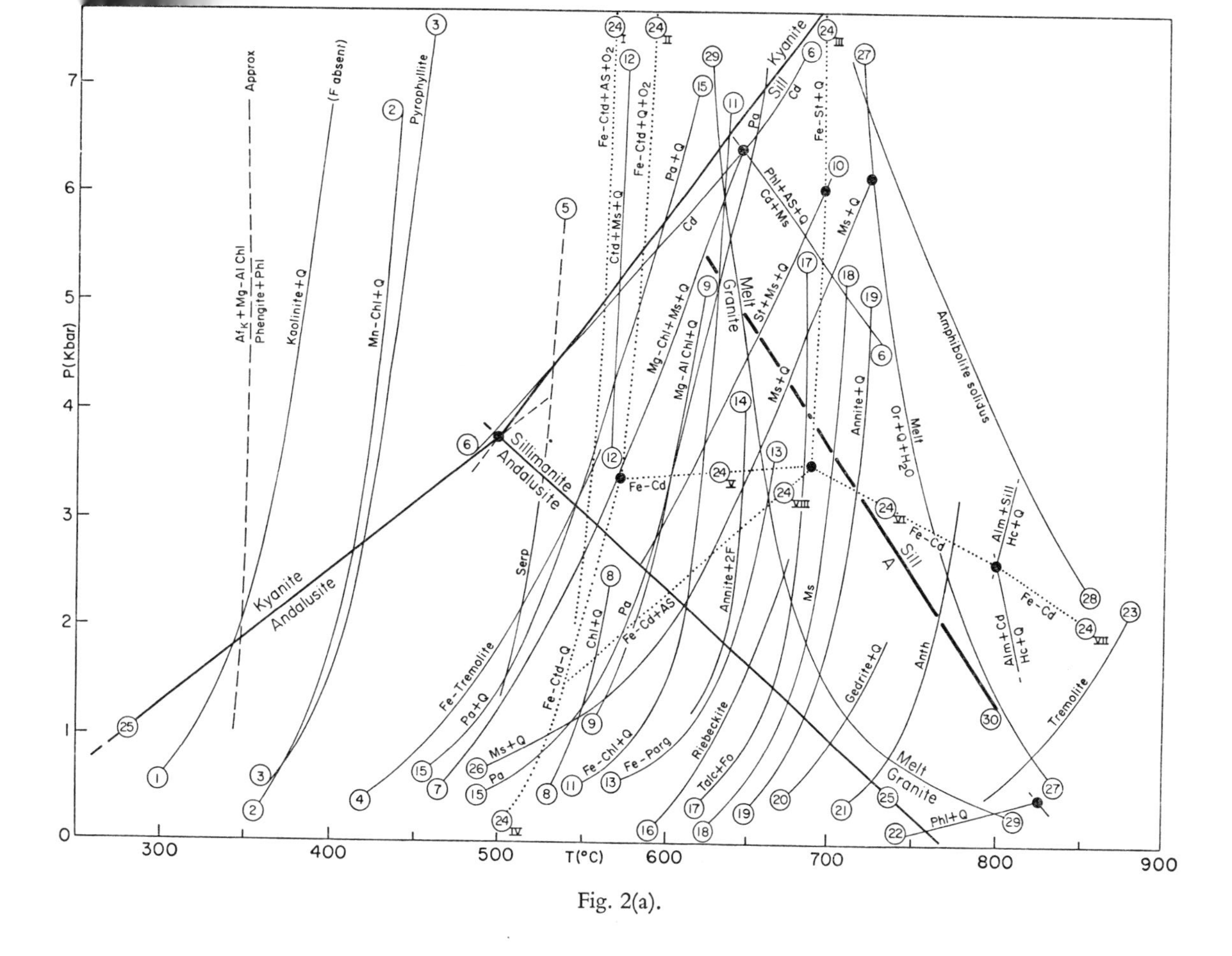

P(Kbar)
T(°C)
0
1
2
3
4
5
6
7
300
400
500
600
700
800
900
Approx
Af_K + Mg-Al Chl
Phengite + Phl
Kaolinite + Q
(F absent)
Mn-Chl + Q
Pyrophyllite
Kyanite
Andalusite
Sillimanite
Sill
Cd
Pa
Pa + Q
Phl + AS + Q
Cd + Ms
Fe-Ctd + AS + O_2
Fe-Ctd + Q + O_2
Ctd + Ms + Q
Fe-St + Q
Ms + Q
St + Ms + Q
Mg-Chl + Ms + Q
Mg-Al Chl + Q
Melt
Granite
Annite + Q
Or + Q + H_2O
Amphibolite solidus
Alm + Sill
Hc + Q
Alm + Cd
Hc + Q
Fe-Cd
Sill
A
Serp
Fe-Tremolite
Fe-Ctd-Q
Chl + Q
Fe-Cd + AS
Annite + 2F
Ms
Gedrite + Q
Anth
Tremolite
Fe-Chl + Q
Fe-Parg
Riebeckite
Talc + Fo
Phl + Q

Fig. 2(a).

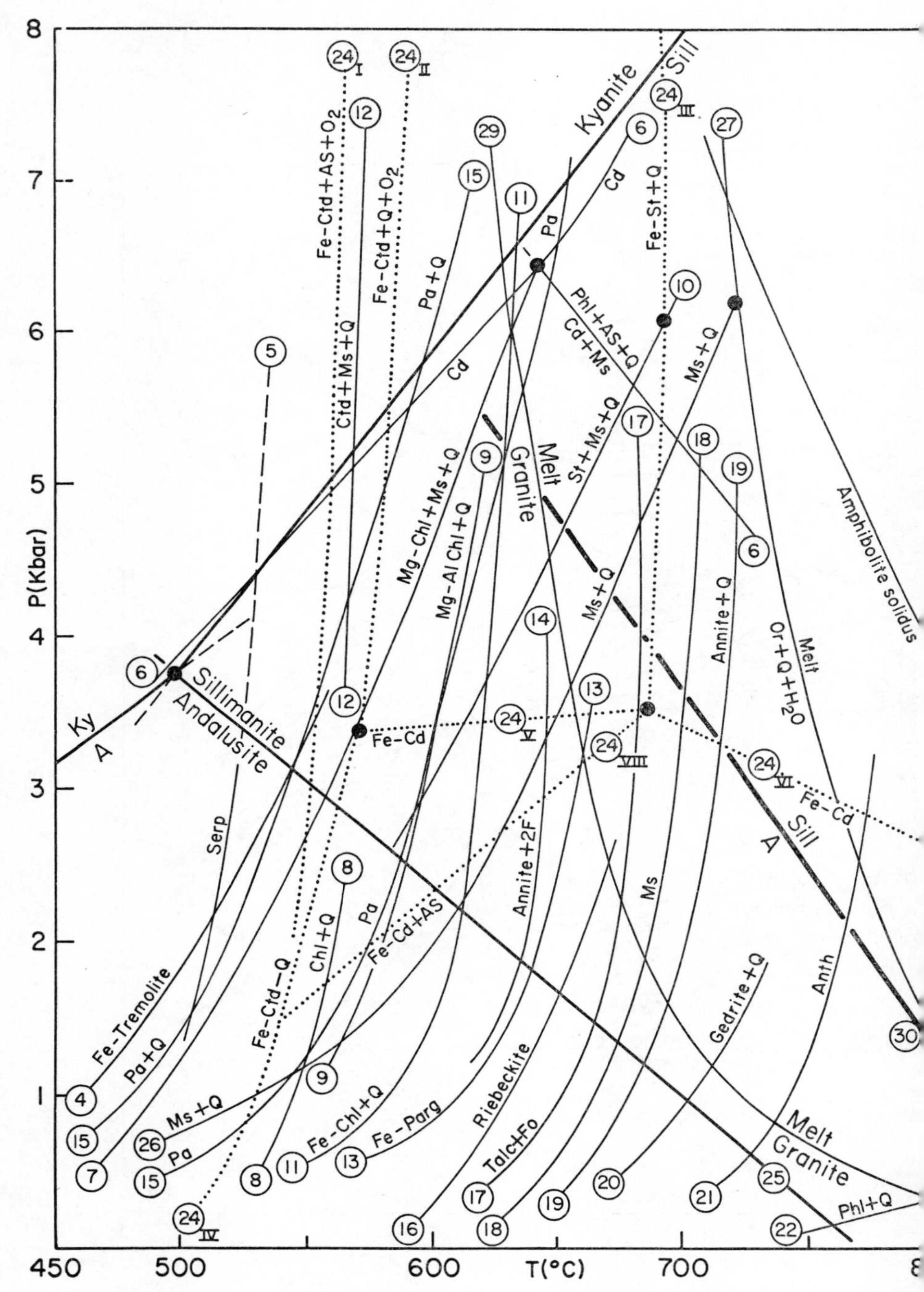

Fig. 2(b). Enlarged portion of 2(a) to show detail in the region of moderate temperatures and pressures.

Fig. 2. Univariant equilibria forming the basis of Fig. 1. These have been replotted from published studies referred to below and detailed in the list of references at the end of the chapter.

1. Thompson, 1970 (Kaolinite + Q = Pyrophyllite + H_2O).
2. Hsu, 1968 (Mn–Chl + Q = Spess + H_2O).
3. Kerrick, 1968 (Pyrophyllite = AS + Q + H_2O).
4. Ernst, 1966 (Fe–Tr = Fa + Q + Hedenbergite + fluid).
5. Bowen and Tuttle, 1949 (Serp = Fo + Tc + H_2O).
6. Schreyer and Seifert, 1969 (Chl + AS + Q = Cd + H_2O).
7. Seifert, 1970 (Mg–Chl + Ms + Q = Mg–Cd + Phl).
8. Akella and Winkler, 1966 (Al–Chl + Q = Gedrite + Cd + H_2O).
9. Fawcett and Yoder, 1966 (Mg–Chl + Q = Tc + Chl + Cd + H_2O).
10. Hoschek, 1969 (St + Ms + Q = AS + Bi + H_2O).
11. Hsu, 1968 (Fe–Chl + Q = Alm).
12. Ganguly, 1968, 1969, 1972; Hoschek, 1969 (Ctd + Ms + Q = St + Bi + Mt).
13. Gilbert, 1966 (Fe–Parg = Mt + Hedenbergite + Gt + Pl + Ne + fluid).
14. Rutherford, 1969 (Na–K–Annite + Q = Fa + Af_{Na} + Af_K + Mt + fluid).
15. Chatterjee, 1970, 1972 (Pa = AS + Ne + Ab + H_2O; and Pa + Q = AS + Ab + H_2O).
16. Ernst, 1962 (Riebeckite + Q = Fa + Mt + Ac + fluid).
17. Greenwood, 1963 (Tc + Fo = Anth + H_2O).
18. Velde, 1966 (Ms = Sa + C + H_2O).
19. Eugster and Wones, 1962 (Annite + Q = Sa + Fa + Mt + V).
20. Akella and Winkler, 1966 (Gedrite + Q = Hypersthene + Cd + H_2O).
21. Greenwood, 1963 (Anth = En + Q + H_2O).
22. Wones and Dodge, 1966 (Phl + Q = Fo + Or + H_2O).
23. Boyd, 1959 (Tr = En + Di + Q + H_2O).
24. Richardson, 1968. Stability relations of Fe–St, Fe–Cd and Fe–Ctd. All curves shown dotted for ease of identification.

    I. Fe–Ctd + AS + O_2 = Fe–St + Mt + Q
    II. Fe–Ctd + Q + O_2 = Fe–St + Alm + Mt + H_2O
    III. Fe–St + Q = Alm + AS + Mt
    IV. Fe–Ctd + Q = Fe–Cd
    V. Fe–Cd = Fe–St + Alm + Q
    VI. Fe–Cd = Alm + AS + Q
    VII. Fe–Cd = Hc + Sill + Q
    VIII. Fe–Cd + AS = Fe–St + Q + Mt

25. Holdaway, 1971 (AS polymorphs).
26. Evans, 1965; Kerrick, 1972 (Ms + Q = Or + AS).
27. Shaw, 1963; Lambert, Robertson and Wyllie, 1969 (Or + Q + H_2O—Melting curve).
28. Yoder and Tilley, 1962 (Tholeiite + H_2O—Solidus).
29. Luth, Jahns and Tuttle, 1964 (Ab + Or + Q + H_2O—Melting curve).
30. Richardson, Gilbert and Bell, 1969 (Sill = A).

mineral facies principle, with full application of field data, microscopic, X-ray, and chemical study. This is then followed by intelligent application of theory to unite the experimental results with the observations on rocks.

(b) APPLICATION OF EXPERIMENTAL RESULTS TO REAL ROCKS

The problems that beset the petrologist who applies experimental results to specific rocks can, with a fair degree of arbitrariness, be considered as problems related to composition, to equilibrium, or to geographic distribution of minerals.

Composition, components, and constituents present a formidable array of problems to the practical application of experimental data to rocks. The problems centre around the application of data for synthetic systems of few components to natural systems of many components. These problems boil down to the question of how additional components will affect the measured equilibria, and how the experimentally determined univariant boundaries are made multi-variant by the addition of new components. The only unequivocal standard that can be applied is the direct determination of the specific equilibrium in question, with all its chemical complexities, and the next best approach is to apply the principles of chemical thermodynamics to estimate the displacement of the equilibria by new components. If thermodynamic data on all the participating phases are available, the procedure is straightforward but involved. More usually the mixing properties are unknown and the only recourse is to assume ideal mixing. The most practicable approach is to express the experimentally determined equilibrium in terms of its equilibrium constant, and to recall that with due regard for standard states the constant is unity at equilibrium. Displacement of the equilibrium due to solid solution can then be computed at constant pressure from the expression

$$\ln K_b - \ln K_a = \frac{-\Delta H}{R}\left[\frac{1}{T_b} - \frac{1}{T_a}\right] \qquad (P\text{ const.}) \qquad \text{(E1)}$$

or, at constant temperature, from

$$\ln K_b - \ln K_a = \frac{-\Delta V}{RT}[P_b - P_a] \qquad (T\text{ const.}) \qquad \text{(E2)}$$

if ΔV is independent of pressure,

or

$$\ln K_b - \ln K_a = -\frac{1}{RT}\int_{P_a}^{P_b} \Delta V\, \mathrm{d}P \qquad \text{(E3)}$$

if ΔV depends on P.

Where:

$\ln K_a$ = natural logarithm of the equilibrium constant for a reaction, at a definite pressure and at temperature T_a° Kelvin.

$\ln K_b$ = similar to $\ln K_a$, but at temperature T_b.

ΔH = enthalpy change for the reaction, determined or calculated for the ambient pressure, P.

R = universal gas constant, in units compatible with the units employed for ΔH.

ΔV = volume change associated with the reaction.

The thermodynamic mixing properties of the minerals enter into these equations through the definition of the equilibrium constant

$$K = \pi_i a_i^{v_i} \tag{E4}$$

Where $\pi_i\, a_i^{v_i}$ is the continued product of the activities of all species involved in the reaction, each raised to the power of its stoichiometric coefficient in the reaction. The results of these calculations will be inexact to the extent that the solid solutions depart from the assumption of ideal mixing, but in many cases the recalculated equilibrium will be closer to the truth than an unmodified application of the simple system determined experimentally.

Another compositional parameter of importance is metasomatism. It is frequently assumed either that all changes that have occurred in metamorphic rocks are essentially isochemical, or that they must have been isochemical before one is entitled to apply either the phase rule or the results of experimental studies made on closed systems. Neither view is correct, nor has the history of the system anything to do with what we can deduce from its final equilibrium state. If equilibrium was reached then all pre-existing equilibrium states must have been obliterated. Of course, many rocks contain textural and compositional evidence of their tectonic and metamorphic history, but one must be wary of making deductions about equilibrium conditions based on a texturally inferred sequence of non-equilibrium steps.

The homogeneity of individual minerals is not the only compositional difficulty. It is common to find rocks in which the approach to equilibrium is very close within limited domains, but which, if taken in their entirety, are grossly out of equilibrium. Such systems are said to be in local equilibrium (Thompson, 1959) and the delimitation of the domain size is an important part of understanding the metamorphic process. An excellent recent contribution to this aspect of metamorphism has been made by Blackburn (1968), who demonstrates that in many rocks the domain of chemical equilibrium is of the order of 10–15 cm, emphasising the need to consider small domains as well as large ones. Within each domain it is possible and reasonable to apply the concepts and consequences of equilibrium thermodynamics, and sometimes adjacent domains of different bulk composition are most useful in determining the mineral facies to which the rock belongs.

3. THE PHASES OF METAMORPHIC ROCKS

Volumes have been written on the individual minerals of metamorphic rocks, and it seems inappropriate to recapitulate here what has been well said elsewhere. Accordingly the only discussion of individual minerals will occur as part of the presentation of data on mineral systems. One important phase is neglected, however, in all mineralogical compilations and in most discussions of the phases present during the crystallisation of metamorphic rocks. The neglected phase is not a mineral, but the fluid phase. Most of the important metamorphic reactions involve components usually thought to be part of a gaseous or at least fluid phase, and where this phase contains several different species and components the effect on individual equilibria can be profound. There follows therefore a brief outline of the fluid phase likely to be present during metamorphism.

(a) THE FLUID PHASE IN METAMORPHIC SYSTEMS

The geological evidence on the existence of a fluid phase during metamorphism is not as scanty as one would assume from noting the number of papers devoted to the subject. The most direct indication comes from study of fluid inclusions from minerals of rocks of undoubted metamorphic origin. The last few years have seen an encouraging resurgence of work in this field. Roedder (1967) reports analyses of fluid inclusions from milky quartz veins in low-grade metamorphic rocks. These inclusions homogenise at temperatures in the range of 150°C–200°C indicating temperatures at least as high as this if the inclusions are primary. Vidale (1969) reports inclusions from high grade metamorphic rocks in Connecticut. Poty and Stalder (1970) have shown that fluid inclusions from greenschist and amphibolite facies rocks in Switzerland are highly saline (1%–12% by weight equiv. NaCl) and are mainly H_2O and CO_2. The results of mass spectrometry on fluid inclusions from quartz, epidote, and fluorite from the Mont Blanc massif, show that these inclusions are mainly H_2O (average approx. 70%), CO_2 (average approx. 30%), with minor H_2 and CH_4 (Zimmermann and Poty, 1970). Fluid inclusions formed during synthesis of calcite homogenise at temperatures within 5°C of their temperatures of synthesis (LeFaucheux, Touray and Guilhamou, 1972). The direct evidence for a fluid phase is thus rather strong.

The indirect and theoretical evidence for the common existence of a fluid phase is also quite strong. Many workers, e.g. Thompson (1955, 1957), point out that the successful graphical analysis of the mineralogy of pelitic schists is consistent with two points of view. One is that the chemical potential of H_2O was imposed on the system from without and the other is that H_2O was present in excess as a separate phase. The choice between the two cannot be made on the basis of graphical analysis of mineral assemblages. Ernst (1963a) has deduced from the mineral relations of phengitic micas from low-grade schists that the water pressure must have been nearly equal to the total pressure. Yoder (1955) presented cogent arguments for supposing the water pressure to be approximately equal to the total pressure during most metamorphic episodes. Fluid inclusions, however, suggest that the metamorphic fluid is usually not pure H_2O, but frequently contains substantial amounts of CO_2, and dissolved salts. For example, Greenwood (1963) and Wones and Eugster (1965) estimate partial pressures of H_2O during Adirondack metamorphism to be in the neighbourhood of 1 to 10 bars, or less than $\frac{1}{2}$% of the total pressure. Wones and Eugster, however, supplement this with an estimate of the CO_2 pressure and find that it must have been very nearly equal to the total pressure, so it seems very likely that a gas phase was present here also. Calculations by French (1966) suggest that coexistence of graphite with any of the usual oxide oxygen buffer pairs will automatically force the gas pressure up to the total pressure. In summary it seems very likely that many metamorphic rocks crystallised in the presence of a gaseous or liquid phase.

(b) THE FLUID PHASE IN SYNTHETIC SYSTEMS

Most phase equilibrium experiments are performed in the presence of a gas phase. The conspicuous exception is the large body of data that has been collected in solid–solid reactions in the complete absence of gas. So far, all experiments with oxygen fugacity buffered by an oxide assemblage contain a gas phase as this is an essential part of the buffering process. It is therefore important to know something of the gases used in experiments.

(1) *Water*

There is no doubt that H_2O is the most abundant and, by that token alone, the most important geological fluid. Most of the hydrothermal studies in the literature have been conducted in the presence of essentially pure H_2O. Ordinarily it departs from stoichiometric H_2O to a small extent by the dissociation of water to form hydrogen and oxygen and by the solution in it of small quantities of the minerals being investigated. To a good first approximation it is pure H_2O. The properties of H_2O that are of the greatest interest to us are the density or molar volume, the free energy, the entropy, and the heat content or enthalpy, all as functions of pressure and temperature. These thermodynamic properties are all derived directly from measurements of density over a range of pressures and temperatures, and such measurements have been reported by a large number of workers. Smoothed calculations have been assembled by Burnham, Holloway, and Davis (1969: 20°C to 1000°C; 0.1 Kbar to 10 Kbar). The latter tabulation contains a great deal of entirely new data, and is the best existing tabulation for the range quoted. Over this range H_2O varies continuously from a state where its fugacity is about one tenth of the pressure to a state near 1000°C and 10 Kbar where its fugacity is nearly twice the pressure. This extreme range in thermodynamic behaviour must obviously be taken into account in any theoretical analysis of high pressure, high temperature dehydration equilibria.

(2) *Carbon Dioxide*

Next to H_2O, in all but strongly reducing environments, CO_2 is the most abundant geological fluid. In strongly reducing environments with H_2 present substantial amounts of CH_4 and CO form. The history of data-collection on the compressibility of CO_2 is almost as long as that of H_2O and will not be listed here. Recent critical compilations of the compressibility and thermodynamic properties of CO_2 appear in an article by Kennedy and Holser (1966) reporting the PVT properties of CO_2 from 0°C to 1000°C and from 0.025 Kbar to 1.4 Kbar in the form of density in gm/cu cm. The free energy of CO_2 over the range 400° to 1200° Kelvin and 0.1 Kbar to 1.4 Kbar, based essentially on the same data as the tables of Kennedy and Holser are given by Robie (1966). These free energy data can readily be converted by standard thermodynamic formulae to fugacities or fugacity coefficients. Over most of the quoted range the fugacity of CO_2 is significantly greater than the pressure. As with H_2O this departure from the ideal gas law must be taken into account in any thermodynamic calculations involving CO_2.

(3) *Hydrogen*

Another gas of importance to geological phase equilibrium studies is hydrogen. This is not so much because of its abundance but rather because of its role in the dissociation of H_2O and the part it plays directly in certain reactions, particularly those involving micas and amphiboles.

Hydrogen is a difficult material with which to work. Being a very small molecule it is difficult to contain, as it diffuses rapidly through the walls of most vessels, including those of steel, particularly at high temperatures. In addition, while it can coexist metastably with high concentrations of oxygen, the reaction of hydrogen and oxygen to H_2O proceeds with explosive violence in the presence of a suitable catalyst or an electric spark. A further difficulty is the tendency of the gas to react with the pressure vessel and form metal hydrides. The consequence of this is twofold: metal hydrides are brittle and high pressure vessels filled with hydrogen can fail suddenly due to this embrittlement; the formation of metal hydrides facilitates the diffusion of the gas out of the pressure vessel by a process of vacancy migration. These awkward attributes of hydrogen have combined to keep the total amount of data on the gas to a modest level. The most recent data have been presented by Presnall (1969). He reports PVT measurement and calculated fugacity coefficients for the range 200°C to 600°C and 0–1800 atm. These data show that the compilation and theoretical extrapolation of earlier work by Shaw and Wones (1964) yield values of the fugacity coefficients that are within about 2% of the new values. These coefficients are needed for the calculation of fugacities of a variety of gas species in geologically important fluids. Reference to Presnall (1969) and to Eugster and Skippen (1967) will indicate the use to which the data may be put.

Most fluids of geological interest are mixtures of several species, most of them falling in the system H-O-C-S-F, and most of them have not been studied as mixtures. Greenwood and Barnes (1966) tabulate the sub-critical phase relations and the super-critical compressibility relations for a variety of geological mixed fluids.

(4) *Water–Carbon Dioxide*

Mixtures of H_2O and CO_2 have been extensively studied in both the sub-critical two phase region and in the supercritical one-phase region. Tabulated by Greenwood and Barnes (op cit.) are data of Takenouchi and Kennedy (1964) on the sub-critical region. These data are reproduced here as Fig. 3, showing the compositions of coexisting H_2O-rich and CO_2-rich fluids, together with the curve of critical mixing. Each of these isobaric curves can be regarded as a sort of "solvus", and could be plotted on any of the isobaric H_2O–CO_2-temperature diagrams presented later for mineral equilibria in mixtures of H_2O and CO_2.

Supercritical data in the system H_2O–CO_2 are tabulated by Greenwood and Barnes (1966). In addition to this tabulation additional data have been presented by Greenwood (1969) on the compressibility of H_2O–CO_2 mixtures between 0 and 0.5 Kbar and 450°C and 800°C. These data are presented in terms of the compressibility factor $Z = PV/RT$ at integral values of pressure, temperature and composition. The fugacities of H_2O and CO_2 in the mixtures, together with the free energies of mixing, have been calculated (Greenwood, 1973). This work indicates that the activity coefficients of both H_2O and CO_2 in the mixtures up to 0.5 Kbar do not differ from unity by more than 15% and for the most part are within 5%. Consequently calculations of equilibria in H_2O–CO_2 mixtures assuming ideal mixing are frequently not grossly in error. A new summary of data on the mixtures up to a pressure of 2 Kbar and 700°C is presented by Barron (1973), in which he fits a regular solution model for the mixtures, making possible exact calculations from 1 atm up to 2 Kbar at temperatures above the 2-phase field illustrated in Fig. 3. Data are still lacking in the region immediately adjacent to the 2-phase boundary, where departures from ideality are expected to be extreme.

(5) *Hydrogen and Water*

Watery fluids that are not completely oxidized to pure stoichiometric H_2O consist of mixtures of H_2O, H_2, and O_2. Their proportions are dictated at equilibrium by the equilibrium constant for dissociation of water to hydrogen and oxygen. The presence of oxygen makes a wholly trivial contribution to the total pressure, so the mixtures can be considered as mixtures of hydrogen and H_2O. These have been studied by Shaw (1963, 1967). A graph is presented by Shaw (1967, op. cit.) summarising the results. The mixing of these gases is far from ideal. Shaw and Wones (1967) give the fugacity coefficients for hydrogen alone up to 1000°C and 3 Kbar.

Other gases that can be important quantitatively in geological fluids are CH_4, CO, H_2S, SO_2, S_2, and COS, and of these CH_4 and SO_2 are by far the

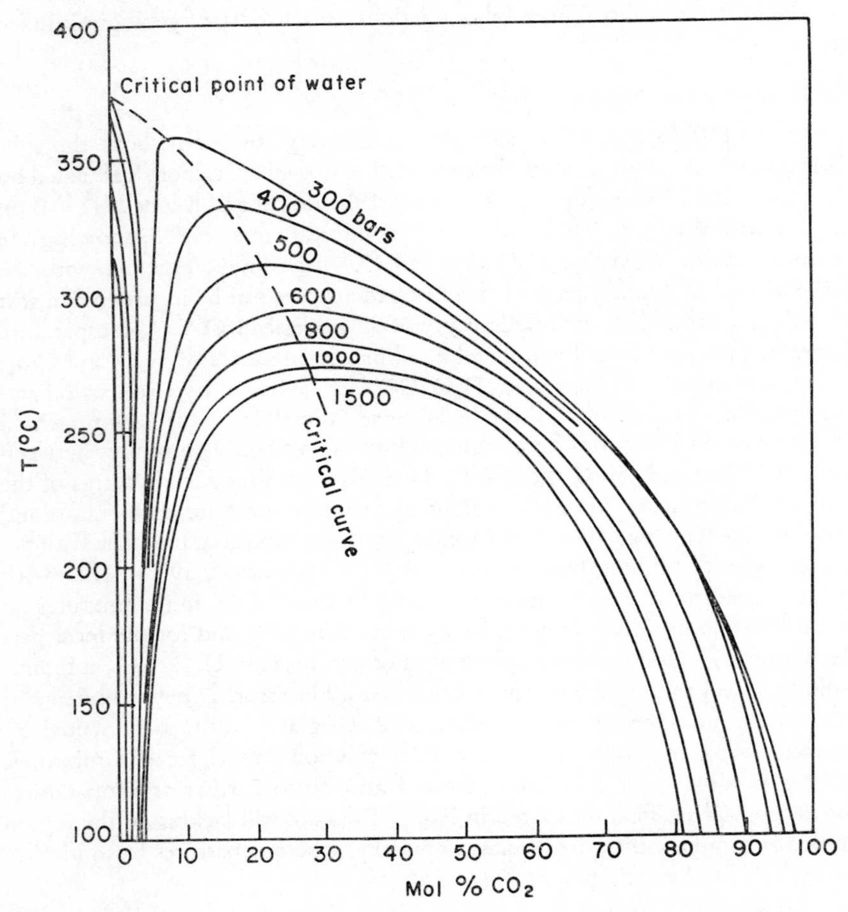

Fig. 3. 2-phase data of Takenouchi and Kennedy (1964) for the system H_2O–CO_2, illustrating the phase separation that characterises this system at high pressures and low to moderate temperatures.

most abundant, lying at opposite ends of the reduction–oxidation spectrum. Estimating the activities of these gases in mixtures with one another is uncertain in the absence of data on the actual mixtures. In the absence of data on the pure gas species the only recourse is to the corresponding state hypothesis, which states, in effect, that all gases obey the same equation of state if they are referred in some way to an appropriate standard, such as their own critical points. This is quite a good approximation for the non-polar molecules, and the tables, graphs, and methods necessary are presented and explained by Pitzer and Brewer (1961). A further application and extension of corresponding state theory has recently been presented by Mel'nik (1972) who shows that the log of the fugacity coefficient is linear with respect to the reduced pressure at constant reduced temperature.

If the properties of the end-members of a gas mixture are known, either from measurements or from corresponding state theory, it is still not possible to calculate with certainty what their individual activities will be in the mixture. In the absence of data to the contrary, the best approximation we can make is that however far the individual gases depart from the ideal gas law when pure, they mix ideally with the other gases, so that in the mixture the fugacity of any gas is just the product of its mole fraction in the gas with its fugacity, pure, at the pressure of the gas mixture.

(c) BUFFERING AND GAS EQUILIBRIA

All gases of interest to petrology are sufficiently complex chemically to consist of several chemical species related to one another by chemical reactions. The resultant gas phase, containing n elements may thus have $n + 1$ degrees of freedom. This means that $n + 1$ independent quantities must be known or specified before the state of the system is fully described. The fact that these gases must be in contact with, and presumably at equilibrium with minerals having definite stability ranges permits independent specification of at least some of these $n + 1$ parameters. For example, any gas at equilibrium with coexisting pyrite and pyrrhotite must have a particular sulphur fugacity (e.g. Fig. 12) and any gas at equilibrium with coexisting pyrite and magnetite must have a particular ratio of the fugacities of sulphur and oxygen. Similarly, the coexistence of hematite and magnetite requires a particular oxygen fugacity which in gases lying wholly within the system H–O automatically specifies ratios of all gas species through the dissociation of H_2O to H_2 and O_2. (The system H–O has two components, one phase, and thus 3 degrees of freedom, of which we have specified pressure, temperature, and the fugacity of oxygen that coexists with hematite and magnetite, thus completely specifying the state of the system.)

Following the contributions of Eugster (1957), Eugster and Skippen (1967), and French (1966) we note that every homogeneous reaction in the gas phase is characterised by an equilibrium constant. This is equal to the continued product of the fugacities of the reacting gas species, each raised to the appropriate power. These constants have the form

$$K_g = \pi_i f_i^{\nu_i} \qquad \text{E5)}$$

where K_g is the equilibrium constant for the gas reaction and ν_i is the stoichiometric coefficient of gas "i" in the reaction. Each buffer assemblage, or heterogeneous reaction is characterised by its equilibrium constant, of the form

$$K_b = \pi_{i,j} f_i^{\nu_i} a_j^{\nu_j} \qquad \text{(E6)}$$

Where

K_b = equilibrium constant for the buffer.
f_i = fugacity of gas species i.
ν_i = stoichiometric coefficient of gas species i in the buffer reaction.
a_j = activity of mineral component j.
ν_j = stoichiometric coefficient of mineral component j in the buffer reaction.

In addition to these equilibrium constants, one for each independent reaction possible in the system, we have the additional constraint that the sum of the partial pressures of the individual species must equal the total pressure. The assumption of ideal mixing of the individual species leads to the statement that the fugacity of each species in the mixture is equal to the product of its fugacity coefficient γ_i with its partial pressure p_i at the temperature and pressure of the mixture. This leads us to the expression:

$$P = \sum_i \frac{f_i}{\gamma_i} \qquad \text{(E7)}$$

This equation is solved simultaneously with all the equilibrium constant equations when $n + 1$ of the parameters have been specified.

These have been solved for a number of special cases of geological interest by French (op. cit.), Eugster and Skippen (op. cit.), and Skippen (1970). Figure 4 (after French, 1966) shows the dependence of gas composition on the fugacity of oxygen in C–O–H gases in the presence of graphite, a common constituent of pelitic rocks. The most significant feature is that at low temperatures and at low oxygen fugacities CH_4 is the dominant gas species, with H_2O and CO_2 remaining relatively unimportant until temperatures over 400°C or oxygen fugacities over about 10^{-22} bars. The exact values depend of course on the temperature and pressure selected for calculation. Thus dehydration equilibria and decarbonation equilibria determined in the presence of the pure reacting gas species will be shifted to lower temperatures due to the reduced activity of CO_2 and H_2O. Theory dealing with such shifts has been presented in some detail by Thompson (1955) and Greenwood (1961).

It may be noted in Fig. 4 that the curves all terminate in dashed lines at high temperatures and at high oxygen fugacity. The explanation for this may be seen in Fig. 5 (French, 1966) which shows the pressure developed by a C–O–H gas phase in equilibrium with graphite and different buffers. For example, consider the curve labelled QFM in Fig. 5. At temperatures less than given by the curve it is possible for a C–O–H gas to coexist stably with graphite and the QFM buffer on the oxygen fugacity defined by the QFM buffer, while at temperatures above the curve it is not possible. If such a gas mixture, with graphite and QFM

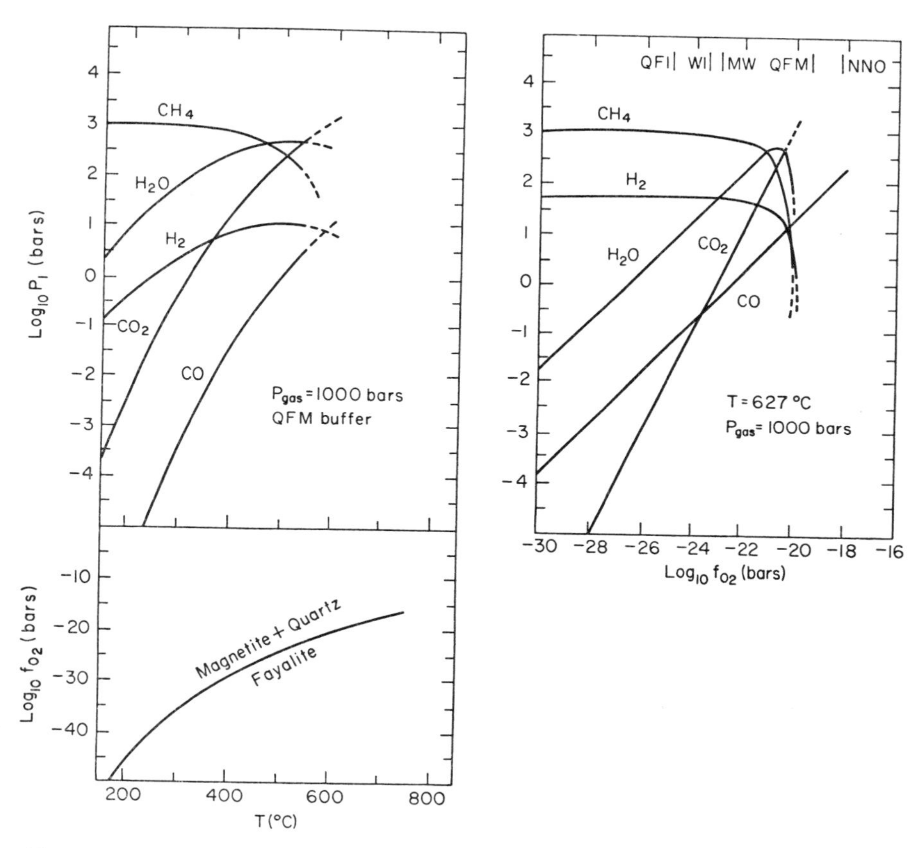

Fig. 4. Proportions of gas species in the system C–O–H in the presence of graphite, after French (1966). The left hand part of the figure illustrates the proportions of the major species at a total gas pressure of 1 Kbar with the fugacity of oxygen fixed by the presence of the QFM buffer (Quartz + Fayalite + Magnetite). The right hand part shows the dependence of the partial pressures on oxygen fugacity at constant temperature. The oxygen fugacities of the different buffers are shown at the top. (QFI = Quartz + Fayalite + Iron; WI = Wüstite + Iron; MW = Magnetite + Wüstite; QFM = Quartz + Fayalite + Magnetite; NNO = Nickel + Nickel Oxide.)

were heated from below the curve at some arbitrary pressure, say 1 Kbar, until the curve were reached, one of two courses would be followed. As reaction progressed either the pressure could be maintained constant and one of the reacting phases would be consumed or if the reaction were taking place inside a rigid container the pressure would increase following the curve until either the container failed or one of the reactants was consumed. If the phase consumed were graphite the system would still be buffered with respect to oxygen activity but if quartz or fayalite were consumed the system would be entirely unbuffered. In any event the potential exists for forcing the pressure of the fluid phase up to that of the confining pressure very rapidly once such a buffer curve is reached. This may be taken as an additional argument in favour of assuming that a gas phase was present during many metamorphic episodes. It has further been pointed out by French (1966) that metamorphism in systems containing

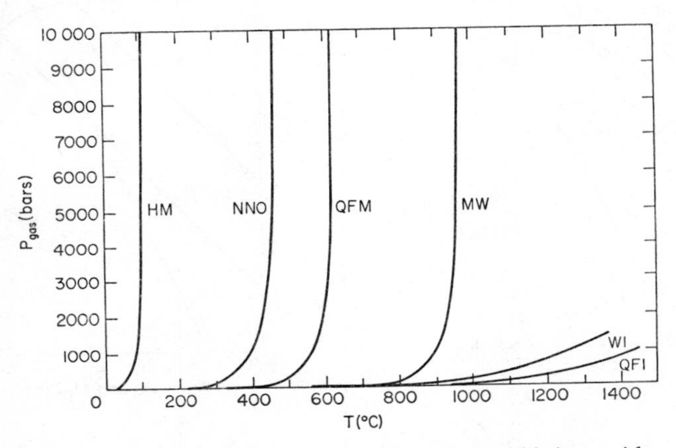

Fig. 5. The total pressure developed by a C–O–H gas at equilibrium with graphite and different buffers (French, 1966).

graphite necessarily follows a path of progressive reduction with increasing temperature at constant pressure, and that the dissappearance of graphite in such oxygen-buffered assemblages could be taken as an isograd.

The experimental details of making buffered experiments have been covered by D. K. Bailey in Part I, and need not be repeated here, except to emphasise the systems that are now susceptible to the buffer technique. Eugster and Skippen (1967) present a clear account of methods of buffering most of the possible species in the system H–C–O–S. Munoz and Eugster (1969) have further extended the method to include treatment of fluorine-bearing compounds, and Frantz and Eugster (1973) show how to buffer the activity of HCl through the use of Ag + AgCl in conjunction with the usual buffers for oxygen (and hydrogen) fugacity. This last advance now makes it possible to study not only minerals containing the transition metals in different oxidation states but also to study acid-base and ion exchange reactions under rigorously controlled conditions.

4. OXIDE AND SULPHIDE EQUILIBRIA IN NATURAL AND SYNTHETIC SYSTEMS

The material in this section is presented in order of increasing complexity, starting with equilibrium data on oxides and sulphides, progressing through equilibria involving only single gas species to equilibria involving several gas species, sulphides, and solid solution in both silicate and non-silicate minerals. Sulphide and oxide equilibria involve fewer components than silicates and do not enter into solid solution with them. Oxides and sulphides thus form chemically simple sub-systems that can be considered without reference to their mineralogical surroundings.

(a) OXIDE EQUILIBRIA

Oxides of the transition metals have long been known to be important in rock-forming processes, and since 1957 they have taken on a new importance to experimental petrology due to Eugster's (Eugster, 1957) discovery of how to use these reactions to buffer the activity of oxygen in hydrothermal experiments.

Data given in Appendix II permit calculation of the equilibrium fugacities of oxygen over various assemblages (Eugster and Wones, 1962; Wones and Gilbert, 1968; Huebner, 1969). Pressure affects all of the equilibria as may be seen from the equation in Appendix II. The $\log_{10} f_{O_2}$ *vs* T curves for these buffers are shown in Fig. 6 for a total pressure of 1 bar. It will be seen that for the whole range of geological conditions oxygen fugacity and hence partial pressure is very low, hence under no conditions can oxygen gas make a substantial fraction of the gas phase.

Figure 6 provides a reference framework for all geological reduction–oxidation reactions. Change in total pressure changes the values but not the relative positions of the curves. Presence or absence in rocks of the assemblages named in Fig. 6 can be used to set limits on the possible values for f_{O_2} and temperature of last equilibration. For example, coexistence of quartz and magnetite indicates that the rock last equilibrated at an f_{O_2} higher than that of the QFM curve for that total pressure and temperature. Fayalite is uncommon in metamorphic rocks for a combination of reasons. First, many rocks appear to have equilibrated in the field of stability of quartz plus magnetite. Second, the presence of potassium feldspar and the availability of H_2O makes possible the reaction:

$$\underset{\text{Fayalite}}{3Fe_2SiO_4} + \underset{\text{Orthoclase}}{2KAlSi_3O_8} + 2H_2O = \underset{\text{Annite (s.s. in biotite)}}{2KFe_3AlSi_3O_{10}(OH)_2} + \underset{\text{Quartz}}{3SiO_2} \qquad \text{(R1)}$$

This reaction has been studied by Eugster and Wones (1962, p. 104) and will be discussed later.

The addition of aluminum to the Fe–O system results in the appearance of hercynite ($FeAl_2O_4$) and the solid solution of Al in hematite and magnetite. Although hercynite has been recorded from silica saturated environments

(Deer, Howie, and Zussman, 1962, p. 67), it is most commonly associated with mafic igneous rocks and in metamorphosed bauxites containing corundum. The compositional relations to corundum and the iron oxides are shown in Fig. 7 (after Turnock and Eugster, 1962), from which it may be seen that hematite and hercynite are not stable together. There is, in addition, a solvus along the join magnetite–hercynite as shown in Fig. 8 (after Turnock and Eugster, op. cit.).

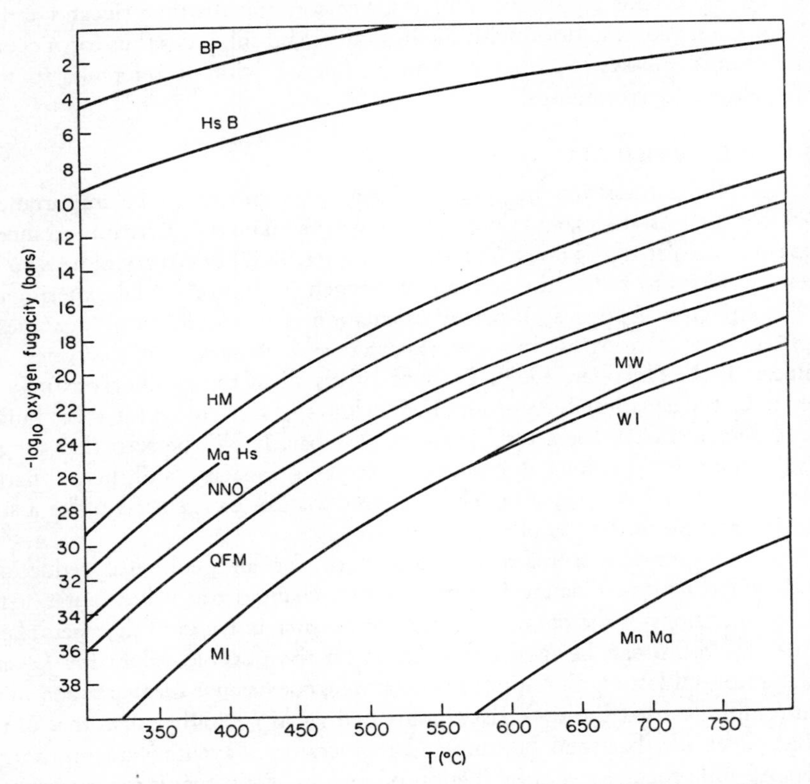

Fig. 6. Curves of oxygen fugacity versus temperature for various buffer assemblages at a total pressure of one bar. Abbreviations for common buffers as in Fig. 4, others from Appendix Ib.

The assemblage is not common, but an example described by Tilley (1924) with coexisting hercynite + magnetite + corundum with Fe-cordierite + biotite could be used to determine both the temperature and the equilibrium oxygen activity.

Reference to Fig. 9 (after Turnock and Eugster, 1962) should make this clear. Suppose for example, in the rocks described by Tilley (1924) the magnetite were known to contain 20 mol % $FeAl_2O_4$ and to be at equilibrium with a more aluminous spinel and with corundum. If this were the case the assemblage must have reached equilibrium at $f_{O_2} = 10^{-17}$ bars and $T = 675°C$. The effect of pressure is negligible in this example. This example, and one presented later on the system Fe–Ti–O emphasise that the compositionally simple oxides can be critical to interpretation of conditions of crystallisation. In addition, they point up a more general fact that given the coexistence of two binary solid solutions related by redox reactions both f_{O_2} and T can be determined. The assemblage hercynite + magnetite + corundum is isobarically univariant, or, as the solvus is essentially independent of pressure, univariant without restriction. Consequently, given Figs 8 and 9 it should be possible to

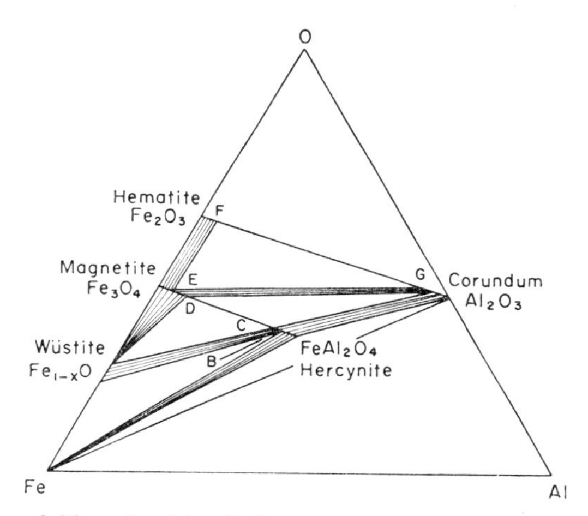

Fig. 7. Coexistence fields and solid solution compositions in the system Fe–Al–O for a temperature of 700°C. Generalised and not to scale, after Turnock and Eugster (1962). The vapour pressure depends on the compositions of the solids, hence changes across the diagram. Points D and C represent the compositions of coexisting magnetite and hercynite, shown as a function of temperature in Fig. 8.

determine all the intensive parameters from a measurement of a single extensive parameter such as the content of hercynite in magnetite solid solution. In this case the accessible intensive parameters are temperature and f_{O_2}, pressure being unavailable because of the insensitivity of the equilibria to total pressure. In principle any pair of coexisting solid solutions can be used in this way, but care must be taken to be sure that no other components have affected the system.

Addition of titanium to the Fe–O system produces, in addition to wüstite, magnetite, and hematite, the phases rutile (TiO_2), pseudobrookite (Fe_2TiO_5)-($FeTi_2O_5$) ilmenite ($FeTiO_3$), and ulvospinel (Fe_2TiO_4). These are shown in Fig. 10 (after Buddington and Lindsley, 1964). The solid solution series are magnetite–ulvospinel, hematite–ilmenite, and pseudobrookite–$FeTi_2O_5$. It is important to notice that there is no solid solution between ilmenite and magnetite, and that the intergrowths of ilmenite and magnetite that can be observed must be due to some process other than *isochemical* exsolution. Lindsley shows conclusively (in Buddington and Lindsley, op. cit.) that these textures must be due to a process of oxidation, for which he coins the descriptive term "oxidation exsolution". It is evident that petrographers must beware of making simple textural interpretations in oxide systems, and probably in many others as well, including sulphides.

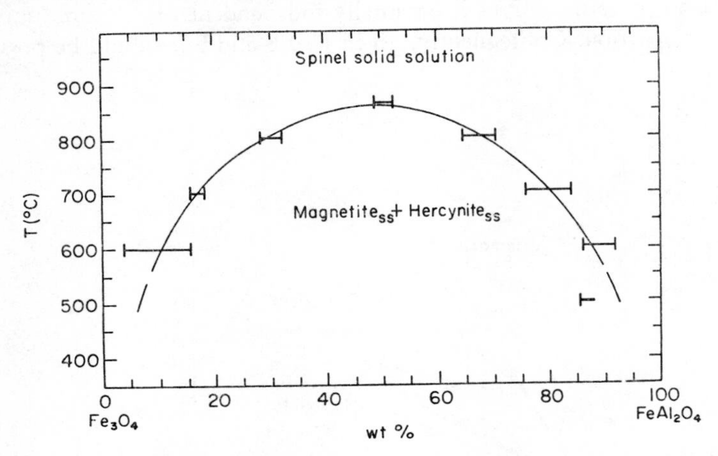

Fig. 8. The magnetite–hercynite solvus at the oxygen fugacity of the QFM buffer, at a total pressure 2 Kbar. (After Turnock and Eugster, 1962. Reversed brackets only are shown.) With increasing temperature the points C and D of Fig. 7 follow upwards along the curve of the solvus until at a temperature of 860°C the two phases become identical in composition and the Fe–Al spinel solid solution is complete.

The compositions of coexisting spinel and rhombohedral phases in the Fe–Ti–O system are sensitive to the activity of oxygen and to temperature, but not to pressure. Consequently there are two equilibrium constants (Buddington and Lindsley, op. cit.)

$$K_{R-2} = \frac{(A_{\mathrm{Mt}})^2(A_{\mathrm{Ilm}})^6}{(A_{\mathrm{Usp}})^6 f_{\mathrm{O}}{}^2} \tag{E8}$$

$$K_{R-3} = \frac{(A_{\mathrm{Mt}})(A_{\mathrm{Ilm}})}{(A_{\mathrm{Usp}})(A_{\mathrm{Hem}})} \tag{E9}$$

corresponding to the reactions

$$\underset{\mathrm{Usp}}{6Fe_2TiO_4} + O_2 = \underset{\mathrm{Mt}}{2Fe_3O_4} + \underset{\mathrm{Ilm}}{6FeTiO_3} \tag{R2}$$

$$\underset{\mathrm{Hem}}{Fe_2O_3} + \underset{\mathrm{Usp}}{Fe_2TiO_4} = \underset{\mathrm{Mt}}{Fe_3O_4} + \underset{\mathrm{Ilm}}{FeTiO_4}. \tag{R3}$$

These two equilibrium constants are functions of temperature and f_{O_2} only, and if the solid solutions are binary within the system Fe–Ti–O the two equations can be solved for the compositions of the phases if the temperature and f_{O_2} are known.

More usefully the equations may be solved for T and f_{O_2} if two of the com-

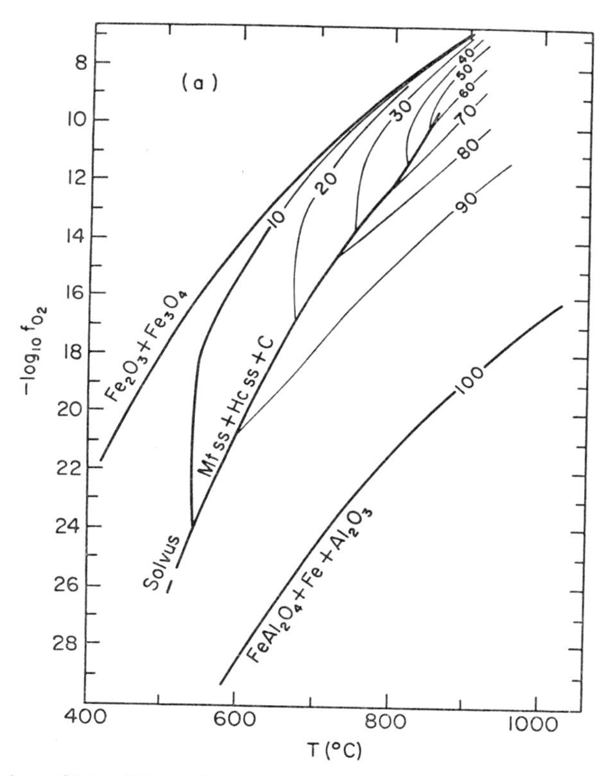

Fig. 9. Projection of Mt_{ss}-Hc_{ss} solvus at equilibrium with corundum (see Fig. 7) onto the $\log_{10} f_{O_2}$ *vs* T plane. (After Turnock and Eugster, 1962, Fig. 12). Along the solvus both f_{O_2} and T are dictated by the coexistence of two spinel phases and corundum. The numbered contours give the compositions of the spinels in terms of mol % $FeAl_2O_4$. At f_{O_2} above the solvus curve only the Fe-rich phase is stable with corundum, the composition being dictated by a combination of f_{O_2} and T. Below the solvus curve only the Al-rich spinel is stable with corundum, the composition again reflecting the combination of f_{O_2} and T. Along the projection of the solvus, which ends at a critical end-point in this projection, two spinels + corundum are stable and the compositions of both and the f_{O_2} are fixed by the temperature of equilibrium.

positions are known. Figure 11 (after Buddington and Lindsley, op. cit.) provide a graphical solution of the equations. Careful application of the data in Fig. 11 can result in a determination of temperature with an uncertainty of about

$\pm 30°C$ and a simultaneous determination of $\log_{10} f_{O_2}$ with an uncertainty of about one $\log_{10}$ unit, or one order of magnitude in the oxygen fugacity. The reader should note that in the system Fe–Ti–O we used a 2-phase assemblage and needed to measure two compositions while in the system Fe–Al–O we used a 3-phase assemblage and measured only one composition. Buddington and Lindsley (1964) have applied this approach to rocks that crystallised in the range from 550°C to 1125°C and with f_{O_2} ranging from 10^{-9} to 10^{-20} bars.

Additional components affect the Fe–Ti–O geothermometer. Buddington and Lindsley give a scheme for the recalculation of analyses which involves computing minor constituents as the end-member formulae and finally normalising in terms of the main components. Anderson (1968) has also made application of the system Fe–Ti–O to the LaBlache Lake titaniferous magnetite

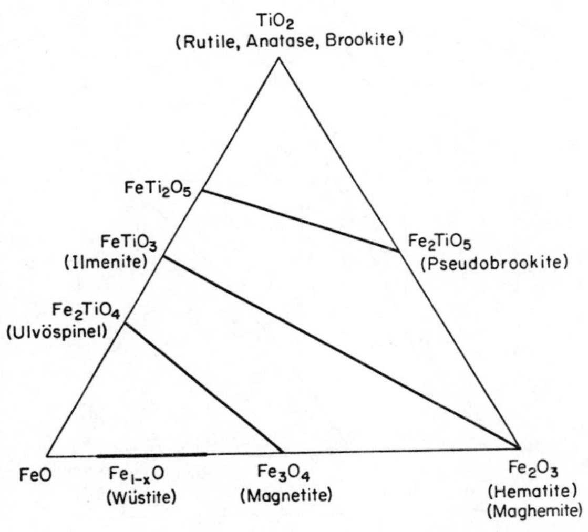

Fig. 10. Phases in the system FeO–Fe_2O_3–TiO_2 showing the major solid solution series magnetite–ulvospinel, hematite–ilmenite, and pseudobrookite–$FeTi_2O_5$ mol %. (From Buddington and Lindsley, 1964.)

deposit. Account was taken of extra components and the recent work of Spiedel (1970) shows that at high temperatures (1100°–1300°C) the addition of Mg to this system shifts the equilibrium f_{O_2} two orders of magnitude. With care, and reference to the original papers, including Anderson (1968) and Spiedel (1970), good estimates may be made of both f_{O_2} and temperature.

(b) SULPHIDE EQUILIBRIA

The sulphide minerals most commonly found in metamorphic rocks are pyrite and pyrrhotite. Others, such as galena, sphalerite, Cu–Fe sulphides, and the other common ore minerals are less abundant and less important in the study of ordinary metamorphic rocks. We must remember, however, that many ore deposits have been metamorphosed, some of them several times. In such cases detailed study of the petrology and phase relations will result at best in an understanding of the conditions of metamorphism and cannot ordinarily be extended to give information on the primary ore-deposition conditions.

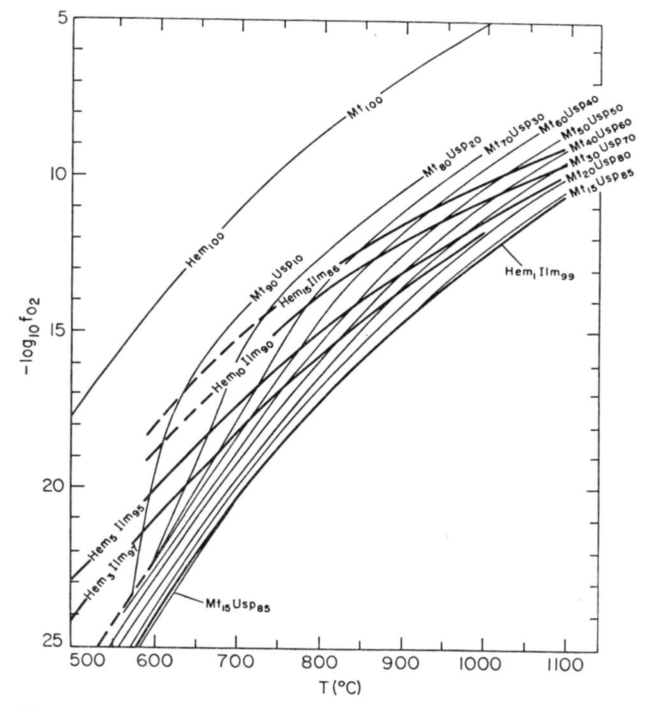

Fig. 11. f_{O_2}–T projection of conjugate composition surfaces from f_{O_2}–T–X space. Fine curved lines are contours of magnetite–ulvospinel solid solutions and heavier curved lines are contours of hematite–ilmenite solid solutions. Intersections of the two families of contours give the compositions of coexisting pairs of solid solutions. (From Buddington and Lindsley, 1964, Fig. 5.)

Excellent accounts of metamorphosed sulphide ore deposits have been presented by Schreyer, Kullerud, and Ramdohr (1964) and by Doe (1962).

Reactions of the kind

$$\text{oxide} + S_2 = \text{sulphide} + O_2 \quad \text{(R4)}$$

$$\text{Mg-Fe silicate} + S_2 = \text{Mg silicate} + \text{sulphide} + O_2 \quad \text{(R5)}$$

tend to run to the right under most geological conditions. Reactions like (R5) above, involving silicates, invariably involve solid solutions and hence are at best only sliding-scale indicators of the ratio of oxygen activity to sulphur activity.

Experimental data on the equilibrium relations and thermodynamic properties of the sulphide minerals are accumulating more rapidly than attempts to use them in connection with metamorphic rocks. We will not attempt a detailed summary of existing sulphide studies, but reference will be made to some of the outstanding work and to recent review articles that will enable the reader to introduce himself to the subject.

A summary of sulphide mineral stabilities has been presented by Barton and Skinner (1967) in which they tabulate a great many sulphide reactions with their

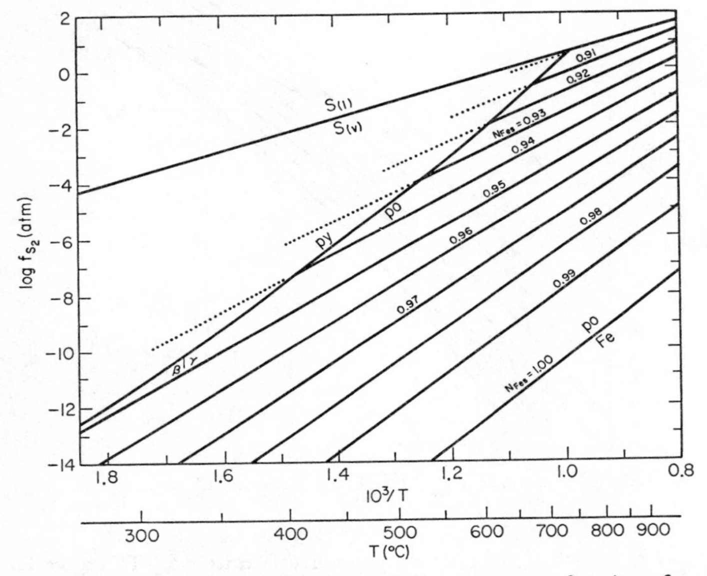

Fig. 12. Composition of pyrrhotite in the iron–sulphur system as a function of temperature and fugacity of S_2. N_{FeS} is the mol fraction of FeS in the system FeS–S_2. (From Toulmin and Barton, 1964.)

equilibrium conditions where these are known. This paper, in addition to a comprehensive treatment of the system Fe–S (Barton and Toulmin, 1964), contains an excellent summary of experimental methods and of the use and interpretation of the phase rule. The number of explanations of the phase rule in the literature presumably reflects an opinion among experts that the phase rule, in spite of its simplicity, is easier to misuse than to use correctly. The explanation by Barton and Skinner (*op. cit.*) is unusually clear and can be recommended. These authors also present diagrams suitable for interpreting the equilibria between sulphides and oxides, in which the co-ordinate axes are given in terms of the activities or the fugacities of oxygen and sulphur. These diagrams emphasise the fact that in most cases the coexistence of a sulphide and an oxide buffers the *ratio* of oxygen and sulphur fugacities, rather than their absolute values even at a specific total pressure and temperature. A useful set of provisional diagrams for more complicated systems has been calculated from theory by Garrells and Christ (1965) for temperatures up to about 300°C.

The common association of pyrite and pyrrhotite may be studied with reference to the work of Toulmin and Barton (1964). Variations in the composition of pyrrhotite permit it to be used as an indicator of sulphur fugacity. Figure 12 (after Toulmin and Barton, 1964) shows the maximum sulphur content in pyrrhotite at the pyrite–pyrrhotite solvus, and further that when the two minerals are at equilibrium at a particular temperature there is but a single value of the sulphur fugacity possible. Contrast, however, the situation where pyrrhotite is the only sulphide. Here the composition of the sulphide is affected by both temperature and sulphur fugacity with f_{S_2} being the more important. The power of this relationship has received too little attention in the literature of metamorphic petrology. Pyrrhotite alone may be of more use than the pair pyrite–pyrrhotite as the 2-phase assemblage is more likely to readjust to falling temperatures.

The fact that the activity of FeS in pyrrhotite is far from unity and not constant means that other equilibria sensitive to the activity of FeS will depend either on the composition of the coexisting pyrrhotite or the sulphur activity (as already remarked, they are not independent). A notable and theoretically useful example of such an equilibrium is the distribution of iron between coexisting iron sulphides and sphalerite. The pioneer work of Kullerud (1953) on the system FeS–ZnS raised hopes that the simple binary solvus between troilite and sphalerite could be used as a geothermometer. However, the variation in sulphur content of pyrrhotite makes it necessary to refer the system pyrrhotite–sphalerite to the system Fe–Zn–S. The system is unavoidably ternary and hence isobaric univariance requires equilibrium coexistence of three solid phases, in particular pyrite, pyrrhotite, and sphalerite. Figure 13 (after Barton and Toulmin, 1966) illustrates the compositional relations involved and shows how a small departure of pyrrhotite from stoichiometric FeS can make a large difference to the composition of the sphalerite that coexists with it. This composition is given in the figure by "point B". Point "A" in the same figure illustrates the composition of sphalerite that could coexist with stoichiometric FeS. In summary, the composition of sphalerite coexisting with pyrrhotite is univariant only if the activity of sulphur is buffered by the presence of pyrite. (Metallic iron would also serve, but is uncommon in rocks.)

5. EQUILIBRIA IN PELITIC ROCKS AND THEIR SYNTHETIC ANALOGUES

Pelitic rocks have been the key to metamorphic "grade" since the first days of systematic description and interpretation of metamorphosed rocks. The importance to metamorphic petrology of these rocks stems from their common occurrence, their limited range of composition, and the large number of minerals to which they can crystallise. In this they are distinct from the meta-morphosed mafic rocks which, although of limited compositional range, show a smaller variety of minerals due to extensive solid solution. In addition the mineralogy of meta-pelites is frequently spectacular even to the unaided eye, making mapping of changes relatively easy.

Their susceptibility to interpretation has been enhanced since 1957 by the appearance (Thompson, 1957) of an adequate graphical representation of the

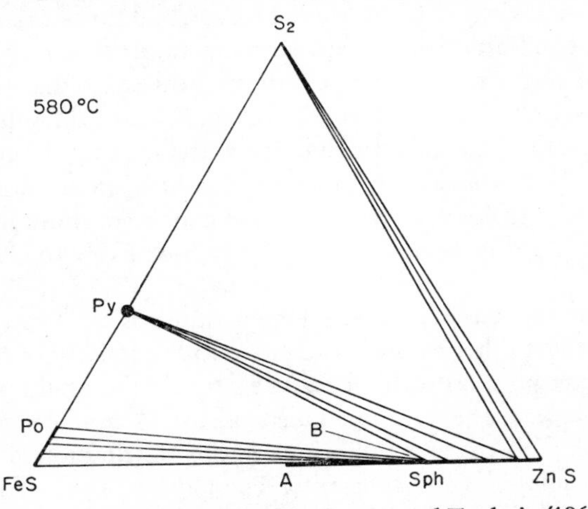

Fig. 13. A portion of the system Fe-S_2-Zn, after Barton and Toulmin (1966, Fig. 11). Point A is the composition of sphalerite that could coexist with FeS at 580°C, and point B is the composition of sphalerite that could coexist at the same temperature with the pair pyrrhotite + pyrite. Small changes in pyrrhotite composition produce large changes in the stably coexisting sphalerite.

phase relations in the system SiO_2-Al_2O_3-MgO-FeO-K_2O-H_2O. The meta-pelites are better known, more studied, and more useful than any of the other common rock groups, with the possible exception of the siliceous magnesian carbonates.

(a) THE SYSTEM SiO_2-Al_2O_3-MgO-FeO-K_2O-H_2O AND ITS SUBSYSTEMS

These six chemical constituents suffice to write formulas for most of the minerals found in meta-pelites, and its subsystems give a good approximation to meta-laterites and ultramafic rocks as well. The metamorphic minerals that crystallise from this system are quartz, kyanite, sillimanite, andalusite, gibbsite, boehmite, diaspore, kaolinite, pyrophyllite, talc, olivine, orthopyroxene, serpentine, anthophyllite, Mg-cummingtonite, gedrite, magnetite, hercynite, orthoclase, muscovite, biotite, chlorite, garnet, staurolite, chloritoid and cordierite. It will be evident that the existence of twenty-six phases in a six component system necessitates the existence of a very large number of reaction-relations between them. In fact, there could be more than 650,000 different univariant reactions! There are data for relatively few of these and for the sake of efficiency in presentation we shall give them by starting with simple sub-systems and progressing to the data for the more complicated whole system.

(1) SiO_2-Al_2O_3

Quartz, kyanite, sillimanite, andalusite, and mullite are the phases of interest. The geologically unknown assemblage quartz + corundum will be dismissed as

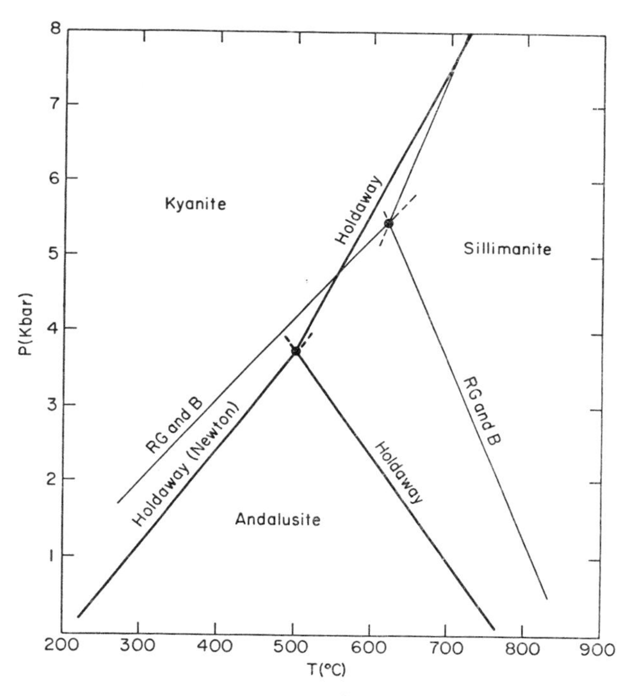

Fig. 14. Stability of the polymorphs of Al_2SiO_5. Heavy lines after Holdaway (1971), light lines after Richardson and Bell (1969). The boundaries after Holdaway are consistent with the experimental data of Newton (1966) on kyanite = andalusite and with those of Richardson Gilbert and Bell (1969) on kyanite = sillimanite. The significant difference is in the location of the boundary andalusite = sillimanite. See text for further discussion.

metastable in spite of the fact that experiments have failed to prove conclusively this antipathy. We are left thus with the equilbrium relations of kyanite, sillimanite, and andalusite in the intermediate temperature range. Mullite is stable only at higher temperatures and is treated by Schreyer in Part II, Section A3 of this volume.

The chemical simplicity of the Al_2SiO_5 minerals would suggest that their phase relations should be freer from ambiguity and easier to apply than the relations of more complicated minerals. Unfortunately, this is far from the case. An entire issue of the *American Journal of Science* (Vol. 269, p. 257–422) was devoted to the Al_2SiO_5 minerals. Since that time other work has appeared and anyone who wants to illustrate the "true" phase relations has a difficult task of selection. The two most recent experimental determinations of the equilibria are shown in Fig. 14. It will be apparent that the experimental results for both sets are in fairly good agreement for the boundaries kyanite = sillimanite and kyanite = andalusite. The main discrepancy lies between the determinations of the boundary andalusite = sillimanite, where the work of Holdaway (1971) places the boundary some 150°C lower than the work of Richardson, Gilbert and Bell (1969). This large disagreement is discussed below.

The difficulty of determining this particular equilibrium stems from the two facts that the free energy of reaction is only a few hundred cal per mol, in contrast with dehydration reactions which have tens of kcal, and that the entropy of reaction is likewise small. Thompson (1955) and numerous others have pointed out that this small entropy change makes the equilibrium temperature very sensitive to perturbations of the free energy of either phase. These perturbations can come from additional components (minor but not negligible), from grain size, strain energy, and Al–Si disorder. Strens (1968) shows that small amounts of contaminants may shift the boundary by more than 100°C. Metastable nucleation of the polymorphs may not be restricted to synthetic studies. Hollister (1969) has presented an interesting account of natural metastable crystallisation of andalusite in the kyanite field, followed by reaction to stable kyanite, and further followed after an increase of temperature by formation of sillimanite.

The effect of small amounts of extra components will depend, naturally, upon the fractionation of these components between the coexisting aluminium silicate polymorphs. Chinner, Smith and Knowles (1969) show that transition-metal content of the polymorphs may reach at high as 1.6 wt % Fe, with Fe greatly exceeding Ti, V, and Cr. In some examples Fe is regularly distributed between the phases, suggesting equilibrium, and in others there is irregular distribution and variation within single grains. Albee and Chodos (1969), however, in a similar study, show little fractionation of Fe, the most abundant contaminant, between coexisting polymorphs, and conclude that extra components that are present could not greatly affect the equilibrium. Preferential solid solution of Fe in sillimanite would expand the sillimanite field at the expense of kyanite and andalusite. This would be consistent with Ganguly's (1969) assertion that the triple point should lie between 500°C and 570°C and 3.5 and 5.5 Kbar, on the basis of his determination of the reactions chloritoid + kyanite = staurolite + almandine + H_2O and chloritoid = almandine + corundum + H_2O, taken together with natural occurrences. In addition to the variability of composition it is possible that the state of ordering of Al and Si in tetrahedral co-ordination in sillimanite may have an equally large effect (Zen, 1969; Holdaway, 1971; Greenwood, 1972).

In Fig. 14 the results of Holdaway (1971) have been given preference over those of Richardson, Gilbert and Bell (1969) for the andalusite = sillimanite reaction. The choice is not easy, as both studies were meticulous. This choice hinges around the observation by Holdaway that in his experiments the presence of fibrolite grossly shifts the apparent boundary to higher temperatures, rather close to that of Richardson, Gilbert, and Bell. It appears that the latter workers may have had substantial amounts of fibrolite in their starting materials leading to an erroneously high reaction temperature. Holdaway demonstrates conclusively that crystals of andalusite decrease in weight when surrounded by powdered sillimanite within the field taken by him to be the stability field of sillimanite and by Richardson, Gilbert, and Bell to be the field of andalusite. He did not demonstrate the corollary that crystals of sillimanite increase in weight when surrounded by powdered andalusite under the same conditions. However, he did demonstrate that fibrolite is progressively converted to sillimanite and that while fibrolite is (metastably) present in the sillimanite field andalusite crystals will *increase* in weight, only to decrease when all the fibrolite is gone. Doubtless more experiments will be done and the results refined further, but for the moment this writer favours the curve of Holdaway.

Additional evidence on the position of the Al_2SiO_5 triple point may be obtained from natural assemblages, although one must be wary of assembling a simple picture based on the coexistence of other very complicated solid solutions. Accordingly it is with some hesitation that this evidence is brought forward. Sillimanite is known in association with paragonite but not with paragonite + quartz (Chatterjee, 1972). Reference to Fig. 2, curves (15), will show that the Richardson, Gilbert, and Bell curve is at too high a temperature for even paragonite alone to be stable with sillimanite except under the narrowest range of conditions. However, the Holdaway curve appears to be at too low a temperature, or paragonite + quartz + sillimanite should have been reported from somewhere. Chlorite has been reported from rocks in a transition zone just before the appearance of sillimanite (Guidotti, 1973) suggesting that in the absence of Al_2SiO_5 as a phase chlorite is stable with quartz at the same temperature as sillimanite. See curves (6), (8), and (9) in Fig. 2. These all suggest that the andalusite = sillimanite curve of Richardson, Gilbert, and Bell lies at too high a temperature, but gives little information on the Holdaway curve. Halferdahl (1961) catalogues chloritoid occurrences and reports that chloritoid + quartz is common in association with kyanite and with andalusite, and rare but not unknown with sillimanite suggesting that curves (12) or (24) may pass close to the triple point. A similar point has been made by Ganguly (1969). Finally, the assemblage quartz, staurolite, andalusite, cordierite, garnet is not uncommon in low-to-moderate pressure metamorphism (Zwart, 1962; Turner, 1968), and the same assemblage except for the substitution of sillimanite for andalusite is equally common. This leads to the suggestion that the andalusite = sillimanite boundary must cross the lower-pressure stability limit of the assemblage Fe-staurolite + almandine + quartz (Fig. 2, Curve 24.v). Summarising the foregoing it appears that the Richardson, Gilbert, and Bell boundary is at too high a temperature, but that the boundary of Holdaway may be at too low a temperature, at least for the appearance of fibrolite. Accordingly, in Fig. 1 liberties have been taken with the experimental data and a curve drawn (dashed) for the andalusite = sillimanite (fibrolitic) boundary 50°C higher than the Holdaway curve and satisfying the petrographic observations noted above. This curve is not experimental and does not agree with either of the best determinations in the laboratory. Rather it lies between them for the reasons cited, and appears to satisfy petrographic experience better than either. It is consistent with a view that the first appearance of fibrolite from either andalusite or micas takes place in the field of sillimanite by a process of metastable nucleation. This is made possible by the large overstepping required for reaction in systems with small free energies of reaction. If this is correct, the first appearance of (fibrolitic) sillimanite at pressures below the triple point is metastable and cannot be taken as an isograd.

(2) SiO_2-Al_2O_3-H_2O

The addition of H_2O to the alumino-silicate system adds the minerals gibbsite, diaspore, boemite, kaolinite, pyrophyllite, and the equilibria between them. The stabilities of kaolinite and pyrophyllite have been studied by several workers. At the time of writing previous inconsistencies seem to be resolved by the work of Kerrick (1968, pyrophyllite) and Thompson (1970, kaolinite). These papers may be taken as an introduction to the literature, in which it may be found that metastable nucleation and persistence of pyrophyllite is a serious problem.

Detailed discussion of the equilibria between the uncommon mineral assemblages noted above is omitted here due to their limited petrologic utility. Turner (1968) ascribes their limited occurrence to the narrow stability range of the assemblage pyrophyllite + quartz + chlorite and to the very similar reaction temperatures for all of the critical reactions. Another equally forceful reason may be metastability of these layer minerals in the presence of alkalies, particularly alkali feldspar. Available data, both experimental and thermodynamic, suggest all the layer minerals without alkalies in their unit cell tend to react spontaneously with feldspar to form micas. Consequently these minerals and their equilibria may only be observed in feldspar-free rocks.

(3) SiO_2-MgO-H_2O (*The ultramafic sub-system*)

This sub-system of the "pelite system" contains none of the common minerals of meta-pelites except quartz. The rocks to which it does apply include chiefly the metamorphosed equivalents of the ultramafic igneous rocks and certain magnesian carbonates. It is included here because of the intensive study the system has received and because stability limits in this system affect many others. This system is further distinguished by the fact that it was the first to be investigated by high pressure hydrothermal methods in the pioneering study by Bowen and Tuttle (1949) in which they established the absence of low temperature melting relations with serpentine. Later studies in this and related more complicated systems have confirmed the early work to a remarkable degree. The upper stability of serpentine is taken from the work of Bowen and Tuttle (1949) and shown in Fig. 2 as curve 5. The reaction talc + forsterite = anthophyllite + H_2O (curve 17, Fig. 2) and the reaction anthophyllite = enstatite + quartz + H_2O (curve 21, Fig. 2) have been taken from the work of Greenwood (1963). Zen (1971) and Greenwood (1971) show that the available thermodynamic data are inadequate to calculate with confidence the relative stabilities of anthophyllite *versus* talc + enstatite. It consequently seems prudent to omit this curve from the present diagrams. Reference to the above papers will introduce the reader to the literature on this system and to the considerable difficulties connected with metastability that have plagued experimental studies.

(4) SiO_2-FeO-H_2O

This sub-system of the "pelite system" has received little direct study, although numerous more complicated systems having this as a bounding system have been examined. The most recent direct study is that of Forbes (1969) into the extent of solid solution between the minerals talc and minnesotaite (the iron analogue). He shows that the solid solution is not one of simple substitution of ferrous iron for magnesium but a complicated exchange substitution of the form $Fe^{+3} + H^{+} \rightleftharpoons Si^{+4}$, coupled with $Fe^{+2} \rightleftharpoons Mg^{+2}$. The most iron-rich member of the series synthesised corresponds to $Fe/Fe + Mg = 0.20$. The earlier work of Flaschen and Osborne (1957), in which the crystallisation of minnesotaite was first reported remains the only other study.

The careful study of Wones and Gilbert (1969) on a re-determination of the stability of the oxygen-buffer assemblage fayalite + magnetite + quartz belongs properly in the system Fe-Si-O_2, but in view of its wide use in hydrous systems and the hydrothermal method used in its redetermination, it is included at this point. Data from this paper have been used to prepare part of Appendix II.

(5) SiO_2-Al_2O_3-MgO-H_2O

Some of the most important work in experimental petrology has been done on this system. The early work of Yoder (1952) and Roy and Roy (1955) has been followed by the efforts of many others in outlining the stability fields of cordierite, gedrite, chlorite, yoderite, serpentine, garnet, and sapphirine, in addition to increasing our knowledge of the bounding ternary systems. All these studies have been done in the presence of a water-rich gaseous phase, hence it is possible to represent graphically all the minerals that can stably coexist with H_2O by projection from H_2O onto the plane SiO_2-Al_2O_3-MgO. (See Fig. 9(b) in Schreyer.)

The lower temperature stability limit of Mg-cordierite has been the subject of a recent careful re-examination by Seifert and Schreyer (1970), superseding earlier studies of this common and important mineral. This study, in addition to

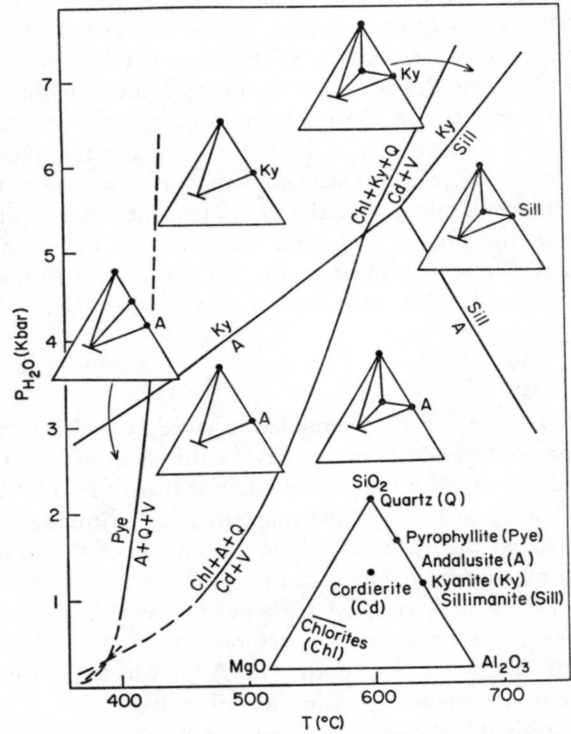

Fig. 15. Pressure–temperature grid of stable associations in part of the system MgO-Al_2O_3-SiO_2-H_2O, after Seifert and Schreyer (1970). The reactions are selected so as to emphasise the lower stability limit of Mg-cordierite in the presence of H_2O.

modifying the accepted position of the cordierite stability, has clarified our understanding of equilibria involving pyrophyllite and the aluminum silicates. The relationship of pyrophyllite to cordierite stability is shown in Fig. 15 (after Seifert and Schreyer, op. cit.). These results are consistent with the study by Seifert (1970) on a part of the system K_2O-MgO-Al_2O_3-SiO_2-H_2O, discussed later, shown as the curves emanating from the invariant point at 6.5 Kbar and 640°C on Fig. 2.

The stability of Mg-Al-chlorite + quartz has been investigated by Fawcett and Yoder (1966) in the absence of muscovite. The results of their work on the reaction chlorite + quartz = talc + cordierite + H_2O appear as curve 9 in Fig. 2. The reaction of chlorite + quartz to gedrite + cordierite + H_2O has been reported by Akella and Winkler (1966) to lie at a lower temperature than the reaction to talc + cordierite reported by Fawcett and Yoder (op. cit.), and this is shown as curve 8 on Fig. 2. The upper stability limit of Mg-Al-gedrite, after Akella and Winkler (1966, op. cit.) is shown as curve 20 on Fig. 2.

(6) SiO_2-Al_2O_3-FeO-H_2O

The iron analogue of the last section has received attention from Turnock (1960), Halferdahl (1961), Hoschek (1967, 1969), Hsu (1968), Hsu and Burnham (1969), Richardson (1968), and Ganguly (1968, 1969, 1972). The minerals concerned include, in addition to those already discussed, staurolite, almandine, chloritoid, and Fe-cordierite. Ganguly (1972) lists and discusses twenty-five separate equilibria in this system, including both experimental data and theoretical derivations and extrapolations. Ganguly's work is in good agreement with that of Richardson and Hsu, and Ganguly's diagrams, embodying earlier work of others, are taken as definitive (Fig. 16, Fig. 2). Reference to Fig. 2 will reveal three main P-T fields for bulk compositions that can crystallise staurolite in the neighbourhood of 600°C: above about 5 Kbar, staurolite + quartz + kyanite but *no cordierite*; below about 3 Kbar, Fe-cordierite + andalusite + quartz but *no staurolite*; between about 5 and 3.5 Kbar staurolite + Fe-cordierite + quartz (curves numbered (24) on Fig. 2). The stability of staurolite alone and of staurolite + quartz is summarised in Fig. 16 (see Ganguly, 1972). It will be seen that the assemblage staurolite + quartz has a much more restricted field than that of staurolite alone, and that the stable field of staurolite is bounded on all sides by reactions, both in terms of P-T and P- $-\log_{10} f_{O_2}$ space. The invariant points shown in Fig. 16A also appear in Fig. 2, at a different scale.

Richardson (1968) notes that in natural assemblages the ratio Mg/(Mg + Fe) follows the scheme cordierite > staurolite > chloritoid > garnet, which leads to the conclusion that the field of cordierite should expand at the expense of that of staurolite with addition of Mg. This is in agreement with the calculations of Ganguly (1972), who finds that addition of Mg to the system shifts the low-pressure limit of staurolite upward in pressure by about 2 Kbar, using the compositions of coexisting natural minerals as a guide. However the work of Seifert and Schreyer (1970) places the stability limits of pure Mg-cordierite close to those for Fe-cordierite, suggesting that the change should be slight. In addition, Schreyer and Seifert (1969) have found a field for pure Mg-staurolite at pressures in excess of 10 Kbar. A consequence of this is that although pure Mg-staurolite is unknown from rocks, substitution of Mg for Fe should become increasingly possible at elevated pressures, and with increasing f_{O_2}.

Almandine is another important mineral in this sub-system of the "pelite system" and has been reported on by Hsu (1968). He shows that almandine stability is very sensitive to variations in oxygen activity, illustrated in Fig. 17 (after Hsu, 1968). Dependence of the equilibria on f_{O_2} makes it necessary to apply a restriction to f_{O_2} before attempting representation on a P–T plot. Accordingly, the polybaric path of invariant point "D" (Fig. 17) is illustrated as curve (11) in Fig. 2. This illustrates the maximum thermal stability of Fe-chlorite in the presence of magnetite and quartz, but chlorite alone should be stable to somewhat higher temperatures.

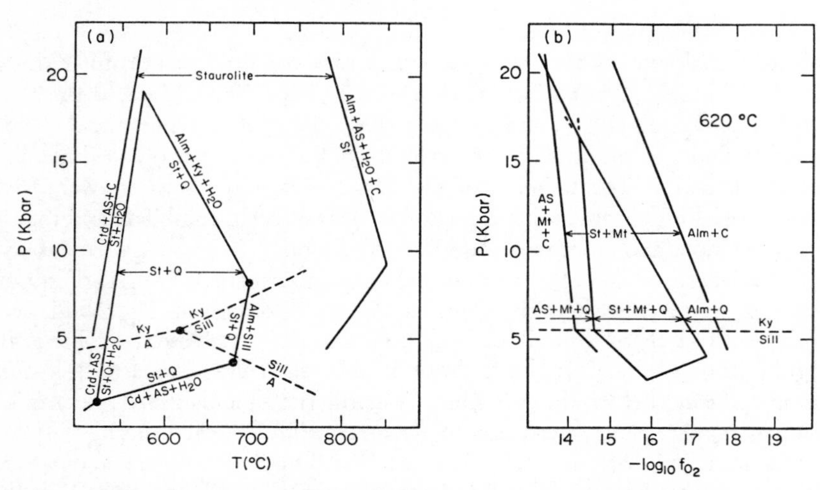

Fig. 16. Staurolite stability, after Ganguly (1972). These curves include some data from the work of Hsu (1968) and Richardson (1968). The dashed stability curves for kyanite, andalusite and sillimanite are from Richardson, Gilbert and Bell (1969). See also Figs. 1, 2, and 14 for Al_2SiO_5 stabilities. A: pressure–temperature projection at about midway between the NNO and QFM buffers. B: pressure- —$\log_{10} f_{O_2}$ section at approximately 620°C through the same fields as shown in A. Abbreviations: AS = andalusite, sillimanite, *or* kyanite, whichever is stable; St = staurolite; Alm = almandine; Ctd = chloritoid; C = corundum.

(7) *Mica Stabilities*

Numerous reactions in pelites involve mica, but before considering these it is necessary first to consider the stabilities of the micas individually.

(i) *Dioctahedral micas, paragonite and muscovite.* These have been studied experimentally by Yoder and Eugster (1955), Eugster and Yoder (1955), Carman (1967), Evans (1965), Munoz and Eugster (1969), Popov (1970), Seifert and Schreyer (1965), Althaus and Johannes (1969), Velde (1965a, 1965b, 1966), Hemley (1959), and Eugster, Albee, Bence, Thompson, and Waldbaum (1972). In addition, others have made theoretical studies and experimental studies in more complicated chemical systems in which the dioctahedral micas participate. We cannot report all this work and hence show only the limiting reactions and boundaries.

Paragonite stability, both with and without quartz, has been presented by Chatterjee (1970, 1972). These two stability boundaries carry the number (15) in Fig. 2, the lower temperature one corresponding to the reaction paragonite + quartz = albite + Al_2SiO_5 + H_2O and the higher temperature one to the reaction paragonite = albite + corundum + H_2O. The thermodynamics of these reactions are discussed by Chatterjee, who shows the two boundaries to be thermodynamically consistent.

Muscovite stability with and without quartz is shown in Fig. 2, the reaction muscovite + quartz = K-feldspar + Al_2SiO_5 + H_2O (Evans, 1965) being

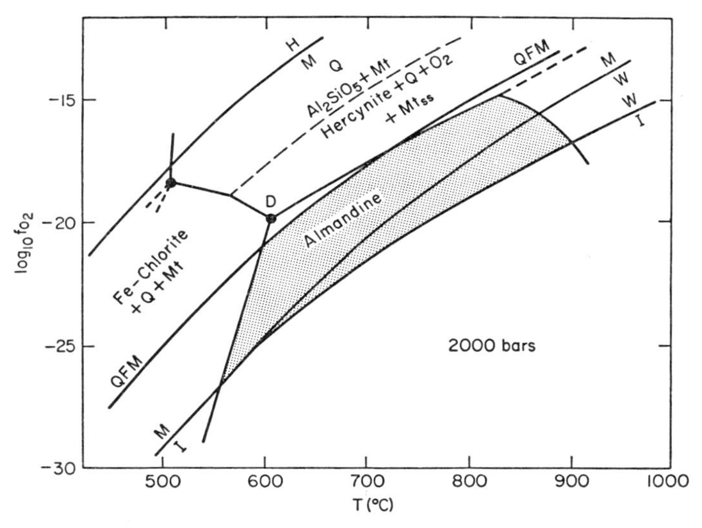

Fig. 17. Stable phase assemblages in the bulk composition of almandine + H–O gas phase. (After Hsu, 1968.) The isobaric invariant point D, representing the highest temperature stability of the assemblage Fe–chlorite + magnetite + quartz is plotted as curve (11) in Fig. 2.

numbered (26) and the reaction muscovite = K-feldspar + corundum + H_2O (Velde, 1966) being numbered (18). It may be noticed from Fig. 2 that the upper stability of muscovite is terminated by intersection with melting curves. In rocks of granitic composition this intersection occurs at 3.7 Kbar and 655°C (curve (29), Fig. 2) and for bulk compositions on the muscovite + quartz join at 6.3 Kbar and 730°C (curve (27), Fig. 2). At temperatures above these muscovite can coexist with other crystals and a more or less granitic melt.

Muscovite and paragonite coexist in some of the more Al-rich bulk compositions of metamorphosed pelitic rocks, and because Na and K are distributed between these minerals in a regular fashion it has long been known that this pair is a potential geothermometer. The history of experimental study goes back to the work of Eugster and Yoder (1955), and presented here is the result of study by Eugster *et al.* (1972), in Fig. 18. The illustrated solvus has been constructed on the basis of numerous experiments and supported by careful attention to a theoretical model of mica solid solutions. The solvus is thermodynamically consistent and can be used to derive the excess thermodynamic properties of mixing. In particular, the excess Gibbs free energy of mixing is

$$G_{ex} = (3082 + .170T + .087P)N_{Pa}\,N_{Ms}^2 + (4164 + .393T + .126P)N_{Ms}N_{Pa}^2$$

with the critical temperature at 2 Kbar equal to 833°C and $N_{crit} = 0.39_{Ms}$. In these expressions "*N*" stands for mol fraction, *P* for pressure in bars, *T* for temperature, °*K*. In Fig. 18 it will be noted that neither paragonite nor muscovite

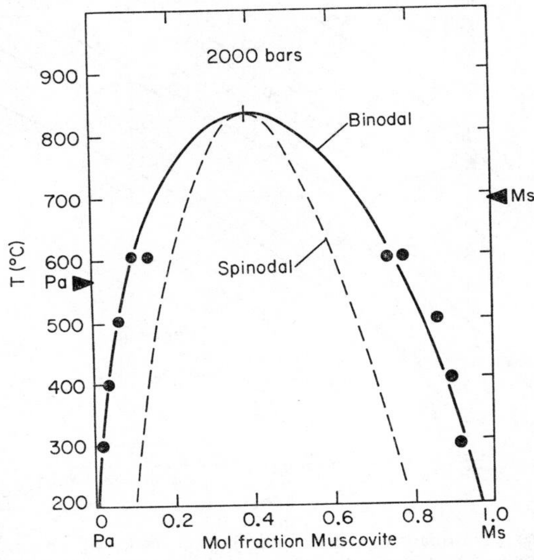

Fig. 18. Paragonite–muscovite solvus after Eugster, Albee, Bence, Thompson and Waldbaum (1972). Inside the binodal loop (solvus) two micas are stable together, while outside a single homogeneous mica is stable. Inside the spinodal the mica phase is unstable with respect to diffusion and will spontaneously unmix to the two stable phases, while between the binodal and spinodal the most stable configuration is a two-phase assemblage but a homogeneous single phase will not spontaneously unmix to the stable composition. The black triangles labelled Pa and Ms represent the upper stabilities of paragonite and muscovite respectively, given by the reactions mica = feldspar + corundum + H_2O. (See Fig. 2, curves (15) and (18) at 2 Kbar.)

is stable up to the temperature of the critical point of the solvus, their breakdown temperatures at 2 Kbar to feldspar + corundum being shown by black triangles. Above the illustrated experimental points the position of the solvus is determined by calculation from the internally consistent thermodynamic parameters.

In a theoretical study of earlier data, Fujii (1966) points out that paragonite takes more Ca into solid solution than muscovite and that consequently the presence of Ca in a natural system will require corrections to be made to the strictly binary solvus before using it as a geothermometer. Further, Munoz and Eugster (1969) show that the micas are very efficient scavengers of fluorine from hydrothermal systems, almost quantitatively removing it from the vapour phase. Because of the great thermal stability of the fluor-micas presence of fluorine in a natural environment will greatly extend the stable field of mica.

(ii) *Trioctahedral micas, phlogopite, phengite, biotite.* Phlogopite, phengite, and celadonite have received study from Wise and Eugster (1964), Wones (1967), Yoder and Eugster (1954), Velde (1965), Seifert and Schreyer (1965), and Wones

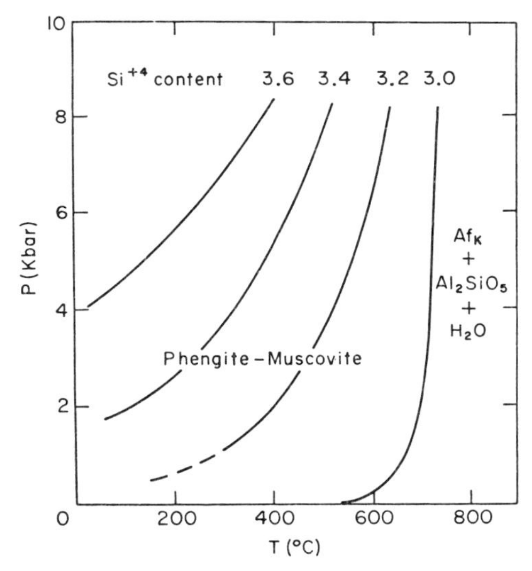

Fig. 19. Si^{+4} content of phengitic muscovite micas as a function of pressure and temperature. The highest temperature curve is the upper stability limit of muscovite in the absence of quartz, and the other curves reflect the higher Si^{+4} contents possible at lower temperatures and higher pressures. After Velde (1967).

and Dodge (1966). The upper stability of phlogopite + quartz is shown in Fig. 2 as curve (27) representing the reaction phlogopite + quartz = K-feldspar + forsterite + H_2O, after Wones and Dodge (1966). As a lower temperature stability limit for either Fe- or Mg-biotite has not been found it seems that some kind of mica should be stable over most of the metamorphic range of conditions, and including the igneous range, in agreement with petrographic experience. The lower stability of coexistence of phlogopitic mica with a muscovite–celadonite solid solution has been approximated at about 350°C on Fig. 2 by combination of the work of Velde (1965) with that of Fawcett (1964). This curve is labelled (approx. K-feld + Mg-Al Chl = Phengite + phlog.).

The phengitic micas have proved difficult to study experimentally due to sluggish kinetics at the low temperatures required. The upper thermal stability of the celadonite end-member of the phengites has been found by Wise and Eugster (1967) to lie at 410°C–430°C at 2 Kbar and to be relatively insensitive to variations in oxygen activity. It thus lies very close to curves (2) and (3) in Fig. 2, and is not plotted for lack of space. The Si^{+4} content of phengite depends upon both pressure and temperature. This dependence is illustrated in Fig. 19 (after Velde, 1967). Muscovite has a minimum Si^{+4} content at its upper stability limit and contains progressively more at lower temperatures and higher pressures

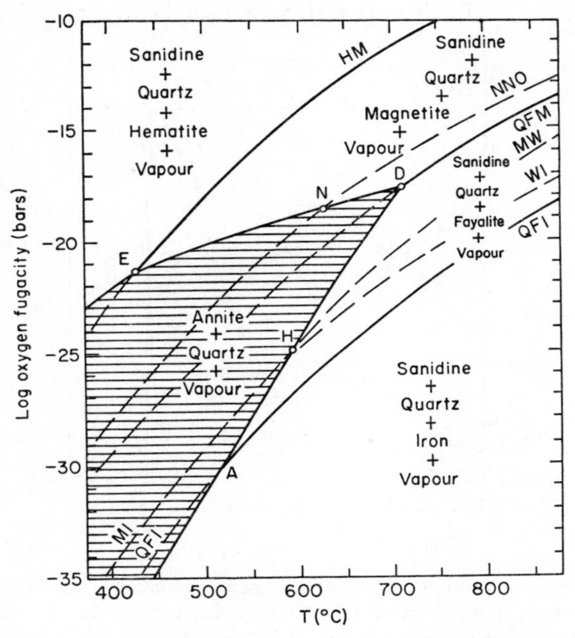

Fig. 20. Phase relations of annite + quartz + vapour as a function of oxygen fugacity and temperature at 2·07 Kbars pressure. After Eugster and Wones (1962). The P–T trajectory of point "D" in this figure is plotted in Fig. 2 as curve (19).

of H_2O, in agreement with compositions of natural phengitic micas. Seifert and Schreyer (1965) have synthesised an unusual mica of the composition $KMg_{2.5}(Si_4O_{10})(OH)_2$, which may form an end-member of the mica solid solution series at high pressures even in the absence of aluminum.

The iron-biotite annite provides the key to making petrologic use of the biotites. This mica, $KFe_3AlSi_3O_{10}(OH)_2$, has been studied by Eugster and Wones (1962), who give in addition a very clear exposition of the theory and practice of making buffered experiments in hydrothermal apparatus. Figure 20 presents their results for annite + quartz. Point "D" in Fig. 20 represents the maximum thermal stability of annite in the presence of quartz at a total pressure of 2.07 Kbar. The locus of all such points, representing the assemblage annite + quartz + sanidine + magnetite + fayalite + vapour, is univariant, and is shown in Fig. 2 as curve (19). It should be noted that the oxygen fugacity along curve (19) is not constant, but buffered by the assemblage quartz + fayalite + magnetite. Wones (1963) has also determined the stability of the ferric iron analogue of annite, ferriannite, $KFe_3{}^{+2}Fe^{+3}Si_3O_{10}(OH)_2$. This end-member has a stability range slightly greater than that of annite. Wones (1963) and Wones

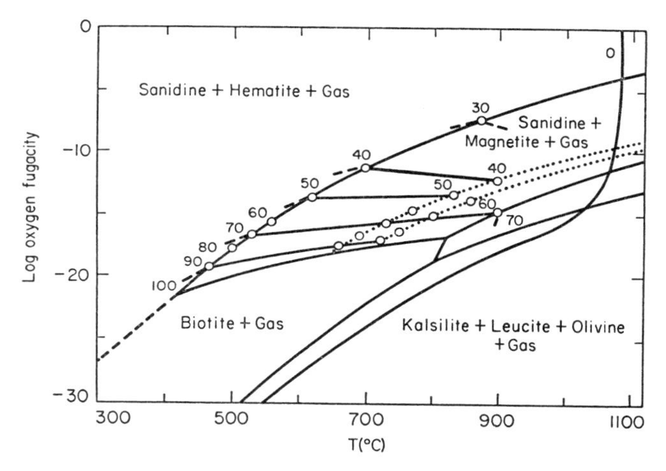

Fig. 21. Equilibrium compositions of the intermediate biotites in terms of oxygen fugacity and temperature at 2.07 Kbars pressure. Numbered contours give Fe/(Fe + Mg) values for the biotites that coexist with K-feldspar + magnetite. The contour marked 100 is the boundary for pure annite and that marked 0 is for pure phlogopite. (After Wones and Eugster, 1965.)

and Eugster (1965) show that biotites synthesised on the join phlogopite–annite are not members of a simple binary solid solution but of a "regular" ternary solution (Guggenheim, 1957, p. 251) of annite, phlogopite, and oxybiotite $KFe_3{}^{+3}AlSi_3O_{12}(H_{-1})$. The latter is a physically unattainable but algebraically convenient "end-member" to use for descriptive and thermodynamic purposes. The Fe/Fe + Mg ratio of these biotites, coexisting with sanidine, magnetite, and gas is sensitive to temperature, pressure, and oxygen fugacity. This dependence is illustrated in Fig. 21.

Practical application of biotite equilibria can be made as follows. Note first that the stability of annite can be written

$$\text{Annite} = \text{sanidine} + \text{magnetite} + H_2,$$

and that in the presence of pure sanidine and magnetite the equilibrium constant is:

$$K = f_{H_2}/a_{Ann}$$

where f_{H_2} = fugacity of H_2 gas, and
a_{Ann} = activity of annite in solid solution.

This can be recast into

$$K = \left\{ \frac{P}{a_{Ann}\left(\frac{1}{\gamma H_2} + \frac{(f_{O_2})^{\frac{1}{2}}}{\gamma H_2O \, . \, K_w}\right)} \right\}$$

where the γ is a fugacity coefficient and K_w is the dissociation constant of H_2O to H_2 and O_2.

At fixed pressure it is thus possible to contour a log f_{O_2} *vs* T diagram in terms of a_{Ann} or composition of biotite and then to use this diagram with the composition of natural biotites coexisting with K-feldspar + magnetite to place one on a univariant curve of f_{O_2}–T, limiting the conditions that were possible for the rock.

Wones and Eugster present equations for calculating the fugacities of hydrogen and of H_2O that can have been at equilibrium with a biotite having a known concentration of annite, in the presence of potassium feldspar and magnetite. These are:

$$\log f_{H_2O} = \frac{3428 - 4212\,(1 - x)^2}{T} + \log x + \tfrac{1}{2} \log f_{O_2} + 8{\cdot}23\,(\pm{\cdot}2)$$

$$\log f_{H_2} = -\frac{9341 + 4212\,(1 - x)^2}{T} + \log x + 11{\cdot}05\,(\pm{\cdot}2)$$

T°K, x = mol fraction annite in biotite.

In addition, they give for the "buffer reaction" annite + sillimanite $+\frac{1}{2}O_2$ = muscovite + magnetite + quartz, the equation:

$$\log_{10} f_{O_2} = -\frac{27000}{T} + 10{\cdot}9\,(\pm 15\%)\ T°K.$$

An application made by the authors indicates that the pressure of H_2O during the Adirondack metamorphism was about three orders of magnitude less than the total pressure, but they deduce on other grounds that the CO_2 pressure could have been approximately equal to the total pressure.

The effect of Na on the equilibrium relations of the intermediate biotites has been studied by Rutherford (1969), who finds that a maximum of about 20 mol % of Na-annite enters into solid solution in mica in the presence of Na-feldspar. As sanidine accepts more albite in solution than annite accepts Na-annite, the net effect is to *reduce* the stability of annite by the addition of Na to the system.

Figure 22 shows some of Rutherford's results, projected on the alkali feldspar binary join. The upper stability of Na–K biotites in the presence of 2 alkali feldspars is shown in Fig. 2 as curve (14).

(8) SiO_2-Al_2O_3-MgO-FeO-K_2O-H_2O

This system, which provides a good model for equilibria in meta-pelites, has received considerable study both from the theoretical standpoint (Thompson, 1957; Albee, 1965) and the experimental standpoint (Hoschek, 1969, 1967; Seifert, 1970; Ganguly, 1969), and innumerable accounts have appeared of natural phase assemblages, among which should be noted Carmichael (1970), Chinner (1960, 1966a, 1966b), Card (1964), Evans and Guidotti (1966), Hollister (1966), Albee (1965a, 1965b), Guidotti (1968, 1963), Zen and Albee (1964), Kretz (1964), Sen and Chakraborty (1968), and Thompson (1957).

Seifert (1970) has determined the stability relations between cordierite, muscovite, chlorite, and aluminosilicate in the system K_2O-MgO-Al_2O_3-SiO_2-H_2O. These results are presented in Fig. 2, as the curves emanating from the invariant point at 6.5 Kbar, 640°C, the curve for the reaction Mg-chlorite + muscovite + quartz = Mg-cordierite + phlogopite + H_2O being numbered (7). It may be seen that Mg-cordierite is restricted to pressures below 6.5 Kbar in the presence of muscovite and quartz. These limits may be compared on the same figure with the stability range of Fe-cordierite, which is similarly limited in pressure, but to a pressure lower than 3.5 Kbar. The upper pressure limit of natural cordierites must lie between these extremes.

Hoschek (1969) has investigated the stability of staurolite and chloritoid in a system approximately within the six-component pelite system, determining the positions of the reactions staurolite + muscovite + quartz = Al_2SiO_5 + biotite + vapour and chlorite + muscovite = staurolite + biotite + quartz + vapour. Ganguly (1968) has deduced that the reaction chloritoid + muscovite + quartz = staurolite + biotite + magnetite + H_2O must lie between the reactions Fe-chloritoid + Al_2SiO_5 = Fe-staurolite + quartz + magnetite and Fe-chloritoid + quartz + O_2 = Fe-staurolite + almandine + magnetite. This is in good agreement with the work of Hoschek (1969) and the boundary according to Hoschek is shown as curve (12) in Fig. 2. Curve (10), Fig. 2, is the reaction staurolite + muscovite + quartz = biotite + Al_2SiO_5 + H_2O + magnetite (after Hoschek, 1969) and is in agreement with the deduction of Ganguly (1968) that this curve should lie at lower temperatures than the reaction Fe-staurolite + quartz + O_2 = almandine + Al_2SiO_5 + magnetite + H_2O. An interesting application of phase equilibrium results to natural assemblages was attempted in 1964 by Schreyer *et al.*, before most of these data were available on staurolite, chloritoid, and Fe-cordierite. The latest data substantiate their conclusion that the rocks at Bodenmais, Germany, crystallised at a pressure of 4.2 $\pm$ .5 Kbar and 710°C $\pm$ 25°C.

(b) THE INFLUENCE OF OTHER COMPONENTS

In spite of the complexity of the diagrams and results so far presented, we have examined systems that are much simpler than natural rocks, and it is of interest to reflect on the effect of adding still more components to the systems covered.

Solution theory such as applied to the micas, with experimental verification is the best way to deal with this difficulty, and at the present state of the art we must proceed with theory even in the absence of direct experimental support.

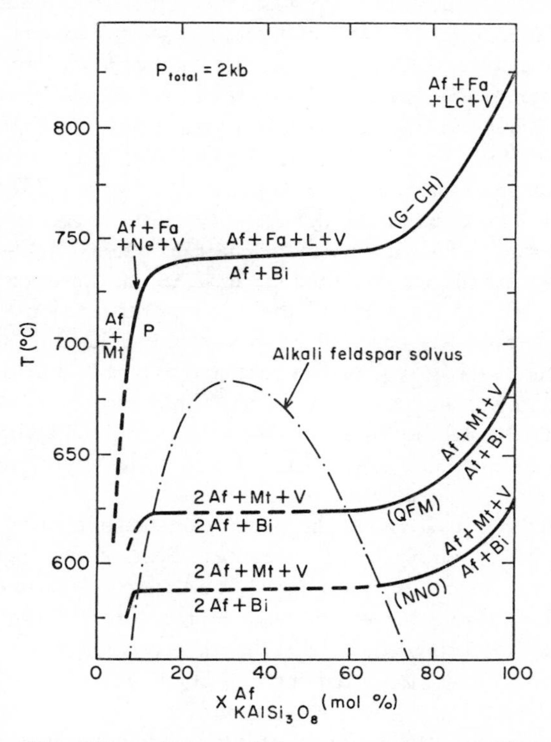

Fig. 22. Stability of Na–K iron biotites at different oxygen fugacities defined by the NNO, QFM, and Graphite–methane (G–CH) buffers, projected on the binary alkali feldspar join. (After Rutherford, 1969.)

Such attempts to treat the effects of extra components cannot be considered accurate but they are better than application of unmodified experimental results on pure systems. We cannot explore this subject in detail here but it is possible at least to note some of the experimental studies that bear on the extension of the "pelite system" to more complex and more realistic compositions.

The influence of Na on the biotite equilibria has been studied by Rutherford. (See Fig. 22 and Fig. 2, curve (14).) The effect of Mn on equilibria involving garnet can be evaluated from the work of Hsu (1968) in which he reports the equilibrium position of the reaction Mn-chlorite + quartz = spessartite + H_2O. This reaction is shown as curve (2) on Fig. 2. The stability of spessartite is practically independent of oxygen fugacity, in contrast to that of almandine. Comparison of curves (2) and (11) of Fig. 2 shows that spessartite becomes stable at a much lower temperature than almandine, in accord with observations from rocks, where it is commonly observed that the first-formed garnets are rich in Mn, becoming richer in Fe as crystallisation progresses. This is probably due to a combination of their relative stabilities as shown in Fig. 2 and Rayleigh fractionation (Hollister, 1966), in which the immediate surroundings are

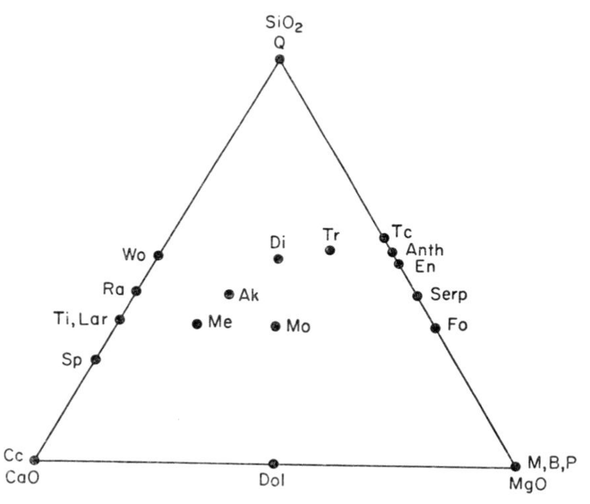

Fig. 23. The phases of metamorphosed Al-free siliceous carbonate rocks. Mol percent. H_2O and C O_2 content not represented.

depleted in Mn by the growing garnet, resulting in zoned crystals. It might be further remarked that it is impossible to obtain garnet stably at temperatures less than curve 2 unless the water pressure is less than the total pressure. This may be accomplished either by contamination with another gas, such as CO_2, CO, or CH_4 or by having the system open to H_2O at an equilibrium pressure less than total pressure. The CH_4 can originate by reaction with graphite at low temperatures, and may be a significant factor in the lower grades of metamorphism.

The addition of Ca to the "pelite system" results in appearance of plagioclase, epidote, zoisite, and solid solution of Ca in other minerals notably garnet and mica. Even in the absence of CO_2 these minerals complicate the phase relations, and unfortunately experimental testing of these effects is unavailable at present.

6. EQUILIBRIA IN SYSTEMS CONTAINING CARBONATES

The metamorphism of carbonate-bearing rocks introduces a new complication, that of a fluid phase made up of major proportions of more than one volatile component. The account of iron-bearing systems at equilibrium with H-O gases touched on the subject, but at all but the most reducing conditions and at the lowest temperatures the gas phase is mostly H_2O in such systems. This is not true in carbonate-bearing systems, where the dilution of reacting volatile components by one another becomes an important cause of displaced equilibrium. Most of the minerals can be represented graphically in terms of a projection onto the triangle CaO–MgO–SiO_2 (Fig. 23), with the fugacities of H_2O and CO_2 represented in other ways, presented later. There follows first an account of carbonate equilibria in the absence of silicates, followed by dehydration equilibria in the absence of CO_2, decarbonation equilibria in the absence of H_2O, and finally the combination of both dehydration and decarbonation reactions in mixtures of H_2O and CO_2.

(a) CARBONATES

The carbonates of Ca, Mg, Mn, Fe, Ba, Sr, and Zn have been studied experimentally. Of these, Ba, Sr, and Zn carbonates are not common in metamorphic rocks although important in some ore deposits and veins. For details refer to Chang (1965, Ba–Sr).

(1) $CaCO_3$-$MgCO_3$

Limited solid solution between calcite and dolomite provides a potential geothermometer. As with all such "solvus temperatures" the temperature that can be deduced is the last temperature of equilibrium, which may in the case of carbonates be well below the maximum temperature that was reached. Figure 24(a) (after Goldsmith and Heard, 1961) shows the range of solid solutions in the Ca-Mg carbonates. The calcite–dolomite solvus is not strongly pressure-dependent, being +0.12 mol % $MgCO_3$ in calcite per Kbar (Goldsmith and Newton, 1969).

(2) $CaCO_3$-$FeCO_3$ *and* $CaCO_3$-$MnCO_3$

This join is important to the study of metamorphosed iron formations where calcite–siderite solid solutions enter into reactions with silicates, forming calcite plus grunerite and minnesotaite as important minerals. The calcite–siderite solvus is illustrated in Fig. 24(b). (Rosenberg, 1963a, 1963b.)

The solid solution relations of the Ca–Mn carbonates are shown in Fig. 24(c) Goldsmith and Graf, 1957). There it may be seen that calcite solid solutions

---

Fig. 24. Solid solution relations in the carbonates of Ca, Mg, Fe and Mn.
(a) Ca–Mg carbonates, after Goldsmith and Heard, 1961.
(b) Ca–Fe carbonates, after Rosenberg, 1963a, 1963b.
(c) Ca–Mn carbonates, after Goldsmith and Graf, 1957.
(d) The ternary Ca–Fe–Mg carbonates at 450°C after Rosenberg, 1967. Tie lines are diagrammatic only, added by H. J. Greenwood.

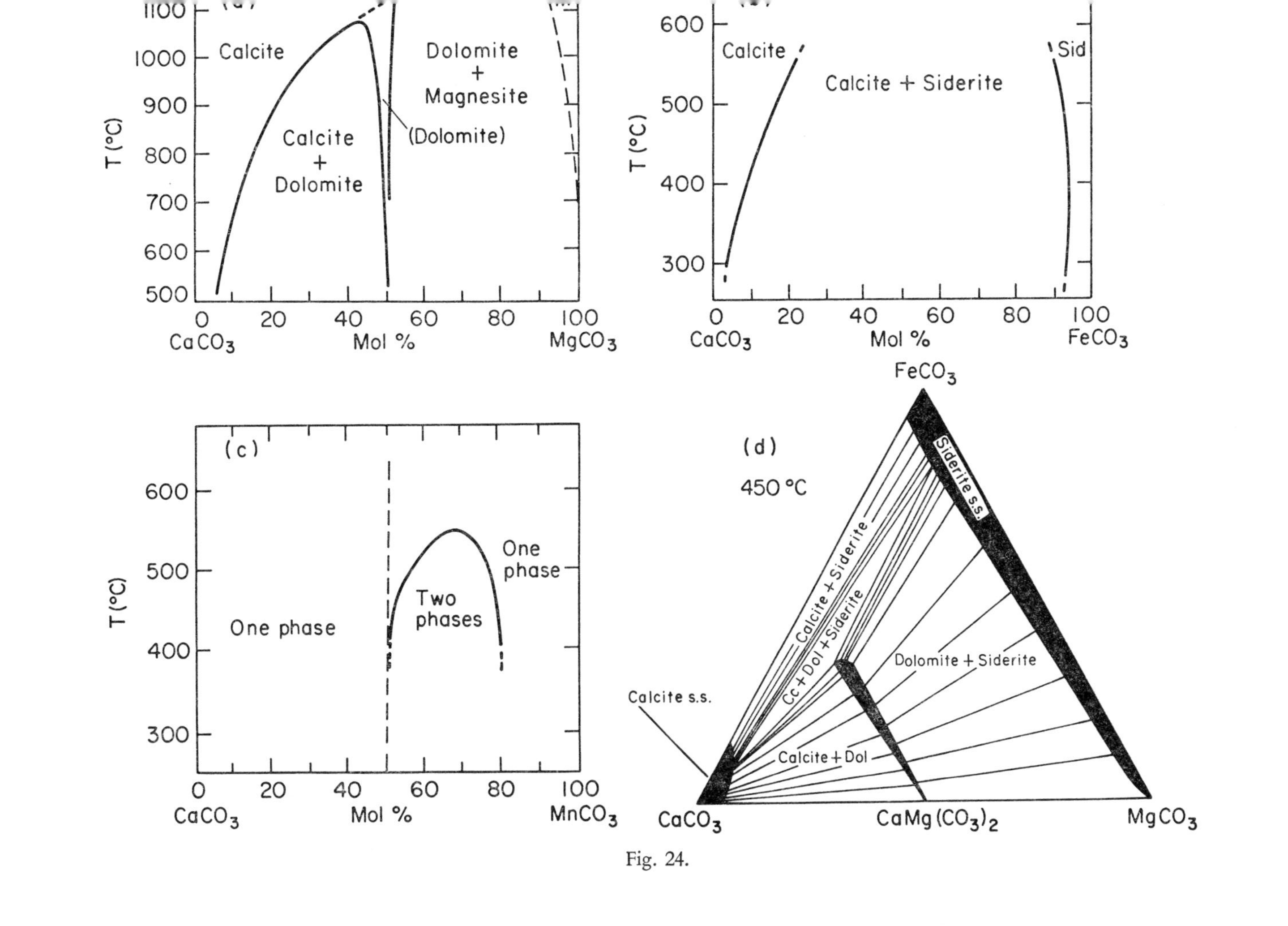

T (°C)
Calcite
Calcite + Dolomite
Dolomite + Magnesite
(Dolomite)
CaCO3
Mol %
MgCO3
Calcite
Calcite + Siderite
Sid
FeCO3
(c)
One phase
Two phases
One phase
MnCO3
(d)
450 °C
Siderite s.s.
Calcite + Siderite
Cc + Dol + Siderite
Dolomite + Siderite
Calcite s.s.
Calcite + Dol
CaMg(CO3)2

Fig. 24.

extend as far as the 50 mol % $MnCO_3$ point before a separate Mn-rich carbonate phase appears. This may have some very pronounced effects both on reactions involving Mn-carbonate and those involving calcite.

An isobaric isothermal (450°C) section through the Ca-Mg-Fe carbonate system is shown in Fig. 24(d) (Rosenberg, 1967). The siderite–magnesite solid solution is complete, the dolomite–ankerite solid solution is extensive but not complete, and the calcite phase is fairly restricted. The tie-lines illustrated are diagrammatic and have not been chemically determined. This system would seem to be a very useful one as a potential geothermometer but Rosenberg

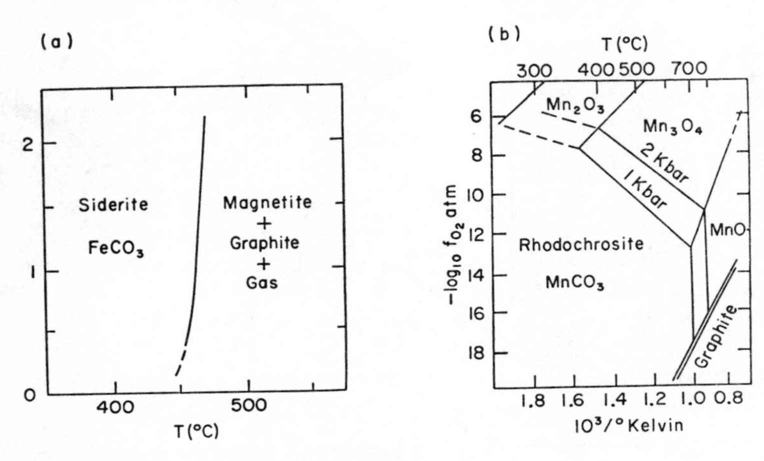

Fig. 25. Upper stabilities of siderite and rhodochrosite.
(a) Stability of siderite in the presence of graphite + gas, after French and Eugster, 1965.
(b) Stability of rhodochrosite in terms of f_{O_2}, temperature, and total gas pressure. At oxygen fugacities less than the double-line labelled "graphite" no vapour phase is stable. (Huebner, 1969.)

(1968) remarks, and reference to Fig. 24(c) will confirm, that $MnCO_3$ is so soluble in calcite that the presence of Mn in a natural system will make a large change in any solvus involving calcite.

Siderite ($FeCO_3$) and rhodochrosite ($MnCO_3$) have stability fields limited not only by the available pressure of CO_2 but also by the oxidation state of the system. Figure 25(a) shows the upper stability of pure siderite in the presence of graphite + gas. (French and Eugster, 1965.) Solid solution of calcite in siderite stabilises the Fe-phase to higher temperatures. (See Fig. 24(b). Figure 25(b) shows the stability of rhodochrosite in terms of oxygen fugacity, temperature, and total gas pressure (Huebner, 1969). As hausmannite is the breakdown product stable over greatest range of f_{O_2}–T, it is not surprising that it is the commonest reaction product found in nature.

(b) CARBONATES IN THE PRESENCE OF H_2O

The addition of H_2O to the carbonate-bearing systems produces melting at modest pressures and temperatures (Wyllie and Tuttle, 1960; Wyllie and Haas, 1965; Wyllie, 1965; Walter, Wyllie and Tuttle, 1962; Franz and Wyllie, 1966). In addition to liquid and gas, $Ca(OH)_2$ and $Mg(OH)_2$ become possible phases. Melting is possible in the presence of a water-rich gas phase at 1 Kbar pressure and temperatures as low as 600°C to 625°C. The solids at equilibrium with liquids include $CaCO_3$, $Ca(OH)_2$, and in the SiO_2-bearing systems, spurrite. It is clear that the possibility of melting during the metamorphism of carbonate rocks must be entertained.

(c) DECARBONATION REACTIONS IN THE ABSENCE OF H_2O

Bowen's (1940) analysis of the progressive metamorphism of siliceous carbonates was predicated on progressive decarbonation, neglecting the progressive dehydration that can accompany such metamorphism. Although we must eventually consider the role of H_2O it is useful at this point to restrict our attention to the H_2O-absent reactions. These are illustrated in Fig. 26, where each curve sets upper temperature limits for each univariant assemblage. These limits are reached only in the presence of free pure CO_2. Dilution of the CO_2 by contaminants such as CH_4 or H_2, or reduction of the ratio of equilibrium pressure of CO_2 to total pressure by osmotic equilibria, will result in lowering of the temperatures given for the curves. In the case of dilution by non-reacting gases the assumption of ideal mixing in the gas phase may be made to estimate the effect, probably without exceeding an error in the displacement of more than 5% or at the most 10%. If however, as is usually the case, CO_2 is diluted with H_2O, equilibria are displaced only until some part of the assemblage becomes metastable due to possible reactions with H_2O. Such reactions are presented later, after examination of simple dehydration reactions.

In Fig. 26 certain points may be noted. Reaction 1, giving the first appearance of diopside, is possible only in very H_2O-poor fluids, hence if petrographic evidence can be found that diopside appeared by this reaction then curve (1) can be used with confidence to set limits on the conditions. Of course dilution of CO_2 by other materials will have to be accounted for. Reactions (3) and (4) cannot both be stable in the relative positions shown, because reaction (3) involves forsterite which is made metastable by reaction (4). Hence the position of one must be in error. Reaction (3) is consistent with five other independent equilibria in the system CaO-MgO-SiO_2-H_2O-CO_2 (Skippen, 1972) and at this writing it seems that reaction (4) probably belongs at a temperature lower than reaction (3). Further experiments are needed.

Sphene is a common constituent of impure metamorphosed carbonate rocks, whereas the assemblage calcite + quartz + rutile is rare. Equilibrium conditions for this reaction have been determined by Schuiling and Vink (1967) (Fig. 26, curve (5)). Applying their data to the occurrence of sphene in the Adirondacks near Emeryville, New York (Engel and Engel, 1958, 1962a, 1962b, 1964) and using a temperature of about 625°C (Fe–Ti–O; Buddington and Lindsley, 1964; Palmer, 1970), we reach the conclusion that the CO_2 pressure was about 5 Kbar. This is consistent with estimates by Engel and Engel of total pressure and is in the same range as calculated for CO_2 by Wones and Eugster (1965).

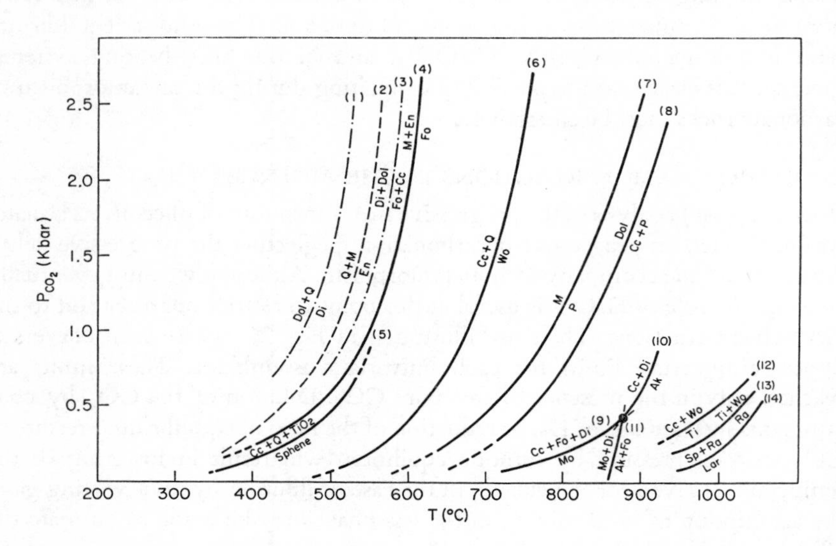

Fig. 26. Decarbonation reactions in the absence of H_2O. Solid curves: experimentally determined by direct experiment; dash–dot curves: calculated from experimental data on closely related reactions; dashes: calculated from standard thermochemical data. Reactions; Abbreviations as in Fig. 23.

(1) Dol + 2Q = Di + $2CO_2$	Skippen, 1972
(2) M + Q = En + CO_2	Greenwood, 1967
(3) Di + 3Dol = 2Fo + 4Cc + $2CO_2$	Skippen, 1972
(4) M + En = Fo + CO_2	Harker and Tuttle, 1955
(5) Cc + Q + Rutile = Sphene + CO_2	Schuiling and Vink, 1967
(6) Cc + Q = Wo + CO_2	Harker and Tuttle, 1956
(7) M = P + CO_2	Harker and Tuttle, 1955
(8) Dol = Cc + P + CO_2	Harker and Tuttle, 1955
(9) 2Cc + Fo + Di = Mo + $2CO_2$	Walter, 1963
(10) Cc + Di = Ak + CO_2	Walter, 1963
(11) Mo + Di = 2Ak + Fo	Walter, 1963
(12) 2Cc + 2Wo = Ti	Zharikov and Shmulovich, 1969
(13) Ti + 2Wo = 2Ra + $2CO_2$	Zharikov and Shmulovich, 1969
(14) Sp + Ra = 4Lar + CO_2	Zharikov and Shmulovich, 1969

Note: Reactions (3) and (4) cannot both be stable in the relative positions shown The good consistency of (3) with other related reactions suggests that the stable location of (4) lies at temperatures *below* (3).

As a corollary, P_{H_2O} must have been low, in agreement with the estimates of Greenwood (1963) and Wones and Eugster (1965).

The minerals akermanite, monticellite, tilleyite, spurrite, and larnite are uncommon minerals of the highest temperature parts of contact metamorphic aureoles. Their equilibrium relations, in the presence of pure CO_2, have been studied by Harker (1959), Walter (1963), and Zharikov and Shmulovich (1969). Walter's results are presented in Fig. 2, curves (9), (10) and (11), and Zharikov and Shmulovich's results on tilleyite, rankinite, larnite, and spurrite as curves (12), (13), and (14). In agreement with petrographic experience, all these minerals belong to very high-temperature parageneses.

(d) DEHYDRATION REACTIONS IN THE ABSENCE OF CO_2

All of the hydrous minerals of metamorphosed carbonates must be stable under some conditions in the presence of pure H_2O. Some of these limiting equilibria are illustrated in Fig. 27. As with Fig. 26, some of the reactions require special mention. The serpentine reactions (1), (2), and (3) are fixed in this diagram by the work of Johannes (1968), Scarfe and Wyllie (1967), and Evans and Trommsdorff (1970, calculation). Earlier work of Bowen and Tuttle (1949) places the reactions nearly 100°C higher, at temperatures more closely in correspondence with natural occurrence. Evans and Trommsdorff (1970) suggest that the explanation may lie in the fact that the serpentine polymorph in regionally metamorphosed ultramafics is antigorite, while the experimental results are for chrysotile. They deduce that antigorite is stable to higher temperatures than shown in Fig. 27 for the chrysotile polymorph. The two dashed tremolite reactions illustrated are both the result of calculation and are consistent with the other curves figured. The reactions involving anthophyllite may possibly be metastable with respect to the reaction talc = 3enstatite + quartz + H_2O, but existing data at present are not adequate to resolve the problem (Zen, 1971; Greenwood, 1971).

(1) CaO-SiO_2-H_2O *and* ZnO-SiO_2-H_2O

In addition to the reactions shown in Fig. 27 the low temperature reactions in the Ca-bearing system have been studied by Buckner, Roy and Roy (1960a), and Harker (1964), and summarised by Roy and Harker (1962). The unfamiliar minerals okenite, nekoite, gyrolite, tobermorite, afwillite, hillebrandite, reyerite, xonotlite, and foshagite all break down in the presence of pure H_2O at temperatures below 400°C. The higher temperatures of most related decarbonation reactions cause the above minerals to be stable only at low temperatures and in the presence of extremely H_2O-rich solutions or vapours. Less than one or two mol % CO_2 in the vapour phase is sufficient to convert all of these minerals to compositionally equivalent carbonate assemblages, providing the explanation for their rare natural occurrence. By this reasoning xonotlite should be, and is, the most common of them.

The zinc minerals sauconite, hemimorphite, and willemite have been studied by Roy and Mumpton (1956) who report that the hydrate hemimorphite reacts to willemite + H_2O at 250°C at a pressure of 1 Kbar. Hemimorphite is thus restricted to low-temperature, hydrous environments.

An interesting feature of Fig. 27 is the low-temperature disappearance of diopside (with serpentine) to form tremolite and forsterite, and the high-temperature reappearance of diopside at the expense of tremolite. Naturally both reactions cannot be exhibited by the same bulk composition.

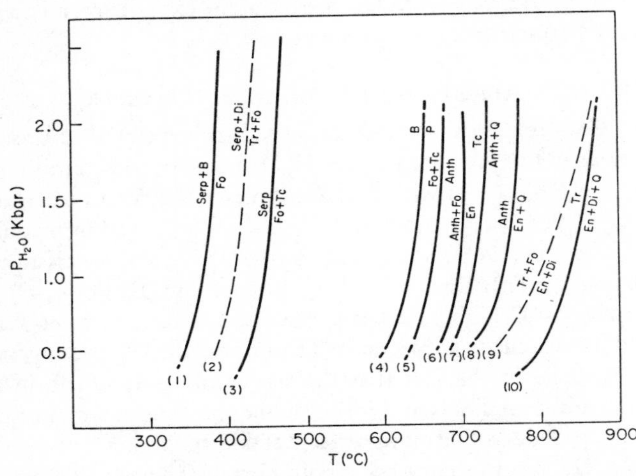

Fig. 27. Dehydration reactions in the absence of CO_2. Solid curves: experimentally determined; dashed curves: calculated from other equilibria, (2) by Evans and Trommsdorff, 1970 and (9) by Skippen, 1972. Reactions, abbreviations as in Fig. 23.

(1) Serp + B = 2Fo + $3H_2O$	Johannes, 1968
(2) 5Serp + 2Di = 6Fo + Tr + $9H_2O$	Evans and Trommsdorff, 1970
(3) 5Serp = 6Fo + Tc + $9H_2O$	Scarfe and Wyllie, 1967
(4) B = P + H_2O	Barnes and Ernst, 1963
(5) 4Fo + 9Tc = 5Anth + $4H_2O$	Greenwood, 1963
(6) Anth + Fo = 9En + H_2O	Greenwood, 1963
(7) 7Tc = 3Anth + 4Q + $2H_2O$	Greenwood, 1963
(8) Anth = 7En + Q + H_2O	Greenwood, 1963
(9) Tr + Fo = 5En + 2Di + H_2O	Skippen, 1972
(10) Tr = 2Di + 3En + Q + H_2O See Fig. 2, curve (23).	Boyd, 1959

Note: It is possible that the curves involving Anth are metastable with respect to Tc = En + Q + H_2O, but existing data are not yet conclusive (Zen, 1971; Greenwood, 1971).

The equilibria presented in Figs 26 and 27 are the limiting reactions for equilibria that occur in the metamorphism of carbonate rocks in the presence of mixtures of H_2O and CO_2. These mixed reactions can be regarded as combinations of the limit-reactions, and will be treated next.

(e) DECARBONATION AND DEHYDRATION REACTIONS IN H_2O-CO_2 MIXTURES

Combining subsystems so as to include H_2O and CO_2 in the gas phase permits a closer approach to petrologic reality and results in numerous reactions impossible in the one-gas-component sub-systems. These reactions can be illustrated in a variety of ways, the most-used of which is the isobaric T-X diagram which plots the proportions of H_2O and CO_2 coexisting with various assemblages versus temperature. The stoichiometry of the reactions possible permit of six types of reaction curve, illustrated in Fig. 28.

Dehydration equilibria can in general extend only part way into a H_2O–CO_2 diagram for the obvious reason that at some point one or other of the reactants

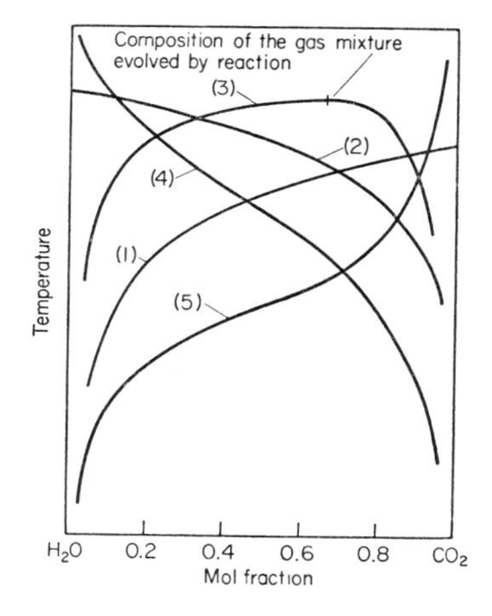

Fig. 28. Shapes of reaction boundaries possible in isobaric T–X sections, in which the proportions of CO_2–H_2O (X) are plotted against temperature at constant gas pressure. (After Greenwood, 1967.)

(1) Reaction that evolves only CO_2. Shape is due to dilution of CO_2 by H_2O.
(2) Reaction that evolves only H_2O. Shape is due to dilution of H_2O by CO_2.
(3) Reaction that evolves both H_2O and CO_2. Such curves pass through a temperature maximum at a gas composition equivalent to the composition of the mixture evolved.
(4) and (5). Reactions that evolve one volatile component and consume the other. Examples of these reactions are found in Fig. 29.

or products will be able to react with the carbon dioxide to form a carbonate, terminating the reaction at an invariant point. A similar statement can be made about the simple decarbonation reactions. Originating from these invariant points are reactions that involve both CO_2 and H_2O, and lead to other invariant points.

(1) *The system* MgO-SiO_2-H_2O-CO_2

This system has been studied in the laboratory by Greenwood (1967) and Johannes (1966a, 1966b, 1967, 1969) over a total range of about 500°C and 7 Kbar. The stable reactions possible in this system under total pressure conditions permitting the stability of anthophyllite are diagramatically shown in Fig. 29. It will be noted that the simple bounding reactions from the H_2O-free and the CO_2-free subsystems cross the T-X field from either side until rendered metastable by intersection with one another. The invariant points so generated give rise to other reactions, most of which use both CO_2 and H_2O as reactants, which in turn cross the T-X field until rendered metastable by other equilibria. Some of the reactions cross without giving rise to invariant points. These are referred to as "indifferent" intersections because the minerals involved occupy non-intersecting parts of the composition diagram and hence have no possible reactions in common.

The topology of Fig. 29 is echoed in Fig. 30, which presents the corresponding 2 Kbar section of the same system. The boundaries here are based on data sources noted in the figure caption, and the parts of the diagram most affected can be seen by reference back to Fig. 29. In Fig. 29 the heavy parts of the reaction boundaries illustrate where the experimental data exist and in the caption of Fig. 30 the sources of the data are acknowledged. The important features to note in Fig. 30 are that serpentine, with or without magnesite, is stable only in fluids very poor in CO_2. (See the positions of the isobaric invariant points MSTQ, MFSB, and MFST) and that anthophyllite + magnesite and enstatite + magnesite are stable only in fluids poor in H_2O and rich in CO_2 (see invariant points MAFE, MFAT, MATQ, and MAQE). Other features worth noting are that the assemblages M + Q, M + T, and M + F are stable over a very wide range of gas compositions and are separated from one another by reaction-boundaries that are more nearly isothermal than the others in this system.

The dependence of reaction temperatures upon gas composition renders these T-X isobaric curves unsusceptible to simple interpretation in rocks. However, provided the proportion of gas to minerals is not too high it is possible for the mineralogy to control the gas composition rather than the converse. A consequence is that the rock may buffer the gas composition by mineralogic reactions until one of the invariant points is reached. At such a point further introduction of heat cannot raise the temperature without also raising the pressure, which must follow the polybaric trace of the isobaric invariant point. Such curves are truly univariant and should serve as the proper boundaries for isograds, rather than T-X curves, which have one more degree of freedom. Such univariant curves, based on the work of Johannes (1969) are presented in Fig. 31.

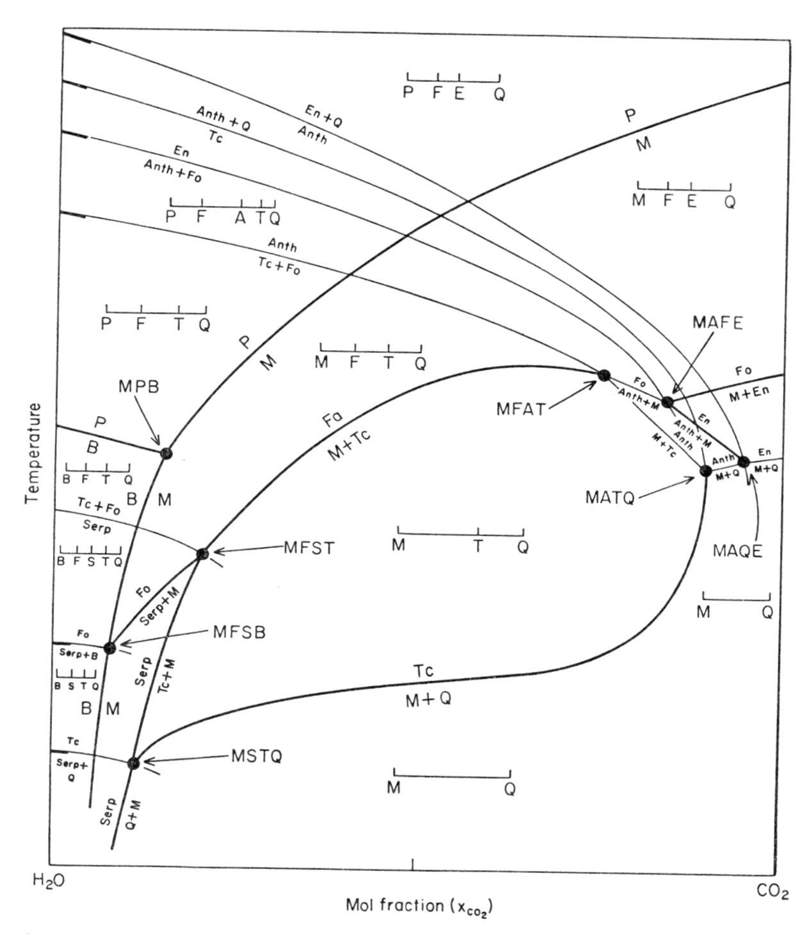

Fig. 29. Isobaric temperature–gas composition (T–X) diagram for the system MgO-SiO_2-H_2O-CO_2 showing all the reactions stable in the system, consistent with the curves shown in Figs. 26 and 27 for the Ca-free system. Isobaric invariant assemblages are indicated by the mineral phases stable together, thus: MPB, MFST, MFSB, MSTQ, MAFE, MFAT, MATQ, MAQE. Heavy lines indicate where experimental data are available and light lines the locations of boundaries fixed by the intersections with experimentally determined reactions. All the types of reaction curve illustrated in Fig. 28 are represented. For example: the brucite dehydration is of type (2), Fig. 28; the magnesite decarbonation is of type (1), Fig. 28; the reaction of magnesite + talc to forsterite + gas is of type (3); and the reaction of magnesite + quartz to talc is of type (5). Abbreviations: P = periclase, M = magnesite, B = brucite, Fo = forsterite, En = enstatite, Anth = anthopyllite, Tc = talc, Serp = serpentine, Q = quartz. Initial letters only used for invariant points and compatibility bars (in which phases are ordered by Mg: Si).

The phase relations in this system have been applied with great success to the metamorphism of ultramafic rocks by Evans and Trommsdorff (1970) and Trommsdorff and Evans (1972) who have been able to deduce temperatures and gas compositions for a wide range of assemblages, in addition to suggesting

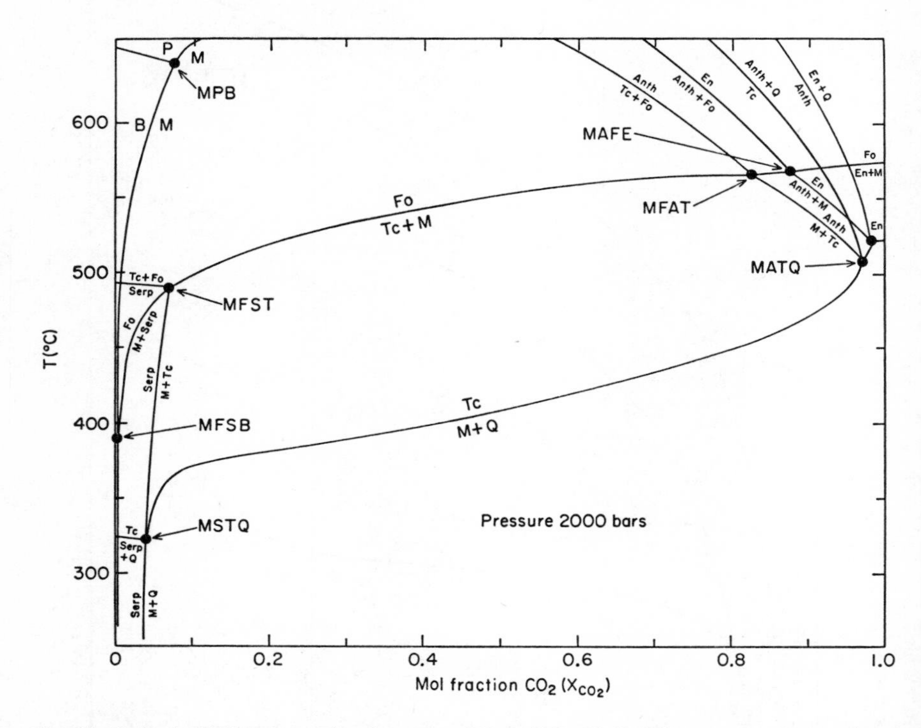

Fig. 30. Isobaric T-X diagram for the system MgO-SiO_2-H_2O-CO_2 at a constant total gas pressure of 2000 bars. Based on the work of Greenwood (1963, anthophyllite reactions in pure H_2O; 1967, Tc + M = Fo at 2000 bars); Walter, Wyllie and Tuttle, (1962, M = P, B = P, B = M); Johannes (1966a, 1966b, 1967, 1969, Q + M = S, Q + M = Tc, Tc + M = Serp, Serp + M = Fo, M + Tc = Fo, Anth + M = En, over a pressure range of 0.5–7 Kbars). Abbreviations as in Fig. 29. Note that the eight isobaric invariant points are confined fairly closely to the margins of the diagram at this pressure.

some alternative topologies for the T-X diagram at different conditions of total pressure. Examination of these papers is recommended.

(f) THE SYSTEM CaO-MgO-SiO_2-H_2O-CO_2

The addition of Ca to the system last discussed permits a close approximation to the phase relations in metamorphosed siliceous dolomites and magnesian limestones. Experimental studies have been reported by Metz and Winkler (1963,

1964, 1965), Metz (1967), Greenwood (1967), Gordon and Greenwood (1970), and Skippen (1972). The shapes of reaction boundaries obey the same rules as those in the Ca-free system and need not be repeated here. Metz and Trommsdorff (1968) summarise pre-existing data on the system and make applications to assemblages in the Swiss Alps.

Figure 32 illustrates reactions in the system that involve calcite. Other reac-

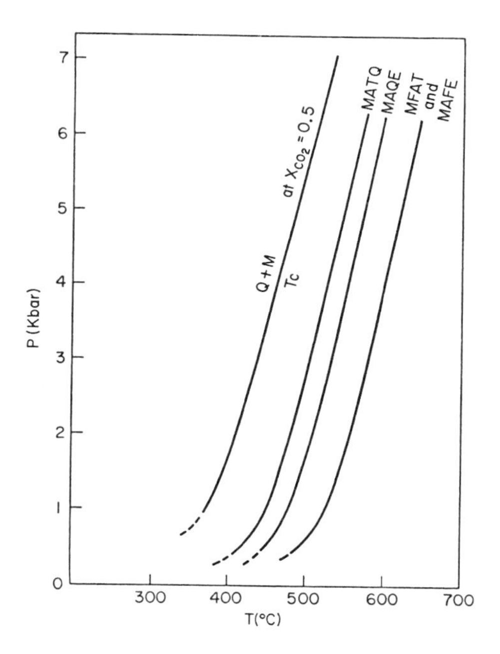

Fig. 31. Pressure–temperature projection of some of the isobaric invariant points from Fig. 30. (These are polybarically univariant.) Pressure variation based on the work of Johannes (1966a, 1966b, 1967, 1969). The lowest temperature curve is univariant only because of the restriction to constant vapour composition of $X_{CO_2} = 0.5$, while the others are strictly univariant. The highest temperature curve represents the positions of two different isobarically invariant assemblages, MFAT and MAFE, because these two are too close together to distinguish at this scale of plotting.

tions, involving one or more of the phases enstatite, magnesite, and anthophyllite have been omitted due to uncertainty about the relative positions of some of the bounding curves. Those that remain are presented on the basis of Skippen's (1972) work, with supporting data from Greenwood (1967) and Gordon and Greenwood (1970). The earlier work of Metz and Winkler on many of the same reactions is in disagreement with Skippen on the location of the (Tc + Cc + Q + Tr + Di) invariant point. Skippen's result is favoured here because of the excellent thermodynamic consistency between all his reactions. Some difficulties still remain, however, as was noted in connection with Fig. 26, where the experimentally determined lower stability of forsterite in the presence of CO_2 lies at a higher temperature than reaction (3) of Skippen, Di + 3Dol = 2Fo + 4Cc + $2CO_2$. (See curves (3) and (4) in Fig. 26.)

Skippen shows that for the phases Gas, Q, Cc, Dol, Tc, Tr, Di, Fo, En, in

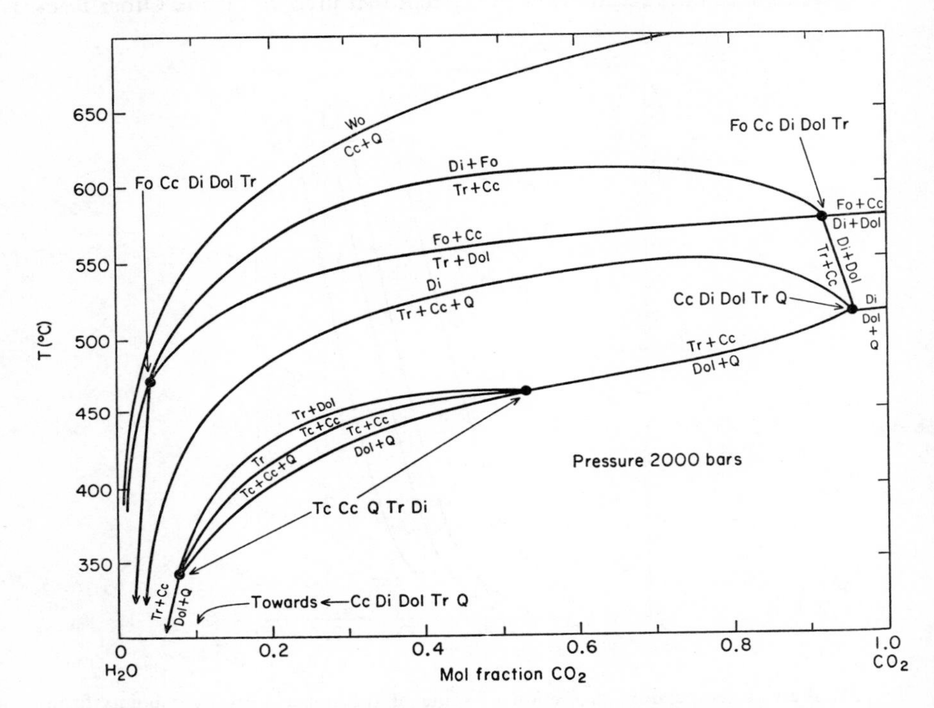

Fig. 32. Isobaric (2 Kbars) T-X diagram for reactions involving calcite in the system CaO-MgO-SiO_2-H_2O-CO_2. Invariant points are referred to by the symbols of the phases *present*, rather than by those absent. Invariant points involving the phases magnesite, enstatite, and anthophyllite are omitted due to uncertainty in the relative positions of some of the bounding curves. Data as follows: with abbreviations as in Fig. 23.

Cc + Q = Wo + CO_2 Greenwood, 1967

3Dol + 4Q + H_2O = Tc + 3Cc + $3CO_2$ Gordon and Greenwood, 1970; Skippen, 1972.

All other reactions, Skippen, 1972.

this 5-component system, exactly 49 reactions can be written between the phases, of which only five can be independent. Of the 49 he deduces that only 25 can be stable under geological conditions due to the metastability of the three assemblages Fo + Q, En + Cc, and Di + Tc. Of the 25 stable reactions five independent ones have been chosen, experimentally determined, and used to calculate the remaining 20. Only ten of the 25 are shown in Fig. 32.

Examination of Fig. 32 will reveal that each of the isobaric invariant points is duplicated on the diagram. Thus there are two with labels (FoCcDiDolTr) and two with the label (TcCcQTrDi). This duplication is not inconsistent with the phase rule requirements for invariance because the assemblages are only *isobarically* invariant. These duplicated points are connected to one another by truly univariant lines in P–T–X space, and for example the points (TcCcQTrDi) approach one another at lower pressures, annihilating the field of Tc + Cc + Tr at pressures below about 1 Kbar.

Application of such diagrams has been made by Trommsdorff (1966, 1968), who notes that the general progression of calc-silicate assemblages in the Bernina–Simplon area of the Alps is talc-tremolite-diopside-forsterite-wollastonite, consistent either with external or internal control of the fluid phase composition and progressively increasing temperature. Greenwood (1967) applies the data on wollastonite stability to a metamorphic aureole, showing that the isograd is controlled by a combination of simple conductive heat-flow and contamination of the CO_2 by H_2O near fissures. Melson shows that in the Granite Peak aureole the mineralogy is strongly controlled by the availability of fissures for the escape of gaseous reaction products. Remote from the fissures high CO_2 concentrations prevailed inhibiting the decarbonation reactions, while within a few centimetres of fissures the reaction went to completion, resulting in mineral zonation parallel to the openings. Carmichael (1969) shows that in the Whetstone Lake area of Ontario, where pelitic rocks of the kyanite–sillimanite zone are developed, pore fluid compositions were more or less homogeneous from unit to unit at each place but differing from one part of the area to another. This has had the effect of reversing the order in which decarbonation and dehydration isograds are crossed in the field, depending on where they are examined.

(g) THE SYSTEM CaO-Al_2O_3-K_2O-SiO_2-H_2O-CO_2

Addition of aluminum to the systems already described makes possible the appearance of many new phases. If we exclude consideration of assemblages including corundum we still must consider anorthite, gehlenite, zoisite, and grossularite. The stability of grossularite with respect to these minerals is shown at a pressure of 2 Kbars in Fig. 33. It will be apparent that both grossularite and zoisite are stable at this pressure only in fluids with a high ratio of H_2O to CO_2 and therefore that the presence of either mineral in metamorphosed carbonate rocks implies that the pore fluid in the rock was water-rich during metamorphism in spite of the abundant carbonates present. Kerrick (1972) and Kerrick, Crawford and Randazzo (1973) give excellent accounts of the application of this system to contact metamorphic rocks in California. In addition to points already made in this chapter, these papers demonstrate that both CO_2 and H_2O

can behave as initial value components and have their activities buffered by the minerals rather than being externally imposed on and controlling the system. Increase of total pressure changes the picture, making grossularite stable in much higher CO_2 concentrations.

The further addition of potassium provides the constituents for muscovite and K-feldspar and results in additional equilibria. Three of these, involving

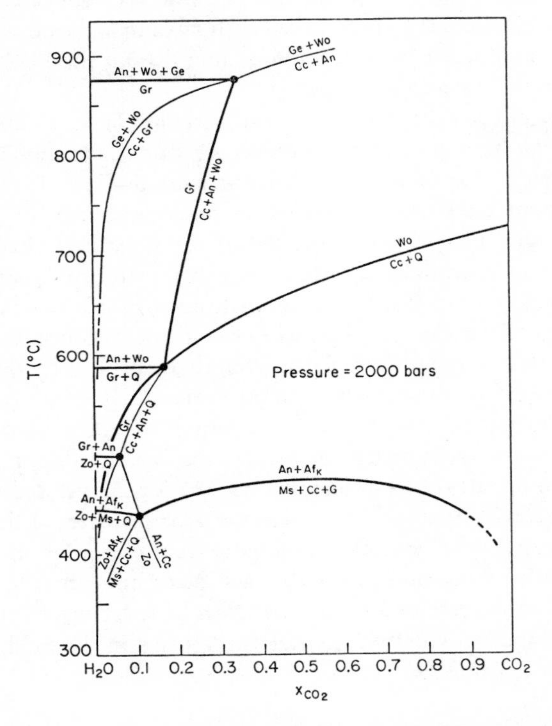

Fig. 33. Isobaric T-X diagram for a portion of the CaO-Al_2O_3-SiO_2-K_2O-H_2O-CO_2 system, illustrating reactions involving grossularite in the absence of corundum (Gordon and Greenwood, 1970b) and the upper stability of muscovite + calcite + quartz (Hewitt, 1973; Johannes and Orville, 1972). Heavy curves show where experimental data exist, light lines calculated from determined equilibria.

calcite, quartz, and muscovite are shown in Fig. 33, following the data of Hewitt (1973) and Johannes and Orville (1972). The determination by Storre and Nitsch (1972) of the illustrated invariant point is not in agreement, lying at higher values of x_{CO_2}. Internal consistency between the illustrated curves gives them the greater weight.

It will be noticed in Fig. 33 that many of the reactions involve anorthite, a mineral that is ordinarily represented in rocks by some intermediate to calcic plagioclase. The effect of adding Na to the system has the effect of enlarging the stability field of plagioclase at the expense of minerals that take little or no Na into their structures. Consequently the fields of zoisite, grossularite, and the assemblage muscovite + calcite + quartz are reduced by the addition of Na to this system.

Phlogopite is stable in carbonate rocks containing MgO, K_2O, and Al_2O_3, under appropriate conditions. Some of the phlogopite equilibria are illustrated

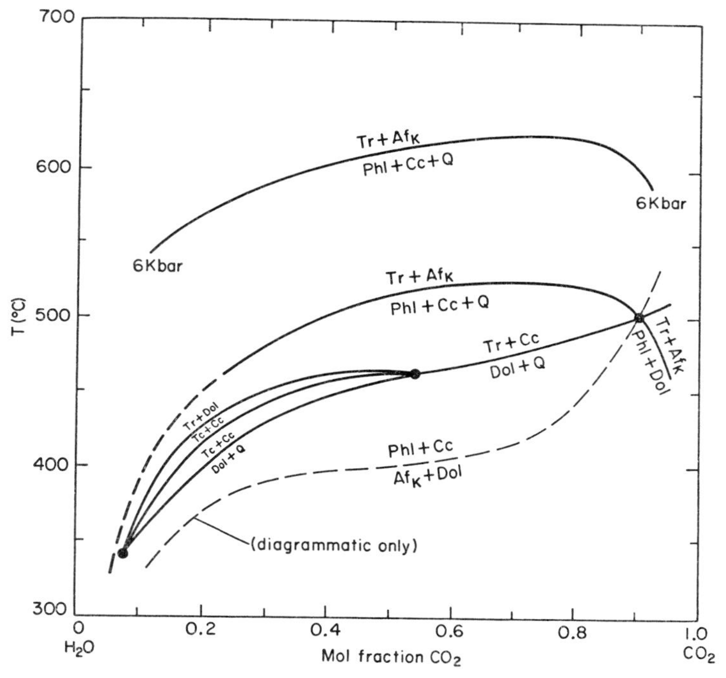

Fig. 34. Isobaric T-X diagram illustrating the stability of the phases tremolite, K-feldspar, phlogopite, calcite, quartz and dolomite. Phlogopite + calcite + quartz stability after Hoschek, 1973, and dolomite + quartz + tremolite + calcite reaction, after Skippen, 1972. (See also Fig. 32.) The curves passing through the invariant points are for 2 Kbars, and the upper curve is for 6 Kbars (Hoschek, 1973).

in Fig. 34, after Hoschek (1973) and including the work of Skippen (1971, see also Fig. 32). The Phlog + Cc + Q = Tr + K-Spar reaction studied by Hoschek can be stable only at temperatures higher than the reaction Dol + Q = Tr + Cc and the intersection generates another equilibrium curve, K-Spar + Dol = Phlog + Cc. This latter curve has been studied by Bailey (1966), who showed it to lie at higher temperatures at 1 Kbar. Bailey's curve is not shown in Fig. 34, but a 2 Kbar curve is sketched in to be consistent with the invariant point implied by the work of Skippen and Hoschek.

The stabilities of the micas muscovite and phlogopite in the presence of calcite and quartz are summarised in a P-T projection in Fig. 35, which shows projection of the upper critical temperatures for the mica stabilities in CO_2–H_2O mixtures. The muscovite curve is from Hewitt (1973, and Fig. 33) with a fluid composition of 50 mol % CO_2. The phlogopite curve is after Hoschek (1973) and passes through a T-X maximum at $x_{CO_2} = 0.75$. Curves of this kind, giving

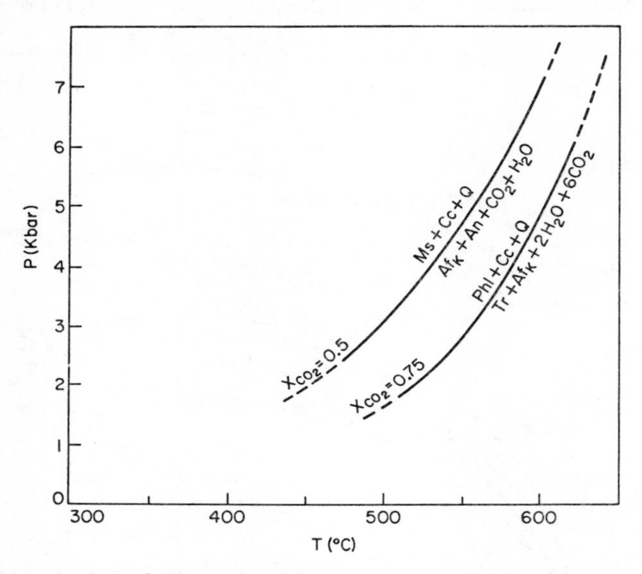

Fig. 35. P-T projection of the upper critical temperatures for the two mica + calcite + quartz reactions. Ms + Cc + Q = Af(K) + An + CO_2 + H_2O after Hewitt (1973). This curve passes through a temperature maximum on a T-X diagram at $x_{CO_2} = 0.5$. See also Fig. 33. Phlogopite + calcite + quartz = Tremolite + K-spar + $2H_2O$ + $6CO_2$ (Hoschek, 1973) passes through a similar maximum at $x_{CO_2} = 0.75$. See Fig. 34 for T-X projection.

P-T projections of upper critical temperatures of reactions evolving both H_2O and CO_2 are univariant. The only other points in T-X diagrams that generate univariant P-T curves are the isobaric invariant points, vapour-absent reactions, and the limiting one-volatile reactions shown in Figs 26 and 27.

(h) METASOMATIC EXCHANGE BETWEEN PELITIC AND CARBONATE ROCKS

There are many accounts in the literature of reactions that have taken place between adjacent pelitic and carbonate rocks. Thompson (1959) has presented an elegant discussion of the theory of open systems applied to such situations, and Vidale (1969) has studied the reactions both in the laboratory and in the field. Figure 36 illustrates the stable phase assemblages (in the presence of excess quartz and fluid) in the system $CaO-Al_2O_3-MgO-KAlO_2$ (from Vidale, 1969). She finds that if wollastonite and the assemblage muscovite + phlogopite + quartz are juxtaposed in a sealed container, together with aqueous chloride solution, reaction takes place according to the scheme shown in Fig. 37. The first step is to equalise the activity of K^+ and SiO_2 by diffusion, eliminating all unstable associations in favour of 5-phase assemblages each of which is at local

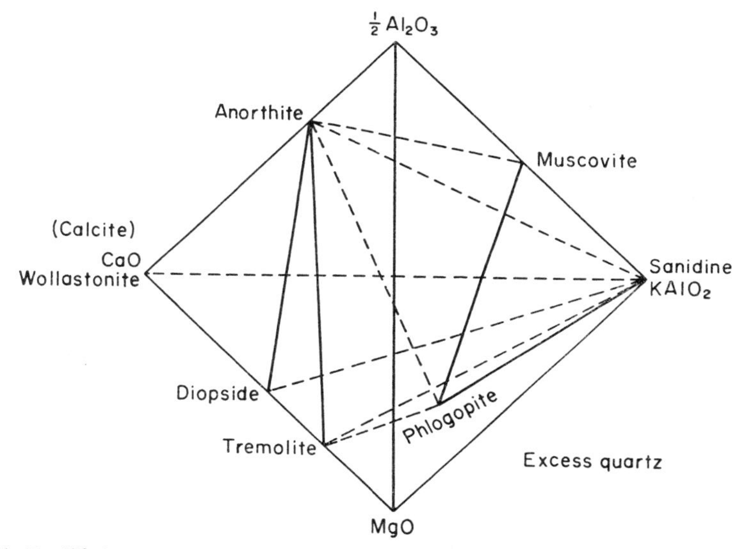

Fig. 36. Equilibrium assemblages in the system $CaO-MgO-KAlO_2-Al_2O_3-H_2O$ at 600°C and 2 Kbars $P_{H_2O} = P_{total}$ (Vidale, 1969).

equilibrium. Further diffusion of components occurs, eliminating some of the zones and simplifying the mineralogy of each zone. The final result is a zoned assemblage in which the number of phases in any zone is less than the number of components of the whole system, and in which each zone is at local equilibrium with respect to all components. These results are in accord with the theoretical considerations presented by Thompson (1957) and with Vidale's petrographic observations. One of the features of the experiments is that zones of tremolite + anorthite + quartz are formed between the carbonate and pelitic layers. If Na were available the feldspar would be some intermediate plagioclase, and the configuration would be much as suggested by Orville (1969), who suggests that many thin-layered amphibolites in metasedimentary sequences owe their origin to just such a process.

SUMMARY

Experimental data and thermodynamic theory have transformed petrology from a descriptive and qualitative subject to at least a semi-quantitative science, with the certain promise of improved quantitative accuracy in the future. The theory and technology of experiments at very high pressure, and of experiments

Stage 1, Gradational Contact

Wollastonite and "Pelite" with Wollastonite decreasing to the right	"Pelite" (Muscovite, Phlogopite, and Quartz)

Stage 2, Local Equilibrium Assemblages at 600°C, 2000 Bars

Wollastonite	Diopside	Tremolite	Muscovite
Diopside	Tremolite	Phlogopite	Phlogopite
Anorthite	Anorthite	Anorthite	Anorthite
Sanidine	Sanidine	Sanidine	Sanidine
Quartz	Quartz	Quartz	Quartz

Stage 3, Some Metasomatism has Occurred

Wollastonite		Di.		Tr		Muscovite
Diopside	Di.	Tr.	Tr.	Phl.	Phl.	Phlogopite
Anorthite	An.	An.	An.	An.	An.	Anorthite
Quartz	Q.	Q.	Q.	Q.	Q.	Sanidine
						Quartz

Stage 4, More Metasomatism has Occurred

Wo.				Muscovite
Di.	Diopside	Tremolite	Phlogopite	Phlogopite
An.	Anorthite	Anorthite	Anorthite	Anorthite
Q.	Quartz	Quartz	Quartz	Sanidine
				Quartz

Fig. 37. Progressive metasomatism in an experimental "carbonate"–"pelite" diffusion couple (after Vidale, 1969). 600°C and 2 Kbars, $P_{H_2O} = P_{total}$.

in the presence of several gas species with full buffer control, are well advanced, and the main effort in the near future will probably be to pursue data on the great number of equilibria still unstudied. Much work remains to be done on testing thermodynamic consistency between existing experimental data.

The exciting prospect for experimental petrology is the possibility of finally being able to calculate the physical and chemical conditions that prevailed during metamorphism on the basis of the excellent accumulation of direct experimental boundaries, some of which are discussed in this volume. The data presented here are for the most part suitable for such estimates although one must be continuously on guard for inconsistencies. Some of these, where obvious, have been pointed out here, but there are doubtless many others still undetected.

This brings us to one of the main challenges to all modern petrologists, but principally to the experimentalist. The challenge is to assure thermodynamic consistency of all newly reported boundary curves and to test consistency with all related pre-existing equilibria and thermodynamic data. It is no longer acceptable to present a diagram that is merely a pictorial view of experiments conducted. These experiments must be demonstrated to be internally consistent and tested against all other available data. The results should be presented in tabular form in the abstract for ease of incorporation in widely-read abstracting services, and in this tabulation should appear not only thermodynamically consistent co-ordinates of reaction boundaries but also specific notation of important agreements and inconsistencies with other work. Only by such a concerted effort can we hope to build a self-consistent framework and establish a justifiable confidence in the results of experimental petrology.

ACKNOWLEDGEMENT

The time and opportunity to assemble material presented here has been provided by financial support for study leave by a Killam Foundation grant, travel funds from the National Research Council of Canada, and the peace and quiet of a place to work by the Eislgenössische Technische Hochschule, Zurich, Institut für Kristallographie und Petrographie. I am grateful for this assistance.

REFERENCES

Akella, J. and Winkler, H. G. F. (1966). *Contr. Miner. Petrol.*, **12,** 1–12.
Albee, A. L. (1965). *Am. J. Sci.*, **263,** 512–536.
Albee, A. L. (1965). *J. Geol.*, **73,** 155–164.
Albee, A. L. (1965). *J. Petrol.*, **6,** 246–301.
Albee, A. L. and Chodos, A. A. (1969). *Am. J. Sci.*, **267,** 310–316.
Althaus, E. and Johannes, W. (1969). *Am. J. Sci.*, **267,** 87–98.
Anderson, A. T. (1968). *J. Geol.*, **76,** 528–547.
Bailey, D. K. (1966). *Min. Soc. of India I.M.A. Volume*, 5–8.
Baker, E. H. (1962). *J. Chem. Soc.*, **87,** 464–470.
Barnes, H. L. and Ernst, W. G. (1963). *Am. J. Sci.*, **261,** 129–150.
Barron, L. M. (1973). *Contr. Mineral. Petrol.*, **39,** 184.
Barton, P. B., Jr. and Toulman, P., III. (1964). *Geochim. et Cosmochim Acta*, **28,** 619–640.

Barton, P. B., Jr. and Toulmin, P., III. (1966). *Econ. Geol.*, **61,** 815–849.

Barton, Paul, B., Jr. and Skinner, Brian J. (1967). Sulfide mineral stabilities; *in* "Geochemistry of Hydrothermal Ore Deposits", Barnes, H. L. (Ed.). Holt, Rinehart, and Winston, 236–333.

Buddington, A. F. and Lindsley, D. H. (1964). *J. Petrol.*, **5,** 310–357.

Burnham, C. Wayne, Holloway and Davis, Nicholas F. (1969). *Geol. Soc. Amer. Spec. Paper 132.*

Blackburn, W. H. (1968). *Contr. Mineral and Petrol.*, **19,** 72–92.

Blake, R. L. (1965). *Am. Miner.*, **50,** 148–169.

Bowen, Norman L. (1940). *J. of Geol.*, **48,** 225–274.

Bowen, Norman L. and Tuttle, O. F. (1949). *Bull. G.S.A.*, **60,** 439–460.

Boyd, Francis R., 1959, Hydrothermal Investigations of Amphiboles *in* "Researches in Geochemistry", Abelson, P. H. (Ed.). Wiley, New York, 377–396.

Buckner, D. A., McRoy, D., and Roy, R. (1960). *Am. J. Sci.*, **258,** 132–147.

Card, K. D. (1964). *Bull. G.S.A.*, **75,** 1011–1030.

Carman, J. H. (1967). *Trans. Am. Geoph. Union*, vol. 48, pp. 224–225.

Carmichael, D. M. (1970). *J. Petrology*, **11,** 147–181.

Carpenter, A. B. (1967). *Am. Min.*, **52,** 1341–1363.

Chakraborty, K. R. and Sen, S. K. (1967). *Contr. Mineral. Petrol.*, **16,** 210–232.

Chatterjee, N. D. (1970). *Contr. Mineral. Petrol.*, **27,** 244–257.

Chatterjee, N. D. (1972). *Contr. Mineral. Petrol.*, **34,** 288–303.

Chang, L. L. Y. (1965). *J. Geol.*, **73,** No. 2, 346–368.

Chinner, G. A. (1960). *J. Petrol.*, **1,** No. 2, Geoph. lab. paper No. 1,327.

Chinner, G. A. (1966). The significance of the aluminum silicates in metamorphism: *Earth Sci. Rev.*, **2,** 111–126.

Chinner, G. A. (1966). *Q. J. Geol. Soc. Lond.*, **122,** 159–186.

Chinner, G. A., Smith, J. V. and Knowles, C. R. (1969). Transition-metal contents of Al_2SiO_5 polymorphs. *Am. Jour. Sci.*, Schairer volume, vol. 267-A, pp. 96–113.

Clark, Sydney P., Jr. (1966). "Handbook of Physical Constants". *Geol. Soc. Amer. Mem.*, **97.**

Deer, W. A., Howie, R. A. and Zussman, J. (1962). "Rock-Forming Minerals: 5". Longmans, London.

Doe, B. R. (1962). *Bull. Geol. Soc. Amer.*, **73,** 833–854.

Engel, A. E. J. and Engel, C. G. (1958). *Bull. G.S.A.*, **69,** 1369–1414.

Engel, A. E. J. and Engel, C. G. (1962). *Bull. G.S.A.*, **73,** 1499–1514.

Engel, A. E. J. and C. G. (1962). *G.S.A. Petrologic Studies, Buddington Volume*, 37–82.

Engel, A. E. J., Engel, C. G. and Havens, R. G. (1964). *J. Geol.*, **72,** 131–156.

Ernst, W. G. (1968). "Amphiboles". Springer-Verlag New York.

Ernst, W. G. (1960). *Geochim. et Cosmichim. Acta*, **19,** 10–40.

Ernst, W. G. (1961). *Am. J. Sci.*, **259,** 735–765.

Ernst, W. G. (1962). *J. Geol.*, **70,** 689–736.

Ernst, W. G. (1963a). *Amer. Mineral.*, **48,** 1357–1373.

Ernst, W. G. (1963b). *Amer. Mineral.*, **48,** 241–260.

Ernst, W. G. (1966). *Am. J. Sci.*, **264,** No. 1, 37–65.

Eskola, P. (1915). *Bull. Comm. Géol. Fin.*, **44.**

Eskola, P. (1920). *Norsk. Géol. Tidsskr.*, **6,** 143–194.

Eugster, Hans P. and Wones, David R. (1962). *J. Petrol.*, **3,** 82–125.

Eugster, H. P. (1957). *J. Chem. Phys.*, **26,** 1,760–1,761.

Eugster, H. P. and Skippen, G. B. (1967). Igneous and metamorphic reactions involving gas equilibria *in* "Researches in Geochemistry, Volume 2", Abelson, P. H. (Ed.), John Wiley and Sons, Inc., N.Y., 492–520.

Eugster, H. P. and Yoder, H. S. (1955). *Carn. Inst. Wash. Yr. Bk.*, **54,** 124–128.

Eugster, H. P., Albee, A. L., Bence, A. E., Thompson, J. B., Jr. and Waldbaum, D. R. (1972). *J. Petrol.*, **13,** 147–179.
Evans, B. W., 1965. *Am. J. Sci.*, **263,** 647–667.
Evans, B. W. and Guidotti, C. V. (1966). *Contr. Mineral. Petrol.*, **12,** 25–62.
Evans, B. W. and Trommsdorff, V. (1970). *Schweiz. Mineral. Petrog. Mitteil.*, **50,** 481–492.
Fawcett, J. J. (1964). *Carn. Inst. Wash. Yr. Bk.*, **63,** 137–141.
Fawcett, J. J. and Yoder, H. S. (1966). *Amer. Miner.*, **51,** 353–380.
Flaschen, S. S. and Osborn, E. F. (1957). *Econ. Geol.*, **52,** 923–943.
Forbes, Warren C. (1969). *Amer. Miner.*, **54,** 1,399–1,408.
Frantz, J. D. and Eugster, H. P. (1973). *Am. J. Sci.*, **273,** 268–286.
Franz, G. W. and Wyllie, P. J. (1966). *Geochim. et Cosmochim. Acta*, **30, 1,** 9–22.
French, Bevan, M. (1966). *Rev. Geophys.*, **4,** 223–253.
French, B. M. and Eugster, H. P. (1965). *J. Geophys. Res.*, **70,** 1529–1539.
French, Bevan M. and Rosenberg, P. E. (1965). *Science*, **147,** 1283–1284.
French, Bevan M. (1971). *Am. J. Sci.*, **271,** 37–78.
Fujii, T. (1966). "Muscovite-paragonite Equilibria", Unpublished doctoral dissertation, Dept. of Geology, Harvard University, Cambridge, Massachusetts.
Ganguly, J. (1968). *Am. J. Sci.*, **266,** 277–298.
Ganguly, J. (1969). *Am. J. Sci.*, **267,** 910–944.
Ganguly, J. (1972). *J. Petrol.*, **13,** 335–365.
Garrels, R. M. and Christ, C. L. (1965). "Solutions, Minerals and Equilibria", Harper and Row, New York.
Gilbert, M. C. (1966). *Am. J. Sci.*, **264,** 698–742.
Goldschmidt, V. M., 1912, *Vidensk. Skr. I, Nat.-Naturv. kl.*, **22.**
Goldsmith, J. R. and Graf, D. L. (1957). *Geochim. et Cosmochim. Acta*, **11,** 310–334.
Goldsmith and Heard, H. C. (1961). *J. Geol.*, **69,** 45–74.
Goldsmith, J. R. and Newton, R. C. (1969). *Am. J. Sci.*, **267-A,** Schairer vol., 160–190.
Gordon, T. M. and Greenwood, H. J. (1970). *Amer. Mineral.*, **56,** 1674–1688.
Gordon, T. M. and Greenwood, H. J. (1970). *Am. J. Sci.*, **268,** 225–242.
Graf, D. L. and Goldsmith, J. R., 1958. *Geochim. et Cosmochim. Acta*, **13,** 218–219.
Greenwood, H. J. (1961). *Jour. Geophys. Res.*, **66,** 3923–3946.
Greenwood, H. J. (1962). *Carn. Inst. Wash. Yr. Bk.*, **61,** 82–85.
Greenwood, H. J. (1963). *J. Petral.*, **4,** 317–351.
Greenwood, H. J. and Barnes, H. L. (1966). *Geol. Soc. Amer. Mem.*, **97,** 385–400.
Greenwood, H. J. (1967). *The Amer. Miner.*, **52.**
Greenwood, H. J. (1967). Mineral Equilibria in the System $MgO-SiO_2-H_2O-CO_2$ *in* "Researches in Geochemistry, Volume 2", Abelson, P. H. (Ed.). John Wiley & Sons, New York.
Greenwood, H. J. (1969). *Am. J. Sci.*, **267-A,** 191–208.
Greenwood, H. J. (1971). *Am. J. Sci.*, **270,** 151–154.
Greenwood, H. J. (1972). *Geol. Soc. Amer. Mem.*, **132,** 553–571.
Greenwood, H. J. (1973). *Am. J. Sci.*, **273,** 561–571.
Guggenheim, E. A. (1957). Thermodynamics, an Advanced Treatment for Chemists and Physicists, North-Holland Pub. Co., Amsterdam.
Guidotti, C. V. (1963). *Amer. Miner.*, **48,** 772–791.
Guidotti, C. V. (1968). *Amer. Miner.*, **53,** 963–74.
Guidotti, C. V. (1973). *Contr. Mineral. Petrol.*, **42,** 33–42.
Halferdahl, L. B. (1961). *J. Petrol.*, **2,** 49–135.
Hariya, Y., Dollase, W. A. and Kennedy, G. C. (1969). *Amer. Miner.*, **54,** 1419–1441.
Harker, A. (1894). *Geology Mag.*, 169–170.
Harker, R. I. and Tuttle, O. F. (1955). *Am. J. Sci.*, **253,** pp. 209–224.
Harker, R. I. and Hutta, J. J. (1956). *Econ. Geol.*, **51,** 375–381.

Harker, R. I. and Tuttle, O. F. (1956). *Am. J. Sci.*, **254**, 239–256.

Harker, R. I. (1959). *Am. J. Sci.*, **257**, 656–667.

Harker, R. I., Roy, D. M. and Tuttle, O. F. (1962). *J. Am. Ceram. Soc.*, **45**, 471–473.

Harker, R. I. (1964). *J. Am. Ceram. Soc.*, **47**, 521–529.

Hellner, E., Hinrichsen, Th., Seifert, F. (1965). The Study of Mixed Crystals of Minerals in Metamorphic Rocks in "Controls of Metamorphism", W. S. Pitcher and G. W. Flinn (Eds.). Oliver and Boyd, Edinburgh.

Hellner, E. and Schurmann, K. (1966). *J. Geol.*, **74**, 322–331.

Helgeson, H. C. (1967). *Economic Geology Monograph*, **3**, 333.

Helgeson, H. C. (1967). *J. Phys. Chem.*, **71**, 3,121.

Hemley, J. J. (1959). *Am. J. Sci.*, **257**, 241–270.

Holdaway, M. J. (1971). *Am. J. Sci.*, **271**, 97–131.

Hewitt, David A. (1973). *Amer. Mineral.*, **58**, 785–791.

Hollister, L. S. (1969). *Am. J. Sci.*, **267**, 352–370.

Hollister, L. S. (1966). *Science*, **154**, 1647–1651.

Holloway, J. R., Burnham, C. W. and Millhollen, G. L. (1968). *J. Geophys. Res.*, **73**.

Hoschek, G. (1967). *Die Naturwissenschaften*, **8**, 200.

Hoschek, G. (1968). *Die Naturwissenschaften*, **5**, 226–227.

Hoschek, G. (1967). *Beifr. Min. Pet.*, **14**, 123–162.

Hoschek, G. (1969). *Contr. Mineral. Petrol.*, **22**, 208–232.

Hoschek, G. (1973). *Contr. Mineral. Petrol.*, **39**, 231–237.

Hsu, L. C. and Burnham, C. W. (1969). *Geol. Soc. Amer. Bull.*, **80**, 2,393–2,408.

Hsu, L. C. (1968). *J. Petrol.*, **9**, Part 1, 40–83.

Huebner, J. S. (1969). *Amer. Miner.*, **54**, 457–481.

Johannes, W. (1966). *Naturwissenschaften*, **3**, 80–81.

Johannes, W. (1966). *N. Jahrb. f. Min. Monatshelfe*, 305–308.

Johannes, W. (1967). *Beitr. Mineral. Petrol.*, **15**, 233–250.

Johannes, W. (1968). *Contr. Mineral. Petrol.*, **19**, 309–315.

Johannes, W. and Orville, P. M. (1972). *Fortschrit. Mineral.*, **50**, 46–47.

Johannes, W. (1969). *Am. Jour. Sci.*, **267**, 1083–1104.

Kennedy, G. C. and Holser, W. (1966). *Geol. Soc. Amer. Mem.*, **97**, 371–383.

Kerrick, Derrill, M. (1968). *Am. J. Sci.*, **266**, 204–14.

Kerrick, D. M. (1972). *Am. Jour. Sci.*, **272**, 946–958.

Kerrick, D. M., Crawford, K. E. and Randazzo, A. F. (1973). *Jour. Petrol.*, **14**, 303–325.

Kopp, O. C. and Harris, L. A. (1967). *Amer. Miner.*, **52**, 1681–1688.

Kretz, R. (1964). *J. Petrol.*, **5**, 1–20.

Kullerud, G. (1953). *Norsk. Geol. Tidsskr.*, **32**, 61–147.

Kullerud, G. and Yoder, H. S., Jr. (1959). *Econ. Geol.*, **54**, 533–572.

Lambert, I. B., Robertson, J. K. and Wyllie, P. J. (1969). *Am. J. Sci.*, **267**, 609–626.

Luth, W. C., Jahns, R. H. and Tuttle, O. F. (1964). *J. Geophys. Res.*, **69**, 759–773.

LeFaucheux, F., Touray, J.-C. and Guilhamou, N. (1972). *Bull. Soc. Fr. Cristall.*, **95**, 620–622.

Matsushima, S., Kennedy, G. C., Aketta, J., Haygarth, J. (1967). *Am. J. Sci.*, **265**, 28–44.

Medaris, L. G. Jr. (1969). *Am. J. Sci.*, **267**, 945–968.

Mel'nik, Yu. P. (1972). *Geochem. Internat.*, **9**, 1–13.

Merril, R. B., Robertson, J. K. and Wyllie, P. J. (1970). *J. Geol.*, **78**, 558–569.

Melson, W. G. (1966). *Amer. Miner.*, **51**, 402–421.

Metz, P. (1967). *Geochim. et Cosmoch. Acta*, **31**, 1517–1532.

Metz, P. and Trommsdorff, V. (1968). *Contr. Mineral. Petrol.*, **18**, 305–309.

Metz, Paul W. and Winkler, Helmut, G. F. (1963). Bildung von talk as kieseligem. *Geochim. et Cosmochim. Acta*, **27**, 431–457.

Metz, P. and Winkler, H. G. F. (1964). *Geochem. International*, **2**, 388–389.

Metz, P. and Winkler, Helmut, G. F. (1965). *Naturwiss.*, **51,** 160.
Misch, P. (1964). *Beitrage zur mineralogie und petrographie*, **10,** 315–356.
Morgan, B. A. (1970). *J. Petrol.*, **11,** 101–145.
Mueller, R. F. (1961). *Am. J. Sci.*, **259,** 460–480.
Munoz, J. L. and Eugster, H. P. (1969). *Amer. Miner.*, **54,** 943–959.
Newton, R. C. (1965). *J. Geol.*, **73,** 431–441.
Newton, R. C. (1966). *Science*, **151,** 1222–1225.
Newton, R. C. (1966). *Science*, **153,** 170–172.
Newton, R. C. (1966). *Am. J. Sci.*, **264,** 204–222.
Orville, P. M. (1963). *Am. J. Sci.*, **261,** 201–237.
Orville, P. M. and Greenwood, H. J. (1965). *Am. Jour. Sci.*, **263,** 678–683.
Orville, P. M. (1969). *Am. J. Sci.*, **267,** 64–86.
Palmer, D. F. (1970). *Econ. Geol.*, **65,** 31–39.
Pitzer, K. S. and Brewer, L. (1961). "Thermodynamics." McGraw-Hill, New York.
Popov, A. A. (1970). *Geochemistry International*, 94–104.
Poty, B. and Stalder, H. A. (1970). *Schweiz. Min. Petr. Mitteil.*, **50,** 141–54.
Presnall, D. C. (1969). *J. Geophys. Res.*, **74,** 6,026–6,033.
Richardson, S. W. (1968). *J. Petrol.*, **9,** 467–468.
Richardson, S. W., Gilbert, M. C. and Bell, P. M. (1969). *Am. J. Sci.*, **267,** 259–272.
Robie, R. A. (1966). *Geol. Soc. Amer. Mem.*, **97.**
Roedder, E. (1969). Fluid Inclusions as Samples of Ore Fluids *in* "Geochemistry of Hydrothermal Ore Deposits", Barnes, H. L. (Ed.), Holt, Rinehart and Winston, New York.
Rosenberg, P. E. (1963). *Amer. Mineral.*, **48,** 1396–1400.
Rosenberg, P. E. (1963). *Am. J. Sci.*, **261,** 683–690.
Rosenberg, P. E. (1967). *Amer. Mineral.*, **52,** 787–796.
Rosenberg, P. E. (1968). *Amer. Mineral.*, **53,** 880.
Rosenbusch (1877). *Abh. Geol. Spezialkarte Elsass-lothr.*, **1,** 80–393 (quoted by Harker, R. I. (1939), p. 24).
Ross, M., Papike, J. J. and Weiblen, P. W. (1968). *Science*, **159,** 1099–1102.
Roy, D. M. and Roy, R. (1955). *Amer. Mineral.*, **40,** 147–178.
Roy, D. M. and Harker, R. I. (1962). *Fourth Internat. Sympos. on the Chemistry of Cement*, **I,** 196–201.
Roy, D. M. and Mumpton, F. A. (1956). *Econ. Geol.*, **51,** 432–443.
Rutherford, M. J. (1969). *J. Petrol.*, **10,** 381–408.
Scarfe, C. M. and Wyllie, P. J. (1967). *Nature*, **215,** 945–946.
Schreyer, W. (1964). *N. Jb. Miner. Abh.*, **102,** 39–67.
Schreyer, W. (1966). *N. Jb. Miner. Abh.*, **105,** 3, 211–244.
Schreyer, W. (1965). *N. Jb. Miner. Abh.*, **103,** 35–79.
Schreyer, W. (1965). *Beitrage zur Min. und. Pet.*, **11,** 297–322.
Schreyer, W. and Seifert, F., 1969. *Am. J. Sci.*, **267,** 371–388.
Schreyer, W. and Seifert, F. (1969). *Am. J. Sci.*, **267-A,** 407–443.
Schreyer, W., Kullerud, G. and Ramdohr, P. (1964). *Neues Jahrb. für Mineral.*, **101,** 1–26.
Schreyer, W. and Siefert, F. (1967). *Contr. Mineral. Petrol.*, **4,** 343–358.
Schreyer, W., Yoder, H. S. Jr. (1964). *Neues. Jahrb. für mineral Abh.*, **101,** 271–342.
Schuiling, R. D. and Vink, B. W. (1967). *Geochimic, et Cosmochic. Acta*, **31,** 2399–2411.
Seifert, F. (1970). *J. Petrol.*, **11,** 73–99.
Seifert, F. and Schreyer, W. (1965). *Amer. Mineral.*, 1114–1118.
Seifert, Von F. and Schreyer, W. (1966). Fluide phasen im system K_2O-MgO-SiO_2-H_2O und ihre mogliche bedeutung fur die entstehung ultrabasischer gesteine: Sonderdruck aus der Zeitschrift: Berichte der Bunsengesellschaft fur physikalische Chemie, Band 70, Heft 9/10.
Seifert, F. and Schreyer, W. (1970). *Contr. Mineral. Petrol.*, **27,** 225–238.

Sen, S. K. and Chakraborty, K. R. (1968). *N. Jb. Miner. Abh.*, **108,** 181–207.
Shaw, Herbert R. (1963). *Science*, **139,** 1220–1222.
Shaw, H. R. (1963). *Amer. Mineral.*, **48,** 883–896.
Shaw, H. R. and Wones, David R. (1964). *Am. Jour. Sci.*, **262,** 918–929.
Shaw, H. R. (1967). Hydrogen Osmosis in Hydrothermal Experiments *in* "Researches in Geochemistry: Vol. 2", Abelson, P. H. (Ed.). John Wiley and Sons, Ltd., New York.
Skippen, G. B. (1972). *J. Geol.*, **79,** 457–481.
Speidel, D. H. (1970). *Am. J. Sci.*, **268,** 341–353.
Storre, B. and Karotke, E. (1971). *Neues Jahrb. Mineral.*, **5,** 237–240.
Storre, B. and Nitsch, K. (1972). *Contr. Mineral. Petrol.*, **35,** 1–10.
Strens, R. G. J. (1968). *Mineral. Mag.*, **36,** 839–849.
Takenouchi, S. and Kennedy, G. C. (1964). *Am. J. Sci.*, **262,** 1055–1074.
Thompson, J. B. Jr. (1957). *Amer. Mineral.*, **42,** 842–858.
Thompson, J. B., Jr. (1955). *Am. J. Sci.*, **253,** 65–103.
Thompson, J. B., Jr. (1959). Local Equilibrium in Metasomatic Processes *in* "Researches in Geochemistry", Abelson, P. H. John Wiley, 427–457.
Thompson, A. B. (1970). *Am. J. Sci.*, **268,** 454–458.
Tilley, C. E. (1948). *Min. Mag.*, vol. XXVIII, 272–276.
Tilley, C. E. (1951). *Geology Mag.*, **88,** 175–178.
Tilley, C. E. (1924). *Quart. J. Geol. Soc.*, LXXX, 22–70.
Toulmin, P., III, and Barton, P. B., Jr. (1964). *Geochim. et Cosmochim. Acta*, **28,** 641–671.
Trommsdorff, Volkmar and Evans, W. (1972). *Am. J. Sci.*, **272,** 423–437.
Trommsdorff, V. (1968). *Schweiz. Mineral. Petrog. Mitteil.*, **48.**
Trommsdorff, V. (1966). *Schweiz. Mineral. Petrog. Mitteil.*, **46,** 431–447.
Turner, F. J. (1967). *Neues Jahrb. Mineral. Monatshefte*, 1–22.
Turnock, A. C. (1960). *Carn. Inst. Wash. Yr. Bk.*, **59,** 98–103.
Turnock, A. C. and Eugester, H. P. (1962). *J. Petrol.*, **3,** 533–565.
Velde, B. (1965a). *Am. J. Sci.*, **263,** 886–913.
Velde, B. (1965b). *Amer. Mineral.*, **50,** 436–449.
Velde, B. (1967). *Contr. Mineral. Petrol.*, **74,** 250–258.
Velde, B. (1966). *Amer. Mineral.*, **51,** 924–928.
Vidale, R. (1969). *Am. J. Sci.*, **267,** 857–874.
Walker, L. S. (1963). *Am. J. Sci.*, 488–500.
Walter, L. S., Wyllie, P. J. and Tuttle, O. F. (1962). *J. Petrol.*, **3,** 49–62.
Winkler, H. G. F. (1967). "Petrogenesis of Metamorphic Rocks", Springer-Verlag, New York.
Winkler, H. G. F. (1957). *Geochim. Cosmochim. Acta*, **13,** 42–69.
Winkler, H. G. F. and von Platen, H. (1958). *Geochim. et Cosmochim. Acta*, **15,** 91–112.
Wise, W. S. and Eugster, H. P. (1964). *Amer. Mineral.*, **49,** 1031–1083.
Weeks, W. F. (1956). *J. Geol.*, **64,** 245–270.
Wones, D. R. (1967). *Geochim. et Cosmoch. Acta*, **31,** 2248–2253.
Wones, D. R. (1963). *Amer. Mineral.*, **48,** 1300–1321.
Wones, D. R. (1963). *Am. J. Sci.*, **261,** 581–596.
Wones, D. R. and Dodge, F. C. W. (1966). *Geol. Soc. Amer. Ann. Meeting Program*, p. 243.
Wones, D. R. and Eugster, H. P. (1965). *Amer. Mineral.*, **50,** 1228–1272.
Wones, D. R. and Gilbert, M. C. (1968). *Ann. Rept. Dir. Geophys. Lab. Carn. Inst. Wash. Yr. Bk.*, **66,** 402–403.
Wyllie, Peter J. (1962). *Min. Mag.*, **33,** 9–25.
Wyllie, P. J. (1965). *J. Petrol.*, **6.**
Wyllie, P. J. and Haas, J. L., Jr. (1965). *Geochim. et Cosmochim. Acta*, **29,** 871–892.
Wyllie, P. J. and Tuttle, O. F. (1960). *J. Petrol.*, **1,** 1–46.
Yoder, H. S., Jr. (1952). *Am. J. Sci.*, **267-A,** 569–627.

Yoder, H. S. (1955). *G.S.A. Sp. Paper*, **62,** 505–524.
Yoder, H. S. (1959). *Proc. Sixth Nat. Conf. Clays*, 42–60. Pergamon, New York.
Yoder, H. S. and Eugster, H. P. (1954). *Geochim. et Cosmochim. Acta*, **6,** 157–185.
Yoder, H. S. and Eugster, H. P. (1955). *Geochim. et Cosmochim. Acta*, **8,** 225–280.
Yoder, H. S., Jr. and Tilley, C. E. (1962). *J. Petrol.*, **3,** 342–532.
Zen, E-an and Albee, A. L. (1964). *Amer. Mineral.*, **49,** 904–925.
Zen, E-an (1969). *Am. J. Sci.*, **267,** 297–309.
Zen, E-an (1969). *Amer. Mineral.*, **54,** 1,592–1,606.
Zen, E-an (1971). *Am. J. Sci.*, **270,** 136–150.
Zimmerman, J. L. and Poty, B. (1970). *Schweiz. Min. Petr. Mitteil.*, **50,** 99–108.
Zwart, H. J. (1962). *Geol. Rundschau*, **52,** 38–65.
Zharikov, V. A. and Shmulovich, K. I. (1969). *Geochem. Internat.*, **6,** 853–868.

3. Experimental Metamorphic Petrology at Low Pressures and High Temperatures

W. Schreyer

1 Introduction: Scope of Subject and Limitations 261
2 Range of Conditions and Rock-types Considered 263
3 Minerals of Low-pressure High-temperature Metamorphism 265
  (a) Natural occurrences 265
  (b) Experimental stability data 265
4 Metamorphic Reactions in Model Systems 283
  (a) The system MgO-SiO_2-H_2O 284
  (b) The system MgO-SiO_2-H_2O-CO_2 285
  (c) The system CaO-MgO-SiO_2-CO_2 289
  (d) The system CaO-MgO-SiO_2-CO_2-H_2O 293
  (e) The system CaO-Al_2O_3-SiO_2-H_2O 298
  (f) The system CaO-Al_2O_3-SiO_2-H_2O-CO_2 300
  (g) The system MgO-Al_2O_3-SiO_2-H_2O 306
  (h) The system K_2O-MgO-Al_2O_3-SiO_2-H_2O 310
  (i) The system Fe-O-Al_2O_3-SiO_2-H_2O 314
  (j) The system Fe-O-MgO-Al_2O_3-SiO_2-H_2O 319
5 Natural Multicomponent Systems and the Problem of Metamorphic Facies 324

1. INTRODUCTION: SCOPE OF SUBJECT AND LIMITATIONS

Metamorphism at high temperatures was the first type of rock alteration under other than weathering conditions to be recognised in the early studies of petrogenesis. Sediments were found to be calcined or partly fused under the influence of hot lavas which extruded at the Earth's surface or cooled rapidly at shallow depths. The glassy metamorphic rocks thus formed were named *buchites*. Very similar processes taking place in burning near-surface coal seams, i.e. in the absence of igneous material, give rise to the so-called *paralavas*. At somewhat greater depths where cooling of magmas in the plutonic cycle is generally much slower the classical environment of *contact metamorphism* occurs. As in the buchites the driving force of this type of metamorphism is, of course, the initial temperature gradient between the hot magma and the cooler country rocks. It has become increasingly evident, however, that mineral assemblages believed to be characteristic of contact metamorphism may also be formed over large volumes of rock in the absence of any intrusive bodies, that is, on a regional scale. Thus the general topic of this chapter on high-temperature, low-pressure metamorphism also includes considerable portions of the overall field of *regional metamorphism*, and there can only be arbitrary limitations against its intermediate pressure and temperature range as discussed in Part II, Section A2 by H. J. Greenwood.

The geological and petrographical aspects of high-temperature, low-pressure metamorphism thus outlined have been aptly described in the literature for many years and are laid down in a number of textbooks on metamorphic petrology. While buchites were studied by, among others, Ramdohr (1919), Thomas (1922), Agrell and Langley (1958), and Wyllie (1961), classical descriptions of the deeper-seated contact metamorphism have been contributed by Rosenbusch (1877), Daly (1903), Goldschmidt (1911), Tilley (1924), Harker (1939), French (1968), Smith (1969) and many others. Typical low-pressure environments of regional metamorphism have been dealt with particularly by Read (1923), Zwart (1958) and Shido (1958). In the modern concept of global tectonics low-pressure regional metamorphic belts play an important role as units of continental crust formerly situated above and strongly contrasted with the high-pressure metamorphic environments of the "downgoing slab" in subduction zones (paired metamorphic belts of Miyashiro, 1973a). Moreover, low-pressure high-temperature regional metamorphism is presumed to have been the characteristic type of rock transformation during early Precambrian times, when the crust was still relatively thin and hot (see Saggerson and Owen, 1968).

In accordance with the general aims of experimental petrology this chapter is to provide a *compilation of quantitative physico-chemical data bearing on the petrogenesis of high-temperature, low-pressure metamorphic rocks*. With the aid of such data modern petrologists should ideally be able to determine accurately the conditions of pressure, temperature and other variables prevailing during the formation of characteristic mineral assemblages in these rocks. In the case of contact metamorphism the results obtained would lend themselves for comparison with conditions similarly deduced for the intruding igneous mass, and might finally allow reconstruction of gradients of temperature and fluid pressure within the thermal aureole. In regional metamorphism a knowledge of the varying pressure-temperature conditions is all the more important for any serious discussions of the ultimate causes of this strange transformation of rocks, which is undoubtedly not only a function of depth of burial.

Unfortunately, however, metamorphic, like all rock-forming processes, are far too complex to be thoroughly imitated by laboratory experiments. Moreover most of the experimental work performed thus far was deliberately confined to relatively simple closed model systems in which there is hope of understanding the basic processes taking place, or of relating them to physical-chemical theory. On the other hand, natural systems such as contact aureoles abound in variables that will necessarily change in the course of time. Whereas total pressure (load pressure, solid pressure) may remain fairly constant, it is quite obvious that temperatures will rise rapidly in the beginning of the metamorphism being followed by longer periods of cooling. Similarly, fluid pressures in wet sediments, probably equalling total pressure during the initial stages, may gradually decrease to virtually nil as water and other volatiles are being driven off from the porous space of the rock as well as through dehydration of hydrous minerals. Thus natural systems will in most cases be open, at least for some of their components; that is the chemical potentials of these components will themselves become variables during the mineralisation processes. Naturally it is not only the subtraction of matter through loss of volatiles, partial melting etc. to be considered for the newly formed metamorphic rock, but also the possible addition of elements. Contact metamorphic environments, in particular, are liable to exhibit extensive metasomatism through transfer of material from the neighbouring igneous body.

In conclusion, therefore, the reader should be warned against too much optimism that the results of modern experimental petrology would immediately solve all the problems of low-pressure metamorphism recognised by generations of petrographic investigators. There are yet severe experimental and especially theoretical limitations inhibiting any rigorous application of laboratory results to natural rocks. It is the author's conviction, however, that such difficulties will gradually be overcome through further systematic studies of progressively more complex synthetic systems.

Antagonists to experimental petrological investigations often hold that the most serious obstacle to applying the results of such studies to petrogenesis in nature is the problem of *equilibration*. Whereas serious laboratory experiments always aim at determining stable equilibrium phase relations, those rocks formed within relatively short periods, particularly of high-temperature low-pressure metamorphism, very often display incompatible mineral assemblages or even metastable crystalline phases. The so-called sanidinite facies is notorious for, if not characterised by, such phase disequilibria. It must be emphasised, however, that it was through comparative experimental equilibrium studies that this fact of natural disequilibria was revealed. Only after their recognition can one also hope to approach the problem of reaction kinetics in such environments.

2. RANGE OF CONDITIONS AND ROCK-TYPES CONSIDERED

As pointed out earlier, the limits of high-temperature and low-pressure metamorphism against the type taking place at intermediate pressures and temperatures can only be chosen arbitrarily. This must be done on the basis of mineral assemblages considered to be characteristic for either type. However, there are obviously also features common to both types. Thus thematic overlaps with Part II Section A2 cannot be prevented. Naturally the fundamental starting materials undergoing metamorphism are identical in both cases: The most common *sediments* will be *pelites, sandstones, graywackes, and pure, or impure carbonates*. Magmatic source rocks can, of course, principally range from *ultrabasic through basic and intermediate to acidic compositions*, almost independently of their origin in the *volcanic* or *plutonic* cycle.

Although in the light of the physical chemistry of the mineralising processes there seem to be no principal differences between the realms of contact and regional metamorphism, petrographic nomenclature does, in some cases, and usually on textural grounds, draw a more or less sharply defined dividing line. Thus massive, fine-grained pelitic *hornfelses of contact metamorphic origin* are distinguished from schistose *gneisses formed by regional metamorphism* despite their possibly identical mineralogy. However, *spotted slates*, i.e. fine-grained pelites showing a pronounced cleavage disturbed by later growth of large porphyroblasts, may be formed by contact as well as regional low-pressure metamorphism; and the same holds for massive *quartzites, marbles* and *lime silicate rocks*. Only *sanidinites* and *buchites*, both formed at the highest temperatures and lowest pressures, are strictly confined to contact metamorphism near the earth's surface, which is sometimes also referred to as *pyrometamorphism*.

Following the classical facies division the present chapter will comprise the *various hornfels facies* (see Winkler, 1967) as well as the *sanidinite facies*. However, because of the overlap with intermediate pressure and temperature metamorphism parts of the *amphibolite facies* will also be touched upon. The temperature range thus covered in this chapter will at very low pressures (less than 1 Kbar) extend from some 400°C to about 1200°C, i.e. the liquidus temperature of basalt, whereas the total pressure range considered will roughly be from 1 atm to 5–6 Kbar at most. At these relatively high pressures the corresponding temperature range of metamorphism will probably be much smaller than the one mentioned previously, namely only about 600°C–900°C. In the search for a common mineral that might be able to characterise this overall pressure temperature range through its exclusive occurrence, one might think of the low-pressure high-temperature phase *cordierite*. It has been shown by experiments, however, that this mineral may, on its own bulk composition, be stable up to pressures of 10–11 Kbar (Schreyer and Yoder, 1964, Schreyer, 1968, Newton, 1972), and indeed it was also found in high-pressure metamorphic rocks of appropriate bulk chemistry, for example in coexistence with kyanite (Wenk, 1968). Conversely, the low-pressure Al-silicate polymorph *andalusite* can only characterise one part, i.e. the low-temperature portion of the PT-range to be discussed. It is proposed, therefore, to draw a convenient though arbitrary dividing line between the mineral assemblages to be discussed here and those of intermediate pressure metamorphism by using *two mineral pairs that are highly characteristic of pelitic rocks:* (1) *cordierite–muscovite* forming at relatively low temperatures; (2) *cordierite-K feldspar* confined to fairly high temperatures. As will be shown in section 4(b) these two assemblages define a *continuous* range of pressure temperature conditions extending from near-surface conditions up to 5–6 Kbar and bordering at high temperatures and all pressures against the realm of partial melting.

The sequence of sections to follow in the present chapter is based on the assumption that gaining knowledge about the behaviour of single mineral phases must be the first step in any systematic petrological study of multimineralic rocks. Only on these grounds can those more complex reactions be deduced, and studied experimentally, which may take place during metamorphism. Finally the discussion will turn to whether or not, with all the reactions taking place in an abundance of different bulk compositions, a meaningful facies classification of high-temperature low-pressure metamorphic rocks is feasible.

3. MINERALS OF LOW-PRESSURE HIGH-TEMPERATURE METAMORPHISM

(a) NATURAL OCCURRENCES

The mineralogy of metamorphic rocks of the types referred to in this chapter has been dealt with by a great many authors (for references see the recent text-books for example by Winkler, 1967, Turner, 1968 and Miyashiro, 1973b). It is most convenient, therefore, to summarise their results in Table 1 which lists the most important minerals found in the various rock types.

(b) EXPERIMENTAL STABILITY DATA

In the following only a very brief survey of experimental data bearing on the stability fields of all but the most important minerals can be given. Such a wealth of data has been collected by experimental petrologists in recent years that for reasonable detail they must be presented for each mineral group in a separate volume. The compilation by Ernst (1968) on amphiboles may serve as an example.

By far the most common rock-forming minerals in low-pressure metamorphism are, of course, silicates. This mineral group will require, therefore, the main attention in this section. In order to allow the reader quick orientation the order of discussion of the various minerals will follow mainly the systematic subdivision of the silicates according to their crystal structure which mineralogists are accustomed to and which is also used in the fine (but by now somewhat out-of-date) compilation of Deer, Howie and Zussmann (1962–63).

(1) *Garnet Group*

The Fe-garnet *almandine*, ideally $Fe_3Al_2Si_3O_{12}$, has been considered by some investigators to be unstable under low-pressure conditions, particularly those of contact metamorphism. Chinner (1962), on the basis of field evidence refuted this general opinion, while Hsu (1968) has shown experimentally that pure almandine is stable at fluid pressures as low as 500 bars. The temperature stability range at this pressure is however dependent on the oxygen fugacity: whereas for the quartz-fayalite-iron buffer it extends from some 500°C to 800°C, it is much narrower for the quartz–fayalite–magnetite (= QFM) buffer (550°C–615°C) and no longer exists at still higher oxygen fugacities. Thus the rarity

TABLE 1

Important non-opaque minerals of high-temperature low-pressure metamorphism.

Rock types	Minerals
Spotted slates and low-pressure mica schists	Chlorite, muscovite, paragonite, biotite, quartz, K feldspar, sodic plagioclase, cordierite, andalusite, epidote-group minerals, almandine-spessartine garnet, staurolite; rarely chloritoid.
Pelitic hornfelses, "hornfels gneisses" and other low-pressure pelitic gneisses	Muscovite, biotite, quartz, K feldspar, plagioclase, cordierite, andalusite, sillimanite, almandine-rich garnet, staurolite, spinel, corundum, cummingtonite, hornblende, orthoamphibole, orthopyroxene; sapphirine in exceptional cases.
Lime silicate hornfelses, siliceous marbles, skarns and tactites	Calcite, dolomite, talc, tremolite, hornblende, epidote-group minerals, diopside, forsterite, grossularite-andradite garnet, vesuvianite, wollastonite, quartz, calcic plagioclase, chlorite, periclase (possibly altered into brucite), humite-group minerals, ilvaite.
Metabasic rocks	Plagioclase, quartz, epidote-group minerals, hornblende, ortho- and clinopyroxenes, garnets.
Ultrabasic rocks	Olivine, pyroxenes, anthophyllite, amphiboles, magnesite, talc.
Metamorphic iron formations	Fayalite, quartz, grunerite, Fe-rich orthopyroxene, spinels, almandine, ilvaite.
Sanidinite-facies rocks of pelitic and calcareous compositions, buchites	Sanidine, plagioclase, tridymite, quartz, mullite, cordierite, osumilite, pyroxenes, spurrite, melilites, merwinite, rankinite, tilleyite, monticellite, (pseudo-)wollastonite, forsterite, periclase.

of almandine in some thermal aureoles could possibly be explained by relatively high oxygen fugacities in these environments that might be introduced from the intruding magma (cf. Hsu, 1968). Conversely there is evidence from certain contact hornfelses that abundant almandine formed at relatively low oxygen fugacities (Okrusch, 1969, 1971).

Relatively small compositional deviations in the direction towards more complicated pyralspite garnets may also have profound influence on the stability relations. For example admixture of *spessartine* component, $Mn_3Al_2Si_3O_{12}$, which on its own composition is stable at 1 atm up to liquidus temperatures (Snow, 1943), will undoubtedly extend the garnet stability field towards higher temperatures, and probably also towards somewhat lower temperatures (cf. Hsu, 1968). On the other hand, the presence of *pyrope component*, $Mg_3Al_2Si_3O_{12}$,

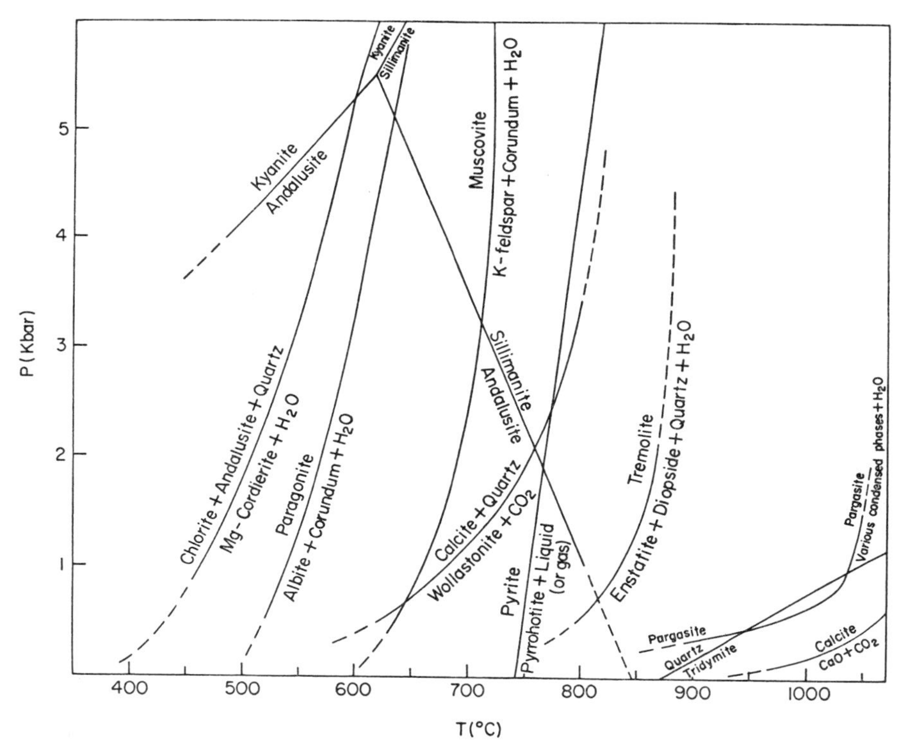

Fig. 1. Pressure–temperature diagram showing stability relations of some important rock-forming minerals. For references see text (section 3(b)).

being a high-pressure phase on its own composition (Boyd and England, 1959) may reduce the pressure stability range in the sense that almandine-pyrope garnets are not stable under the lowest pressures.

The stability relations of Ca-rich garnets containing *grossularite* ($Ca_3Al_2Si_3O_{12}$), and *andradite* ($Ca_3Fe_2Si_3O_{12}$), have long been a matter of debate, although it was quite clear from their occurrence in contact metamorphosed lime silicates that they must be stable at low pressures. For pure *andradite* Huckenholz and Yoder (1971) determined the upper temperature stability limit at 1 atm in the presence of excess oxygen as 1137°C. *Grossularite* had been synthesised by Yoder (1954) at a water pressure as low as 2 Kbar at 800°C, and Roy and Roy (1957) reported its breakdown at 1 Kbar and 850°C. Extrapolation of reversed high-pressure stability data by Hays (1967) to lower pressures, and more recent work at 1 Kbar by Boettcher (1970), indicate that the upper temperature stability limit of grossularite lies at 1 Kbar near 840°C and at 1 atm near 800°C (see Fig. 7). Thus grossularite-rich grandites are considerably less stable in high-temperature metamorphism than andradite-rich ones, Fe^{3+} in this case stabilising the garnet structure. An additional stabilising effect at high temperatures is introduced through incorporation of Ti in andradite: "Ti-andradites" containing some 50% of the theoretical end member $Ca_3Fe_2Ti_3O_{12}$ are stable at 1 atm up to liquidus temperatures near 1300°C (Huckenholz, 1969). This is in agreement with the occurrence of Ti-rich garnets (melanites) in alkali igneous rocks.

The lower temperature stability limits of the Ca-garnets are strongly dependent on the nature of the coexisting gas phase. This complication will be dealt with in section 4(f).

(2) *Aluminium Silicates*

Of the three polymorphs of Al_2SiO_5, *kyanite*, *andalusite* and *sillimanite*, only the latter two will be considered here. Their stability fields, which had been a matter of debate for many years, have been investigated extensively in recent years by several authors (e.g. Althaus, 1967, Richardson *et al.* 1968, 1969, and Holdaway, 1971) and fair agreement between the various results has finally been attained. The univariant curves shown in Fig. 1 are those preferred by Richardson *et al.* (1969). The high-temperature phase sillimanite forms with increasing degree of metamorphism at the expense of andalusite, and this transition takes place with rising pressures at successively lower temperatures. The highest pressure under which the low-pressure phase andalusite can form stably is that of the triple point, i.e. about 5.5 Kbar according to Richardson *et al.* (1969), which coincides approximately with the upper limit of low-pressure metamorphism as proposed in this chapter. This pressure would be lowered to a value of only 3.75 Kbar if the experimental data of Holdaway (1971) are used. If compared with field and paragenetic studies the results of Richardson *et al.* (1969) seem more likely. However, small variations in the amounts of impurities (Althaus, 1969) as well as in the structural state of sillimanite (Greenwood, 1972) may have considerable effects on the polymorphic transition.

Unfortunately, there is still some uncertainty concerning the stability of the aluminium silicate *mullite*, which exhibits more aluminous compositions such as $3Al_2O_3 \cdot 2SiO_2$. It forms at 1 atm at temperatures near or above the upper limit of the andalusite field and is a prominent phase in buchites. Because of the compositional difference between mullite and Al_2SiO_5 the stability field of pure mullite cannot strictly adjoin the fields of andalusite and sillimanite but should overlap with them, whereas the breakdown reactions of both andalusite and sillimanite should be $Al_2SiO_5 \rightarrow$ mullite + silica. One theoretically possible configuration of the stability diagram implying instability of the phase sillimanite at atmospheric pressure is shown in Fig. 2(a). On the other hand, Hariya, Dollase and Kennedy (1969) have recently reported on potential solid solutions between sillimanite and the structurally similar mullite which would result in a stability diagram of the type of Fig. 2(b). Here, with decreasing pressure, sillimanite should become more and more aluminous and approach true mullite composition near atmospheric pressure. If this were correct, there should be no distinction between the sillimanite and mullite fields, and mullite–sillimanite solid solutions should form in natural rocks at moderate depths as opposed to the true mullites of buchites and the true sillimanites of high-pressure high-temperature metamorphism. As the sillimanites of contact aureoles and other low-pressure environments are usually extremely fine-grained (fibrolites) defying mineral separation their bulk chemistry is not known very accurately. The recent investigations of Moore and Best (1969) on fibrolites from two contact aureoles have shown, however, that they have the X-ray properties of normal sillimanites. In addition Kwak (1971) found through microprobe work that sillimanites formed at low pressures in coexistence with Al_2O_3 were practically pure Al_2SiO_5 thus casting doubts on the validity of the interpretation given in Fig. 2(b). A third possibility supported by the thermochemical calculations of Holm and Kleppa (1966) as well as Weill (1966) involves stable conversion of andalusite into sillimanite even at atmospheric pressure and the breakdown of the latter into mullite + SiO_2 at higher temperatures. This simplest interpretation of the stability relations is shown in Fig. 2(c).

According to Weill's (1966) calculation the reaction of sillimanite to form mullite and tridymite takes place at temperatures above 1300°C, which would be geologically irrelevant. Unpublished data of F. Seifert (1972, see Dasgupta *et al.*, 1974, Fig. 7) reduce the breakdown temperature to about 1080°C for pressures below 1 Kbar. In addition, however, preliminary experimental results by Seifert (personal communication, 1973; Sahl and Seifert, 1973; Abraham and Seifert, 1973) obtained at pressures of several Kbars indicate that at temperatures above some 1050° Al-silicate phases structurally and compositionally intermediate between mullite and sillimanite can be synthesised in the presence of excess SiO_2. Thus some combination of the phase diagrams labelled (b) and (c) in Fig. 2 will probably represent the equilibrium relationships. Far more experimental work including careful X-ray studies and microprobe analyses will be necessary to resolve this important mineralogical problem of low-pressure metamorphism at very high temperatures.

(3) *Staurolite*

Staurolite was long believed to be restricted to relatively high pressures of metamorphism, but its occurrence in contact aureoles (e.g. Compton, 1960)

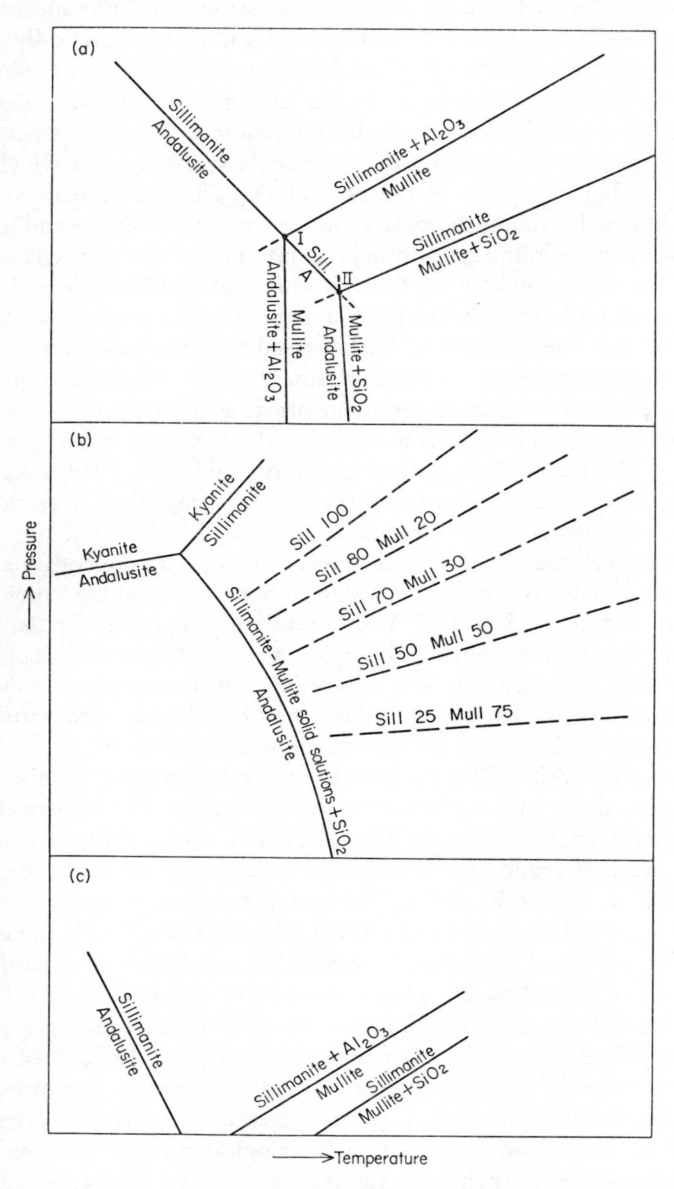

Fig. 2. Possible pressure–temperature stability limits of the two Al_2SiO_5 polymorphs sillimanite and andalusite in relation to the phase mullite. Whereas in cases (a) and (c) mullite is presumed to be of a different, i.e. more aluminous, bulk composition than Al_2SiO_5, case (b) envisages the possibility of complete solid solution between sillimanite and mullite as a function of pressure and temperature. For details see text.

and andalusite-bearing schists clearly indicates that this mineral also has to be considered in low-pressure metamorphism. Recent experimental studies by Richardson (1967, 1968) on the pure Fe-staurolite end member are in good agreement with these observations, because this author could synthesise pure Fe-staurolite at pressures as low as 2 Kbar. The upper temperature stability limit for oxygen fugacities controlled by the QFM buffer as given by Richardson (1967) is, however, not in agreement with his later results (Richardson, 1968), which are discussed in section 4(i) and summarised in Fig. 13. The influence of increasing amounts of Mg in staurolite solid solutions is not known acurately, but it is likely that they will first introduce, and then raise, a minimum pressure necessary for staurolite stability, the pure Mg end member only being stable above 10 Kbar (Schreyer and Seifert, 1969b). A low-temperature stability limit for pure staurolite has not yet been defined. However, Richardson (1968) determined the stability limit of staurolite + quartz at 520°C–540°C for pressures greater than 1.5 Kbar. The significance of the stability relations of this quartz-bearing assemblage for low-pressure metamorphism will be discussed in section 4(i). A recent summary has been given by Ganguly (1972).

(4) *Chloritoid*

Although perhaps more common in high-pressure metamorphic rocks, chloritoid does also occur as a low-temperature mineral in low-pressure regional metamorphism and also in the outer parts of contact aureoles (cf. Halferdahl, 1961). In the pressure range considered here, the upper temperature stability of chloritoid is about 500°C–550°C according to Halferdahl (1961), but may vary somewhat with the prevailing oxygen fugacity (Ganguly, 1969). Grieve and Fawcett (1970a) were able to synthesise chloritoid at a pressure as low as 1 Kbar up to 525°C using the Ni–NiO buffer. The curve for the breakdown to Fe-cordierite + hercynite included in Fig. 13 lies at somewhat higher temperatures (QFM buffer) (Grieve and Fawcett, 1970b).

(5) *Epidote Group*

The monoclinic *clinozoisite-epidote* minerals, $Ca_2(Al,Fe^{+3})\ Al_2Si_3O_{12}(OH)$, and, more rarely, orthorhombic *zoisite*, $Ca_2Al_3Si_3O_{12}(OH)$, occur in relatively low grades of the type of metamorphism discussed here. Fe-rich epidotes are particularly common in metasomatic skarns formed at limestone contacts, but also occur in relatively low-grade metabasites. Experimental stability investigations are hampered by sluggish reaction rates, which may be one reason for the incompatible results presented by different authors. The stability of Fe-bearing clinozoisite-epidotes was studied by Merrin (1962) and more recently by Holdaway (1972). For the upper stability limit of pure, Fe-free zoisite Newton (1965) and later Boettcher (1970), reported values of about 560°C at 2 Kbar and 725°C at 5 Kbar thus indicating a rather flat positive slope of the univariant breakdown curve (Fig. 7). At low pressures the breakdown curves of the monoclinic phases generally lie at somewhat higher temperatures, the thermal stability increasing with the Fe^{3+}-contents of the epidotes (Holdaway, 1972).

Additional experimental work was aimed at determining the upper stability limits of the parageneses of epidote minerals with free quartz, which must necessarily lie at somewhat lower temperatures (e.g. Nitsch and Winkler, 1965). Of special importance seem to be the results of Holdaway (1966) indicating that, in the presence of quartz, relatively Fe-poor clinozoisites are replaced at increasing temperatures by the more stable Fe-rich epidotes. The latter phases may coexist with quartz to temperatures some 50°C above those determined for the upper stability limit of zoisite–quartz as determined by Newton (1966) and Boettcher (1970) included here in Fig. 7.

Although on the basis of experimental work the relative stability of zoisites versus clinozoisites of identical composition is not known as yet, it appears from the natural occurrences that zoisite formation is favoured by relatively high pressures. Possibly Fe-free clinozoisite is not a stable phase at all. At any rate the monoclinic series can accommodate higher amounts of Fe^{+3}. Therefore the stable coexistence of both zoisite and clinozoisite having different bulk chemistry should be possible and has been observed in nature (Myer, 1966; Ackermand and Raase, 1973). Since the assemblage zoisite-hematite has apparently not been found in rocks, one is inclined to conclude that the large field at high temperatures in which monoclinic Fe-bearing epidote is unstable does not overlap with the stability field of the most ferrian zoisite.

(6) *Vesuvianite*

The complex hydrous CaMgAl-silicate vesuvianite (*idocrase*) occurs chiefly in contact marbles and lime silicates, but has also been found in regional metamorphic calcareous schists formed at relatively low temperatures (Chatterjee, 1962). Although this mineral has been synthesised in the laboratory under various conditions (e.g. Coes, 1955; Walter, 1966), its growth field has only recently been determined (Ito and Arem, 1970). In agreement with the field evidence, the temperature range is rather extensive, from some 400°C up to about 700°C. However, compositional variations complicate the issue so that the true equilibrium stability limits of possible solid solutions are not known as yet. Clearly vesuvianites break down at higher temperatures into melilite–monticellite assemblages.

(7) *Melilite Group*

Melilites of variable composition may occur in lime silicate rocks formed through high-temperature contact metamorphism. In addition to the two well-known end members gehlenite, $Ca_2Al_2SiO_7$, and akermanite, $Ca_2MgSi_2O_7$, complex melilite solid solutions may also contain appreciable amounts of iron akermanite, $Ca_2Fe^{+2}Si_2O_7$, and soda melilite, $NaCaAlSi_2O_7$. Of these end members only the last mentioned is not stable as a pure phase at atmospheric pressure. Yoder (1964) found that it breaks down at pressures below some 5 Kbar into wollastonite and nepheline solid solution. An upper pressure limit of melilite solid solutions lies certainly well above 6 Kbar (cf. data for akermanite of Kushiro (1964a) and Yoder (1968); preliminary experiments on gehlenite by Kushiro (1964b)) and outside the range of metamorphism considered in this section. It is interesting to note, however, that ideally pure akermanite of stoichiometric composition is at 1 atm only stable above 1385°C: at lower temperatures the oxide ratio $2CaO \cdot MgO \cdot 2SiO_2$ is slightly changed and is variable (Schairer, Yoder and Tilley, 1967).

A lower temperature stability limit for melilites has so far only been determined for akermanite. According to Harker and Tuttle (1956b) and Yoder (1968) this phase breaks down in the absence of CO_2 at pressures between 1 and 6 Kbar and at temperatures below 700°C–750°C to form wollastonite + monticellite. In the presence of CO_2 the lower temperature limit of akermanite stability is raised considerably above this temperature provided the confining pressure is greater than about 900 p.s.i. (Walter, 1963a). These relationships are depicted in Fig. 5 and will be discussed in section 4(c). Gehlenite was obtained at a temperature of 800°C at 1 Kbar water pressure by Yoder (1950b). Gehlenite-forming reactions will be discussed in sections 4(f) and 4(g). All melilites are stable up to liquidus temperatures. The stability relations of akermanite- and soda-rich melilites were recently summarised by Yoder (1973).

(8) *High-temperature Lime-silicate and Lime-magnesia-silicate Minerals*

All the mineral phases to be considered under this heading occur exclusively in carbonate-bearing rocks of sanidinite facies environments characterised by *very* low pressures but rather high temperatures. These minerals are: *Monticellite*, $CaMgSiO_4$; *spurrite*, $Ca_5(SiO_4)_2CO_3$; *tilleyite*, $Ca_5Si_2O_7(CO_3)_2$; *merwinite*, $Ca_3Mg(SiO_4)_2$; *rankinite* and its low-temperature polymorph *kilchoanite*, $Ca_3Si_2O_7$; *larnite*, β-Ca_2SiO_4. In addition the melilite mineral *akermanite* (see 3(b)(7)) may, on chemical grounds, be also listed in this group. The stability relations of these minerals which all have rather similar compositions exhibit considerable complications, mainly because they vary strongly with the prevailing CO_2 pressure. For details the following original papers may be consulted: monticellite (Walter, 1963a); spurrite and tilleyite (Harker, 1959); merwinite (Walter, 1965; and Smulovich, 1969); rankinite, kilchoanite and larnite (Roy, D. M., 1958; and Agrell, 1965); akermanite (Walter, 1963a). Some relationships are also discussed in section 4(c) on metamorphic reactions.

(9) *Sapphirine*

Although most natural sapphirines, which have the approximate composition $(Mg,Fe)_2Al_4SiO_{10}$, occur in high-pressure granulite facies environments, the mineral has also been found in emery deposits formed by contact metamorphism (Friedman, 1952).

The stability of a pure Mg end-member sapphirine to very high temperatures above 1400°C was demonstrated by Forster (1950), but recent studies by Seifert (1974 and Fig. 11) indicate a surprisingly low breakdown temperature near 650°C at 1 Kbar water pressure for this phase. These latter results pose the problem of why sapphirines do not appear much more often in high-temperature, low-pressure metamorphic rocks. Because of the strong preference of Mg over Fe^{+2} in the sapphirine structure it seems probable that its rarity is due to compositional rather than to environmental (pressure–temperature) limitations (Seifert, 1974).

(10) *Pyroxene Group*

Among the pyroxenes of low-pressure metamorphism the chemically relatively simple *orthopyroxenes* $(Mg, Fe^{+2})_2Si_2O_6$, and *diopsidic clinopyroxenes* are predominant. Whereas the latter minerals are stable up to liquidus temperatures, orthopyroxenes show polymorphic transitions at high temperatures to proto- and clino-forms (cf. Foster, 1951, Boyd *et al.*, 1964, Lindsley and Munoz, 1969, Ernst and Schwab, 1970 and others) that might be useful for geothermometry.

The lower stability limit of pure *diopside*, $CaMgSi_2O_6$, is, for the natural environments in carbonate-bearing rocks, largely dependent on the mole fraction of CO_2 in the fluid phase. According to Metz and Winkler (1964) it may vary at 1 Kbar pressure from some 400°C to 540°C. However, in more recent work to be discussed in section 4(d) Metz (1970) specified the diopside stability in much more detail (see Fig. 6). The pure Mg-orthopyroxene *enstatite*, $Mg_2Si_2O_6$, has according to Greenwood (1963) a lower temperature stability limit at 2 Kbar water pressure near 700°C (Fig. 3). Incorporation of iron to form hypersthene will probably lower this temperature. Although the pure Fe end member ferrosilite, $FeSiO_3$, is not stable under the conditions of low-pressure metamorphism (Lindsley, 1965, Smith, 1971), orthopyroxenes (eulites) containing between 70 and 90 mol % $FeSiO_3$ occur in contact metamorphosed iron formations (e.g. Tsuru and Henry, 1937).

Whereas appreciable incorporation of Al into orthopyroxenes has formerly been expected to occur only at high pressures (Boyd and England, 1964), more recent experimental work by Anastasiou and Seifert (1972) has shown that even in the pressure range of 1–5 Kbar aluminous enstatites containing up to 9 wt % Al_2O_3 may be stable. Indeed in this pressure range the temperature dependence of solid solubility was found to be much more pronounced than the pressure dependence. Therefore, orthopyroxenes occurring for example in high-temperature hornfelses or sanidinites may contain Al_2O_3 and thus lend themselves for use as geothermometers, provided they coexist with more aluminous phases.

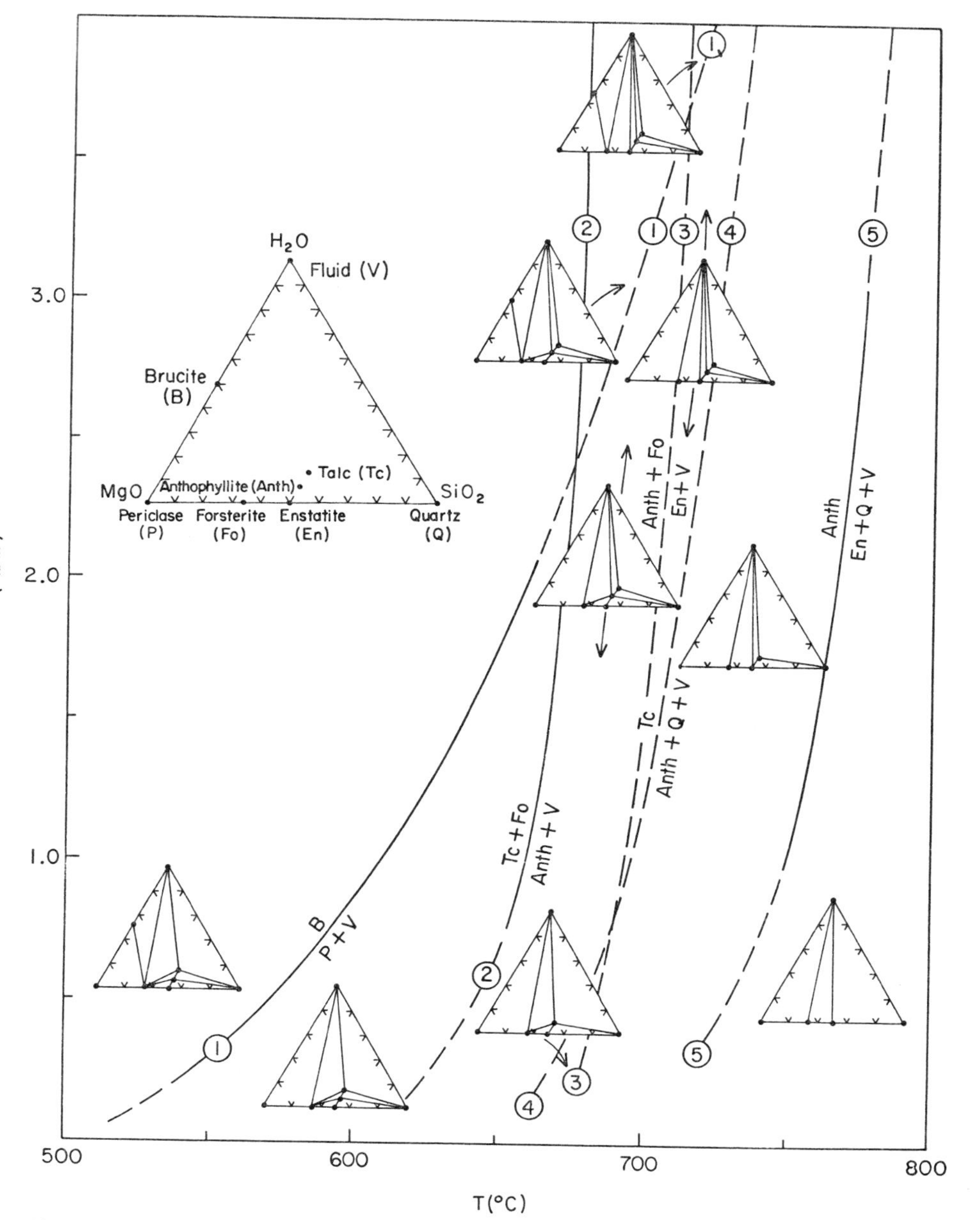

Fig. 3. Pressure–temperature diagram demonstrating important compatibility relations in the system MgO-SiO_2-H_2O at elevated temperatures and low pressures. Locations of univariant curves are based on the experimental data by Greenwood (1963) and Barnes and Ernst (1963). The phases involved in the reactions, and their abbreviations, are shown in an enlarged compositional triangle (upper left side).

(11) *Wollastonite*

As a typical mineral of lime silicate contact rocks and skarns, wollastonite, $CaSiO_3$, is stable only at relatively high temperatures especially at the higher confining pressures. In the presence of pure CO_2, wollastonite does not become stable until a temperature of some 670°C at 1 Kbar is attained (Harker and Tuttle 1956a and Fig. 1). Greenwood (1967a), however, has determined that wollastonite formation may take place at successively lower temperatures down to some 400°C, when the mol fraction of H_2O in the gas phase increases. This is partly included in Fig. 8 and is further discussed in section 4(f).

Although it is known from synthesis experiments that considerable amounts of Ca in wollastonite can be replaced by Fe^{+2}, the influence of this substitution on the lower stability limit of the wollastonite stability field has not been determined. However, the petrological significance may be limited because natural wollastonites are rarely rich in Fe.

The high-temperature polymorph of $CaSiO_3$, pseudowollastonite, is stable at low pressures only above some 1120°C to 1200°C (Buckner and Roy, 1955; Kushiro, 1964c) and can, therefore, only be expected in highest-temperature sanidinite facies environments.

(12) *Amphibole Group*

The most widespread amphiboles of low-pressure metamorphic rocks are *common hornblendes* of very complicated and variable compositions. Because of their chemical complexity little experimental work has been done to determine their stability ranges. As a first approximation Yoder and Tilley (1962) have determined the upper temperature stability limits of undefined hornblendes coexisting with other minerals in several basalts, and obtained values at 2 Kbar P_{H_2O} of some 920°C–1010°C. These high temperatures must still be regarded as minimum temperatures for the stability of the pure hornblende phases of complex compositions.

Lime silicate rocks usually contain chemically much simpler amphiboles such as members of the *tremolite-actinolite* series, $oCa_2(Mg,Fe^{+2})Si_8O_{22}(OH)_2$. The upper stability limit of pure Mg–tremolite as determined by Boyd (1959) is about 830°C at 1 Kbar (Fig. 1). For pure *Fe–tremolite* it is drastically lowered to some 465°C only at 1 Kbar for the QFM-buffer (Ernst, 1966). In addition, the appearance of Fe–tremolite is confined to oxygen fugacities defined by the QFM-buffer, or lower. Introduction of Al into these amphiboles usually causes expansion of their stability ranges towards higher temperatures. Thus pure *pargasite*, $NaCa_2Mg_4AlSi_6Al_2O_{22}(OH)_2$, is stable at 1 Kbar up to 1040°C (Boyd 1959 and Fig. 1), and its Fe^{+2}-analogue, *ferropargasite*, remains stable to considerably higher temperatures and even oxygen fugacities than Fe–tremolite (Gilbert, 1966). When the amphiboles in the lime silicate rocks coexist with carbonate minerals, the rock fluid will certainly contain considerable amounts of

CO_2. Under these circumstances the fugacity or mole fraction of CO_2 in the gas phase will strongly influence the upper and lower stability limits of the amphiboles. This problem will be discussed in more detail in section 4(d).

Orthoamphiboles of the *anthophyllite–gedrite* series may occur in metamorphosed ultrabasic rocks and, more importantly, in Al-rich hornfelses and gneisses together with cordierite (e.g. Eskola, 1914; Bugge, 1943; Abraham and Schreyer 1973). According to the recent analytical investigations of Robinson *et al.* (1971), natural gedrites always contain some Na although this element cannot be regarded as a necessary constituent of aluminous orthorhombic amphiboles. Thus in the nomenclature a distinction should be made between the Na-bearing gedrites and Na-free aluminous anthophyllites,

$$(Mg, Fe^{2+})_{6-5}Al_{1-2}[Si_6(Si,Al)_2O_{22}](OH,F)_2.$$

Pure anthophyllite, $oMg_7Si_8O_{22}(OH_2)$, has a narrow temperature stability range extending at 1 Kbar P_{H_2O} from some 660°C to 750°C (Greenwood 1963 and Fig. 3). With incorporation of iron this stability range is probably gradually shifted to lower temperatures (Hinrichsen, 1967). The stability relations of the aluminous anthophyllites are not very well understood as yet. Pure *MgAl-anthophyllites* have so far only been synthesised at high pressures, near 10 Kbar, and their stability field under these conditions lies at generally higher temperatures than that of pure anthophyllite (Schreyer and Seifert, 1969b). It is uncertain whether or not these pure, Fe-free synthetic orthoamphiboles would also be stable in the range of low-pressure metamorphism. However, with the introduction of Fe^{2+} alone Hinrichsen (1974) was able to synthesise Fe-bearing Al-anthophyllites with $Mg/(Fe^{2+} + Mg) = 0.5$ at water pressures as low as 5 Kbar. In addition, Hsu and Burnham (1969) working in a Na-free system obtained ferrous Al-anthophyllites even at 2 Kbar.

It is quite clear that Fe-bearing, Na-containing *gedrites* are stable under low pressure conditions. Using natural materials Akella and Winkler (1966) have determined a temperature stability range for an intermediate MgFe-gedrite ($Mg_{56}Fe_{44}$ in atomic proportions) extending at 1 Kbar P_{H_2O} from about 550°C to 700°C. Moreover, Hinrichsen (1968) has synthesised at this same pressure Na-bearing Fe-rich gedrites above temperatures ranging from 580°C (for Mg-free members) up to 680°C (for $Mg_{40}Fe_{60}$-gedrites). Amphibole synthesis was achieved, at 1 Kbar, only when 0.3 Na were substituted for R^{+2}, and, even with this provision, the Mg-rich members ($Mg/(Mg + Fe) > 0.30$) were not obtained.

Unfortunately the stability relations of the predominantly Fe-rich, monoclinic *cummingtonite–grunerite* series, $o(Mg,Fe^{+2})_7Si_8O_{22}(OH)_2$, are not yet known, because these clinoamphiboles could not be synthesised in the pure system, i.e. without the deliberate addition of impurities such as Ca (cf. Schürmann, 1967). It seems possible from their natural occurrences that grunerites are stable to relatively low pressures, whereas successively higher pressures are required for more Mg-rich members.

(13) *Mica Group*

Among the white micas pure *muscovite*, $KAl_2[AlSi_3O_{10}](OH)_2$, has the highest temperature stability limit (Velde, 1966), whereas *paragonite*, $NaAl_2[AlSi_3O_{10}]$-(OH_2), according to Chatterjee (1970) breaks down some hundred degrees lower (Fig. 1), and *margarite*, $CaAl_2[Al_2Si_2O_{10}](OH)_2$ is only stable up to about 510°C at 2 Kbar (Chatterjee, 1974). When coexisting with quartz the upper stability limits of all these micas will be lowered further (Evans, 1965; Chatterjee, 1972, and unpublished data on margarite–quartz). Therefore, only muscovite will be expected in the low-pressure metamorphic rocks formed at elevated temperatures, though there are some cases in which andalusite slates do contain paragonite, probably as a retrograde product (Harder, 1956).

In the biotite group *phlogopite*, $KMg_3[AlSi_3O_{10}](OH)_2$, is by far the most stable mica. At 2 Kbar P_{H_2O} it does not break down below 1180°C, i.e. outside the range depicted in Fig. 1, and under conditions of P_{H_2O} less than P_{tot} its upper temperature stability limit is still higher (Yoder and Kushiro, 1969). Incorporation of iron to form *biotites* on the join phlogopite-annite, $KFe^{+2}[AlSi_3O_{10}](OH)_2$, reduces these temperatures drastically, especially at higher oxygen fugacities (Wones and Eugster, 1965). This explains why biotites, although very common in pelitic hornfelses and gneisses, are often found to be replaced by anhydrous phases; for example in buchites they are altered mainly to cordierite (Ramdohr, 1919). Biotites more aluminous than those of the phlogopite-annite join can be expected to be stable to higher temperatures. Rutherford (1973) could show that Mg-free biotites along a join between annite and a theoretical pure aluminium biotite, $K_2Al_6[Al_2Al_6O_{20}](OH)_4$ are stabilised to higher temperatures with increasing Al-contents, the most aluminous synthetic biotite containing 25% of the theoretical end member.

(14) *Chlorites*

Judging from the rarity of chlorites in high-temperature low-pressure metamorphic rocks it is rather surprising to note that their upper temperature stability limit lies at fairly high temperatures. At 2 Kbar water pressure the most stable member of the chlorite series, which according to Nelson and Roy (1958) is pure *clinochlore*, $Mg_5Al[AlSi_3O_{10}](OH)_2$, does not break down until a temperature near 720°C is attained (Fawcett and Yoder, 1966). Its upper temperature stability limit is included in Fig. 9 (see section 4(g)). Chlorites with different Al contents and, in particular, those incorporating iron have smaller stability ranges (Turnock, 1960). Yet the virtual absence of chlorite in the metamorphic rocks considered here, except for their lowest grades (spotted slates), must be explained on the basis of more complex metamorphic reactions that consume chlorite (see sections 4(g) and 4(h)).

(15) *Talc*

The appearance of talc, $Mg_3[Si_4O_{10}](OH)_2$, in contact and other lower-pressure types of metamorphism is restricted to siliceous dolomites and ultrabasic rocks of rather low grades. It is surprising, therefore, that the upper temperature stability limit of the pure phase lies as high as about 700°C (Greenwood 1963; and Fig. 3 in section 4(a)). As for the chlorites this apparent discrepancy is explained by the influence of the bulk chemistry of the rock, as will be discussed in section 4(d).

(16) SiO_2-*Polymorphs*

Most of the rocks discussed in this section have formed in the stability field of quartz, in particular of *high-quartz*, i.e. above some 575°C–590°C depending on pressure (Yoder, 1950a). The occurrence of *tridymite* in buchites and sanidinites may be regarded as evidence for very high temperatures in combination with relatively low pressures. Although tridymite could not be synthesised from ideally pure SiO_2 (e.g. Flörke, 1955), the amounts of impurities or fluxes possibly necessary to stabilise this phase are trifling and probably present everywhere in natural environments. However, the nature and quantities of such impurities may influence the stable inversion temperature quartz $\rightleftharpoons$ tridymite usually given as 870°C for 1 atm (e.g. Flörke and Langer, 1972). Thus the univariant pressure-temperature curve for this inversion as determined in the system SiO_2–H_2O (Tuttle and England, 1955; Tuttle and Bowen, 1958) and reproduced in the pressure–temperature plot in Fig. 1 should be applied with some caution to the natural case. Moreover tridymite and, in particular, cristobalite believed to be stable only above 1470°C at 1 atm are notorious for their metastable formation under much lower temperatures. Clearly metastable cristobalite is found in diatomites contact metamorphosed by rhyolites (e.g. Novak, 1967), and both tridymite and cristobalite occur frequently in late vesicles of volcanic rocks (e.g. Larsen *et al.*, 1936), and in fritted sediments (porcellanites) within the igneous rock (e.g. Agrell and Langley, 1958).

(17) *Feldspars*

The pressure–temperature stability relations of the feldspars, especially in the presence of H_2O, are still not very well known. For the highest temperature environments considered, i.e. above some 700°C, there is complete solid solution within the *alkali feldspar series* ($NaAlSi_3O_8$–$KAlSi_3O_8$) and, for all practical purposes, within the *plagioclase series* ($NaAlSi_3O_8$–$CaAl_2Si_2O_8$). However, there is only very limited solid solubility between these two series as found by analyses of natural materials (e.g. Rahman and MacKenzie, 1969) as well as by experiment (Seck, 1971a and b). With decreasing temperatures *alkali feldspars* unmix into albite- and K feldspar-rich portions (perthites) following the solvus curve in that system (Orville, 1963; Luth and Tuttle, 1966; Thompson and

Waldbaum, 1969; Seck, 1972). These two feldspars are stable down to room temperatures. Pure *anorthite* forms only at elevated temperatures (300°C–400°C), but is also stable over most of the low-pressure high-temperature region to be considered in this chapter (cf. Boettcher, 1970, Nitsch, 1972; see also section 4(e)). However, practically nothing is known experimentally about the low-temperature stability limits of the intermediate plagioclases as a function of (water) pressure. Mainly on the basis of reported observations of plagioclase exsolutions and coexistences in medium grade metamorphic rocks as well as of crystallographic considerations, Smith (1972) presented a hypothetical TX-phase diagram for plagioclases for an undefined pressure. If this is correct, there should be, even above 700°C, a miscibility gap in the range An_{70-90} which extends up to liquidus temperatures under conditions of truly stable equilibrium. This diagram does not account for Wenk's (1962) finding of an increase in the anorthite content of plagioclases coexisting with calcite with rising grade of regional metamorphism. The recent study by Frey and Orville (1974) suggests that the relations may actually be more complicated in nature with other Ca-bearing phases appearing.

More complexities displayed by the feldspars are related to the extreme variability of their structural states exhibiting different degrees of Al/Si ordering as a function of environmental conditions. Despite extensive experimental work the *stable* transition temperatures of the many feldspar polymorphs encountered are largely unknown, the main reason being that during the relatively short durations of the experiments equilibrium could only rarely be attained. The observed structural states of natural feldspars will thus also be time-dependent and probably reflect the cooling history of the rock rather than its environmental conditions during the peak of metamorphism. Only for alkali feldspars containing less than 65% albite can it be stated with some confidence that the polymorph stable in the realm of high-temperature low-pressure metamorphism is a *monoclinic sanidine*, or *Na-sanidine*. Goldsmith and Laves (1954) were able to convert the triclinic form of K-feldspar, microcline, into this form at a temperature as low as 525°C. The monoclinic/triclinic inversion temperature is, however, raised with increasing Na-content, monoclinic albite (= monalbite) being stable only above 980°C, and inverting again on cooling (Laves, 1960).

(18) *Cordierites*

Because of its extensive stability field at low pressures (Schreyer and Yoder, 1964) the framework silicate cordierite is one of the prominent phases of the type of metamorphism considered here. At 2 Kbar water pressure the most stable end member, pure Mg-cordierite, $Mg_2Al_4Si_5O_{18} \cdot xH_2O$, has its lower stability limit near 500°C (Seifert and Schreyer 1970 and Fig. 1) and persists up to its incongruent melting point. The lower temperature stability limit of Fe-cordierite appears to be similar to that of Mg-cordierite (Schreyer, 1965b), though its variation with changing oxygen fugacities has not been investigated as yet. It is important to note that there exists an upper pressure stability limit of Fe-cordierite near 3.5 Kbar for the oxygen fugacity of the quartz–fayalite–magnetite (= QFM) buffer (Richardson 1968 and Fig. 13 in section 4(i)). This limit is well within the pressure range discussed in this section. The oxygen fugacities indicated for regional metamorphic rocks are similar to those created by the QFM buffer, which would require that iron-rich cordierites occur only in low pressure environments as suggested by Chinner (1959). Further details will be given in section 4(i).

On the other hand, it should be emphasised once again that pure Mg-cordierite must be stable up to pressures near 10 Kbar, since it is common in the granulite facies and other high-pressure rocks (Schreyer, 1968, Schreyer and Seifert, 1969a). Newton (1972) has pointed out, however, that hydrous cordierites containing H_2O in structural channels are pressure–stabilised with respect to cordierites having open channels.

The stability field of the theoretical end member Mn-cordierite was recently found by Dasgupta *et al.* (1974) to be restricted to very low pressures well below 1 Kbar. Thus, assuming that Mg,Fe^{+2}, and Mn^{+2} would be equally available in natural environments, the decreasing pressure stabilities of the end member phases in the order Mg-cordierite, Fe-cordierite, Mn-cordierite would also be reflected in the composition of the average natural cordierite in which $Mg > Fe^{+2} > Mn^{+2}$ (Schreyer, 1965a). These relations also strongly influence the distribution of Mg, Fe^{+2}, and Mn^{+2} among cordierite and other coexisting ferromagnesian minerals such as biotite and garnet, a subject investigated long ago by Eskola (1915). Changing $Mg/(Mg + Fe^{+2})$-ratios of cordierites and garnets as a function of pressure, temperature and the remaining mineral assemblages have been investigated by Hensen and Green (1973). As previously recognised by Chinner (1959), Mg–Fe-distribution may be used for determining the physical conditions of formation of cordierite-garnet bearing rocks that appear in the realm of low-pressure, high-temperature metamorphism (e.g. Chinner, 1962, Okrusch, 1969).

Cordierites, regardless of composition, are stable phases at liquidus temperatures. Therefore they occur in buchites, paralavas and other rocks formed by partial melting, and, of course, in igneous rocks of appropriate compositions.

At equilibrium even the cordierites formed under the highest temperatures in nature are orthorhombic "low"-cordierites, hexagonal high-cordierites being metastable phases for all natural conditions (Schreyer and Schairer, 1961). Therefore, different structural states of natural cordierites, produced by Al/Si order–disorder as in the feldspars, cannot be used as geothermometers but are a reflection of the complete thermal history of these minerals. This may lead to the seeming irregularity that "higher" structural states are found in cordierites formed under conditions of lower metamorphic grades (Schreyer, 1966; Harwood and Larson, 1969).

(19) *Osumilite*

Although very rare the mineral osumilite,

$$(K,Na,Ca)(Mg,Fe^{+2})_2(Al,Fe^{+3},Fe^{+2})_3(Si,Al)_{12}O_3 \cdot H_2O$$

is clearly confined to high-temperature, low-pressure environments. In addition to its occurrence in vesicles in volcanic rocks (Miyashiro, 1956) osumilite has recently been found in buchites (Chinner and Dixon, 1973). On the basis of an experimental study of the synthetic end member $KMg_2Al_3Si_{10}Al_2O_{30} \cdot xH_2O$, Schreyer and Seifert (1967) concluded that this osumilite is metastable at all temperatures and pressures relative to such common mineral assemblages as cordierite + K feldspar + quartz, or biotite + muscovite + cordierite + quartz. Osumilite persistence in nature may be favoured by the relatively short periods of heating typical for many natural sanidinite-facies, high-temperature environments. Nevertheless, it cannot be ruled out that some complex osumilite solid solutions may have small stability fields at very low pressures(< 1 Kbar) and high temperatures. Clarifying experiments in this range will be difficult due to slow reaction rates.

(20) *Carbonates*

Only the most common carbonates *calcite*, $CaCO_3$, and *dolomite*, $CaMg(CO_3)_2$ need to be considered here. Whereas calcite usually remains stable up to the highest temperatures attained in low-pressure contact metamorphism (cf. Harker and Tuttle, 1955a and Fig. 1), dolomite breaks down into the assemblage calcite + *periclase* (MgO). According to Harker and Tuttle (1955a) this disintegration of dolomite which is accompanied by loss of CO_2 takes place in the presence of excess CO_2 between about 600° and 900°C with temperatures rising the higher the pressures. A portion of this reaction curve is shown in Fig. 5.

It is also of considerable interest that Mg may enter the calcite structure at elevated temperatures up to a maximum amount of about 25 wt % $MgCO_3$ to form *magnesian calcites* (Harker and Tuttle, 1955b). Since the pressure effect on this solid solubility is only small, and also known quantitatively (Goldsmith and Newton, 1969), assemblages of calcite with dolomite or periclase may be used as geothermometers (Carpenter, 1967). The stability relations of the pure Mg-carbonate *magnesite* will be dealt with in section 4(b).

(21) *Periclase and Brucite*

The periclase formed in dolomite marbles through the mechanism just outlined is in many natural rocks altered retrogressively through introduction of water into *brucite*, $Mg(OH)_2$. The equilibrium curve between brucite and periclase in the presence of excess water has been determined by many investigators with considerably different results. The latest study which also evaluates earlier results, is that of Weber and Roy (1965). In Fig. 3 the equilibrium curve determined by Barnes and Ernst (1963) is used, because it was followed to higher pressures.

(22) *Sulphides*

Only the most important iron sulphides, *pyrite* and *pyrrhotite*, will be considered here briefly. For high-temperature metamorphism it is important that pyrite, FeS_2, decomposes in the presence of excess sulphur at temperatures between roughly 750°C and 800°C depending on pressure (Fig. 1) to form pyrrhotite and a sulphur-rich liquid or gas (Kullerud and Yoder, 1959). At lower sulphur fugacities this breakdown will occur at even lower temperatures (cf. Toulmin and Barton, 1964). Therefore, pyrrhotite is the dominant Fe-sulphide in shallow high-temperature contact rocks. Since the degree of metal deficiency in their general formula $Fe_{1-x}S$ is, in the presence of excess S, a function of temperature (Arnold, 1962) pyrrhotites can be used as geologic thermometers, providing at least minimum temperatures of metamorphism.

(23) *Temperature-pressure indifferent Minerals*

In contrast to all the mineral phases discussed previously there are some that are stable over the entire pressure–temperature range considered in this chapter, independently of the availability of water, and, with one exception, also of CO_2. These minerals are *corundum*, *spinel* and the *olivines*, which are stable up to the highest liquidus temperatures and which break down in the presence of water into hydrous assemblages at temperatures in the order of only 250°C–400°C. Experimental data have been provided for corundum by Kennedy (1959), for spinel by Roy, Roy and Osborn (1953), for forsterite by Scarfe and Wyllie (1967), and for fayalite by Flaschen and Osborn (1957). Under the influence of high fugacities of CO_2 only forsterite becomes unstable at temperatures below some 550°C at 2 Kbar total pressure, giving rise to carbonate assemblages of magnesite with enstatite, anthophyllite, talc or even quartz (Johannes, 1969). Some more aspects of this problem are discussed in section 4(b) and depicted in Fig. 4.

4. METAMORPHIC REACTIONS IN MODEL SYSTEMS

Knowledge of the experimentally determined *stability relations* of rock-forming minerals under varying physical and chemical conditions is undoubtedly one of the fundamentals of modern metamorphic petrology. Merely from the overlap or lack of overlap of stability fields in the pressure-temperature diagram (e.g. Fig. 1) it can be decided whether or not coexistence of certain minerals in a rock is, *in principle*, possible under conditions of equilibrium. However, having established this possibility stable coexistence is by no means *certain* and more experimental work concerning the *compatibility relations* of the requisite minerals is necessary. The result of this work will be the delineation of stability fields of characteristic *mineral assemblages* of metamorphic rocks. As in the case of single minerals these stability fields are limited by *chemical reactions* leading to breakdown, consumption, or incompatibility of some of the mineral phases.

The basic law underlying the chemical reactions and compatibility relations of minerals in metamorphic rocks is, of course, the *phase rule*

$$P = C - F + 2$$

which controls the number of coexisting phases (P) as a function of the number of components (C) available in the chemical system, and of the degrees of freedom (F) for the intensive variables, that is, most importantly, for pressure and temperature. As an example five phases may co-exist in a three-component system at one particular pressure and temperature. This *invariant* situation, expressed as a uniquely defined *point* in the pressure-temperature diagram, is changed into a *univariant curve* through gaining one degree of freedom and thus through the loss of one of the phases. Consequently in the *divariant fields* of the pressure-temperature diagram, characterised by two degrees of freedom, three phases may coexist in the same system. Stability fields of single minerals as well as of most mineral assemblages representing equilibrium conditions in nature are divariant or have still higher variances. Reactions between minerals often occur along univariant curves separating the various stability fields, but they may also be multivariant and thus proceed over a whole range of pressure-temperature conditions forming an at least divariant field on their own.

In order to acquaint the reader with the most important principles of metamorphic reactions, mineral compatibility and reaction relationships are discussed in ten model systems of silicates which are of particular relevance to natural rocks. The reader will notice, however, that none of these relatively simple systems (six components at most) represents an adequate model for the large group of metabasic rocks, which consist of the oxides CaO, Na_2O, MgO, FeO, Fe_2O_3, Al_2O_3, SiO_2, and H_2O as major components. Therefore, the metabasites, although particularly sensitive in their mineralogy to metamorphic changes, are only partly covered by this chapter.

(a) *The System* MgO-SiO_2-H_2O

Phase relations in this system may be applied to most dunitic ultrabasic rocks low in Ca and Al, but also to talc schists and anthophyllite bodies. The experimental data were compiled by Bowen and Tuttle (1949) and later supplemented by Greenwood (1963), who introduced anthophyllite as a stable phase even in the presence of excess water.

The most important reactions for high-temperature low-pressure metamorphism taking place under the condition $P_{fluid} = P_{tot}$ are summarised in Fig. 3. They involve the phases brucite, periclase, talc, forsterite, anthophyllite, enstatite, quartz and a water-rich gas phase (fluid). At the lowest temperatures and pressures considered in the present context only brucite, forsterite, talc and quartz may coexist with the gas. The phases periclase, enstatite, and anthophyllite occur only in the "water-deficient region" (Yoder, 1952) of the system, meaning that these phases are incompatible with fluid under these conditions. With increasing temperature the coexistence of periclase with fluid is established first through dehydration of brucite (reaction curve *1* of Fig. 3) provided the pressure is below some 2.5 Kbar. At still higher temperatures the coexistence of talc and forsterite is discontinued through reaction *2* to form anthophyllite plus water-rich gas. With further increase in temperature and at pressures above some 900 bars forsterite can no longer coexist with anthophyllite, because their connecting tie line disappears for the sake of the new tie line enstatite-gas (reaction *3*). Because of its relatively flat slope the univariant curve of reaction *1* intersects that of reaction *2* and probably also that of reaction *3*, thus resulting in a reversal of the reaction sequence at higher fluid pressures. A similar reversal of reactions may possibly occur at lower pressures, where the curves of reactions *3* and *4* appear to intersect. Through this measure talc would become unstable relative to anthophyllite + quartz + fluid *before* enstatite coexists with gas. With reaction *5* anthophyllite is finally dehydrated to form the stable assemblage enstatite plus quartz co-existing with gas.

Independently of the various intersections it is clear that reactions *1* through *5* provide a *step-wise dehydration of the stable mineral assemblages with rising temperatures*. This is one of the *fundamental principles of metamorphism*, especially at low pressures. The univariant curves of the dehydration reactions commonly have relatively steep dP/dT slopes, that is the reactions are mainly temperature-dependent, and to a much lesser extent pressure-dependent.

Although more recent thermodynamical calculations (Zen, 1971; Greenwood, 1971) introduce new uncertainties concerning the stability of the phase anthophyllite, the experimental data by Greenwood (1963) strongly suggest that its stability field extends at least to a pressure of 3 Kbar. For higher pressures, however, preliminary experimental results by Seifert (personal communication, 1974) indicate that the lower temperature stability limit of anthophyllite will be given by a new breakdown reaction to form enstatite + talc not involving any fluid. Thus the relations outlined in Fig. 3 will be changed drastically at pressures in excess of about 5 Kbar. The slope of the new reaction curve must, for chemographic reasons, be flatter than that of curve *2* in Fig. 3; therefore, the stable appearance of anthophyllite in the presence of excess H_2O will be confined to a still narrower temperature range the higher the pressure.

(b) *The System* MgO-SiO_2-H_2O-CO_2

This system is of interest in the present context for two reasons:

(i) The phase relations in this system can be compared with those discussed in the last paragraph for the CO_2-free system, and thus the principal behaviour of rocks under the influence of *two volatile components* may be demonstrated.

(ii) When undergoing metamorphism some ultrabasic rocks may be subjected to the influence of a gas phase rich in CO_2 thus leading to mineral assemblages carrying magnesite, $MgCO_3$. For example the unusual enstatite–magnesite rocks ("sagvandites") of Northern Norway (cf. Barth, 1930; Schreyer *et al.*, 1972) or talc-magnesite rocks from the Alps and elsewhere may be cited (cf. Johannes, 1969, 1970).

Experimental data and theoretical deductions on this system were reported by Greenwood (1967b) and Johannes (1969). The result is a very complicated series of reaction relationships. For the sake of simplicity Fig. 4 gives only a small selection of typical reaction curves relevant to high-temperature metamorphism. For a more complete view of the phase relations the reader is referred to the original literature or to Part II, Section A, Chapter 2 of this book.

The main difference between the present CO_2-bearing system and the purely hydrous one of section 4(a) is that the fluid is no longer of relatively constant, water-rich composition (Fig. 3), but may vary, essentially within the two-component system H_2O–CO_2. Thus a new extensive variable, the composition of the co-existing fluid, defined as its mole fraction of CO_2,

$$X_{CO_2} = \frac{CO_2}{CO_2 + H_2O},$$

may be introduced to describe the phase relations. One convenient way of description as used in Fig. 4 is to plot X_{CO_2} against temperature with total fluid pressure held constant. Only with this provision do reactions take place along (isobarically) *univariant* curves as previously discussed. However, if one used a simple fluid-pressure temperature diagram as in Fig. 3 thus disregarding the variable fluid compositions, all reactions would become *divariant*, that is they could proceed over the more or less wide temperature ranges depicted in Fig. 4 even at constant pressure. Because of this relationship *no geothermometry based on any mineral reactions involving a multi-component gas phase* can be applied to natural rocks *unless there is independent evidence about the composition of this gas phase* as well as about total pressure. The composition of the co-existing gas phase in turn may be governed by large "outside" reservoirs such as in hydrothermal activity causing metasomatism, or else it may be buffered by a specific mineral assemblage present in sufficiently large amounts. An example of the latter behaviour will be given in section 4(d).

It may be seen from Fig. 4 that pure dehydration reactions such as reaction *5* of Fig. 3,

$$\text{anthophyllite} \rightleftharpoons \text{enstatite} + \text{quartz} + H_2O$$

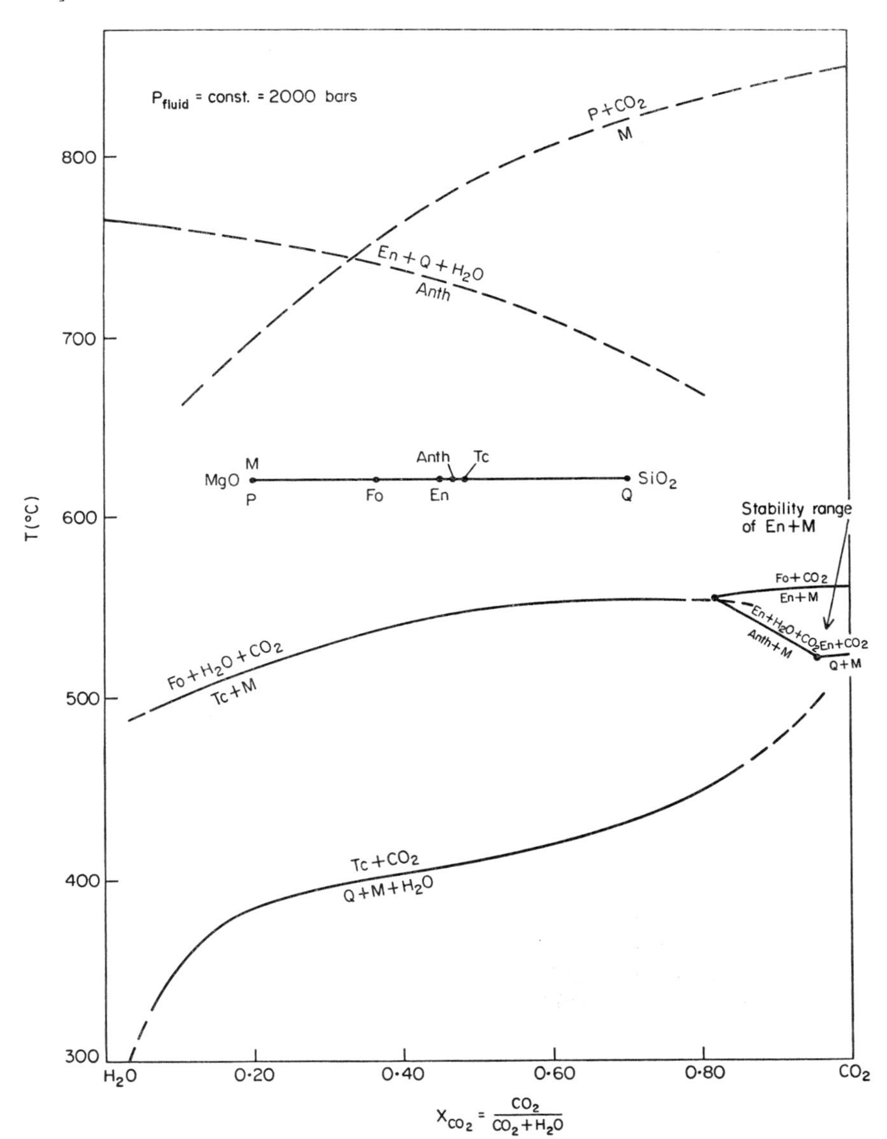

Fig. 4. Some selected reaction relationships within the system MgO-SiO_2-H_2O-CO_2 depicted in an isobaric TX_{CO_2} section for 2 Kbar fluid pressure. Curve locations are based on experimental data compiled by Johannes (1969). The crystalline phases involved in the reactions are projected into the binary system MgO-SiO_2 shown in the central portion of the figure. For discussion see text.

take place at gradually decreasing temperature the more the fluid phase is enriched in CO_2. Similarly, pure decarbonation reactions such as the breakdown of magnesite,

$$\text{magnesite} \rightleftharpoons \text{periclase} + CO_2$$

also take place at drastically lower temperatures when the CO_2-gas is diluted by water. The relationship becomes more complicated for reactions involving dehydration and decarbonation at the same time. An example in Fig. 4 is the reaction

$$1\text{talc} + 5\text{magnesite} \rightleftharpoons 4\text{forsterite} + 5CO_2 + 1H_2O.$$

In this case the reaction curve exhibits a temperature maximum at an intermediate value of X_{CO_2}, which represents the molar fractions of H_2O and CO_2 produced by the reaction itself. This value may even lie along the metastable extension of the curve which is true for the case cited in Fig. 4 with an X_{CO_2} of 5/6 according to the coefficients of the above equation. Finally there is the possibility that a reaction may consume CO_2 simultaneously with releasing H_2O or vice versa. Such cases exemplified by the reaction

$$\text{magnesite} + \text{quartz} + H_2O \rightleftharpoons \text{talc} + CO_2$$

of Fig. 4 are characterised by univariant curves with a turning point. For the theory underlying these relationships in two-volatile systems see Greenwood (1962) and (1967b).

It is also important to note from Fig. 4 that certain *mineral assemblages* may be *confined to special ranges of compositions of the coexisting fluid* in addition to a limited temperature stability range. The enstatite–magnesite assemblage of the "sagvandites" previously mentioned provides an impressive example. This assemblage can only become stable when the gas contains more than about 85 mol $\%CO_2$. Such X_{CO_2}-critical mineral assemblages may then provide the independent evidence about fluid compositions as previously demanded. However, the limiting X_{CO_2}-value given for the assemblage enstatite–magnesite is only valid for the isobaric condition at 2 Kbar as depicted in Fig. 4. With increasing total pressure the enstatite–magnesite field will expand towards considerably lower X_{CO_2} values (cf. Schreyer *et al.*, 1972). In addition, due to the different phase relations in the limiting system MgO-SiO_2-H_2O above 5 Kbar (section 4(a)) the entire topology of the TX_{CO_2}-sections will be changed from that given by Greenwood (1967b) and Johannes (1969) through several steps at intermediate pressures (Evans and Trommsdorff, 1974). Thus the compatibility relations of the ultramafic minerals forsterite, enstatite, anthophyllite, talc, and magnesite in CO_2-metasomatic ultrabasites are very sensitive not only to gas composition but also total gas pressure.

Instead of using the composition of the coexisting gas phase, X_{CO_2}, as a variable influencing the phase relations in two-volatile systems one can, of course, also refer to partial pressures of the gas species involved, i.e. P_{CO_2} and P_{H_2O}. For non-ideal gases such as those considered here it would be preferable to use their fugacities f_{CO_2} and f_{H_2O}, or, even better, their chemical potentials μ_{CO_2} (= log f_{CO_2}) and μ_{H_2O} (= log f_{H_2O}). A diagram, in which μ_{CO_2} is plotted versus μ_{H_2O} for the system MgO-SiO_2-H_2O-CO_2 has recently been worked out by Finger and Burt (1972). The main advantage of this way of representation is that, for a particular set of total pressure and temperature, not only can the mineral assemblages coexisting with the gas phase along the saturation line be seen but also those present in the "gas-deficient region", in which H_2O and CO_2 may be present as mobile components, but not as a gas phase.

(c) *The System* CaO-MgO-SiO_2-CO_2

This system provides the simplest model for metamorphic reactions in siliceous carbonate rocks, comprising both limestones and dolomites, as demonstrated already by Bowen (1940) on the basis of natural mineral assemblages. Whereas at relatively low temperatures water must also be regarded as a necessary component (see 4(d)), the water-free system discussed here may directly be applied to sanidinite-facies reactions that take place at temperatures above those of most dehydration reactions. These sanidinite-facies reactions involve such rare minerals as monticellite, merwinite and others named in section 3(b)(8).

The most important compatibility relations were determined experimentally by Walter (1963a, b; 1965). The results of his work and of additional phase theoretical deductions by the present author are summarised in the P_{CO_2}-T plot of Fig. 5. The phase relations depicted involve only the minerals dolomite, calcite, periclase, akermanite, wollastonite, diopside, forsterite, monticellite, merwinite, spurrite, and a CO_2-rich gas phase (fluid). They are only valid for the condition that CO_2 is present in excess amounts of those necessary to form the most carbonated mineral assemblages stable at the various pressures and temperatures. The additional minerals tilleyite, rankinite and larnite, although stable in some portions of the PT-range of Fig. 5, had been disregarded in the experimental studies of Walter (1963a, b; 1965) and were, therefore, not included here. They will, however, certainly cause further complications of the phase relationships.

The univariant reaction curves shown in Fig. 5 intersect to form four invariant points I_α through I_δ. I_α, I_β and I_γ were recognised by Walter (1965); I_δ is a necessary consequence of the intersection of reaction curve monticellite + wollastonite = akermanite with reaction curve calcite + wollastonite = spurrite + CO_2. According to a Schreinemakers analysis (see Zen, 1966) of the phase relations around I_δ there is one additional reaction curve originating in I_δ, viz. calcite + akermanite = spurrite + monticellite + CO_2, which had been listed by Walter (1965) as reaction 10. Since this reaction is also involved in the phase relations around invariant point I_γ (Walter, 1965), its univariant curve in fact connects the two invariant points I_δ and I_γ.

The main petrological applications of Fig. 5 to the natural mineral assemblages of siliceous carbonate rocks are:

(1) All univariant curves with relatively flat positive dP/dT slopes at these very low pressures characterise reactions involving *decarbonation* with increasing temperatures. This is in agreement with Bowen's (1940) observation of a discontinuous decarbonation series in prograde metamorphism of impure carbonates.

(2) There are three reaction curves originating from the invariant points I_α, I_β and I_γ, respectively, which are characterised by remarkably steep

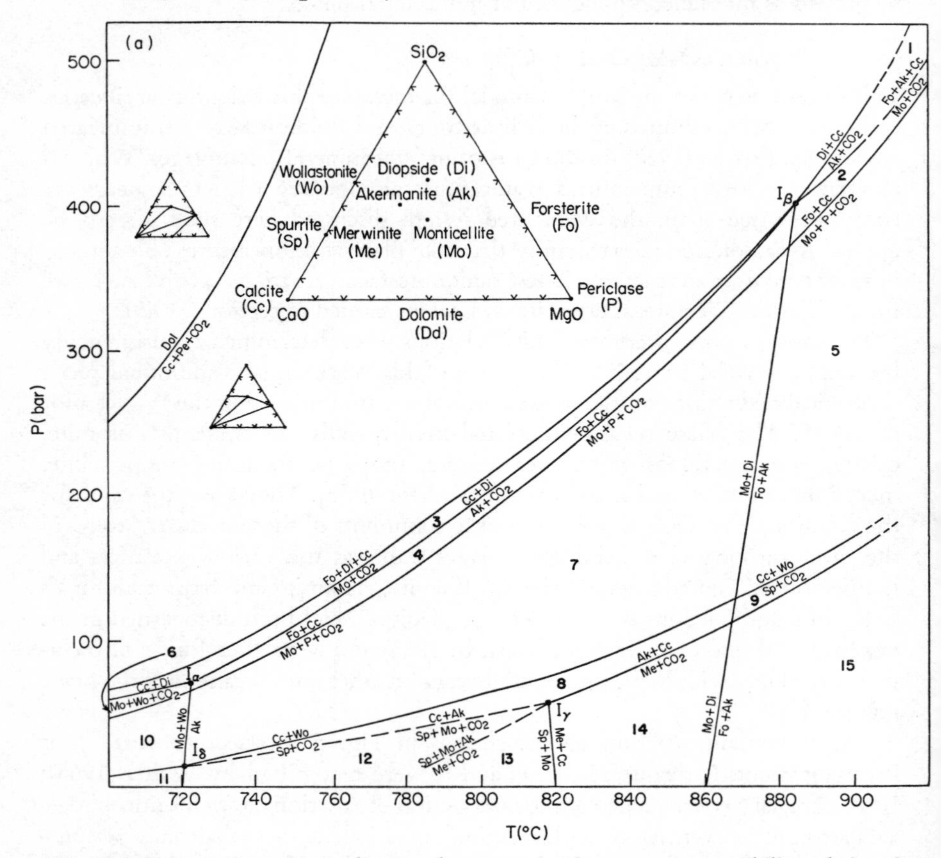

Fig. 5(a). Pressure–temperature diagram demonstrating important compatibility relations in the system CaO-MgO-SiO_2-CO_2 at high temperatures and very low pressures for the limiting conditions of fluid pressure equalling total pressure. Locations of univariant curves are based on the experimental data of Walter (1963a, b; 1965). The crystalline phases involved in the reactions, and their abbreviations, are shown in an enlarged compositional triangle representing a projection of the quaternary system onto its CO_2-free base (upper left).

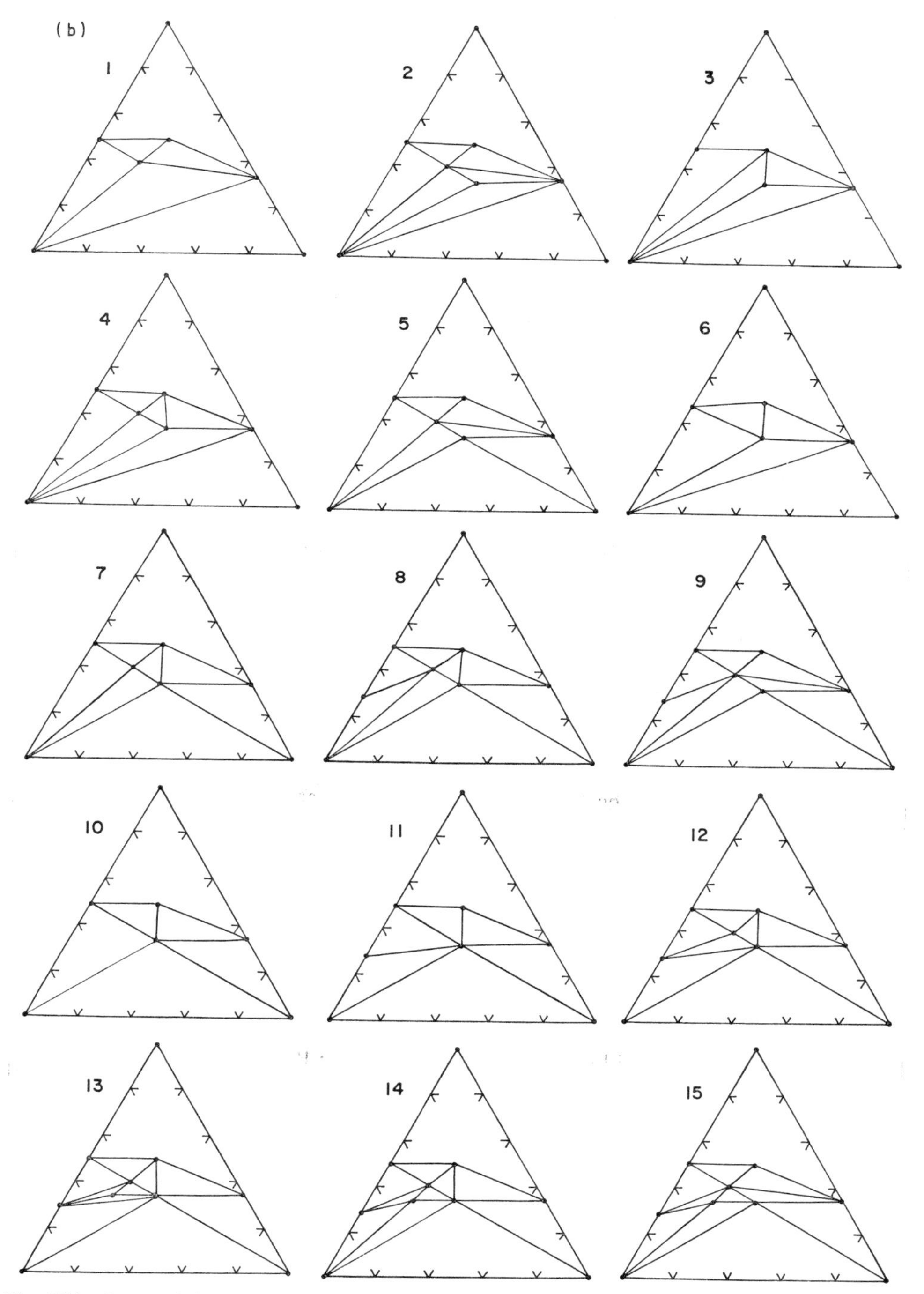

Fig. 5(b). Compatibility triangles for the numbered fields shown in (a).

slopes. They apply to reactions taking place *among solid phases only*, i.e. without interaction between solids and the gas phase. The steep slopes indicate that these reactions are accompanied by only small volume, but considerable entropy, changes.

(3) Mineral assemblages involving the phases monticellite, akermanite, spurrite, and merwinite are generally *limited to very low pressures* and relatively high temperatures, provided the CO_2 pressure equals total pressure. These conditions obtain in shallow intrusive contacts such as Scawt Hill, Northern Ireland (Tilley, 1925) or Crestmore, California (Burnham, 1959). With increasing temperature and/or decreasing pressure of metamorphism the sequence of appearance of individual minerals within the contact aureole is monticellite–akermanite–spurrite–merwinite.

(4) The mineral paragenesis *forsterite + calcite*, which is often found in nature as spectacular intergrowths sometimes called "ophicalcite" (Rost and Hochstetter, 1964) is confined to relatively high pressures and/or low temperatures. At lower pressures and/or higher temperatures this assemblage is decarbonated to form the pair monticellite + periclase.

(5) As a consequence of the four invariant points as well as of the various other intersections of reaction curves without formation of invariant points the *sequence* of metamorphic reactions taking place under isobaric conditions upon rise of temperature is strongly dependent on the magnitude of CO_2-pressure and may thus also be a function of the depth of burial. As an example the three-phase assemblage forsterite–akermanite–monticellite may be considered, which is stable at 1 atm only above some 860°C. At pressures up to that of the invariant point I_β ($\sim$400 bars) this assemblage forms through a reaction involving the pre-existing paragenesis diopside–monticellite. At pressures greater than 400 bars the same assemblage forms through reaction of forsterite + akermanite with additional calcite to produce the monticellite.

(6) Over most of the pressure-temperature range depicted in Fig. 5 the carbonate mineral dolomite is not a stable phase. It is replaced by the assemblage of periclase plus calcite. Therefore, even pure dolomite rocks will be altered under the influence of this type of low-pressure contact metamorphism to form contact marbles consisting of calcite and Mg-oxide and/or, because of its frequent retrograde hydration, of brucite, $Mg(OH)_2$. This marble variety is often referred to as *predazzite* after the historically famous locality Predazzo, Southern Alps, Italy, in which contact metamorphism of fossiliferous carbonate rocks (Triassic) by a "granite" enabled the young age of the igneous rock to be first recognised.

(d) *The System* CaO-MgO-SiO_2-CO_2-H_2O

Bowen (1940), Tilley (1948), Turner (1967) and many others have pointed out that low-pressure metamorphism of siliceous limestones and dolomites at moderate temperatures may also produce hydrous minerals such as talc and tremolite. For this reason water must be considered as an additional component. It introduces, of course, further complexities like those discussed in section 4(b) for the MgO-SiO_2-CO_2-H_2O system.

Although a considerable amount of experimental data is available for this system (Metz and Winkler, 1963; Metz, 1967, 1970; Metz and Trommsdorff, 1968; Gordon and Greenwood, 1970; Skippen, 1971), the problem of stable phase relations as a function of gas composition has not yet been solved in a fully satisfactory fashion. The reason for this state of affairs is that there are two sets of experimental data which, in addition to major agreements, show conflicting results in several aspects. This contrast is best seen by comparing the TX_{CO_2}-section of the system derived by Metz and Trommsdorff (1968) on one hand with that published by Skippen (1971) on the other. Since the discrepancy is outside the range of experimental error, it is possible that it is due to the presence of impurities in the natural starting materials used by one of the investigators (Metz, 1970). Because the experimental results of Skippen (1971) are discussed here in Chapter IIA2 by Greenwood, the present writer will deal mainly with the data of Metz on the basis of his unpublished thesis (Habilitationsschrift of 1970), which have already found their way into textbooks (e.g. Miyashiro, 1973b). No matter what the final outcome for the pure, synthetic system will be, the mere possibility that impurities present in the natural environment may have profound influences on the phase relations deserves the special attention of the petrologist.

In the following only some salient features of the compatibility relations in the system are reported. Figure 6 shows solely the CO_2-rich portion of an isobaric TX_{CO_2}-diagram at $P_{fluid} = 1$ Kbar for the system discussed, and only in the narrow temperature range between 460°C and 560°C, in which according to Metz (1970) most of the metamorphic reactions involving hydrous minerals and the two carbonates calcite and dolomite occur. Figure 6 involves only the minerals calcite, dolomite, quartz, diopside, forsterite, talc and tremolite. It was shown through the experimental work of Metz (1970) that the various decarbonation and dehydration reactions take place without any significant changes in the reaction sequence over a wide range of X_{CO_2} extending from at least 0.1 to about 0.90. Only at the still higher values of X_{CO_2} considered in Fig. 6 are the phase relationships drastically changed through intersection of isobarically univariant points I_1–I_4 (cf. Fig. 6). The main differences in the results of Skippen (1971) are that the location of invariant point I_1 is shifted to an X_{CO_2}-value as low as 0.5 and, partly as a consequence of this, the invariant point I_3 does not exist. Moreover, the intersection of curves creating invariant point I_4 takes place

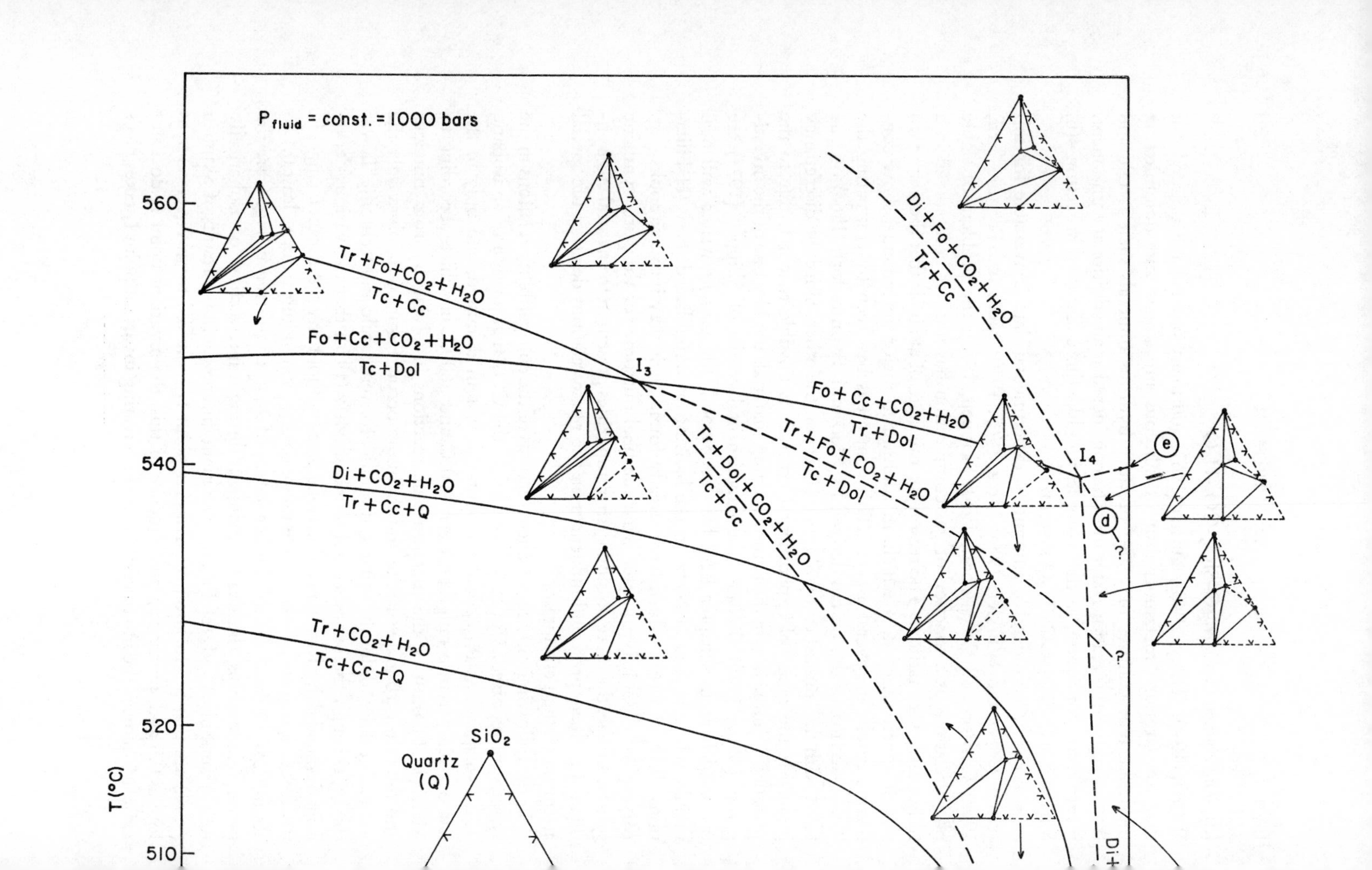

P_{fluid} = const. = 1000 bars
560
540
520
510
T (°C)
$Tr + Fo + CO_2 + H_2O$
$Tc + Cc$
$Fo + Cc + CO_2 + H_2O$
$Tc + Dol$
I_3
$Fo + Cc + CO_2 + H_2O$
$Tr + Dol$
$Tr + Fo + CO_2 + H_2O$
$Tc + Dol$
$Tr + Dol + CO_2 + H_2O$
$Tc + Cc$
$Di + Fo + CO_2 + H_2O$
$Tr + Cc$
I_4
e
d
?
?
$Di + CO_2 + H_2O$
$Tr + Cc + Q$
$Tr + CO_2 + H_2O$
$Tc + Cc + Q$
SiO_2
Quartz
(Q)
Di+

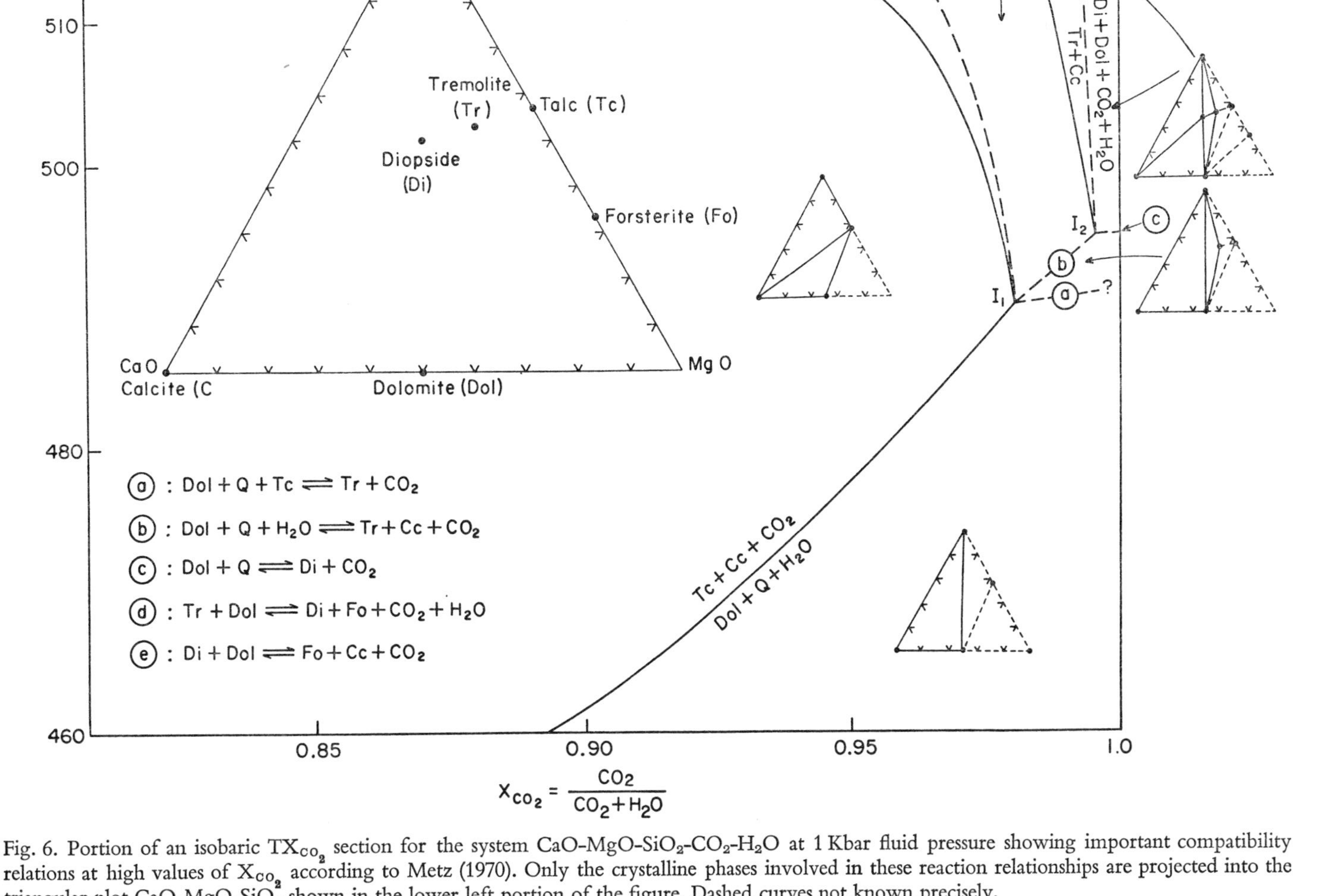

Fig. 6. Portion of an isobaric TX_{CO_2} section for the system CaO-MgO-SiO_2-CO_2-H_2O at 1 Kbar fluid pressure showing important compatibility relations at high values of X_{CO_2} according to Metz (1970). Only the crystalline phases involved in these reaction relationships are projected into the triangular plot CaO-MgO-SiO_2 shown in the lower left portion of the figure. Dashed curves not known precisely.

twice in the entire TX_{CO_2}-section. Thus this invariant point reappears once again at a very low X_{CO_2}-value.

The main petrological applications of the compatibility relations as given in Fig. 6 may be summarised as follows:

(1) Except for very high X_{CO_2} values above I_1 the first metamorphic reaction of a siliceous dolomite (sedimentary dolomite–quartz rock) is the formation of the *assemblage talc–calcite* (which is in agreement with Tilley's (1948) field observations). With a coexisting gas phase of nearly pure CO_2 the dolomite–quartz assemblage persists to temperatures near 500°C and subsequently breaks down to form directly the paragenesis *tremolite–calcite*. With more water present in the gas phase this latter paragenesis does not become stable until temperatures appreciably above 500°C are attained, by reaction of the previously formed talc–calcite assemblage with quartz. It is clear that the different location of I_1 proposed by Skippen (1971) would allow tremolite to form at much lower temperatures and lower CO_2-contents of the gas phase, and even as the first reaction product of the dolomite–quartz rock.

(2) A similar X_{CO_2}-dependence exists for the formation of *diopside*. Whereas in the presence of water diopside forms at the expense of the assemblage tremolite–calcite–quartz at relatively high temperatures, extremely water-poor or water-free gas phases allow diopside formation directly from dolomite + quartz at a temperature near 500°C, i.e. above the temperature of I_2.

(3) At temperatures near 550°C and nearly independently of X_{CO_2} the *ophicalcite-paragenesis* forsterite + calcite becomes stable. An upper temperature limit for this assemblage has already been discussed in section 4(c) and Fig. 5. It is only the reaction path by which this assemblage forms that differs as a function of X_{CO_2}. Under relatively hydrous conditions with X_{CO_2} values lower than that of the invariant point I_3 *talc* and dolomite react to form forsterite + calcite; at X_{CO_2} between those of I_3 and I_4 *tremolite* and dolomite represent the stable starting assemblage, and at the highest X_{CO_2} values, calcite and forsterite are formed at the expense of *diopside* and dolomite. The reader will realise that this trend of reactions with rising X_{CO_2} involves a gradual dehydration of the stable low-temperature reactants of the "ophicalcite" assemblage. These relationships are simplified in the diagram of Skippen (1971), where I_3 does not exist and the assemblage tremolite–dolomite is stable towards much lower X_{CO_2}-values.

(4) Whereas the assemblage *tremolite–forsterite* is indicative of relatively high temperatures of metamorphism only at intermediate X_{CO_2} values, the assemblage *diopside–forsterite* persists up to temperatures of the so-called sanidinite-facies (cf. Fig. 5). It is to be noted, too, that at those high

temperatures extending into the "magmatic" range the diopside-forsterite assemblage is stable independently of the availability and composition of a fluid (e.g. Osborn and Tait, 1952).

In Fig. 6 for the sake of simplicity the phases enstatite, anthophyllite, and magnesite and their compatibility relations with the other phases have been neglected, although they are certainly stable at least under the high X_{CO_2} values of the diagram. However, in natural siliceous marbles these phases are usually absent. For more information on the stability relations reference should be made to the TX_{CO_2}-diagrams of Johannes (1969) on the system MgO-SiO_2-H_2O-CO_2 of which only a few excerpts are reproduced as Fig. 4 in section 4(b). Through comparison of Johannes' diagrams with Fig. 6 careful observers may also determine under which limiting TX_{CO_2}-conditions the phases talc and forsterite can only be involved in the reactions with calcium-bearing minerals shown in Fig. 6. The phase *wollastonite* is not stable under the conditions considered in Fig. 6 (cf. Fig. 1), but will be at higher temperatures, where it will simply act as an additional phase that can co-exist with quartz, diopside, or calcite in the system discussed here (Skippen, 1971).

The influence of total pressure on the phase relations has not been determined by experiment as yet, but can be calculated using thermochemical data such as those derived by Skippen (1971). Certainly, the differences to be expected between the isobaric TX_{CO_2}-sections of 1 Kbar (given by Metz) and 2 Kbar (given by Skippen) will not suffice to account for the discrepancies of the experimental results of these authors as discussed previously. It should also be pointed out that the theoretical treatment of the system in isobaric X_{CO_2}-diagrams as verified above is, by itself, only a limiting case valid under certain prerequisites. As French (1965) has pointed out H_2O and CO_2 are by no means always the only possible gas species appearing in the system C–H–O, which should basically be considered for the metamorphism of siliceous dolomites and many other rocks. Under the low oxygen fugacities imposed by the presence of graphite, for example, the additional gas species H_2, CO and CH_4 will also be stable and thus exert their partial pressures on the solid phases (see also Eugster and Skippen, 1967).

In contrast to the system discussed in section 4(c) the main applications of the phase relations of the system CaO-MgO-SiO_2-CO_2-H_2O will undoubtedly be in the realm of *regional* metamorphism, where siliceous carbonate rocks may crop out over large areas showing gradual changes in metamorphic conditions. A very impressive example is offered in the Central Alps, where Trommsdorff (1972) was even able to construct a polybaric phase diagram from the assemblages observed in the progressive metamorphic sequence. The frequent occurrence of four-phase mineral assemblages in the system discussed, over large volumes of rock, indicates that the PTX_{CO_2}-conditions must have changed during progressive metamorphism along divariant surfaces, or in comparison to the isobaric section of Fig. 6, along univariant curves. This means that the solid mineral assemblages were able to buffer their coexisting fluid phases to the X_{CO_2}-values requisite for stable equilibrium at a given temperature and total pressure.

(e) *The System* $CaO-Al_2O_3-SiO_2-H_2O$

This system presents a simple though seldom applicable model for lime silicate rocks which originated from pelitic limestones, limey marls, and calcareous clays. If one neglects the phase margarite (see section 3(b)13) and Chatterjee, 1974), the mineral equilibria of interest for low-pressure metamorphism are those compiled in Fig. 7, which was drawn on the basis of the work by Newton (1966) and Boettcher (1970). They involve only the phases wollastonite, quartz, grossularite, anorthite, zoisite, gehlenite and corundum. Since zoisite is the only hydrous mineral amongst them, only two of the reactions involve its dehydration breakdown with increasing temperatures, whereas the others are related to the *instability of the garnet phase grossularite* at high temperatures and low pressures. Although two of the reaction curves intersect, there is no invariant point creating a new set of reactions.

These compatibility data of the 7 phases named above are incomplete, and more experimental work is required. As an example the reaction leading to the first appearance of the high-temperature phase gehlenite in this model system is not known. One might expect that gehlenite forms in the pure system at the expense of wollastonite and a phase $CaAl_2O_4$ at temperatures below those of the grossularite reaction with corundum. However this mechanism involving a phase not known as a mineral is probably insignificant for natural rocks. It is more likely that the carbonate mineral calcite takes part in the natural reaction relationships and thus the system discussed in section 4(f) must be applied.

The compatibility relations are further complicated and even changed at low temperatures by the incoming of additional phases such as margarite, $CaAl_2[Al_2Si_2O_{10}](OH)_2$, and prehnite, $Ca_2Al_2Si_3O_{10}(OH)_2$. Thus margarite will, through its mere existence, prohibit the coexistence of anorthite and corundum at temperatures below some 500°C at 2 Kbar (see section 3(b)13). The phase prehnite, which becomes stable at temperatures below about 400°C for the pressure range of interest here, will replace the three-phase assemblage zoisite–grossularite–quartz shown in Fig. 7 (Liou, 1971). In addition the low-temperature stability limit of the feldspar anorthite (Boettcher, 1970; Nitsch, 1972) may be marked at low pressures by the breakdown into the three phases margarite + zoisite + quartz (Chatterjee, personal communication, 1974). Yet at this stage of the experimental and theoretical investigations it is not possible to present a stable grid of the complex overall phase relations at low temperatures. The most recent work by Storre and Nitsch (1974) on margarite cannot be accepted as a contribution to the pure system as it was done on natural material containing 25% of impurities.

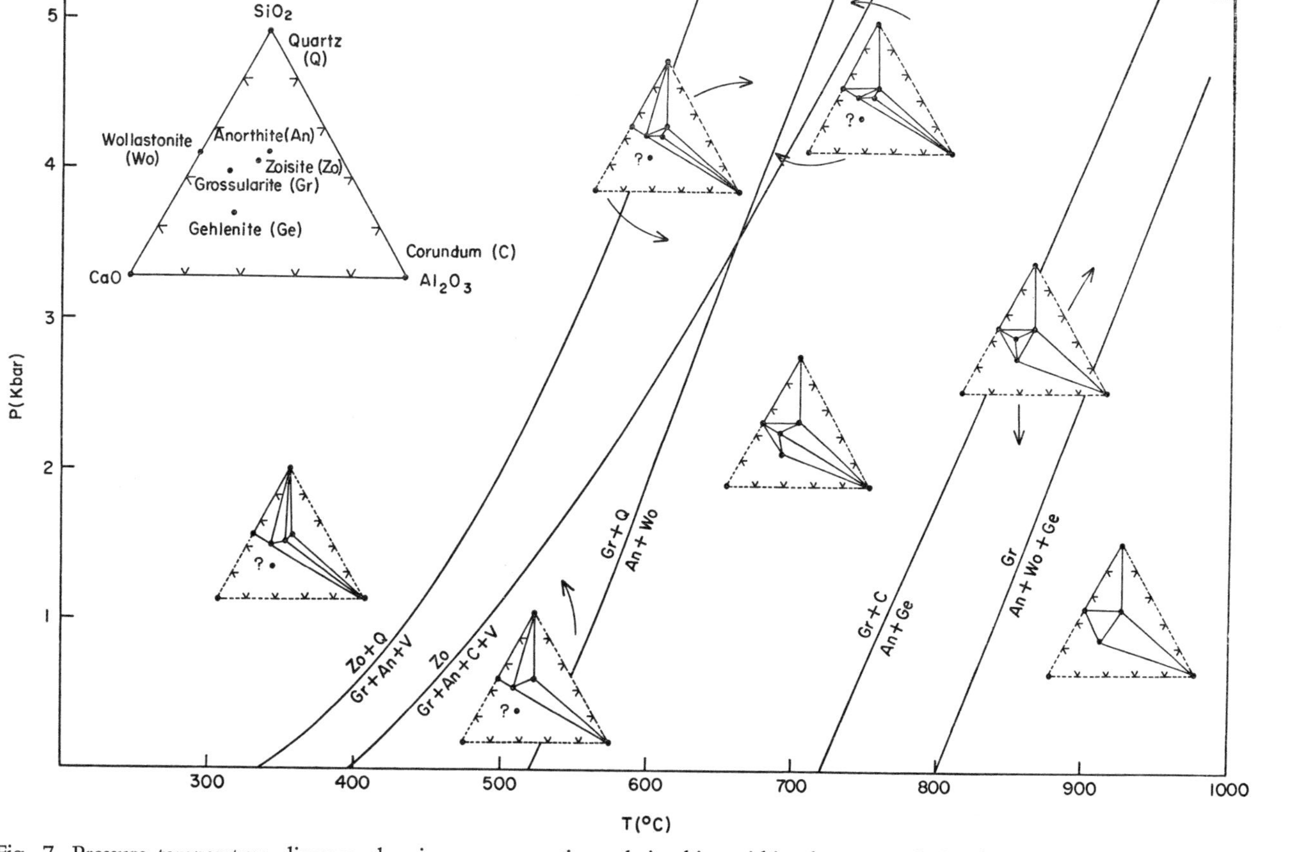

Fig. 7. Pressure–temperature diagram showing some reaction relationships within the system CaO-Al_2O_3-SiO_2-H_2O after Newton (1966) and Boettcher (1970). They involve solely those crystalline phases depicted in the triangular CaO-Al_2O_3-SiO_2 plot (upper left). The univariant curves for the two reactions involving zoisite are valid only for the limiting condition that the pressure of the hydrous fluid (V) equals total pressure. The other three reactions are independent of the magnitude of fluid pressure versus total pressure.

(f) *The System* $CaO-Al_2O_3-SiO_2-H_2O-CO_2$

Compared to the system $CaO-Al_2O_3-SiO_2-H_2O$ (section 4(e)) the present CO_2-bearing system is more applicable to natural calcareous pelites. Experimental data partly aided by thermodynamical calculations were only recently obtained by Storre (1970) and Gordon and Greenwood (1971). The latter work is already based on the important low-temperature study by Liou (1971) on the stability of prehnite, $Ca_2Al_2Si_3O_{10}(OH)_2$, and does, therefore, not involve any metastable equilibria. Figure 8 is a combined version of Figs. 3 and 5 given by Gordon and Greenwood (1971). The phase relations depicted are valid for a total fluid pressure of 2 Kbar and involve the phases grossularite, quartz, wollastonite, anorthite, calcite, gehlenite, zoisite and corundum. For the sake of simplicity the compatibility relations of the phases margarite and prehnite significant below about 500°C have been omitted. Thus the phase relations shown are of main interest for the appearance of both grossularite and gehlenite as a function of the CO_2-content of the coexisting fluid. The critical finding was that the univariant lines of the CO_2-free system already discussed in section 4(e) and Fig. 7 extend isothermally, or very nearly so, into the TX_{CO_2}-diagram where they intersect with strongly X_{CO_2}-dependent curves such as the wollastonite versus calcite + quartz equilibrium. Thus isobaric invariant points are created, which cause drastic changes in the phase relations through elimination of three out of four of the ternary CaAl-silicate phases.

The geological significance of the phase relations of Fig. 8 may be summarised as follows:

(1) The stability of the assemblage *grossularite + quartz* already limited in temperature to a relatively narrow range (cf. Fig. 7) is also severely limited with regard to the CO_2 content of the coexisting gas phase. This assemblage can only be present in environments poor in CO_2, i.e. $X_{CO_2} < 0.16$, the location of invariant point I_3.

(2) A similar restriction to relatively water-rich gas phases exists for the mineral *grossularite* itself. Its stability is limited by virtue of I_5 to X_{CO_2} values of less than about 0.37 for a total pressure of 2 Kbar. Whereas its upper temperature stability remains constant, within this range of X_{CO_2}, there is a dramatic increase in lower temperature stability from very low values (<500°C) up to "magmatic" temperatures as a function of increasing CO_2-content in the gas. Thus, even provided the proper composition of solids is available in a rock, grossularite formation in calcareous pelites must be considered as a rather haphazard process strongly influenced by both gas composition and temperature of metamorphism. Over large TX_{CO_2} ranges it will be substituted by the equivalent high-temperature carbonate assemblage anorthite + wollastonite + calcite. On the other hand, in the rather CO_2-poor environment of peridotite hydration to form serpentinites (cf. Johannes, 1969) grossularites, or even hydro-

grossularites, may form within rodingite bodies at very low temperatures (e.g. Coleman, 1967).

(3) Due to the distribution of reaction curves around invariant point I_3 the order of appearance of the minerals *grossularite and wollastonite* with increasing temperature and thus grade of metamorphism depends primarily on the CO_2-contents of the gas phase: whereas at low X_{CO_2}-values grossularite will form as the earlier phase even at relatively low temperatures, wollastonite may crystallise at X_{CO_2} in excess of about 0.17 at temperatures as much as 200°C below those necessary for grossularite growth.

(4) The reactions leading to the formation of the phase *gehlenite* found to be uncertain in the CO_2-free system (section 4(e)) can now be specified in a way that is also realistic for natural environments. At low X_{CO_2}-values (lower than that of $I_4 \approx 0.17$) gehlenites form at the expense of a pre-existing assemblage grossularite + calcite + corundum, but at higher CO_2-contents of the gas phase the assemblage anorthite + calcite + corundum precedes the growth of gehlenite. Over most of the X_{CO_2}-range considered, with the exception of nearly CO_2-free fluids, the temperatures of gehlenite formation are very high (about 700°C–850°C). This is similar to the conditions of formation of akermanite (see Fig. 5 and section 4(c)) thus explaining why melilites as a mineral group are confined to very high-temperature contact metamorphism of impure limestones.

(5) The mineral *zoisite*, and even more so its coexistence with free quartz, is limited to relatively low temperatures (Fig. 7). As is normal for a hydrous phase this limitation is still more pronounced with increasing X_{CO_2} the two-phase assemblage anorthite–calcite appearing in its place. In fact, on the basis of more recent experimentation with a natural zoisite, Storre and Nitsch (1972) depict the curve of the breakdown reaction of zoisite to calcite + anorthite as originating in I_2 and passing through I_1 (Fig. 8) as a vertical line. This would indicate that this reaction is independent of temperature with zoisite being restricted to very hydrous gas phases with $X_{CO_2} < 0.02$ at $P_{fluid} = 2$ Kbar.

Despite the elucidating conclusions made possible by the work of Gordon and Greenwood (1971) it should be borne in mind that their direct application to natural assemblages is often problematical. As an example, Kerrick (1970) and Kerrick *et al.* (1973), studying contact-metamorphosed calcareous rocks, have pointed out that solid solution in plagioclase, garnet, and (clino-)zoisite must have appreciable effects on the phase relations. This may also be an explanation for the discrepancy in the zoisite stability as indicated by Gordon and Greenwood (1971) compared with that by Storre and Nitsch (1972).

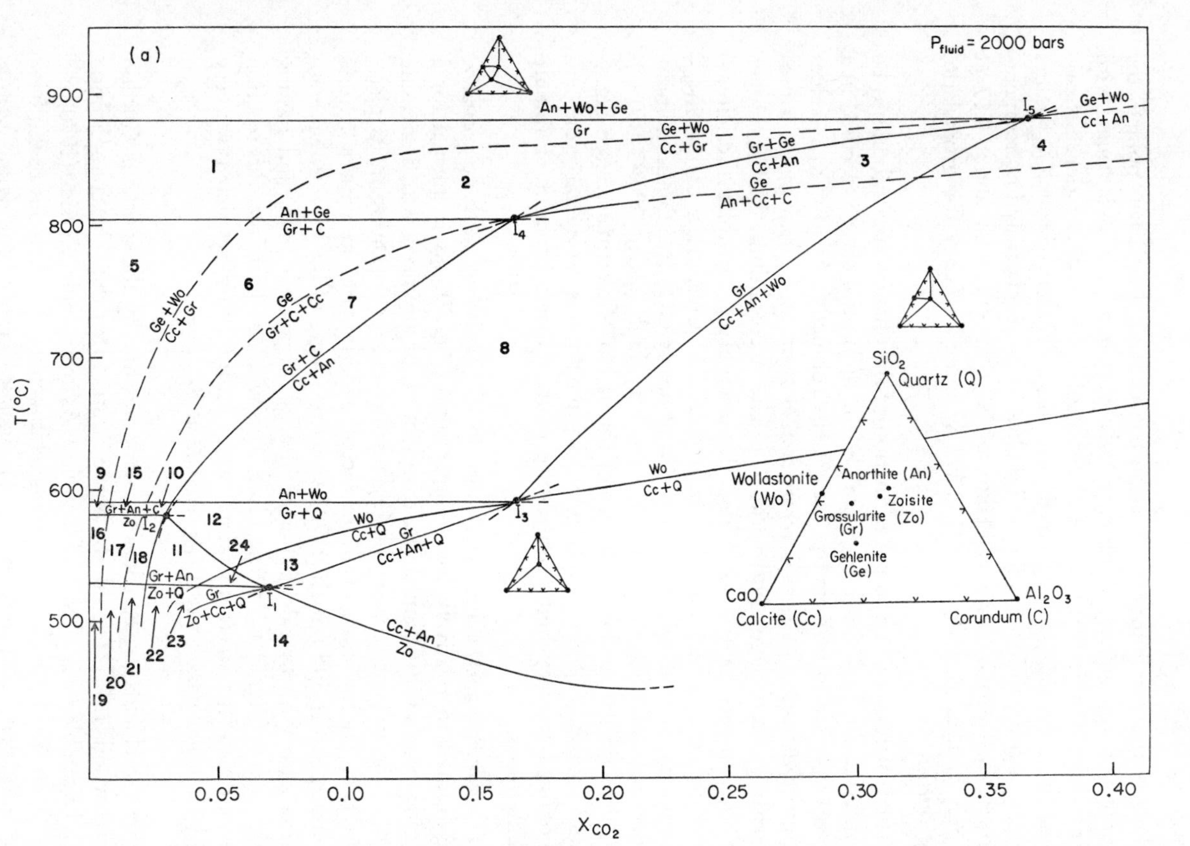

Fig. 8(a). Portion of an isobaric TX_{CO_2} section for the system CaO-Al_2O_3-SiO_2-H_2O-CO_2 at 2 Kbar fluid pressure showing compatibility relations at relatively low values of X_{CO_2} according to Gordon and Greenwood (1971). Only the eight crystalline phases involved in these reaction relationships are projected into the triangular CaO-Al_2O_3-SiO_2 plot shown in the lower right portion of the figure. The line separating the fields 18 and 11 represents the reaction Gr + C = Zo + Cc

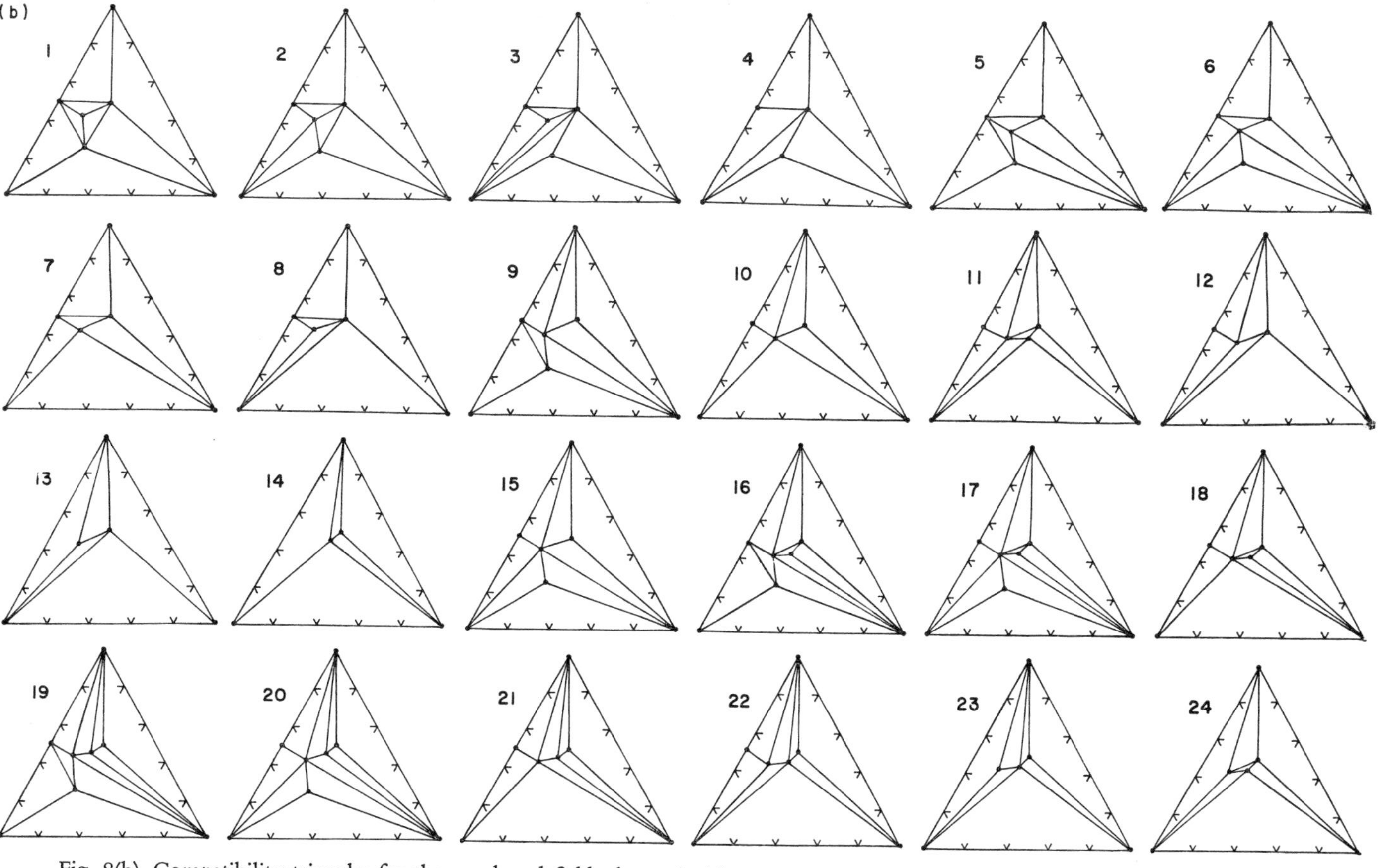

Fig. 8(b). Compatibility triangles for the numbered fields shown in (a).

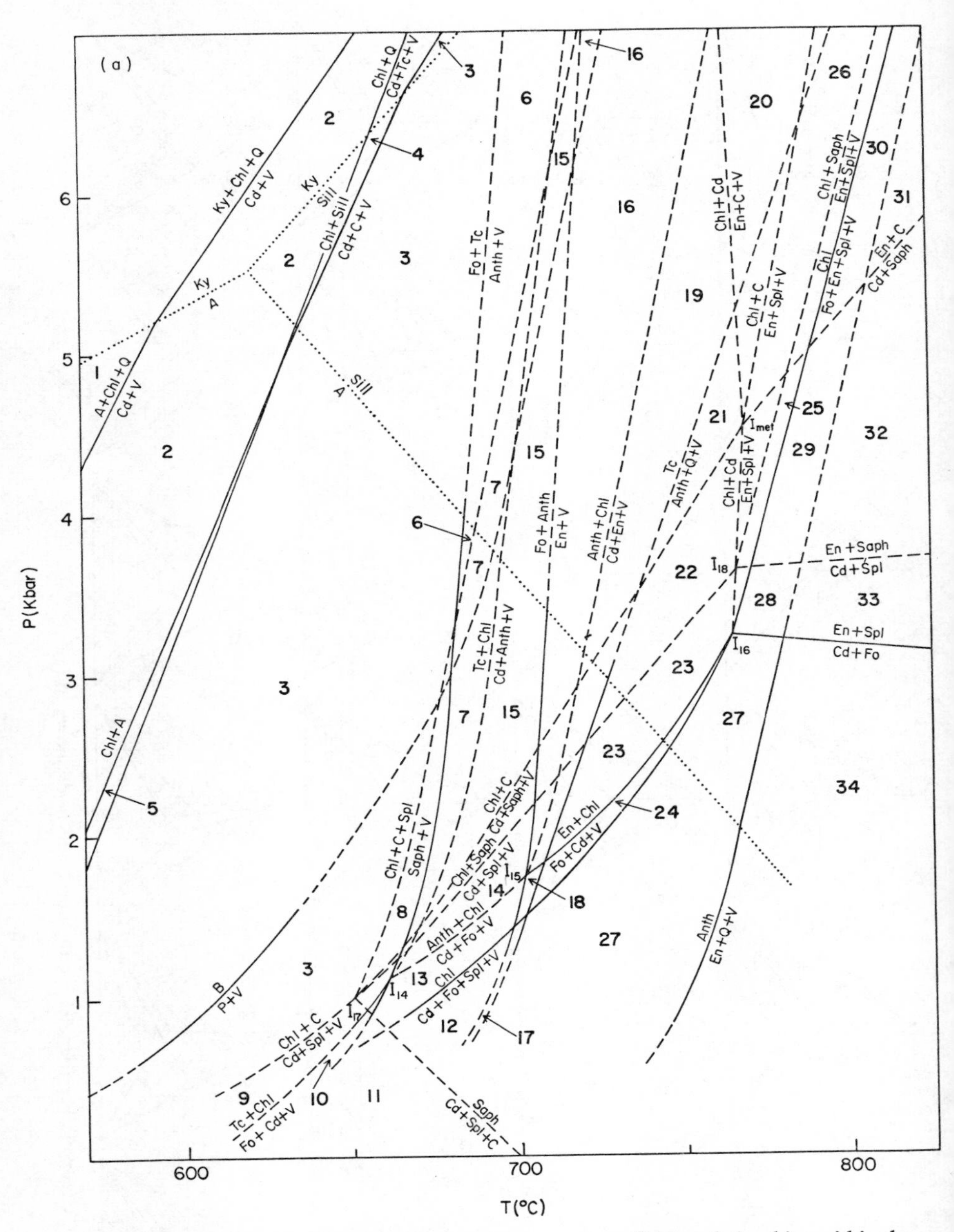

Fig. 9(a). Pressure–temperature diagram summarising compatibility relationships within the system $MgO-Al_2O_3-SiO_2-H_2O$ for the range 0–7 Kbar, 580°C–820°C on the basis of experimental data by Seifert (1970b). The crystalline phases involved in the reactions, and their abbreviations, are shown in an enlarged compositional triangle representing a projection of the quaternary system onto its water-free base (Fig. 9(b)). Dashed curves not known precisely. For phase relations near the invariant point I_{met}, see Fig. 10. The line joining I_{16} to I_{18} represents the reaction: Chlorite + Cordierite = Enstatite + Spinel + Vapour.

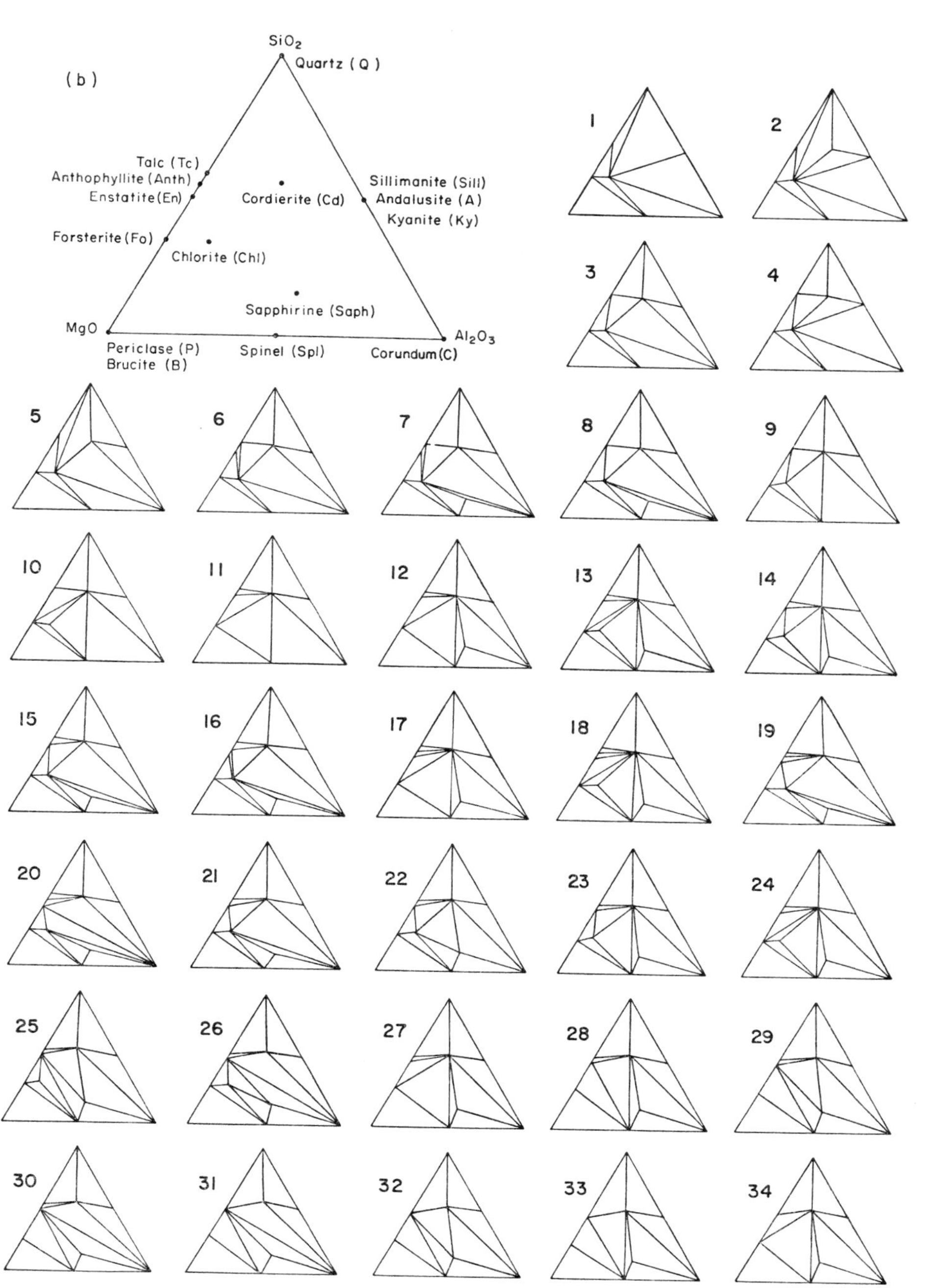

Fig. 9(b). Phase composition triangle, and compatibility triangles for the numbered fields shown in (a).

(g) *The System* $MgO-Al_2O_3-SiO_2-H_2O$

As a first approximation this system may be regarded as a model system for the lime-free pelitic rocks which are among the most common sediments. The significance of this system has been recognised for many years, and the first experimental studies to resolve its phase relations were made by Yoder (1952) and Roy and Roy (1955). After several detailed studies on the stability and compatibility relations of a limited number of phases by Schreyer and Yoder (1964), Fawcett and Yoder (1966), Schreyer (1968), and Seifert and Schreyer (1970), another attempt at summarising phase relationships was made by Schreyer (1970). A still more recent, but not yet fully published, study is that of Seifert (1970b) which, because of its particular relevance to low-pressure metamorphism, serves as the basis of the present review. Some portions of this study have since been refined and are published (Seifert, 1973, 1974, 1975).

On the basis of his experimental results Seifert (1970b) has drawn a P_{H_2O}–T diagram of the system for the range 1 atm–7 Kbar, 400°C–1250°C. From uniquely defined compatibility relations some 80 divariant fields were distinguished in the subsolidus range alone. In Fig. 9 Seifert's diagram is reproduced for the critical temperature range 580°C–820°C in which most of the reactions important for low-pressure high-temperature metamorphism occur. Although the complications of this plot may seem overwhelming at first glance for a "simple quaternary system", it should be remarked that they are minor as compared to those at higher pressures, where instead of just the three quaternary or anhydrous ternary phases chlorite, cordierite, and sapphirine there are eight (Schreyer and Seifert, 1969b). One of these quaternary high-pressure phases, boron-free kornerupine with a composition close to

$$3.67MgO \cdot 3.3Al_2O_3 \cdot 3.67SiO_2 \cdot H_2O,$$

has since been found, however, to remain stable down to water pressures near 5 Kbar (Seifert, 1975), and will thus affect and complicate the phase relations of Fig. 9 in their high-pressure portion. The necessary modification of the phase diagram due to the appearance of kornerupine is, for the sake of simplicity, shown in a separate plot involving only 7 crystalline phases (Fig. 10). In an additional diagram (Fig. 11a, b) some of the recent, more precise determinations by Seifert (1974), mainly concerning the phase sapphirine are compiled. Finally, it must be remembered from section 4(a) that a modification of the anthophyllite stability in the limiting system $MgO-SiO_2-H_2O$ near 5 Kbar (Seifert, unpublished data 1973) will be necessary when more precise data are available.

Combining Figs. 9–11 there are 8 invariant points in the system $MgO-Al_2O_3-SiO_2-H_2O$ within the pressure-temperature range considered. They were numbered I_{14}–I_{21} in continuation of those previously defined at higher pressure (cf. Schreyer and Seifert, 1969a). It should be noted that the invariant point I_{met} shown in Fig. 9 at about 770°C, 4.5 Kbar must be considered metastable. With the incoming of kornerupine it is stably replaced by the three invariant points I_{19}–I_{21} of Fig. 10.

Although Figs. 9–11 are self-explanatory, some of their salient features relevant to the high-temperature metamorphism of pelites should be noted:

(1) The *stable reaction to form cordierite* at pressures below 5 Kbar (Fig. 9) is, contrary to the earlier findings of Schreyer and Yoder (1964),

$$\text{andalusite} + \text{chlorite} + \text{quartz} \rightleftharpoons \text{cordierite} + H_2O.$$

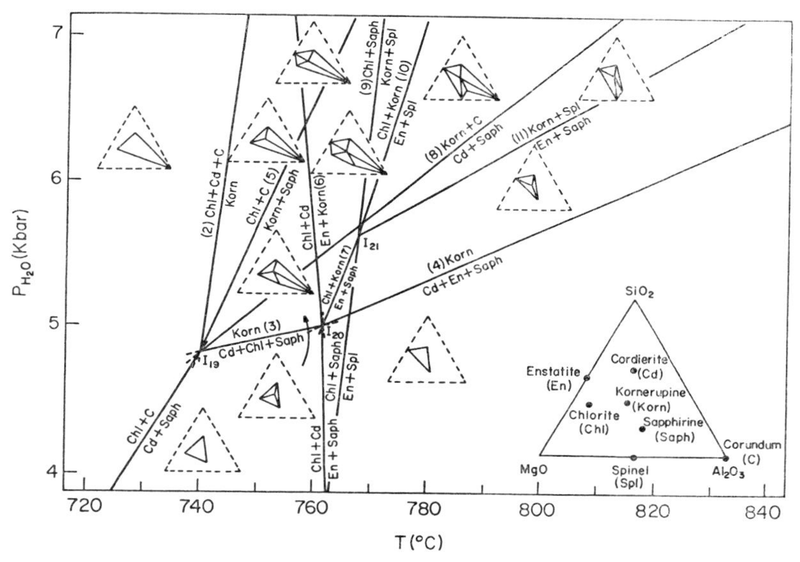

Fig. 10. Pressure–temperature diagram exhibiting some phase relations near the lower pressure stability of the phase boron-free kornerupine within the system MgO-Al_2O_3-SiO_2-H_2O according to Seifert (1975). This diagram supersedes some of the relations shown in Fig. 9 within the requisite pressure–temperature range. For example, the invariant point I_{met} of Fig. 9 which is to be considered metastable is substituted by the three invariant points I_{19}–I_{21}. Numbers in parentheses apply to the original paper by Seifert.

This explains why in some spotted slates of the outermost portions of contact aureoles the assemblage andalusite + chlorite + quartz is commonly observed, but no cordierite appears (cf. Seifert and Schreyer, 1970).

(2) In agreement with the compatibility relations given by Roy and Roy (1955), but at variance with those of Yoder (1952), *cordierite and corundum represent a stable assemblage* under most conditions of Fig. 9. This assemblage forms at the expense of chlorite + Al_2SiO_5 above some 550°C to

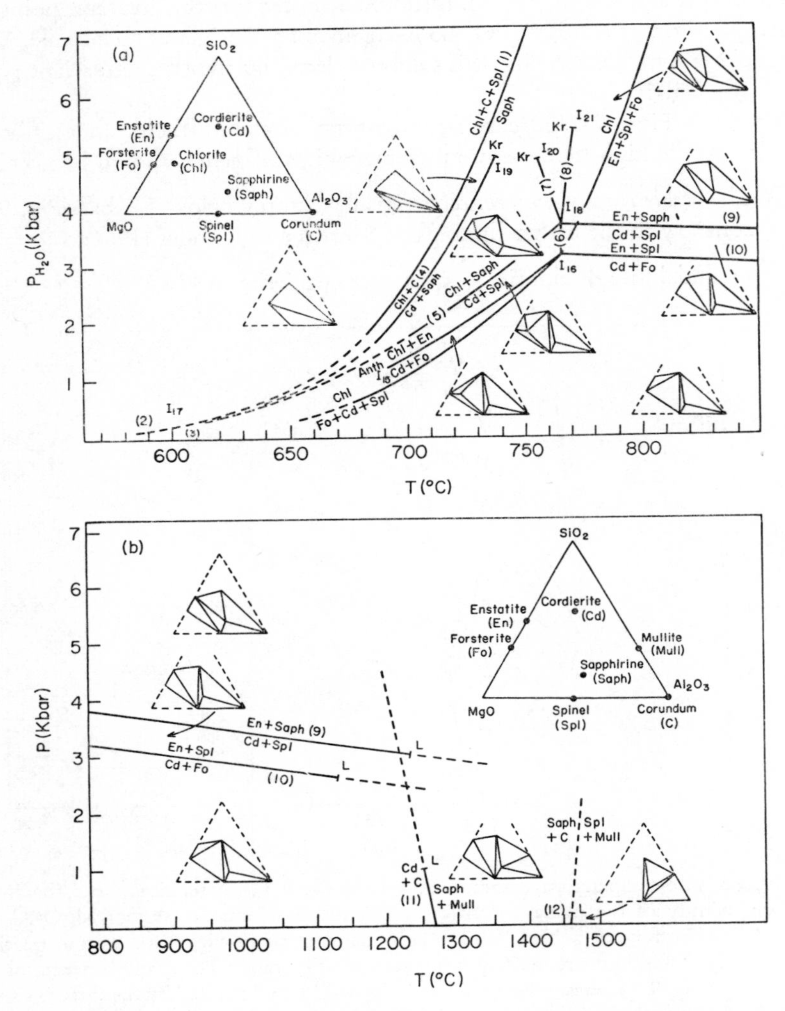

Fig. 11. Pressure–temperature grids showing reaction relationships at relatively low (a) and high (b) temperatures involving sapphirine and other silica-poor phases of the system MgO-Al_2O_3-SiO_2-H_2O as given by Seifert (1974). Diagram (a) shows more precise locations of some of the curves already included in Fig. 9. Abbreviations not included in the compositional triangle (upper left), i.e. Kr = boron-free kornerupine (see Fig. 10), and Anth = anthophyllite, mark invariant points involving these two phases in addition to others shown in the triangle. In diagram (b) the dashed curves show equilibria that are only stable for conditions of $P_{H_2O} < P_{tot}$. L (= liquid) marks invariant points stable at $P_{H_2O} = P_{tot}$ involving hydrous melts as additional phases. Numbers in parentheses apply to the original paper by Seifert.

650°C. The cordierite-corundum pair becomes unstable only above some 1100°C, the assemblage mullite + sapphirine forming instead (Fig. 11b). For an assemblage spinel-aluminium silicate there is no stability range except very near liquidus temperatures, i.e. above 1460°C (Foster, 1950) as shown in Fig. 11b. Therefore, natural cordierite-corundum assemblages as described from pelitic hornfelses (e.g. Tilley, 1923) probably do not indicate disequilibrium as suggested by Schreyer and Yoder (1964). The natural occurrence of the assemblage Al–silicate + spinel, by itself, or in addition to cordierite + corundum (e.g. Tilley, 1924) must then be attributed to additional components such as FeO (Seifert, 1973).

(3) The paragenesis *chlorite–quartz* is limited to relatively low temperatures below about 550°C–650°C (Fig. 9) in accordance with the results of Fawcett and Yoder (1966). Its high-temperature equivalent, cordierite-talc, has recently been found in highly magnesian rocks (Kulke and Schreyer, 1973), but is generally very rare in nature due to the appearance of garnet in the usually Fe^{+2}-bearing bulk chemistry of rocks.

(4) Other than chlorite–quartz the assemblage *chlorite–corundum* has a relatively wide stability range extending to some 600°C–750°C depending on pressure (Figs. 9, 10, 11a). These temperatures are close to the upper stability limit of chlorite alone. A chlorite–corundum rock was described from South Australia by Oliver and Jones (1965). Due to the additional presence of sapphirine, however, the temperature of formation of this rock can, for the conditions $P_{H_2O} = P_{tot}$, be determined as lying within the very narrow range between the upper limit of chlorite–corundum and the lower limit of sapphirine (reactions (4) and (1) in Fig. 11a).

(5) Most of the reaction curves shown in Fig. 9, which include those of the limiting system MgO-SiO_2-H_2O discussed in section 4(a) and depicted in Fig. 3, apply to *dehydration reactions* thus exhibiting steep slopes and typical curvatures (i.e. convex towards higher temperatures). Progressive metamorphism of pelites is, therefore, mainly a process of successive dehydration.

(6) There are, however, a few curves with notably different, flatter or even negative dP/dT-slopes, which apply to reactions not involving a gas phase (V). In addition to the polymorphic transitions of the compound Al_2SiO_5 (cf. Fig. 1) these reactions mark the solid state breakdown of the low-pressure phase cordierite, if it occurs in assemblages with the phases spinel (± corundum) and forsterite. The requisite reaction equations arranged in the order of increasing pressure are:

cordierite + spinel + corundum ⇌ sapphirine (Fig. 9)
cordierite + forsterite ⇌ enstatite + spinel (Fig. 11a, b)
cordierite + spinel ⇌ enstatite + sapphirine (Fig. 11a, b).

In all cases the cordierite-bearing assemblages are stable on the low-pressure side of the diagram, because they represent the larger-volume assemblages. These reactions are, therefore, *compression reactions*, which are mainly susceptible to pressure and to a lesser degree only to temperature. It is particularly noteworthy that such compression reactions make their appearance in the realm of low-pressure metamorphism discussed here.

(7) For a better appreciation of the compatibility relations it is suggested that the reader trace all the limiting reaction curves of the stability fields of some particular two- or three-phase assemblages. Doing this he will find, for example, that the assemblage *cordierite + spinel + corundum* is limited to very low pressures below that of the invariant point I_{17} (see Fig. 9) from which the two limiting reaction curves originate. According to Seifert (1974a) the location of I_{17} is shifted to a still lower pressure and temperature (Fig. 11a). Thus it is very unlikely that an assemblage cordierite + corundum + spinel will ever be found in natural rocks unless both cordierite and spinel contain FeO as an additional component. In highly magnesian rocks the mineral sapphirine will occur instead of the three-phase assemblage. Similarly, the assemblage *cordierite + forsterite* is limited to pressures below about 3.25 Kbar at temperatures above some 650° to 750°C (Figs. 9 and 11a, b). Its field is bordered by reaction curves connecting the invariant points I_{16} (Fawcett and Yoder, 1966), I_{15}, and I_{14}. The latter two invariant points are located upon two reaction curves of the limiting ternary system MgO-SiO_2-H_2O (cf. Fig. 3) which represent degenerate curves in the quaternary system discussed here. In nature an assemblage of cordierite with an olivine has so far been found only in ferruginous hornfelses (Abraham and Schreyer, 1973) and will, therefore, be discussed in section 4(j). The assemblage *cordierite + spinel* can occur only at pressures below about 4 Kbar (Fig. 11a, b) and is limited towards lower temperatures by a curve connecting the invariant points I_{18} and I_{17}. Cordierite-spinel assemblages are well known from many pelitic hornfelses and low-pressure gneisses.

(8) Because of its location at pressures above some 5 Kbar the stability field of the phase boron-free kornerupine (Fig. 10) has apparently little bearing on the type of metamorphism discussed in this chapter. It is possible, however, that the usual incorporation of boron and often of sodium into the structure stabilises the mineral towards lower pressures. Nevertheless kornerupine and its sodian variety prismatine are rare minerals in nature confined to certain pegmatitic veins (e.g. Scheumann, 1960) and some granulite-type metamorphic rocks (Vogt, 1947). This rarity still awaits explanation (Seifert, 1975).

(h) *The System* K_2O-MgO-Al_2O_3-SiO_2-H_2O

For most pelitic rocks this system is of greater importance than the system MgO-Al_2O_3-SiO_2-H_2O already discussed in 4(g), because it contains among its phases two of the most common micas, muscovite and phlogopite (as a model for biotite), as well as K feldspar. Phase relationships in this system were determined, among others, by Luth (1967) and Seifert (1970a and unpublished data). Luth's work was mainly concerned with the melting relations in a relatively Al-poor part of the system, while that of Seifert concentrated on the Al-rich, Si-saturated portion of the system and has, therefore, special relevance for the low-pressure high-temperature metamorphism of natural pelites. Indeed, it is on the basis of Seifert's results that an admittedly arbitrary border between low-pressure and intermediate-pressure metamorphism may be defined.

In addition to the detailed publication by Seifert (1970a) dealing merely with the compatibility relations at relatively low temperatures, a shorter summary of the larger-scale phase relations was given by Schreyer and Seifert (1969a). Figure 12 offers a survey of the whole picture up to the beginning of melting; it includes a number of unpublished results. It must be emphasised that it shows only reactions within the SiO_2-rich portion of the quinary system occurring exclusively in the presence of free quartz. It is for this reason that, graphically, SiO_2 can be treated as an excess component and the classical triangular AKF-plot (Eskola, 1915) can be used to demonstrate the stable phase relations. The main features of the relatively simple phase relations are:

(1) The dehydration reaction of the low temperature assemblage chlorite + muscovite + quartz to form *cordierite* + *phlogopite* takes place at just slightly higher temperatures than the initial cordierite-forming reaction in the limiting system MgO-Al_2O_3-SiO_2-H_2O (cf. 4(g) and Fig. 9). Thus the growth of cordierites in the common potash-bearing pelitic rocks of contact aureoles (e.g. spotted slates) may practically occur near the lower stability limit of that phase. These cordierites can coexist stably with two micas or with chlorite and just a trioctahedral mica ("biotite").

(2) With rising temperatures at pressures greater than about 2 Kbar *chlorite* disappears next from the AKF diagram through its reaction with quartz to form cordierite + talc, which occurs again in the limiting system MgO-Al_2O_3-SiO_2-H_2O (cf. 4(g) and Fig. 9).

(3) Continued rise in temperature at these pressures leads to the *instability of the two-mica assemblage*, which reacts with quartz to form the practically anhydrous paragenesis *cordierite* + *K feldspar*. It is this reaction relationship which prompted Winkler (1967) to rename the "pyroxene-hornfels facies" as "cordierite-K feldspar hornfels facies".

(4) At higher temperatures, though still below the beginning of melting *muscovite* disappears from the AKF diagram through its reaction with quartz to form Al-silicate and K feldspar. This reaction takes place in the limiting system K_2O-Al_2O_3-SiO_2-H_2O (cf. Evans, 1965) and is, therefore, degenerate in the quinary system.

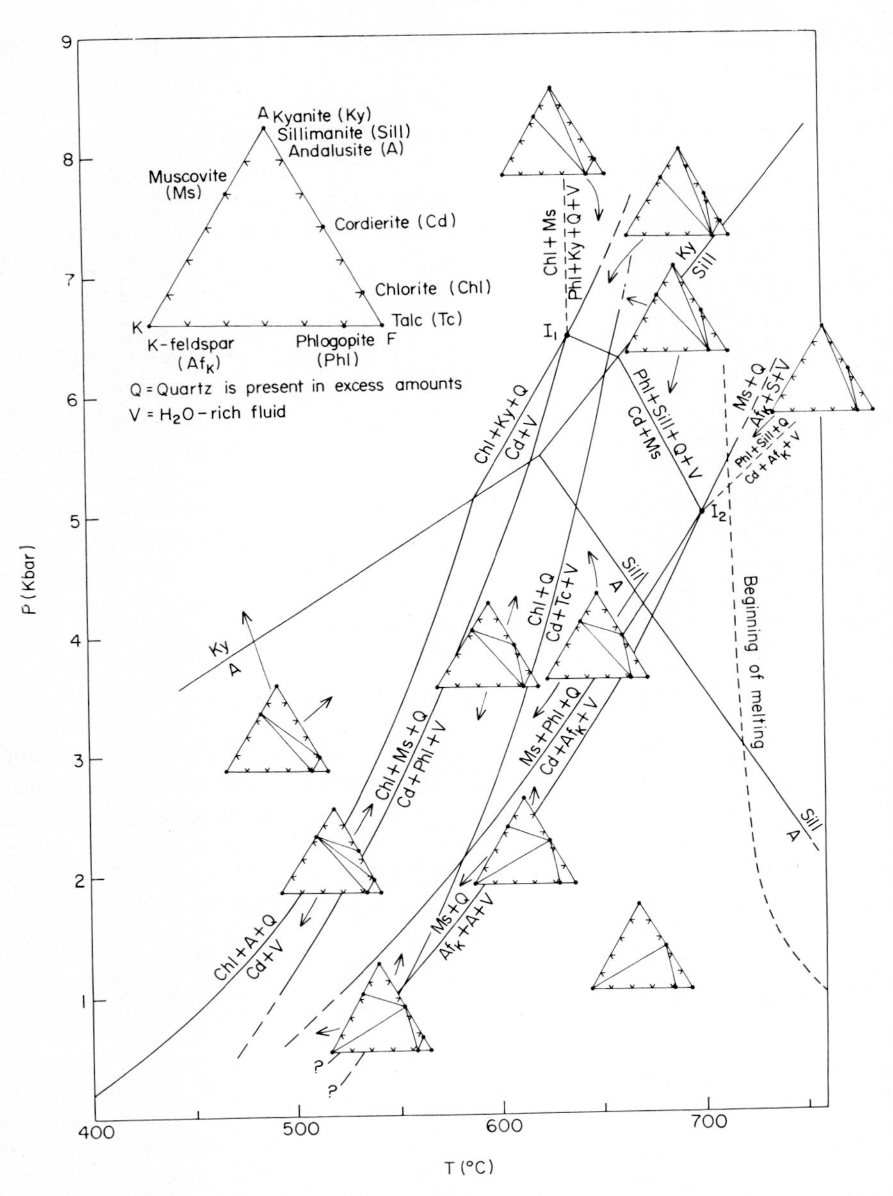

Fig. 12. Pressure–temperature diagram summarising the most important compatibility relations within the quartz-saturated portion of the system K_2O-MgO-Al_2O_3-SiO_2-H_2O at elevated temperatures and pressures between 0 and 7 Kbar based on the experimental work by Seifert (1970a and unpublished data). Except for quartz the crystalline phases involved in the reactions, and their abbreviations, are shown in the enlarged AKF-triangle (upper left) representing a projection of the H_2O- and SiO_2-saturated portion of the quinary system into a ternary subsystem Al_2O_3 (=A) — $K_2O \cdot Al_2O_3$ (=K) — MgO (=F).

(5) The apparent intersection of the upper temperature limit of the chlorite-quartz assemblage with the lower temperature limit of the cordierite + K feldspar paragenesis near 2 Kbar does not lead to any drastically new phase relations at still lower pressures. It is interesting to note the implication, however, that in relatively shallow contact aureoles the commonly accepted high-temperature paragenesis cordierite-K feldspar may make its appearance while chlorite and quartz may still coexist stably in K-poor portions of the rock suite. Thus the often invoked retrograde nature of chlorite assemblages in natural hornfelses should be taken with some caution.

(6) The dehydration reaction relationships described in paragraphs (1)–(5) as being typical for natural low-pressure environments such as hornfels aureoles, are, according to Fig. 12, indeed limited to relatively low pressures. Intersection of the chlorite + muscovite reaction with the lower stability curve of cordierite, and of the two-mica reaction with the upper stability curve of muscovite and quartz at elevated pressures creates *two invariant points* I_1 *and* I_2, which are interconnected by a reaction curve

$$\text{cordierite} + \text{muscovite} \rightleftharpoons \text{phlogopite} + Al_2SiO_5 + \text{quartz}.$$

Through this reaction as well as through a second one originating from I_2,

$$\text{cordierite} + \text{K feldspar} \rightleftharpoons \text{phlogopite} + \text{sillimanite} + \text{quartz},$$

the occurrence of the phase cordierite is restricted to bulk compositions of pelites unusually poor in K_2O. Thus at elevated pressures and intermediate temperatures *normal* pelites will no longer contain cordierite because of the stable coexistence of "biotite" and an aluminium silicate. Cordierite can only appear at pressures near that of I_2 in very high-grade rocks (gneisses) which show signs of partial melting (migmatites). It is to be noted that the curve "beginning of melting" dashed in Fig. 12 applies to the eutectic Mg-cordierite + K feldspar + SiO_2 first defined by Schairer (1954); granite melting will take place at still lower temperatures.

Although in natural rocks the conditions of these two limiting reactions originating from I_1 and I_2 will undoubtedly vary with the Fe-content, it is expected that they will, as in the model system discussed here, principally be a function of pressure and only less importantly of temperature. Simply on the basis of the lower pressure stability limit of Fe-bearing cordierites (cf. section 3(b)(18)) it can be predicted that these reactions, in addition to becoming divariant, will take place at successively lower pressures the higher the Fe^{+2} content in the rock. Because of the profound mineralogical changes endured by the rocks through the disappearance of cordierite from its assemblages with muscovite and K feldspar *one may use the two above-named reactions as a qualitative criterion for the distinction between low-pressure and intermediate-pressure metamorphism.*

Finally, it should be recalled that within the system K_2O-MgO-Al_2O_3-SiO_2-H_2O a quinary osumilite-type phase $KMg_2Al_3Si_{10}Al_2O_{30} \cdot xH_2O$ could be synthesised (Schreyer and Seifert, 1967). It was not included in the phase relations discussed here, because it proved to be metastable under all conditions of synthesis so far applied (see also section 3(b)(19)).

(i) *The System* Fe-O-Al_2O_3-SiO_2-H_2O

This system may be considered as a model system for ferruginous pelites as well as for rocks of metamorphic iron formations. Like all iron-bearing systems it is complicated due to the variable oxidation state of iron as a function of *oxygen fugacity*, f_{O_2}, which must be regarded as an additional variable besides pressure and temperature. Thus it was only after the oxygen buffering technique was developed by Eugster (1957) that a quantitative experimental approach was possible and reproducible results could be obtained. Despite considerable progress in recent years knowledge of the phase relationships within the system Fe-O-Al_2O_3-SiO_2-H_2O is still rather scanty being either limited to the f_{O_2}-range of one particular buffering system (Richardson, 1968) or else to isobaric sections with temperature and oxygen fugacity as variables (Hsu, 1968). However, more and more data are becoming available (Grieve and Fawcett, 1970b; Rutherford, 1970) so that it may be possible to establish a PTf_{O_2}-diagram in the near future.

In the present context the schematic pressure–temperature plot of Richardson (1968) valid only for the quartz–fayalite–magnetite (= QFM) buffer, has been slightly modified using the more recent data on the upper chloritoid stability by Grieve and Fawcett (1970b) and is shown here as Fig. 13. In spite of its incomplete and partly theoretical nature there are a number of interesting reaction relationships involving the seven phases Fe-chloritoid, Fe-staurolite, almandine, Fe-cordierite, hercynite solid solution, aluminium silicate, and quartz, which seem to be of particular relevance to low-pressure metamorphism. The main featues and their petrological implications are:

(1) Through intersection of reaction curves six invariant points I_1–I_6 are created which fall into the approximate pressure range between 1 and 4 Kbar and cause drastic changes in the compatibility relations at high pressures versus those at relatively low pressures.

(2) As one of the most obvious features the appearance of the phase *Fe-cordierite* is limited to pressures below some 3.5 Kbar. Its various breakdown reactions form a continuous series of curves interconnecting the invariant points $I_6 \leftrightarrow I_5 \leftrightarrow I_4 \leftrightarrow I_1$ thus defining a low-pressure stability field. Starting at the lowest temperatures Fe-cordierite is first formed through dehydration of the paragenesis chloritoid + quartz, the reaction curve exhibiting the typically curved steep positive slope. At higher temperatures and pressures Fe-cordierite breaks down first into a hydrous, but

dense assemblage (staurolite + almandine + quartz), and then into the anhydrous and denser assemblages, almandine + aluminium silicate + quartz, and hercynite + aluminium silicate + quartz. The latter assemblage is particularly intriguing as it involves the stable coexistence of free silica with an aluminous spinel phase; such compatibility is impossible in the related system MgO-Al_2O_3-SiO_2-H_2O (cf. 4(g) and Fig. 9). It must be emphasised that the hercynite stable under these conditions is not of the ideal composition $FeAl_2O_4$, but forms a solid solution towards magnetite, Fe_3O_4 (Turnock and Eugster, 1962). Since the reaction curve of the Fe-cordierite breakdown into the hercynite-bearing assemblage exhibits a negative slope, it follows, at least for the QFM-buffer, that Fe-cordierite is not a stable phase at very high temperatures, unless the pressure is extremely low. Extrapolation of the breakdown curve given by Richardson (1968) to higher temperatures would indicate a 1 atm upper stability limit of Fe-cordierite of the order of 1070°C, which lies probably still at subsolidus temperatures. On the other hand, Schairer and Yagi (1952) working in the presence of metallic iron found Fe-cordierite to be a stable liquidus phase up to a temperature of about 1210°C. This seems to indicate that the breakdown of Fe-cordierite is considerably influenced by the prevailing oxygen fugacity with high f_{O_2} values favouring the hercynite-quartz assemblage. The latter assemblage formed at temperatures above that of I_1 at the expense of almandine + aluminium silicate is apparently stable up to rather high pressures (Keesmann *et al.*, 1971, Fig. 2). The instability of Fe-rich cordierites at elevated pressures is also sufficiently supported by observations made on natural rocks (Eskola, 1915). Such cordierites have so far been found only in low-pressure pegmatites (Schreyer, 1965a).

(3) Roughly paralleling the upper pressure stability limit of pure Fe-cordierite but consistently displaced towards lower pressures there is a series of reaction curves interconnecting the invariant points $I_5 \leftrightarrow I_3 \leftrightarrow I_2 \leftrightarrow I_1$. In these reactions Fe-cordierite is expelled from all assemblages involving the more aluminous phases Fe-chloritoid and hercynite. They are essentially comparable to the *compression reactions* which were discussed in the related Mg-system (cf. 4(g) and Fig. 9) and which involve the phase Mg-cordierite. Consequently they lead to denser mineral assemblages, i.e. in the present case almandine–staurolite and almandine–aluminium silicate. Figure 14 may serve as an example that such reaction relationships are also encountered in natural rocks. The spatial distribution of the four phases garnet (almandine-rich), cordierite, green to opaque spinel, and sillimanite clearly indicates that a former assemblage of directly adjacent garnet and sillimanite had reacted—upon pressure release—to form a narrow zone of cordierite with spinel as a by-product concentrated in small grains along the sillimanite–cordierite boundary. Thus this observa-

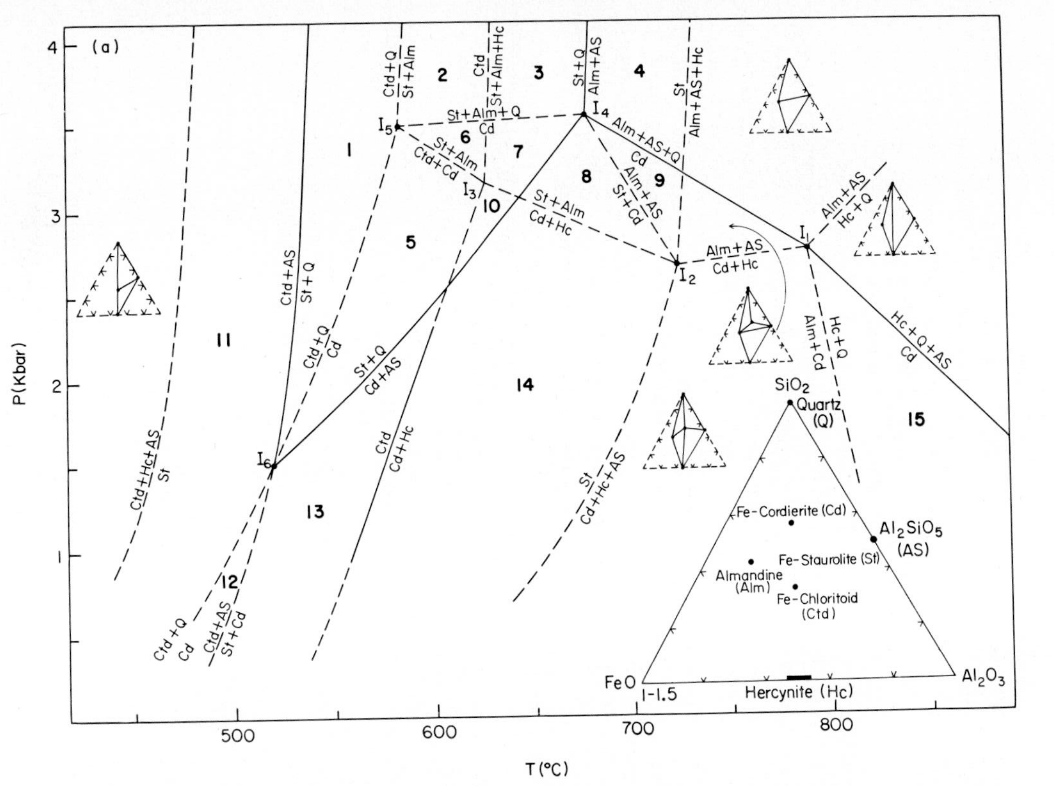

Fig. 13(a). Pressure–temperature diagram demonstrating schematically important compatibility relations within the system Fe-O-Al_2O_3-SiO_2-H_2O at elevated temperatures and relatively low pressures, modified after Richardson (1968). The diagram is meant to be valid for one particular range of oxygen fugacities as defined by the quartz–fayalite–magnetite buffer. The crystalline phases involved in the reactions, and their abbreviations, are shown in an enlarged compositional triangle (lower right) representing a projection of the quinary system into a water-free, pseudoternary subsystem $FeO_{1-1.5}$(=$FeO + Fe_2O_3$)-Al_2O_3-SiO_2.

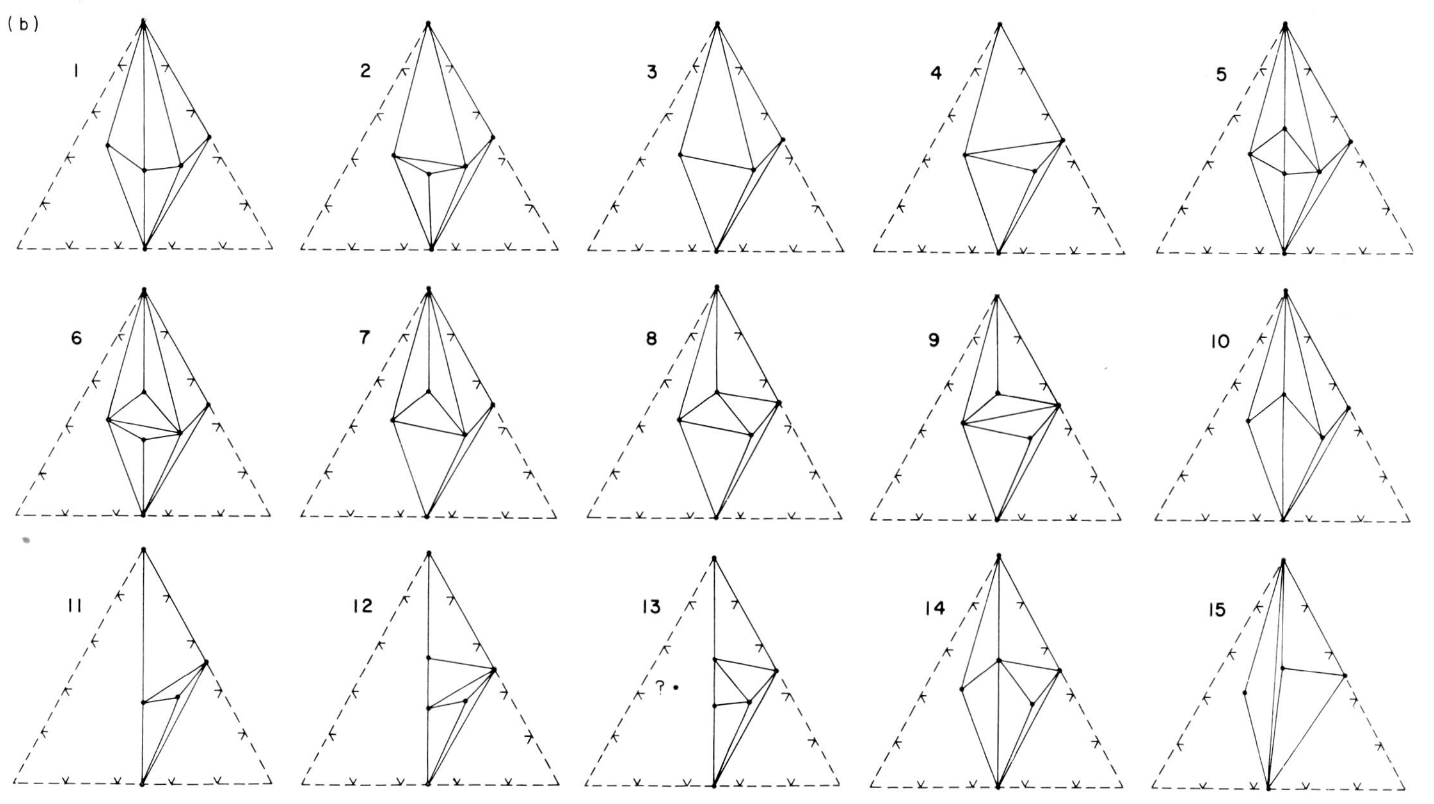

Fig. 13(b). Compatibility triangles for the numbered fields shown in (a).

tion provides direct evidence for the reaction shown in Fig. 13 connecting invariant points I_1 and I_2. It seems possible that the earlier garnet–sillimanite assemblage belongs to a regional metamorphic higher pressure stage followed by low-pressure contact metamorphism.

(4) Contrary to the stability of assemblages involving Fe-cordierite, the stability range of the assemblage *Fe-staurolite + quartz* is restricted to relatively *high* pressure above some 1.5 Kbar. The dehydration reactions

Fig. 14. Photomicrograph of a thin section showing reaction relationship between garnet (corroded crystal in upper portion) and sillimanite (needle-like crystals in lower portion) to form cordierite (white channel in centre) and Fe-rich spinel (black). Nowhere in the rock are garnet and sillimanite in mutual contact. Gneissose pelitic hornfels from granite contact, Gold Butte, Clark Co., Nevada, U.S.A.

causing the breakdown of this assemblage originate from and interconnect the invariant points I_6 and I_4, exhibiting relatively steep slopes. The location of I_6 at a pressure near 1.5 Kbar implies that the staurolite-quartz assemblage common in even relatively low-pressure regional metamorphism is not likely to be encountered in shallow contact metamorphism.

(5) Compared with the staurolite–quartz stability field that of pure *Fe-staurolite* is considerably larger regarding both temperature and pressure. It is likely that it will extend nearly to atmospheric pressure, in a similar way to pure Fe-chloritoid.

Some of the limitations of the compatibility relations shown in Fig. 13 should also be mentioned. Due to the omission of other phases in the system Fe-O-Al_2O_3-SiO_2-H_2O such as Fe-chlorite and fayalite the breakdown reactions of the phase almandine are not shown in Fig. 13 although they do occur in the PT-range discussed (cf. Hsu, 1968). The compatibility of the phase corundum with the Fe-bearing phases chloritoid, almandine, and staurolite has been disregarded although Ganguly (1969) has proposed a chloritoid breakdown into almandine + corundum at higher pressures. It is likely that the limiting tie line Al_2SiO_5-hercynite is not stable over the entire pressure-temperature range of Fig. 13. Grieve and Fawcett (1970b) confirmed the existence of a phase aluminous ferro-anthophyllite, previously synthesised by Schreyer (1965b) at pressures as low as 5.5 Kbar. Its possible appearance at lower pressures, which would further complicate the phase relations of Fig. 13, cannot be excluded. The hercynite solid solution towards magnetite (Turnock and Eugster, 1962) has been considered only for the phase relations around invariant point I_1, whereas for the hercynite taking part in reactions around I_2 and I_3 the ideal composition $FeAl_2O_4$ was assumed. This leads to the degenerate breakdown reaction of chloritoid into just the two phases Fe-cordierite and hercynite, which is also reported by Grieve and Fawcett (1970b) for pressures of 1–2 Kbar.

It is clear that completely different compatilility relations will result when the oxygen fugacity is no longer controlled by the QFM buffer. The abstract by Rutherford (1970) describing work at 2 Kbar may just give a few hints. At the higher f_{O_2} values defined by the hematite–magnetite buffer the high-temperature breakdown assemblage of Fe-staurolite + quartz is magnetite + Al_2SiO_5, that of pure Fe-staurolite is magnetite + Al_2SiO_5 + corundum. The phase Fe-cordierite oxidises at f_{O_2} values slightly higher than those given by the Ni-NiO buffer to form the assemblage staurolite + magnetite + quartz. Thus it is clear that there are severe complications introduced by iron as a component in rock-forming processes. They are, of course, increased further when iron appears *in addition to* magnesium, which is the case in practically all natural rocks (cf. 4(j)).

(j) *The System* Fe-O-MgO-Al_2O_3-SiO_2-H_2O

As a final example of a synthetic system a six-component system is cited which because of its combination of Fe and Mg is a more adequate model for alkali-free pelites than its two limiting systems Fe-O-Al_2O_3-SiO_2-H_2O and MgO-Al_2O_3-SiO_2-H_2O already discussed in sections 4(g) and 4(i), respectively.

Experimental studies within this system have thus far principally been restricted to the synthesis and stability of some of its crystalline phases, which usually exhibit extensive Mg $\rightleftharpoons$ Fe^{+2} solid solution (Yoder and Chinner, 1960; Hellner *et al.*, 1965; Hinrichsen, 1974). A theoretical treatment of phase relations involving cordierite and garnet coexisting with anhydrous phases, mainly applicable to high pressures, was published by Hensen (1971). In the course of experimental work on the almandine-pyrope series, however, some of the stable phase relations within the system at low pressures and high temperatures could also be deduced (Hsu and Burnham, 1969). The following excerpts from these results are intended as a first introduction into the problem.

Under the total pressure of 2 Kbar applied by Hsu and Burnham (1969) and with oxygen fugacities defined by the QFM-buffer there is a complete $Fe^{+2} \rightleftharpoons Mg$ substitution in the phases cordierite, olivine, spinel and chlorite, whereas the orthopyroxene series is not quite complete, ferrosilite-rich members being

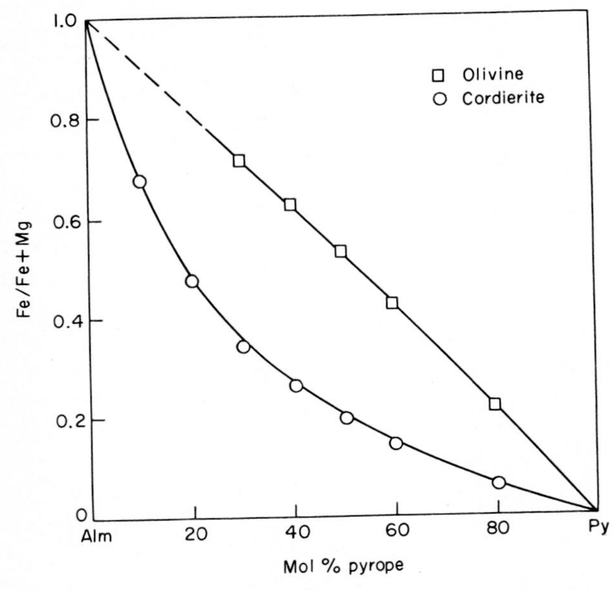

Fig. 15. Distribution of Mg and Fe^{+2} in coexisting olivines and cordierites synthesised from garnet bulk compositions within the system Fe-O-MgO-Al_2O_3-SiO_2-H_2O at 2 Kbar and temperatures between 775°C and 810°C with oxygen fugacities controlled by the quartz–fayalite–magnetite buffer. (After Hsu and Burnham, 1969.) Alm = almandine, Py = pyrope.

unstable. Despite these extensive miscibilities the distribution of Fe^{+2} and Mg in coexisting Fe–Mg-minerals is by no means homogeneous. As an example Fig. 15 shows that in coexisting olivines and cordierites the latter phases are strongly enriched in Mg, and a similar result was obtained for orthopyroxene–cordierite pairs. Since the quantitative relationships change as a function of the PTf_{O_2}-conditions of synthesis, the *Mg/Fe distribution coefficient for stably co-existing mineral phases is a new critical parameter* to be used in the study of the crystallisation history of natural rocks.

Figure 16 is a reproduction of the isobaric TX projection of the partial system $Fe_3Al_2Si_3O_{12}$-$Mg_3Al_2Si_3O_{12}$-H_2O valid for the QFM buffer at relatively high temperatures for a total fluid pressure of 2 Kbar, i.e. for conditions to be expected in shallow contact metamorphism. It provides also an excellent example of the general features, as well as complexities, of the compatibility relations of phases exhibiting extensive solid solutions. The projection involves the solid phases almandine-rich garnet, quartz, chlorite$_{ss}$ (ss = solid solutions), cordierite$_{ss}$, orthoamphibole$_{ss}$, olivine$_{ss}$, orthopyroxene$_{ss}$, talc with very limited Fe contents, and two spinel phases, an Al-rich one close to the spinel-hercynite

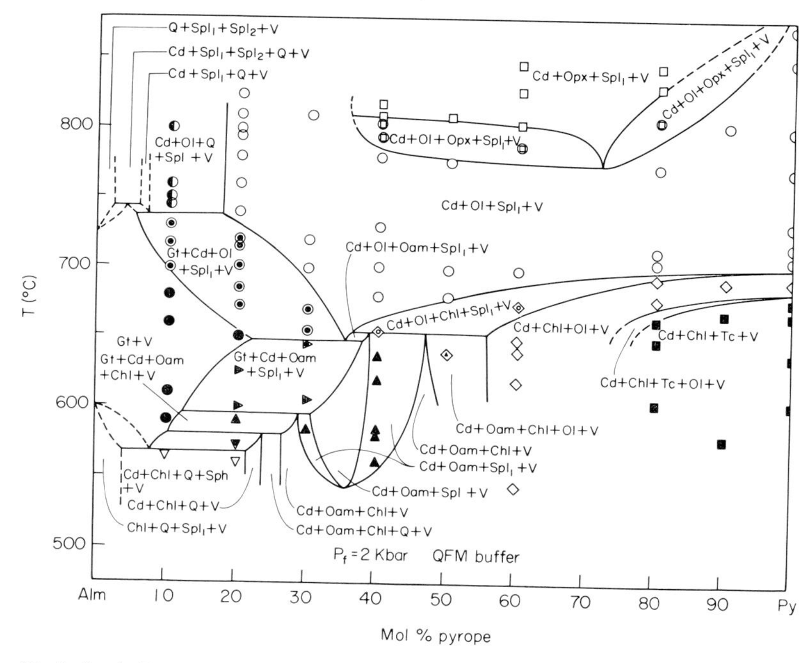

Fig. 16. Isobaric TX-projection of the system $Fe_3Al_2Si_3O_{12}$ (almandine, Alm)-$Mg_3Al_2Si_3O_{12}$ (pyrope, Py)-H_2O at 2 Kbar fluid pressure with oxygen fugacities defined by the fayalite–quartz–magnetite buffer. Reproduced from Hsu and Burnham (1969). Abbreviations of phases: Chl = chlorite, Cd = cordierite, Gt = garnet, V = fluid, Oam = orthoamphibole, Ol = olivine, Opx = orthopyroxene, Q = quartz, Spl_1 = Al-rich spinel solid solution, Spl_2 = Fe^{+3}-rich spinel solid solution, Tc = talc.

join, and an Fe^{+3}-rich one close to magnetite. The incoming of pure Mg-anthophyllite in the limiting system MgO-Al_2O_3-SiO_2-H_2O which prohibits the reaction chlorite + *talc* ⇌ cordierite + olivine at pressures greater than about 1 Kbar (cf. Fig. 9) has not been taken into account. Attention is drawn to the following main features:

(1) Garnet as a single phase is stable over only a limited temperature interval and a compositional range far to the Fe-rich side of the diagram. In more

Mg-bearing compositions, containing up to about 40% of the pyrope component, garnets with variable maximum pyrope contents coexist stably with three other solid phases, the nature of which change as a function of temperature.

(2) Assemblages consisting of *four solid phases* such as garnet + cordierite + orthoamphibole + spinel, and garnet + cordierite + olivine + spinel (both adjacent to the garnet single-phase field) are divariant. They are stable over a range of temperatures and total compositions, and, within this range, the individual compositions of the phases involved will vary. Thus these divariant ranges represent actually the locus where the divariant *Fe–Mg exchange reactions* take place. Such four-solid-phase assemblages may be related to each other through simple univariant reactions. For the example given the reaction taking place at 648°C is:

$$\text{garnet}_{ss} + \text{orthoamphibole}_{ss} \rightleftharpoons \text{cordierite}_{ss} + \text{olivine}_{ss} + \text{spinel} + H_2O.$$

For univariant reactions the Fe/Mg-ratios of the participating phases are uniquely defined.

(3) On the other hand, some of the high-temperature assemblages consisting of only *three solid phases* such as cordierite + olivine + spinel, and cordierite + orthopyroxene + spinel, are necessarily trivariant. Thus they cannot be related to each other through univariant reactions, but only through divariant reactions. Consequently the two trivariant fields mentioned are separated in the projection by divariant fields containing the four solid phases cordierite + olivine + orthopyroxene + spinel, which will again be all of variable composition within these fields. Only for the invariant condition, in which the compositions of all the phases of both of the three-solid-phase-assemblages are unique, are the two fields directly adjoining. This takes place for the composition pyrope_{72} at about 780°C.

Application of these phase relations to natural rocks is manifold. One may think of the *cordierite–gedrite–garnet rocks*, believed by many authors to be of metasomatic origin, for which the Mg/Fe distribution in coexisting minerals may yield information concerning their physical conditions of formation. A particularly rewarding example was found to be a ferruginous hornfels to which the phase diagram of Fig. 16 could almost directly be applied thus yielding elucidating results concerning its metamorphic history (Abraham and Schreyer, 1973). In the present context two more examples of the variable mineralogy of lime- and alkali-poor *ferruginous hornfelses* may be added.

The two photomicrographs of Figs. 17 and 18 are from hornfelses both formed within gabbro contact aureoles of the Harz Mountains and the Stillwater Complex, respectively. However, in Fig. 17 the critical three-phase assemblage is fayalitic olivine + cordierite + spinel, whereas it is hypersthene + cordierite + spinel in Fig. 18. According to Fig. 16 the olivine-bearing assemblage is formed at relatively low temperatures between some 650°C and 730°C. For all compositions with Mg/(Mg + Fe)-ratios higher than about 0.35 this assemblage is being changed with temperatures rising beyond some 800°C into the orthopyroxene-bearing paragenesis. Thus under the assumption that the two rocks have similar Mg/(Mg + Fe)-ratios, for example near 0.4 in Fig. 16, the Harz hornfels would indicate a lower temperature of

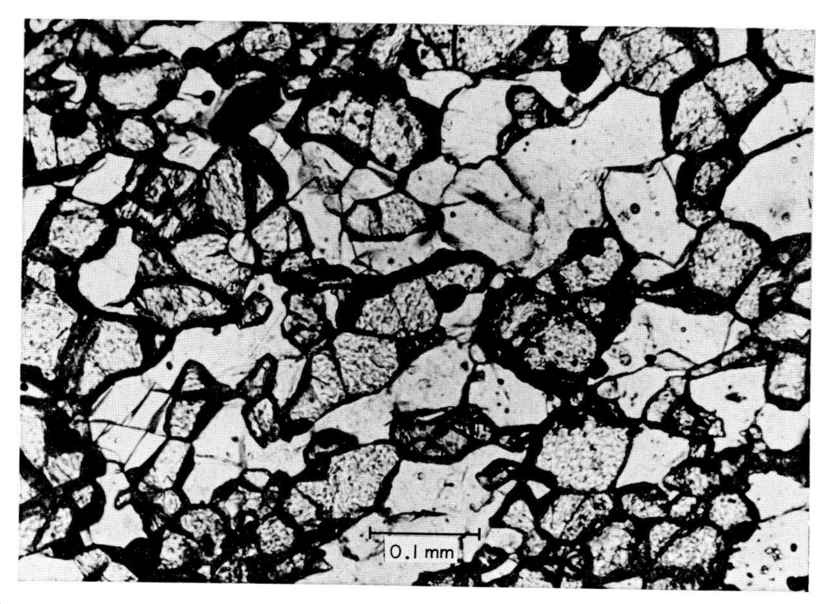

Fig. 17. Photomicrograph of a thin section exhibiting the mineral assemblage fayalitic olivine (high relief)—cordierite (white)—hercynite-rich spinel (black). Ferruginous hornfels, Riekensglück, Bad Harzburg, Germany.

metamorphism than the Stillwater rock. Indeed, the striking myrmekitic intergrowths of hypersthene and spinel in the Stillwater rock (Fig. 18) may be considered evidence for the previous existence of the olivine + cordierite + spinel assemblage during a lower-temperature episode of its thermal history. This olivine-bearing assemblage was apparently replaced by the present one through the divariant reaction

$$\text{olivine}_{ss} + \text{cordierite}_{ss} \rightleftharpoons \text{orthopyroxene}_{ss} + \text{spinel}_{ss},$$

whereby the hypersthene-spinel myrmekites probably mark those volumes of the rock which were formerly predominantly occupied by olivine. It should also be emphasised that the divariant reaction just mentioned must be strongly pressure–dependent as it represents a compression reaction with the high temperature assemblage also being stable at the higher pressures. For the iron-free system $MgO-Al_2O_3-SiO_2-H_2O$ these relations involving a negative dP/dT slope are depicted in Figs. 9 and 11 and were discussed in section 4(g). Therefore, the hypersthene-bearing mineralogy of the Stillwater rock could also be explained by higher confining pressures as compared to those operative in the Harz hornfels.

5. NATURAL MULTICOMPONENT SYSTEMS AND THE PROBLEM OF METAMORPHIC FACIES

In the foregoing sections it has been pointed out repeatedly that the metamorphic reactions investigated in model systems are only of limited value to the

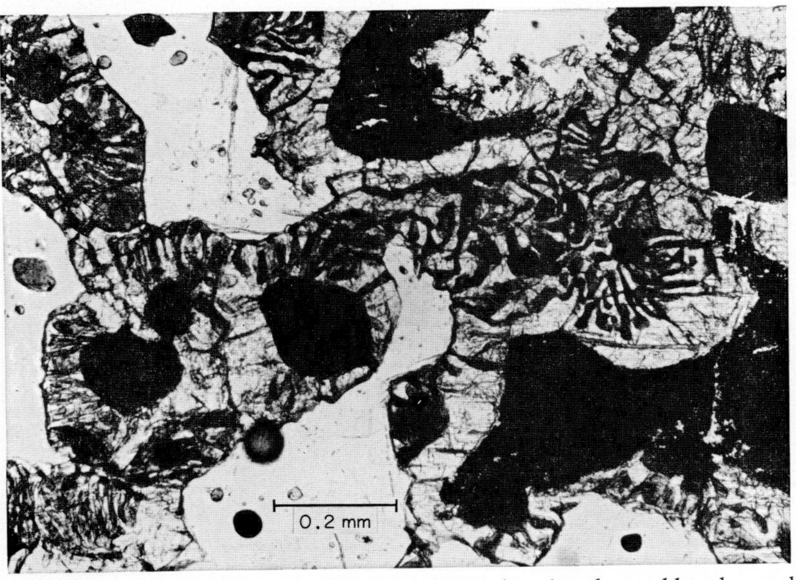

Fig. 18. Photomicrograph of a thin section exhibiting the mineral assemblage hypersthene (high relief)—cordierite (white)—hercynite-rich spinel (black). The spinel phase appears in two generations, the earlier one forming large crystals, the later one appearing as "worms" in myrmekite-like intergrowth within the large hypersthene crystals. This latter fabric is interpreted to have formed at the expense of a pre-existing olivine-cordierite assemblage such as shown in Fig. 17. For further explanations see text and Fig. 16. Ferruginous hornfels, Iron Mountain, Stillwater Igneous Complex, Montana, U.S.A.

understanding of metamorphism in natural rocks, simply because the latter involve so many more components than the model systems. Thus even for carbonate-poor *pelites* a relevant system would have to include both K_2O and Na_2O in addition to the six components combined as $Fe-O-MgO-Al_2O_3-SiO_2-H_2O$ in the final section just discussed. If there are carbonates present, as in many geosynclinal pelites, CaO and CO_2 have to be added as further components. Attempts at determining phase relations in more realistic synthetic pelitic compositions involving 8 components have been made by Hensen and Green (1971, 1972, 1973), but only at high pressures. Similar difficulties exist for the study of phase relations in rocks of *basaltic* composition, which would necessitate the treatment of at least nine components. In many texts on metamorphic petrology the description of metabasite phase relations has been attempted by means of the classical ACF diagram (Eskola, 1915) thus combining FeO and MgO and neglecting K_2O and Na_2O. It is indicative of the youthfulness of experimental petrology that a systematic subsolidus study of a model ACF system such as the SiO_2-saturated portion of the system CaO-MgO-Al_2O_3-SiO_2-H_2O, has not yet been undertaken.

Thus the present status of systematic experimental research in metamorphic petrology hardly allows any unequivocal, final petrogenetic conclusions concerning natural multicomponent rock systems. The experimental studies, however, provide quantitative knowledge of a great number of limiting conditions confining particular cases of rock formation in nature. Most importantly these studies have allowed elucidating insight into the fundamental physical-chemical laws governing rock-forming processes within the Earth's interior. As just one example of this type, but, at the same time, as one of the basic problems of metamorphic petrology, the question of *metamorphic facies* will be considered with special emphasis on high-temperature, low-pressure metamorphism.

In many textbooks of metamorphic petrology, including those of Winkler (1967) and Turner (1968), the realm of very low-pressure metamorphism is subdivided into four metamorphic facies: the albite–epidote–hornfels facies; the hornblende-hornfels facies; the pyroxene-hornfels facies (renamed by Winkler the cordierite-K feldspar-hornfels facies); and the sanidinite facies. The characterisation of these facies by means of critical mineral asssmblages in ACF, AKF and AFM diagrams turned out to be a difficult task, because natural rocks of one particular facies often show several compatibility pairs which should exclude each other in the above mentioned diagrams. Earlier proposals to solve this problem by introducing a variety of subfacies (e.g. Turner and Verhoogen, 1960) have led to very complex relations causing even greater confusion and have, therefore, been abandoned (Fyfe and Turner, 1966). Thus Turner (1968, Figs. 6–10) accepts the crossing of particular tie lines in AKF diagrams by as many as three other tie lines within one metamorphic facies. In addition, he accepts appearance as well as complete absence of particular mineral phases within the same facies (cf. Turner, 1968, Figs. 6–17). On the other hand, he admits that the borders between individual metamorphic facies are by no means sharp, which implies that they involve a number of different mineral reactions.

Under these circumstances, and with the results of experimental petrology in "simple" model systems (cf. section 4) in mind one is led to ask whether or not the metamorphic facies classification is indeed of any critical value for detailed, modern studies of rock metamorphism.

Experimental petrologists working in model systems encompassing only three or four components are deeply impressed by the severe complications of phase relations which they encounter in the course of their studies. It is hoped that this impression has also been conveyed to the readers of this chapter. For a quantitative appraisal the model system MgO-Al_2O_3-SiO_2-H_2O may be cited again, for which Seifert (1970b) has found some 80 divariant fields in the range 400°C–1250°C, and 1 atm–7 Kbar, each defined by unique compatibility relations. Since, on the basis of its original definition, the metamorphic facies principle has to be valid for *all* bulk compositions encountered in a rock suite, the divariant fields of a model system are to be compared with individual metamorphic facies as defined on a rigorous mineralogical basis. Thus, for the variety of natural rocks, at least all the divariant fields depicted in the figures for the various independent model systems must be counted as metamorphic facies in this sense. Taking into account the much greater chemical complexity of natural rocks, however, the number of metamorphic facies becomes so enormous that distinction of individual facies is impractical, if not impossible.

Thus, there is no way to a rigorous classification of multicomponent natural metamorphic rocks on a mineralogical basis. On the other hand, classification does not seem to be the prime goal of modern petrology. The author believes that, compared with classification, quantitative connotations of the magnitudes of pressure, temperature, fugacities of oxygen, water, carbon dioxide, etc. during metamorphism represent much more valuable information, revealing the conditions generated and maintained within the Earth's interior. For this purpose more detailed work in experimental petrology, on increasingly complicated systems, will be necessary in the future.

REFERENCES

Abraham, K. and Schreyer, W. (1973). *Contr. Miner. Petrol.*, **40,** 275–292.
Abraham, K. and Seifert, F. (1973). *Forsch. Ber. des Landes Nordrh.-Westf. Nr.* **2374,** 21.
Ackermand, D. and Raase, P. (1973). *Contr. Miner. Petrol.*, **42,** 333–341.
Agrell, S. O. (1965). *Min. Mag.*, **34,** 1–15.
Agrell, S. O. and Langley, J. M. (1958). *Proc. R. Ir. Acad.*, **B59,** 93–127.
Akella, J. and Winkler, H. G. F. (1966). *Contr. Miner. Petrol.*, **12,** 1–12.
Althaus, E. (1967). *Contr. Miner. Petrol.*, **16,** 29–44.
Althaus, E. (1969). *Am. J. Sci.*, **267,** 273–277.
Anastasiou, P. and Seifert, F. (1972). *Contr. Miner. Petrol.*, **34,** 272–287.
Arnold, R. G. (1962). *Econ. Geol.*, **57,** 72–90.
Barnes, H. L. and Ernst, W. G. (1963). *Am. J. Sci.*, **261,** 129–150.
Barth, T. F. W. (1930). *Min. Pet. Mitt.*, **40,** 221–234.
Boettcher, A. L. (1970). *J. Petrol.*, **11,** 337–379.
Bowen, N. L. (1940). *J. Geol.*, **48,** 225–274.
Bowen, N. L. and Tuttle, O. F. (1949). *Bull. Geol. Soc. Amer.*, **60,** 439–460.

Boyd, F. R. (1959). Hydrothermal Investigations of Amphiboles, *in* "Researches in Geochemistry", P. H. Abelson (Ed.), 377–396. John Wiley, New York.
Boyd, F. R. and England, J. L. (1959). *Carn. Inst. Wash. Yr. Bk.*, **58,** 83–87.
Boyd, F. R. and England, J. L. (1964). *Carn. Inst. Wash. Yr. Bk.*, **63,** 157–161.
Boyd, F. R., England, J. L. and Davis, B. T. C. (1964). *J. Geophys. Res.*, **69,** 2,102–2,109 (1964).
Buckner, D. A. and Roy, R. (1955). *Bull. Geol. Soc. Amer.*, **66,** 1,536.
Bugge, J. A. W. (1943). *Norg. Geol. Unders.*, **160,** 150.
Burnham, C. W. (1959). *Geol. Soc. Amer. Bull.*, **70,** 879–920.
Carpenter, A. B. (1967). *Amer. Miner.*, **52,** 1,341–1,363.
Chatterjee, N. D. (1962). *Beitr. Miner. Petrogr.*, **8,** 432–439.
Chatterjee, N. D. (1970). *Contr. Miner. Petrol.*, **27,** 244–257.
Chatterjee, N. D. (1972). *Contr. Miner. Petrol.*, **34,** 288–303.
Chatterjee, N. D. (1974). *Schweiz. Min. Petrogr. Mitt.*, **54,** 753–767.
Chinner, G. A. (1959). *Carn. Inst. Wash. Yr. Bk.*, **58,** 112–113.
Chinner, G. A. (1962). *J. Petrol.*, **3,** 316–340.
Chinner, G. A. and Dixon, P. D. (1973). *Min. Mag.*, **39,** 189–192.
Coes, L., Jr. (1955). *J. Amer. Ceram. Soc.*, **38,** 298.
Coleman, R. G. (1967). *U.S. Geol. Surv. Bull.*, **1247,** 49.
Compton, R. R. (1960). *Bull. Geol. Soc. Amer.*, **71,** 1,383–1,416.
Daly, R. A. (1903). *U.S. Geol. Surv. Bull.*, **209.**
Dasgupta, H. C., Seifert, F. and Schreyer, W. (1974). *Contr. Miner. and Petrol.* **43,** 275–294.
Deer, W. A., Howie, R. A. and Zussman, J. (1962–63). "Rock-forming Minerals, Vol. 1–5. Longmans, London.
Ernst, Th. and Schwab, R. (1970). *Phys. Earth Planet. Int.*, **3,** 451–455.
Ernst, W. G. (1966). *Am. J. Sci.*, **264,** 37–65.
Ernst, W. G. (1968). "Amphiboles". Springer, New York.
Eskola, P. (1914). *Bull. Comm. géol. Finlande*, **40.**
Eskola, P. (1915). *Bull. Comm. géol. Finlande*, **44.**
Eugster, H. P. (1957). *J. Chem. Phys.*, **26,** 1,760.
Eugster, H. P. and Skippen, G. B. (1967). Igneous and Metamorphic Reactions Involving Gas Equilibria, *in* "Researches in Geochemistry, Vol. 2", P. H. Abelson (Ed.). John Wiley, New York.
Evans, B. W. (1965). *Am. J. Sci.*, **263,** 647–667.
Evans, B. W. and Trommsdorff, V. (1974). *Am. J. Sci.*, **274,** 274–296.
Fawcett, J. J. and Yoder, H. S., Jr. (1966). *Amer. Miner.*, **51,** 353–380.
Finger, L. W. and Burt, D. M. (1972). *Carn. Inst. Wash. Yr. Bk.*, **71,** 616–620.
Flaschen, S. S. and Osborn, E. F. (1957). *Econ. Geol.*, **52,** 923–943.
Flörke, O. W. (1955). *Ber. Dtsch. Keram. Ges.*, **32,** 369–381.
Flörke, O. W. and Langer, K. (1972). *Contr. Miner. Petrol.*, **36,** 221–230.
Foster, W. R. (1950). *J. Geol.*, **58,** 135–151.
Foster, W. R. (1951). *J. Amer. Ceram. Soc.*, **34,** 255–259.
French, B. M. (1965). "Some Geological Implications of Equilibrium Between Graphite and a C–H–O Gas Phase at High Temperatures and Pressures". National Aeronautics and Space Administration, Goddard Space Flight Center, pub. X-641-65-324, Greenbelt, Md., U.S.A.
French, B. M. (1968). *Minn. Geol. Surv. Bull.*, **45,** 103.
Frey, M. and Orville, P. M. (1974). *Am. J. Sci.*, **274,** 31–47.
Friedman, G. M. (1952). *Amer. Miner.*, **37,** 244–249.
Fyfe, W. S. and Turner, F. J. (1966). *Contr. Miner. Petrol.*, **12,** 354–364.
Ganguly, J. (1969). *Am. J. Sci.*, **267,** 910–944.
Ganguly, J. (1972). *J. Petrol.*, **13,** 335–365.

Gilbert, M. C. (1966). *Am. J. Sci.*, **264,** 698–742.

Goldschmidt, V. M. (1911). *Vidensk. Skrifter. I. Mat.-Naturv. K.*, **11.**

Goldsmith, J. R. and Laves, F. (1954). *Geochim. et Cosmochim. Acta*, **5,** 1–19.

Goldsmith, J. R. and Newton, R. C. (1969). *Am. J. Sci.*, **267-A** (Schairer vol.), 160–190.

Gordon, T. M. and Greenwood, H. J. (1970). *Am. J. Sci.*, **268,** 225–242.

Gordon, T. M. and Greenwood, H. J. (1971). *Amer. Miner.*, **56,** 1,674–1,688.

Greenwood, H. J. (1962). *Carn. Inst. Wash. Yr. Bk.*, **61,** 82–85.

Greenwood, H. J. (1963). *J. Petrol.*, **4,** 317–351.

Greenwood, H. J. (1967a). *Amer. Miner.*, **52,** 1,669–1,680.

Greenwood, H. J. (1967b). Mineral equilibria in the system MgO-SiO_2-CO_2-H_2O. *in* "Researches in Geochemistry, Vol. 2", Abelson, P. H. (Ed.). John Wiley, New York.

Greenwood, H. J. (1971). *Am. J. Sci.*, **274,** 151–154.

Greenwood, H. J. (1972). *Geol. Soc. Am. Mem.*, **132,** 553–571.

Grieve, R. A. F. and Fawcett, J. J. (1971a). *Amer. Miner.*, **55,** 517–521.

Grieve, R. A. F. and Fawcett, J. J. (1971b). *Trans. Amer. Geophy. Union*, **51,** 437.

Halferdahl, L. B. (1961). *J. Petrol.*, **2,** 49–135.

Harder, H. (1956). *Beitr. Miner. Petrogr.*, **4,** 227–269.

Hariya, Y., Dollase, W. A., and Kennedy, G. C. (1969). *Am. Miner.*, **54,** 1,419–1,441.

Harker, A. (1939). "Metamorphism". Methuen, London, 362 p.

Harker, R. I. (1959). *Am. J. Sci.*, **257,** 656–667.

Harker, R. I. and Tuttle, O. F. (1955a). *Am. J. Sci.*, **253,** 209–224.

Harker, R. I. and Tuttle, O. F. (1955b). *Am. J. Sci.*, **253,** 274–282.

Harker, R. I. and Tuttle, O. F. (1956a). *Am. J. Sci.*, **254,** 239–256.

Harker, R. I. and Tuttle, O. F. (1956b). *Am. J. Sci.*, **254,** 468–478.

Harwood, D. S. and Larson, R. R. (1969). *Amer. Miner.*, **54,** 896–908.

Hays, J. F. (1967). *Carn. Inst. Wash. Yr. Bk.*, **65,** 234–239.

Hellner, E., Hinrichsen, Th. and Seifert, F. (1965). The study of mixed crystals of minerals in metamorphic rocks, *in* "Controls of Metamorphism". W. S. Pitcher and G. W. Flinn (Ed.), Oliver and Boyd, London.

Hensen, B. J. (1971). *Contr. Miner. and Petrol.*, **33,** 191–214.

Hensen, B. J. and Green, D. H. (1971). *Contr. Miner., Petrol.*, **33,** 309–330.

Hensen, B. J. and Green, D. H. (1972). *Contr. Miner. Petrol.*, **35,** 331–354.

Hensen, B. J. and Green, D. H. (1973). *Contr. Miner. and Petrol.*, **38,** 151–166.

Hinrichsen, Th. (1967). *Neues Jahrb. Min. Monatshefte*, 257–270.

Hinrichsen, Th. (1968). *Neues Jahrb. Min. Monatshefte*, 41–57.

Hinrichsen, Th. (1974). *Fortschr. Min.*, **51,** Beih. 1, 19–20.

Holdaway, M. J. (1966). *Am. J. Sci.*, **264,** 643–667.

Holdaway, M. J. (1971). *Am. J. Sci.*, **271,** 97–131.

Holdaway, M. J. (1972). *Contr. Miner. Petrol.*, **37,** 307–340.

Holm, J. L. and Kleppa, O. J. (1966). *Amer. Mineral.*, **51,** 1,608–1,622.

Hsu, L. C. (1968). *J. Petrol.*, **9,** 40–83.

Hsu, L. C. and Burnham, C. W. (1969). *Geol. Soc. Am. Bull.*, **80,** 2,393–2,408.

Huckenholz, H. G. (1969). *Am. J. Sci.*, **267-A,** 209–232.

Huckenholz, H. G. and Yoder, H. S., Jr. (1971). *Neues Jahrb. Min. Abh.*, **114,** 246–280.

Ito, J. and Arem, J. E. (1970). *Am. Miner.*, **55,** 880–912.

Johannes, W. (1969). *Am. J. Sci.*, **267,** 1,083–1,104.

Johannes, W. (1970). *Neues. Jahrb. Min. Abh.*, **113,** 274–325.

Keesmann, I., Matthes, S., Schreyer, W. and Seifert, W. (1971). *Contr. Miner. Petrol.*, **31,** 132–144.

Kennedy, G. C. (1959). *Am. J. Sci.*, **257,** 563–573.

Kerrick, D. M. (1970). *Bull. Geol. Soc. Amer.*, **81,** 2,913–2,938.

Kerrick, D. M., Crawford, K. E. and Randazzo, A. F. (1973). *J. Petrol.*, **14,** 303–325.

Kulke, H. and Schreyer, W. (1973). *Earth Planet. Sci. Letters*, **18,** 324–328.
Kullerud, G. and Yoder, H. S., Jr. (1959). *Econ. Geol.*, **54,** 533–572.
Kushiro, I. (1964a). *Carn. Inst. Wash. Yr. Bk.*, **63,** 84–86.
Kushiro, I. (1964b). *Carn. Inst. Wash. Yr. Bk.*, **63,** 90–92.
Kushiro, I. (1964c). *Carn Inst. Wash. Yr. Bk.*, **63,** 83–84.
Kwak, T. A. P. (1971). *Amer. Miner.*, **56,** 1,750–1,759.
Larsen, E. S., Irving, J., Gouyer, F. A. and Larsen, E. S. 3rd (1936). *Amer. Miner.*, **21,** 679.
Laves, F. (1960). *Z. Krist*, **113,** 265–296.
Lindsley, D. H. (1965). *Carn. Inst. Wash. Yr. Bk.*, **64,** 148–149.
Lindsley, D. H. and Munoz, J. L. (1969). *Carn. Inst. Wash. Yr. Bk.*, **67,** 86–88.
Liou, J. G. (1971). *Amer. Miner.*, **56,** 507–531.
Luth, W. C. (1967). *J. Petrol.*, **8,** 372–416.
Luth, W. C. and Tuttle, O. F. (1966). *Amer. Miner.*, **51,** 1,359–1,373.
Merrin, S. (1962). "Experimental Investigations of Epidote Paragenesis". Ph.D. thesis, College of Mineral Industries, The Pennsylvania State University.
Metz, P. (1967). *Geochim. et Cosmochim. Acta*, **31,** 1,517–1,532.
Metz, P. (1970). "Experimentelle Untersuchung der Metamorphose von kieselig dolomitischen Sedimenten". Habilitationsschrift Universität Göttingen.
Metz, P. and Trommsdorff, V. (1968). *Contr. Miner. Petrol.*, **18,** 305–309.
Metz, P. and Winkler, H. G. F. (1963). *Geochim. et Cosmochim. Acta*, **27,** 431–457.
Metz, P. and Winkler, H. G. F. (1964). *Naturwissenschaften*, **51,** 460.
Miyashiro, A. (1956). *Amer. Miner.*, **41,** 104–116.
Miyashiro, A. (1958). *Tokyo Univ. Fac. Sci. J., Sec. 2,* **11,** 219–272.
Miyashiro, A. (1973a). *Tectonophysics*, **17,** 241–254.
Miyashiro A. (1973b). "Metamorphism and Metamorphic Belts". Allen and Unwin London.
Moore J. M. and Best M. G. (1969). *Amer. Miner.* **54,** 975–979.
Myer G. H. (1966). *Am. J. Sci.* **264,** 364–385.
Nelson B. W. and Roy R. (1958). *Amer. Miner.* **43,** 707–725.
Newton R. C. (1965). *J. Geol.* **73,** 431–441.
Newton R. C. (1966). *Am. J. Sci.* **264,** 204–222.
Newton R. C. (1972). *J. Geol.* **80,** 398–420.
Nitsch K. H. (1972). *Contr. Miner. Petrol.*, **34,** 116–134.
Nitsch K. H. and Winkler, H. G. F. (1965). *Beitr. Miner. Petrol.*, **11,** 470–486.
Novak, J. (1967). *Acta Univ. Carolinae-Geol.*, **2,** 123–131.
Okrusch, M. (1969). *Contr. Miner. Petrol.*, **22,** 32–72.
Okrusch, M. (1971). *Contr. Miner. Petrol.*, **32,** 1–23.
Oliver, R. L. and Jones, J. B. (1965). *Min. Mag.*, **35,** 140–145.
Orville, P. M. (1963). *Am. J. Sci.*, **261,** 201–237.
Osborn, E. F. and Tait, D. B. (1952). *Am. J. Sci.*, Bowen Vol., 413–433.
Rahman, S. and MacKenzie, W. S. (1969). *Am. J. Sci.*, **267-A,** 391–406.
Ramdohr, P. (1919). *Jb. Preuß. geol. Landesanst.*, **40,** 284–433.
Read, H. H. (1923). *Mem. Geol. Surv. Scot.* (Expl. Sheets 86 and 96).
Richardson, S. W. (1967). *Carn. Inst. Wash. Yr. Bk.*, **65,** 248–252.
Richardson, S. W. (1968). *J. Petrol.*, **9,** 468–488.
Richardson, S. W., Bell, P. M. and Gilbert, M. C. (1968). *Am. J. Sci.*, **266,** 513–541.
Richardson, S. W., Gilbert, H. C. and Bell, P. M. (1969). *Am. J. Sci.*, **267,** 259–272.
Robinson, P., Ross, M. and Jaffe, H. W. (1971). *Amer. Miner.*, **56,** 1,005–1,041.
Rosenbusch, H. (1877). *Abh. zur geol. Special-karte von Elsass-Lothringen*, **1,** 79–393.
Rost, F. and Hochstetter, R. (1964). *Neues Jahrb. Min. Abh.*, **101,** 173–194.
Roy, D. M. (1958). *Amer. Miner.*, **43,** 1,009–1,028.
Roy, D. M. and Roy, R. (1955). *Amer. Miner.*, **40,** 147–178.

Roy, D. M. and Roy, R. (1957). *Bull. Geol. Soc. Amer.*, **68,** 1,788–1,789.
Roy, D. M., Roy, R. and Osborn, E. F. (1953). *Am. J. Sci.*, **251,** 337–361.
Rutherford, M. J. (1970). *Trans. Amer. Geoph. Union*, **51,** 437.
Rutherford, M. J. (1973). *J. Petrol.*, **14,** 159–180.
Saggerson, E. P. and Owen, L. M. (1969). *Geol. Soc. South Afr., Spec. Publ.*, **2,** 335–349.
Sahl, K. and Seifert, F. (1973). *Nature Physical Science*, **241,** 46–47.
Scarfe, C. M. and Wyllie, P. J. (1967). *Nature*, **215,** 945–946.
Schairer, J. F. (1954). *J. Amer. Ceram. Soc.*, **37,** 501–533.
Schairer, J. F. and Yagi, K. (1952). *Am. J. Sci., Bowen Vol.*, 471–512.
Schairer, J. F., Yoder, H. S., Jr. and Tilley, C. E. (1967). *Carn. Inst. Wash. Yr. Bk.*, **64,** 95–100.
Scheumann, K. H. (1960). *Abh. sächs. Akad. Wiss.*, **47,** No. 2, 23.
Schreyer, W. (1965a). *Neues Jahrb. Min. Abh.*, **103,** 35–79.
Schreyer, W. (1965b). *Beitr. Mineral. Petrogr.*, **11,** 297–322.
Schreyer, W. (1966). *Neues Jahrb. Min. Abh.*, **105,** 211–244.
Schreyer, W. (1968). *Carn. Inst. Wash. Yr. Bk.*, **66,** 380–392.
Schreyer, W. (1970). *Fortschr. Miner.*, **47,** 124–165.
Schreyer, W., Ohnmacht, W. and Mannchen, J. (1972). *Lithos*, **5,** 345–364.
Schreyer, W. and Schairer, J. F. (1961). *J. Petrol.*, **2,** 324–406.
Schreyer, W. and Seifert, F. (1967). *Contr. Miner. Petrol.*, **14,** 343–358.
Schreyer, W. and Seifert, F. (1969a). *Am. J. Sci.*, **267,** 371–388.
Schreyer, W. and Seifert, F. (1969b). *Am. J. Sci.*, **267-A,** 407–443.
Schreyer, W. and Yoder, H. S., Jr. (1964). *Neues Jahrb. Min. Abh.*, **101,** 271–342.
Schürmann, K. (1967). *Neues Jahrb. Min. Monatsh.*, 270–284.
Seck, H. (1971a). *Neues Jahrb. Min. Abh.*, **115,** 315–345.
Seck, H. (1971b). *Contr. Miner. Petrol.*, **31,** 67–86.
Seck, H. (1972). *Fortschr. Miner.*, **49,** 31–49.
Seifert, F. (1970a). *J. Petrol.*, **11,** 73–99.
Seifert, F. (1970b). "Das petrogenetische P_{H_2O}-T-Netz des Systems MgO-Al_2O_3-SiO_2-H_2O im Druck-Temperatur-Bereich 0-7 kb, 400–1,250°C. Habilitationsschrift, Ruhr-Universität Bochum.
Seifert, F. (1973). *Contr. Miner. Petrol.*, **41,** 171–178.
Seifert, F. (1974). *J. Geol.*, **82,** 173–204.
Seifert, F. (1975). *Am. J. Sci.*, **275,** 57–87.
Seifert, F. and Schreyer, W. (1970). *Contr. Miner. Petrol.*, **27,** 225–238.
Shido, F. (1958). *Tokyo Univ. Fac. Sci. J.*, Sec. 2, **11,** 131–217.
Skippen, G. B. (1971). *J. Geol.*, **79,** 457–481.
Smith, D. (1971). *Am. J. Sci.*, **271,** 370–382.
Smith, D. G. W. (1969). *J. Petrol.*, **10,** 20–55.
Smith, J. V. (1972). *J. Geol.*, **80,** 505–525.
Smulovich, K. I. (1969). *Doklady Acad. Nauk. USSR*, **184,** 1,177–1,179.
Snow, R. B. (1943). *J. Amer. Ceram. Soc.*, **26,** 11–20.
Storre, B. (1970). *Contr. Mineral. Petrol.*, **29,** 145–162.
Storre, B. and Nitsch, K. H. (1972). *Contr. Mineral. Petrol.*, **35,** 1–10.
Storre, B. and Nitsch, K. H. (1974). *Contr. Mineral. Petrol.*, **43,** 1–24.
Thomas, H. H. (1922). *Q.J. Geol. Soc. Lond.*, **78,** 229–259.
Thompson, J. B. Jr. and Waldbaum, D. R. (1969). *Amer. Miner.*, **54,** 811–838.
Tilley, C. E. (1923). *Geol. Mag.*, **60,** 101–107.
Tilley, C. E. (1924). *Q.J. Geol. Soc. Lond.*, **80,** 22–70
Tilley, C. E. (1925). *Min. Mag.*, **22,** 78–86.
Tilley, C. E. (1948). *Min. Mag.*, **28,** 272–276.
Toulmin, P., III and Barton, P. B. Jr. (1964). *Geochim. et Cosmochim. Acta*, **28,** 641–671.

Trommsdorff, V. (1972). *Schweiz. Miner. Petrogr. Mitt.*, **52,** 567–571.
Tsuru, K. and Henry, N. F. M. (1937). *Min. Mag.*, **24,** 527.
Turner, F. J. (1967). *Neues Jahrb. Min. Monatsh.*, 1–22.
Turner, F. J. (1968). Metamorphic Petrology: Mineralogical and Field Aspects. McGraw-Hill, New York.
Turner, F. J. and Verhoogen, J. (1960). "Igneous and Metamorphic Petrology", McGraw-Hill, New York.
Turnock, A. C. (1960). *Carn. Inst. Wash. Yr. Bk.*, **59,** 98–103.
Turnock, A. C. and Eugster, H. P. (1962). *J. Petrol.*, **3,** 533–565.
Tuttle, O. F. and Bowen, N. L. (1958). *Mem. Geol. Soc. Am.*, **74,** 153.
Tuttle, O. F. and England, J. L. (1955). *Bull. Geol. Soc. Am.*, **66,** 149–152.
Velde, B. (1966). *Amer. Miner.*, **51,** 924–929.
Vogt, T. (1947). *Comm. géol. Finlande Bull.* **140,** 15–24.
Walter, L. S. (1963a). *Am. J. Sci.*, **261,** 488–500.
Walter, L. S. (1963b). *Am. J. Sci.*, **261,** 773–779.
Walter, L. S. (1965). *Am. J. Sci.*, **263,** 64–77.
Walter, L. S. (1966). *Geol. Soc. Amer. Ann. Meet. Progr. Abstr.*, 235.
Weber, J. N. and Roy, R. (1965). *Am. J. Sci.*, **263,** 668–677.
Weill, D. F. (1966). *Geochim. et Cosmochim. Acta*, **30,** 223–237.
Wenk, E. (1962). *Schweiz. Mineral. petrogr. Mitt.*, **42,** 139–152.
Wenk, E. (1968). *Schweiz. Mineral. petrogr. Mitt.*, **48,** 455–457.
Winkler, H. G. F. (1967). "Petrogenesis of Metamorphic Rocks". Springer-Verlag, New York.
Wones, D. R. and Eugster, H. P. (1965). *Amer. Miner.*, **50,** 1,228–1,272.
Wyllie, P. J. (1961). *J. Petrol.*, **2,** 1–37.
Yoder, H. S., Jr. (1950a). *Trans. Amer. Geophys. Union*, **31,** 827–835.
Yoder, H. S. Jr. (1950b). *J. Geol.*, **58,** 221–253.
Yoder, H. S. Jr. (1952). *Am. J. .Sci.*, **267-A,** 569–627.
Yoder, H. S., Jr. (1954). *Carn. Inst. Wash. Yr. Bk.*, **53,** 121.
Yoder, H. S. Jr. (1964). *Carn. Inst. Wash. Yr. Bk.*, **62,** 86–89.
Yoder, H. S. Jr. (1968). *Carn. Inst. Wash. Yr. Bk.*, **66,** 471–477.
Yoder, H. S., Jr. (1973). *Fortschr. Min.*, **50,** 140–173.
Yoder, H. S. Jr. and Chinner, G. A. (1960). *Carn. Inst. Wash. Yr. Bk.*, **59,** 81–84.
Yoder, H. S. Jr. and Kushiro, I. (1969). *Am. J. Sci.*, **267-A,** 558–582.
Yoder, H. S. Jr. and Tilley, C. E. (1962). *J. Petrol.*, **3,** 342–532.
Zen, E-an (1966). *U.S. Geol. Survey Bull.*, **1225.**
Zen, E-an (1971). *Am. J. Sci.*, **270,** 136–150.
Zwart, H. J. (1958). *Geol. en Mijnbouw*, **20,** 18–30.

Part II B

Experimental Petrology: Igneous Rocks

SUMMARY

This section is devoted to reviews of experimental petrology in granitic and alkaline compositions. The first intention is to provide an account of two areas covering the felsic igneous rocks, which also form a link with the previous section, in that the onset of melting (and the formation of felsic magma) constitute the upper limit of metamorphism in the crust. The beginning of melting in the deep mantle is also touched on when considering the conditions of formation of strongly undersaturated mafic alkaline rocks.

Although there has been much new experimental work on the granite system since Tuttle and Bowen's classic memoir in 1958, there has been no comprehensive review of the field since that time. In the first chapter Luth brings together the chief lines of progress in the experimental study of granites, building his review on the foundation laid by Tuttle and Bowen, namely the haplogranite system ($NaAlSi_3O_8$-$KAlSi_3O_8$-SiO_2-H_2O). Firstly he outlines the changes in the water-saturated system as the range of physical conditions has been extended, and he then draws attention to the most relevant new development, the effects of water undersaturation on crystallisation and melting. He then explores the consequences of adding new components to the haplogranite system, leading to some modern experimental studies on natural granitic compositions. A salutary discussion is provided of the difficulties and dangers of interpreting and applying the experimental results in complex systems to the even more complex natural condition. The chapter concludes with some fascinating glimpses of new studies in non-equilibrium crystallisation.

The second chapter reviews the applications of modern experimental results to alkaline rocks, including oversaturated and undersaturated felsic varieties and the strongly undersaturated mafic rocks such as the nephelinites and melilitites. Again, the themes are developed from the earlier studies in which Bowen figured so prominently. The experimental systems most relevant to alkaline magmas have recently been reviewed by Edgar (1974), whose prime aim was to present the more important phase diagrams with a short commentary on each. The present review, however, follows the lines set by Bowen in that it tries to pinpoint the main problems in each group of rocks and looks to see how experimental petrology has helped to elucidate them. A separate discussion of the phase relations of the alkali feldspars is provided, in view of their essential status in the felsic rocks. This topic, and that of peralkaline oversaturated rocks, link the alkaline rocks and the granite system of the previous chapter. The strongly undersaturated mafic alkaline rocks extend the enquiry directly into mantle conditions, compositions, and melt extracts, and in this way the two chapters together deal with igneous conditions relating directly to the melting boundaries in crust and mantle.

One general theme running through the alkaline chapter is the balance between alumina and total alkalis in magmas. This is the topic of the concluding section, which examines the development of peralkalinity. Alkaline igneous activity in many regions has this distinctive character, standing in marked contrast to the aluminous composition of the continental crust, and experimental petrology offers perhaps the best line of attack on this problem.

Consideration of alkaline rocks of all systems leads inevitably to the question of alkali transfer. The experimental evidence of alkali mobility (backed by the alkali metasomatism in and around alkaline intrusions) leads to the difficult subject of open-system magmatism. Designing experiments relevant to open systems, and communicating the results, is one of the exciting new challenges of experimental petrology.

1. Granitic Rocks

W. C. Luth

1 Introduction 335
2 Equilibrium Studies 336
(a) The haplogranite system: $NaAlSi_3O_8$-$KAlSi_3O_8$-SiO_2-H_2O 338
(b) The haplogranite system in the presence of additional components 368
(c) Natural systems 394
3 Non-Equilibrium Studies 404
(a) Crystal growth 405
4 Conclusions 412

1. INTRODUCTION

The classic work of Tuttle and Bowen (1958) has served as the base for many additional experimental studies bearing on the origin and development of the granitic rocks. Of perhaps greater importance than the number and quality of these contributions has been the widespread use of experimentally obtained data by the petrologist in defining and resolving petrologic problems. Ready acceptance of the experimental approach has too often led to the uncritical use of experimental results by geologists. In general, the granitic rock as seen in outcrop, hand specimen, or thin section, is the result of all processes which have acted upon it subsequent to its original formation. The working hypothesis employed in this review is that granitic rocks are products of multi-stage processes and that attempts to treat such rocks (other than in exceptional cases) as products of single-stage magmatic or metamorphic processes, are subject to severe limitations. It is the author's impression that attempts to resolve questions bearing on the magnitude, and variability, of intensive thermodynamic parameters such as pressure, temperature, fugacity or partial pressure of various species, which ignore the multi-stage nature of the overall process are not likely to be successful.

Questions of the type: "At what P, T, f_{O_2}, f_{H_2O}, etc. did this rock form?" are quite analagous to questions of the type: "What is the age of this rock?" In the latter case it is well-established that different techniques provide different "age" values, each of which may be meaningful to an event in the over-all history of the development of the particular rock. In the case of geochronology the "misfit" of the analytic data is frequently used to develop meaningful information. However, in the case of application of experimental data to petrologic problems involving the granitic rocks we still see many efforts to force the chemical and mineralogic data to "fit" an experimentally based model which is essentially single-stage in style. A meaningful approach involving the application of experimental data to granitic rocks should include an attempt to resolve discrepancies, or "misfits" between different methods of estimating intensive parameters. This is particularly necessary in view of the potential information content of the discrepancies relative to the temporal development of the granitic rock body. Because a review of literature on a subject such as this is inherently biased in terms of the reviewer's prejudices, it seems desirable to provide the reader with a statement of my key prejudices at the outset. Forewarned is forearmed!

1. Most granitic rocks have an ultimate magmatic origin and early history.
2. Most granitic rocks exhibit textural and mineralogic features which are less related to the ultimate magmatic origin and early history than to subsequent sub-solidus recrystallisation.
3. Most granitic rocks display an array of non-equilibrium textural and mineralogic features which must be evaluated relative to a simplistic equilibrium model derived from experimental studies.
4. Vastly more information on the kinetics of nucleation, crystallisation, and recrystallisation is required before a high level of predictability can be attained in the interpretation of natural occurrences on the basis of experimental studies.

This review is not intended to be an exhaustive annotated bibliography of the literature related to experimental studies bearing on the origin and development of the granitic rocks. Rather, the intention is to survey some of what the reviewer considers to be the major developments since the pioneering work of Tuttle and Bowen (1958). Careful examination of this contribution suggests that most of these more recent studies represent extensions of, and elaborations upon, the pioneering study.

2. EQUILIBRIUM STUDIES

Fundamentally different purposes and goals are involved in experimental studies which employ synthetic, as contrasted with natural, starting materials. Each approach has advantages, depending on the objectives of the experimentalist. Synthetic starting materials are generally required when the investigation is directed toward systematic studies in systems which involve limited chemical variability. Typically these studies are concerned with phase relations appropriate to a wide variety of compositions over a relatively large pressure–temperature regime. In general, such studies are based on an analogue (or model) approach where the applicability to a specific granitic rock is directly related to the quality of the model. Experimental studies which employ natural rock starting materials are generally concerned with the behaviour of a specific silicate composition, or suite of related compositions, as a response to changing pressure–temperature conditions. Typically these studies involve a restricted range of compositions, with a much higher degree of chemical variability. When such starting materials are employed the analogy with the natural occurrence is obviously much closer. However, in attempting to correlate with the specific natural occurrence, regarding its magmatic history, there is an implicit assumption that the rock composition chosen as a starting material has undergone no sub-solidus bulk composition changes. That is, the composition (or compositions) are the same as at the completion, and in some cases at the onset, of magmatic crystallisation. Under the best of circumstances it seems that experimental studies using synthetic starting materials, for systematic studies, are most suitable for evaluation and analysis of processes, while those which employ natural starting materials are of greatest importance when considering details related to specific cases.

A large number of problems exists regardless of whether synthetic or natural starting materials are used in experimental studies on quartz–feldspathic compositions. The discussions of Fyfe (1960), Piwinskii (1967) and Piwinskii and Martin (1970) are particularly useful in this context. Although by far the majority of the experimental studies discussed in this review are referred to as "equilibrium" or "phase equilibrium" studies it is quite rare to have an unequivocal demonstration of the attainment of equilibrium. In the case of experimental studies using synthetic starting materials, as concerned with systematic investigations, it is possible to evaluate the results in terms of internal consistency. Of course this factor is related to the relatively low level of chemical variability of the systems being studied. In analysing experimental studies on natural starting materials the application of these familiar "phase rule" related techniques (e.g. Ricci, 1951; Schreinemakers, 1912–1925) is of less value in view of the uncertainty associated with evaluating the number of independent chemical variables necessary to completely represent the compositional variations of all the phases encountered.

In the analyses of experimental data with reference to natural or synthetic starting materials and attainment of equilibrium by means of "reversal" techniques, it is necessary to distinguish clearly between the state of the starting materials when the experiment is being prepared, and the state of the starting materials which are present at the beginning of the experiments. Most often, in the literature, the term "starting materials" refers to the former. Unfortunately, we do not have experimental apparatus which permits an instantaneous change in pressure-temperature conditions from ambient to those of the experiment. In general, a finite time is required, and in the case of gels, glasses, and finely ground ($<5\ \mu m$) crystalline material of granitic composition, the starting material at the initial attainment of the experimental pressure and temperature may be quite different from what was initially prepared. This phenomenon is to be expected in the case of such "high-energy" materials in view of the well-known maxima in nucleation density and rate as functions of temperature.

(a) THE HAPLOGRANITE SYSTEM: $NaAlSi_3O_8$-$KAlSi_3O_8$-SiO_2-H_2O

The systematic studies by Tuttle and Bowen (1958) on phase relations in this system provide the basis for a variety of more recent studies. Given the inherent complexity of liquidus relationships in quaternary systems, and considering that their experimental study (conducted between 1949 and 1954) was the first such investigation at high pressure, the scope and magnitude of this study becomes even more significant. The major portion of their experimental study dealt with determination of phase relations in this quaternary system, and the ternary subsystems $NaAlSi_3O_8$-SiO_2-H_2O; $KAlSi_3O_8$-SiO_2-H_2O; and $NaAlSi_3O_8$-$KAlSi_3O_8$-H_2O at pressures at and below 3000 kg/cm^2 (1000 kg/cm^2 = 14 223 lb/in^2 = 980.7 bars), although some data at 4000 kg/cm^2 were presented. The principal aspect of phase equilibria in this study was determination of liquidus relations involving a liquid saturated with (coexisting with) an aqueous vapour* phase in both ternary and quaternary systems. Tuttle and Bowen were clearly aware of the importance of other aspects of the phase relations in these systems, particularly those where the silicate liquids are not saturated with respect to an aqueous vapour phase (*op cit*. 47–48, 50–53). Much of the experimental work on this quaternary system following that of Tuttle and Bowen has been concerned with expanding the pressure–temperature–composition (P–T–X) regime studied by these authors. However, a firm understanding of the contribution of Tuttle and Bowen remains the starting point for all workers seriously concerned with the application of experimental studies to the origin and development of the granitic and related rocks. Indeed, in the following treatment it is assumed that such is the case and this review will not include a detailed treatment of the original work but will concentrate on developments since the "Granite Memoir".

The subsequent studies in this system have been devoted primarily to extending and elaborating on the Tuttle–Bowen treatment in two related areas: higher pressure and more detailed treatment of the subsystem relative to the solidus reactions and phase equilibria involving liquids not saturated with respect to an aqueous vapour phase.

Solidus Reactions: P–T Coordinates

Of particular importance is the set of univariant reactions which defines the solidus† temperature as a function of pressure in the quaternary system and its

* In order to avoid a lengthy discussion, I will employ the term "vapour phase" to mean the relatively low density, low viscosity, aqueous phase which may coexist with a silicate liquid. "Fluid phase" or "gas phase" are equally viable terms.

† Solidus temperature in the present context is used to refer to the temperature at a given pressure, below which a silicate liquid is not stable although an aqueous vapour phase is capable of stable existence.

subsystems. The experimental results for the solidus reactions are summarised. in Table 1, Fig. 1, 3 and 4 for the systems $NaAlSi_3O_8$-$KAlSi_3O_8$-SiO_2-H_2O (Fig. 1); $NaAlSi_3O_8$-$KAlSi_3O_8$-H_2O (Fig. 1); $NaAlSi_3O_8$-SiO_2-H_2O (Fig. 3); $KAlSi_3O_8$-SiO_2-H_2O (Fig. 3); $NaAlSi_3O_8$-H_2O (Fig. 4); $KAlSi_3O_8$-H_2O (Fig.

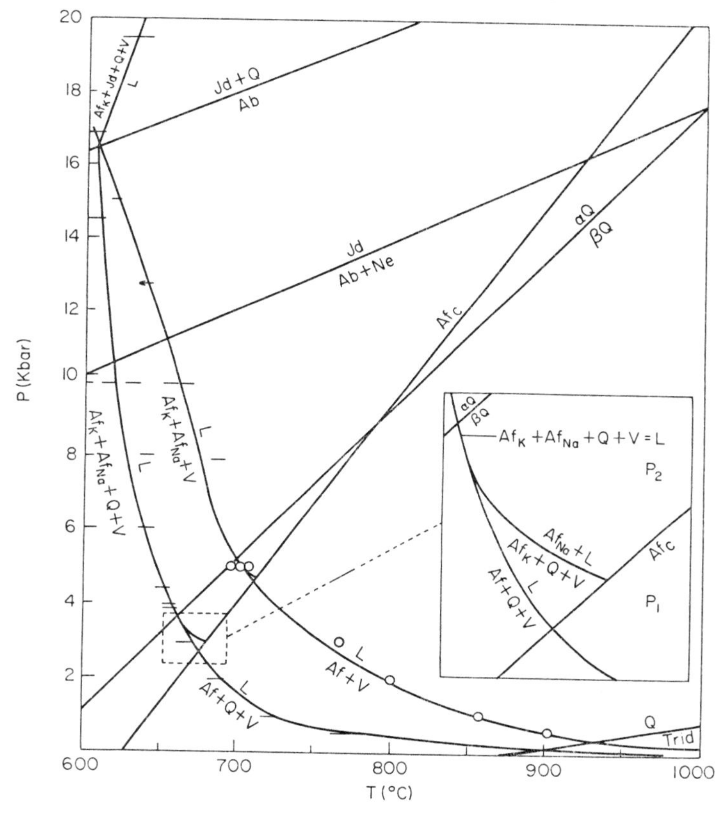

Fig. 1. Pressure–Temperature Projection of Invariant and Univariant Equilibria: $NaAlSi_3O_8$-$KAlSi_3O_8$-SiO_2-H_2O and $NaAlSi_3O_8$-$KAlSi_3O_8$-H_2O.
Coordinates of solidus reactions from Table 1, data indicated by horizontal bars and open circles. Reactions involving jadeite from Boettcher and Wyllie (1968), other sources discussed in the text. Inset refers to invariant and singular quaternary equilibria, in schematic fashion, related to the transition from the low pressure minimum-type solidus reaction to the high pressure eutectic-type solidus reaction (see text). Designation of the phases:

Af	alkali feldspar crystalline solution				
Af_K	K-rich alkali feldspar crystalline solution				
Af_{Na}	Na-rich alkali feldspar crystalline solution				
Af_C	Alkali feldspar critical temperature and composition (see Fig. 9)				
Ab	albite, $NaAlSi_3O_8$	Jd	jadeite, $NaAlSi_2O_6$	Ne	nepheline, $NaAlSiO_4$
Q	quartz	αQ	low-quartz	βQ	high-quartz
L	silicate liquid	V	aqueous vapour phase	Trid	tridymite

4); and SiO_2-H_2O (Fig. 4). In each case the smooth curves were constructed by the reviewer and represent his (current!) interpretation of the available data.

Luth *et al.* (1964) extended the study of Tuttle and Bowen (1958) to 10 Kbar presenting both solidus data and vapour-saturated liquidus data on the quaternary system and some information on the bounding ternary and binary hydrous systems. On the basis of more recent work by Boettcher and Wyllie (1969)

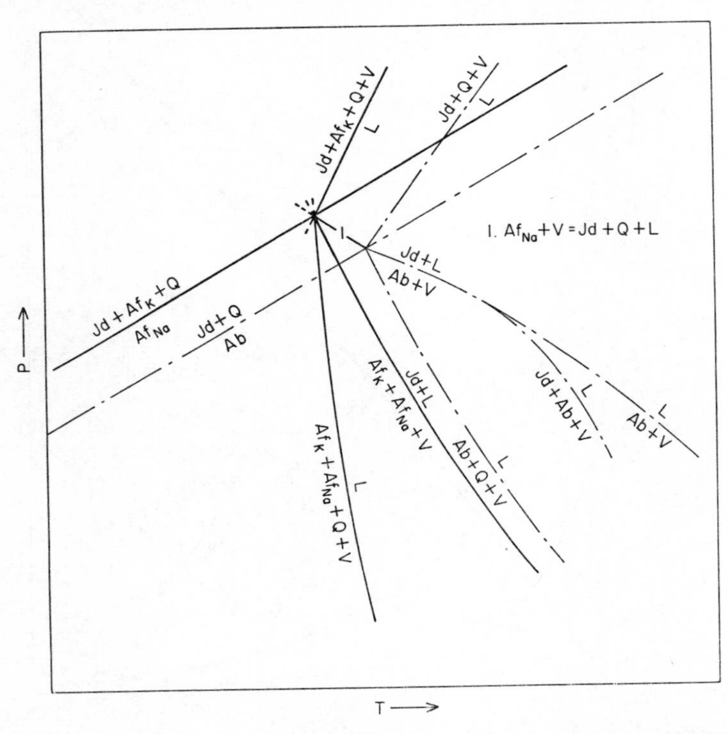

Fig. 2. Pressure–Temperature Projection of Invariant and Univariant Equilibria: Schematic Treatment of Equilibria Involving Jadeite in the Quaternary System. Quaternary reactions indicated by solid lines. Ternary reactions in the $NaAlSiO_4$-SiO_2-H_2O system shown as dash–dot broken lines.

on the system $NaAlSiO_4$-SiO_2-H_2O; Lambert *et al.* (1969) on the join $95KAlSi_3O_8 \cdot 5NaAlSi_3O_8$-$SiO_2$-$H_2O$; and Merrill *et al.* (1970) on the quaternary system, it appears that temperatures given by Luth *et al.* (1964) may be 10°C (Merrill *et al.*, 1970), 20°C (Lambert *et al.*, 1969), or even 30°C (Boettcher and Wyllie, 1969) high at 10 Kbar. Although not directly pertinent to this review, other related studies support the correction of the Luth *et al.* (1964) 8–10 Kbar values by 10–15°C. Peters *et al.* (1966) and Scarfe *et al.* (1966) used the same apparatus (Luth and Tuttle, 1963) and experimental method as used by Luth *et al.* (1964) in their studies of the systems $NaAlSiO_4$-SiO_2-H_2O and $KAlSiO_4$-SiO_2-H_2O at pressures to 10 Kbar. Kim and Burley (1971) also reported on the system $NaAlSiO_4$-SiO_2-H_2O and found the solidus reaction (analcite + albite + vapour = liquid) in the silica-undersaturated portion of the system 625 ± 6°C as contrasted to the Peters *et al.* (1966) value of 640 ± 5°C at 10 Kbar. Kim and Burley (1971) located the invariant point involving the

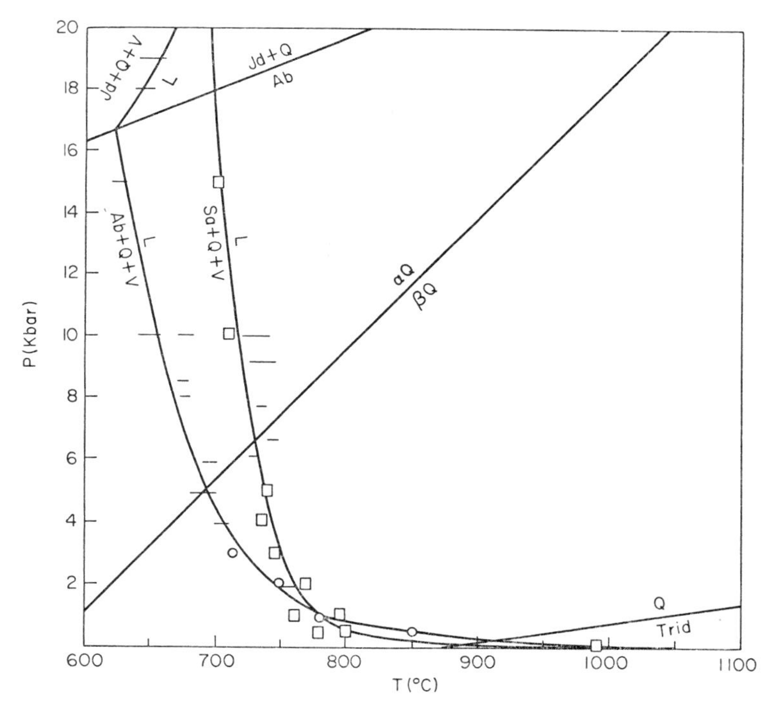

Fig. 3. Pressure–Temperature Projection of Invariant and Univariant Equilibria: Solidus reactions in the $NaAlSi_3O_8$-SiO_2-H_2O and $KAlSi_3O_8$-SiO_2-H_2O Systems. Data, indicated by horizontal bars, open circles and squares, from Table 1. Phase designations as in Fig. 1, and Sa = sanidine, $KAlSi_3O_8$.

coexistence of analcite + albite + nepheline + liquid + vapour at 657°C, 5.15 Kbar as contrasted to the Peters *et al.* (1966) value of 665°C, 4.75 Kbar. The Kim and Burley (1971) results do not support the approach taken by Boettcher and Wyllie (1969) of subtracting 30°C from the value of the invariant point and other temperatures obtained by Luth *et al.* (1964). Boettcher and Wyllie (1969) obtained a temperature of 600°C at 10 Kbar for the solidus reaction also studied by Kim and Burley (1971) and Peters *et al.* (1966). In a similar sense the study of Scarfe *et al.* (1966) on the system $KAlSiO_4$-SiO_2-H_2O on the univariant reaction leucite = kalsilite + sanidine, is supported by Lindsley's (1967) study

TABLE 1

Pressure–temperature coordinates of solidus reactions in the haplogranite system.

Pressure (bars)	Temperature (°C)	Reference see below	Pressure (bars)	Temperature (°C)	Reference see below	Pressure (bars)	Temperature (°C)	Reference see below	Pressure (bars)	Temperature (°C)	Reference see below
$NaAlSi_3O_8$–$KAlSi_3O_8$–SiO_2–H_2O			$NaAlSi_3O_8$–SiO_2–H_2O			$NaAlSi_3O_8$–H_2O			$KAlSi_3O_8$–H_2O (*contd.*)		
1	950–970	1	1	1062	8	1	1 118	8	5 000	855	16
490	760–780	1	490	850	1	490	950–960	1	6 500	845	16
981	715–725	1	981	780	1	510	956–960	14	10 000	825	13
1 961	680–690	1	1 961	750	1	887	905– ?	14	15 000	795	13
2 942	660–670	1	2 942	712	1	981	900–920	1	20 000	775	13
3 923	650–660	1	3 923	700–710	1	1 450	? –910	14			
4 000	650–660	2	4 900	680–700	3	1 567	849– ?	14	SiO_2–H_2O		
4 400	645–655	3	5 850	690–700	3	1 700	847–860	14	1	1 723	5, 6
6 000	635–645	3	8 000	?–680	3	1 961	840–850	1	500	1 400–1 420	17
8 000	635–645	3	8 500	670– ?	3	2 414	? –830	14	800	1 290–1 307	17
10 000	625–635	3	10 000	670–680	3	2 942	790–800	1	1 100	? –1 210	17
10 000	610–620	4	10 000	640–657	9	3 923	745–755	1	1 500	1 155–1 165	17
14 500	600–610	4	15 000	620–630	9	5 000	745	6	1 750	1 105–1 140	17
19 500	620–640	4				5 000	758	7	2 000	1 120–1 130	17
			$KAlSi_3O_8$–SiO_2–H_2O			4 910	? –760	1	2 300	1 104–1 130	17
$NaAlSi_3O_8$–$KAlSi_3O_8$–H_2O			1	990	10	5 219	727– ?	1	2 600	1 100–1 115	17
1	1 063	5	490	800	1	6 100	? –723	3	3 000	1 092–1 101	17
490	900	1	500	825	11	6 700	? –740	3	3 100	1 100–1 110	17
981	? –860	1	981	760	1	8 050	722–735	3	3 500	1 090–1 100	17
1 961	800	1	1 000	795	11	10 000	680–690	9	4 000	1 090–1 099	17
2 942	765	1	1 000	760	12	10 000	? –690	15	5 000	1 089–1 095	17

5 000	? –700	3	1 961	750–760	1	10 150	695–720	3	5 000	1 055–1 070	18
5 000	695	6	2 000	767	11	15 000	650–670	9	6 000	1 077–1 088	17
5 000	703	7	2 942	745	1	17 000	620–625	9	7 000	1 073–1 081	17
7 900	680–690	3	4 000	735	11	$KAlSi_3O_8$-H_2O			8 000	1 069–1 085	17
10 000	660–670	3	5 000	740	13				9 000	1 080–1 088	17
10 000	650–660	4	6 100	723– ?	3	1	1 150	10	9 500	1 079–1 086	17
12 250	610– ?	4	6 600	? –748	3	500	1 035	14	10 000	1 040–1 050	18
12 750	? –640	4	7 700	? –732	3	1 000	? –1 000	14			
13 250	? –630	4	8 100	722– ?	3	1 550	965– ?	14			
15 000	613–620	4	9 150	724–745	3	2 725	950–960	14			
16 750	600–610	4	10 000	720–740	3	3 582	? –940	14			
19 500	610– ?	4	10 000	710	13	3 636	934– ?	14			
			15 000	700	13	4 000	880	16			
			20 000	697	13	5 000	875	13			
						5 000	876	6			

References Cited

1. Tuttle and Bowen (1958)
2. Steiner *et al.* (1974)
3. Luth *et al.* (1964)
4. Merrill *et al.* (1970)
5. Schairer and Bowen (1950)
6. Yoder *et al.* (1956)
7. Morse (1970)
8. Schairer and Bowen (1956)
9. Boettcher and Wyllie (1969)
10. Schairer and Bowen (1955)
11. Shaw (1963)
12. Carmichael and MacKenzie (1963)
13. Lambert *et al.* (1969)
14. Goranson (1938)
15. Burnham and Jahns (1962)
16. Spengler (1965)
17. Kennedy *et al.* (1962)
18. Stewart (1967)
19. Schairer and Bowen (1935)
20. Fenn (1973)

of the same reaction at higher pressures. These arguments are not definitive but taken in conjunction with other data on solidus reactions generally support a reduction of about 10–15°C in the temperatures obtained by Luth *et al.* (1964) at pressures of 8 to 10 Kbar.

$NaAlSi_3O_8$-$KAlSi_3O_8$-SiO_2-H_2O. The quaternary solidus curve extends from the temperature of the ternary minimum in the anhydrous system at 1 bar

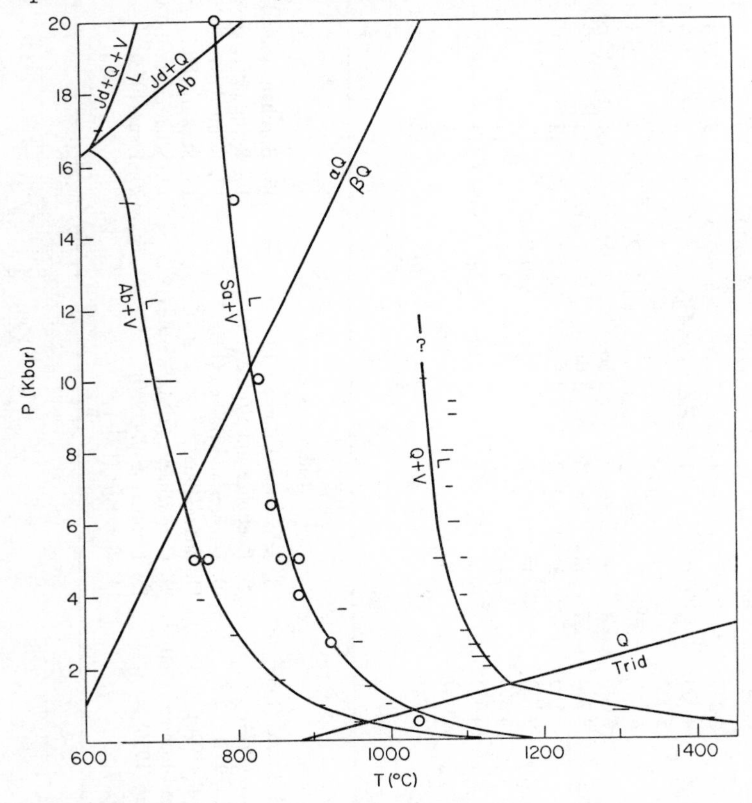

Fig. 4. Pressure–Temperature Projections of Invariant and Univariant Equilibria: Solidus Reactions in the $NaAlSi_3O_8$-H_2O, $KAlSi_3O_8$-H_2O, and SiO_2-H_2O Systems. Data, indicated by horizontal bars and open circles, from Table 1. Phase designations as given previously.

(960°, Tuttle and Bowen, 1958) through invariant points related to the tridymite-βquartz and α quartz-β quartz transformations and a singular point involving the change from a minimum type to eutectic type reaction (estimated at 3500 ± 200 bars, 660 ± 10°C) to the invariant point (Merrill, *et al.*, 1970) involving jadeite and the instability of albite. To facilitate discussion of these equilibrium relations the curves for the α–β quartz transformation from Yoder (1950), the alkali feldspar critical curve from Luth *et al.* (1974) and the corrected (Johannes *et al.*, 1971) curve for the albite = jadeite + quartz reaction are also shown in Fig. 1. The effects of the α–β quartz transformation and the tridymite-β quartz reaction are familiar (Shaw, 1963, Tuttle and Bowen, 1958), and need not be discussed in detail in this review. If both reactions are assumed to be first order, then there is a discontinuity in the slope of the solidus reaction at the point of intersection of the solidus reaction curve with the unary reactions. In each case the solidus reaction curve is steeper on the high pressure side of the tinersection. Luth (1974) has recently discussed the data and problems related to location of the alkali feldspar critical curve mapped out by the maxima on the solvus. In the present context it is sufficient to note that the alkali feldspar solvus critical curve shown in Fig. 1 (subject to an uncertainty of nearly ±25°C) intersects the solidus curve at about 2.75 Kbar, 675°C. However, the diagrams (and data) given by Tuttle and Bowen (1958) for pressures of 3000 kg/cm² (2942 bars) and 4000 kg/cm² (3923 bars), as well as more recent data obtained by Steiner (1970), indicate a change from minimum to eutectic behaviour at some pressure between 3 and 4 Kbar. This apparent inconsistency is related to the fact that the intersection referred to above is an intersection in P–T projection only and is not an invariant point. A schematic (since definitive data are not available) treatment of the phase relations in the vicinity of this intersection point is presented in the inset to Fig. 1. Although the intersection is at 2.75 Kbar a higher pressure, equivalent to P_1 (Fig. 1), must be attained before the critical feldspar (temperature and composition at the crest of the solvus) can coexist with both liquid and vapour. This invariant point is defined by the coexistence of four phases in the quaternary system. True invariance is achieved in view of the n–1 independent variations in state of which a critical phase is capable while remaining a critical phase (Gibbs, 1906, p. 131). Extending from this invariant (degenerate by a critical restriction) point to higher pressure and lower temperature is a univariant reaction curve which involves the coexistence of five phases: sodic alkali feldspar (Af_{Na}), potassic alkali feldspar (Af_K), quartz (Q), liquid (L), and vapour (V). Although it is not certain, in view of the lack of sufficient data, it appears that this reaction is: $Af_{Na} + Q + V = Af_K + L$. This interpretation is based on the composition of the liquid pertinent to the minimum solidus at 3 Kbar (Tuttle and Bowen, 1958), the composition of the eutectic liquid at 4 Kbar (Steiner, 1970), and the estimated composition of the critical phase (Luth *et al.*, 1974). The K/Na ratio of the liquids referred to and the critical phase appear to be sufficiently close so that the alternative reaction: $Af_K + Q + V = Af_{Na} + L$ (shown inset in Fig. 1) is only slightly less likely. As pressure is increased above P_1 (Fig. 1) at some particular pressure (P_2) the plane in the composition tetrahedron defined by three of the univariant phases, Af_{Na}–Q–V (alternatively Af_K–Q–V) becomes coincident with the plane defined by the three coplanar (in the composition tetrahedron) phases which participate in the melting relationship. Then at $P > P_2$ the univariant reaction becomes $Af_K + Af_{Na} + Q + V = L$. P_2–T_2 of Fig. 1 is a singular point where the quaternary univariant reaction changes from $Af_K + Q + V = Af_{Na} + L$ (alternatively $Af_{Na} + Q + V = Af_K + L$) to $Af_K + Af_{Na} + Q + V = L$ and marks the high pressure termination of the minimum-type solidus reaction $Af + Q + V = L$. The quaternary solidus reaction continues to higher pressure and is terminated at an invariant point involving the coexistence of Af_K, Af_{Na}, Jadeite (Jd), Q, L, and V at 16.8 Kbar and 605°C as located by Merrill *et al.* (1970). These authors assumed that the Na-rich alkali feldspar participating in the invariant equilibria was pure albite, thus resulting in a degenerate quaternary invariant point where three (Af_{Na} = albite, Q and jadeite) of the six coexisting phases are colinear in the composition tetrahedron. This assumption of pure albite is not consistent with the data of Goldsmith and Newton (1974) or extrapolations from Luth *et al.* (1974). Both of these studies indicate two feldspars ($ab_{95}or_5$ and about $ab_{20}or_{80}$) coexisting at this temperature and a pressure of 16.3 Kbar. (See Johannes *et al.*, 1971 for the pressure correction of 16.8 to 16.3 Kbar.) If the sodium-rich alkali feldspar participating in the invariant equilibria is not pure albite, then the invariant point is not degenerate and does not lie on the Ab = Jd + Q reaction curve. A schematic representation is given in Fig. 2. Data on the composition of the coexisting phases, or the P–T co-ordinates of the reactions are not sufficient to warrant treating this aspect of the equilibria in any detail. The silica-undersaturated portion of the system $NaAlSiO_4$-$KAlSiO_4$-SiO_2-H_2O although of vital importance in dealing with the nepheline syenites is not of primary concern in this review. It is worthy of note that the reaction curves; $Af_K + Af_{Na} + Q + V = L$ and $Af_K + Af_{Na} + V = L$ do not intersect (as indicated by Merrill *et al.*, 1970) if the Na rich alkali feldspar is not pure albite. The singular equilibria illustrated on the $Jd + Af_K + Af_{Na} + L + V$ curve are particularly schematic in view of the total lack of data on the P–T co-ordinates of the invariant point involving $Jd + Af_K + Af_{Na}$ + Nepheline + L + V and the possible complexities introduced by the existence of analcite as a liquidus phase. The experimental data on the quaternary solidus reactions (Table 1, Fig. 1) seem to be in reasonable agreement if the temperatures obtained by Luth *et al.* (1964) are reduced by about 10–15°C at 8 and 10 Kbar. Although it would be desirable to have more data at $P < 500$ bars, the solidus reaction curves seem reasonably well defined at higher pressures.

$NaAlSi_3O_8$-$KAlSiO_8$-H_2O. The ternary solidus reaction curve in this system extends from the temperature of the minimum in the anhydrous system at 1 bar to a singular equilibria at about 5 Kbar, 700°C as a ternary minimum-type reaction. At pressures above this singular equilibria the solidus reaction is $Af_K + Af_{Na} + V = L$ which is terminated by some high pressure equilibria involving jadeite. The singular equilibria on the solidus reaction curve is analogous in form to that previously discussed in connection with the quaternary solidus reaction. The alkali feldspar critical curve intersects the Af + V = L minimum-type reaction at about 715°C, 4.8 Kbars (Fig. 1). This intersection is not invariant, and the invariant point lies on the alkali feldspar critical curve at some higher pressure. Extending to higher pressure and lower temperature from this invariant (by a critical restriction) point is the four-phase univariant curve: $Af_{Na} + V = Af_K + L$ as discussed by Yoder *et al.* (1956). The singular point on the solidus reaction curve is generated at some higher pressure when the liquid participating in the minimum reaction becomes colinear with the Af_{Na} and V participating in the four phase univariant reaction. The eutectic-type univariant reaction, extending to higher pressure from this singular equilibria, has been briefly discussed in the previous section regarding its high pressure determination (Fig. 1). The P–T co-ordinates of the solidus reaction seem to be in reasonable accord, particularly when the data of Luth *et al.* (1964) at 8 and 10 Kbar are corrected by about 10–15°C. Determinations of a number of workers at 5 Kbar are all in agreement at 700 ± 10°C regarding the solidus temperature. Yoder *et al.* (1956) indicated that the singular point was above 5 Kbar, Morse (1970) indicated that the singular point was below 5 Kbar, and the results of Luth *et al.* (1964) were not definitive regarding the location of the singular equilibria.

$NaAlSi_3O_8$-SiO_2-H_2O. The solidus reaction in this system at low pressure is albite (Ab) + tridymite (Tr) + V = L, and changes to Ab + βQ + V = L on intersection with the βQ–Tr reaction curve at about 230 ± 50 bars, 915 ± 10°C. A second invariant point exists at about 5 Kbar, 705°C where the solidus reaction changes to Ab + αQ + V = L. These relations have been discussed by Tuttle and Bowen (1958) and need not be elaborated on here. Boettcher and Wyllie (1969) found that the solidus reaction Ab + αQ + V = L terminates at an invariant point involving jadeite, albite, alpha quartz, liquid and vapour. The solidus reaction extending to higher pressure from this invariant point is Jd + αQ + V = L and has a positive P–T slope (dP/dT) in contrast to the case at pressures below this invariant point. The data are given in Table 1 and shown in Fig. 3. For reasons outlined in previous pages, the suggestion offered by Boettcher and Wyllie (1969) that the temperatures of Luth *et al.* (1964) and Peters *et al.* (1966) were some 30–35°C too high is rejected in favour of a —10 to —15°C correction at 8 and 10 Kbar of the older data. In view of the differences in experimental apparatus and method the agreement of the data on the solidus reaction seems satisfactory.

$KAlSi_3O_8$-SiO_2-H_2O. The solidus reactions in this system (Table 1, Fig. 3) involve the βQ–Tr and βQ–αQ reactions in a manner similar to that discussed previously for $NaAlSi_3O_8$-SiO_2-H_2O. Shaw (1963) has provided a discussion of the βQ + Tr + sanidine (Sa) + L + V invariant point in connection with his redetermination of the solidus and liquidus relations in this system. The determinations by Shaw (1963) are about 30°C higher than those of Tuttle and Bowen (1958) at 0.5 and 1 Kbar, though the agreement is reasonable from 2 to 4 Kbars. Lambert *et al.* (1970) studied the solidus reaction at higher pressure than Luth *et al.* (1964) and previous workers by extrapolating from results on the $95KAlSi_3O_8 \cdot 5NaAlSi_3O_8$-$SiO_2$-$H_2O$ join. In discussing their results they noted the problem encountered by workers who have attempted to study Na-free systems using the gel method as described by Luth and Ingamells (1963) and Roy (1956). Even using the ammonium stabilised variety of Ludox both Shaw (1963) and Luth *et al.* (1964) had sufficient Na_2O in the starting materials to provide between 1 and 0.5% albite in the potassium end member on the join reported as $KAlSi_3O_8$-SiO_2, as pointed out by Lambert *et al.* (1970). This results in the notable melting interval found by all workers since Tuttle and Bowen (1958), and has been discussed fully by Lambert *et al.* (1970). Luth *et al.* (1964) mentioned the problem and noted that their gels on the "$KAlSi_3O_8$"—SiO_2 join were not on composition but contained excess alkali over the 1:1 molar ratio of Na + K/Al. All other gels in this system used by Luth *et al.* (1964) and prepared by the method of Luth and Ingamells (1963) were prepared for stoichiometry in this ratio by considering the Na_2O content of the Ludox in the $NaAlSi_3O_8$ end member. At present it appears that only the data of Tuttle and Bowen (1958) directly pertain to the $KAlSi_3O_8$-SiO_2-H_2O system, and that all other results depend largely on interpretation. Even with this in mind, the agreement on the location of the P–T co-ordinates of the solidus reaction proposed for this system is not unreasonable, particularly if the temperatures of Luth *et al.* (1964) are reduced by some 10–15°C. At present it seems desirable to favour the results of Tuttle and Bowen (1958) at 0.5 and 1 Kbar over those of Shaw (1963) for reasons related to the melting interval as discussed by Lambert *et al.* (1970).

$KAlSi_3O_8$-H_2O, $NaAlSi_3O_8$-H_2O, *and* SiO_2-H_2O. The data for solidus reactions in these systems are given in Table 1 and illustrated in Fig. 4. The workers referenced in Table 1 have all provided detailed discussions of the phase relations in these systems and no extended treatment is necessary for the purposes of this review. The solidus reactions for $KAlSi_3O_8$-H_2O, including the incongruent melting of leucite, have been thoroughly discussed by Goranson (1938), Scarfe *et al.* (1966), Lambert *et al.* (1970) and others. Lambert *et al.* (1970) found no experimental evidence to indicate the presence of a primary field of muscovite on the $KAlSi_3O_8$-H_2O join at pressures above 10 Kbar as found by Seki and Kennedy (1965). Indeed, Lambert *et al.* (1970) suggest that their "results can be interpreted as if Or–H_2O remained a binary system to 18.5 Kbar" (*op. cit.*, p. 619). The solidus reaction in the system SiO_2-H_2O has been studied by Kennedy *et al.* (1962) and Stewart (1967). Stewart's (1967) results at 5 and 10 Kbar are at considerably lower temperatures than those of Kennedy *et al.* (1962) as indicated in Table 1 and Fig. 4. Stewart (1967) also did not observe the second critical endpoint at 9.5 Kbar and 1080°C reported by Kennedy *et al.* (1962). Kushiro's (1969) results on the system SiO_2-H_2O are not definitive, representing only a minor portion of his study, but he suggests a value of 1020 ± 15°C at 20 Kbar for the solidus. Additional work on this system seems required before a satisfactory conclusion can be drawn.

In summary, it appears to me that the P–T co-ordinates of the solidus reactions in the quaternary haplogranite system and its bounding hydrous binary and ternary systems are reasonably well-known. If the Luth *et al.* (1964) temperature values at 8 and 10 Kbar are adjusted downward by about 10–15°C the agreement becomes excellent. The experimental studies performed by Wyllie and co-workers using piston–cylinder apparatus yield values that are rather consistently lower in temperature by about 20–30°C than those of Luth *et al.* (1964) obtained in gas apparatus. However, the excellent agreement of Peters *et al.* (1966) and Kim and Burley (1971) on the system $NaAlSiO_4$-SiO_2-H_2O and comparison of the results of Scarfe *et al.* (1966) with those of Lindsley (1967) on the leucite = sanidine + kalsilite reaction suggests a correction of −10 to −15°C of the Luth *et al.* (1964) results on solidus reactions in the haplogranite system at 8 and 10 Kbar.

Studies at pressures less than 15 Kbar are most applicable to problems related to the generation of granitic magmas as there is little geophysical or geologic information which suggests the existence of granitic compositions at pressures of 15–30 Kbar. If granitic compositions do not exist in this depth regime, then the melting and crystallisation relations of these granitic compositions (which are very interesting from the phase equilibria point of view) are not directly pertinent to an analysis of the origin and development of the granitic rocks. Nonetheless studies at these pressures are of vital importance in providing information on limiting conditions.

The change to a positive P–T slope (dP/dT) for the solidus reaction curves in systems involving sodic feldspar and quartz at high pressure first demonstrated by Boettcher and Wyllie (1969) is a phenomenon of major significance. Prior to this work, which has been supported by the study of Merrill *et al.* (1970) for the haplogranite system, it was assumed that either the solidus reaction curves continued with negative slope to higher pressure, or changed slope in a continuous fashion through a point where dP/dT becomes infinite. These high pressure studies by Wyllie and co-workers referred to previously have shown that for quartz-alkali feldspar systems involving an albitic phase and H_2O, there is a minimum temperature value at about 16–17 Kbar for the formation of the granitic liquid. At pressures above and below this value, the solidus temperature takes on higher values.

The P–T co-ordinates of the solidus reactions are of major importance in considering melting and crystallisation relations of the granitic compositions. Equally important is knowledge of the variation in composition of the solution phases (crystalline, liquid, or vapour) which participate in the solidus reactions.

Solidus Reactions: Compositional Relations

Liquid Phase Data on the compositions of the phases which coexist at the solidus temperature are subject both to greater experimental uncertainties and are less readily available. The composition of the liquid participating in the quaternary solidus reaction (Table 2, Fig. 5) is based on data obtained by Tuttle and Bowen (1958), Steiner (1970), and Luth *et al.* (1964). The experimental study

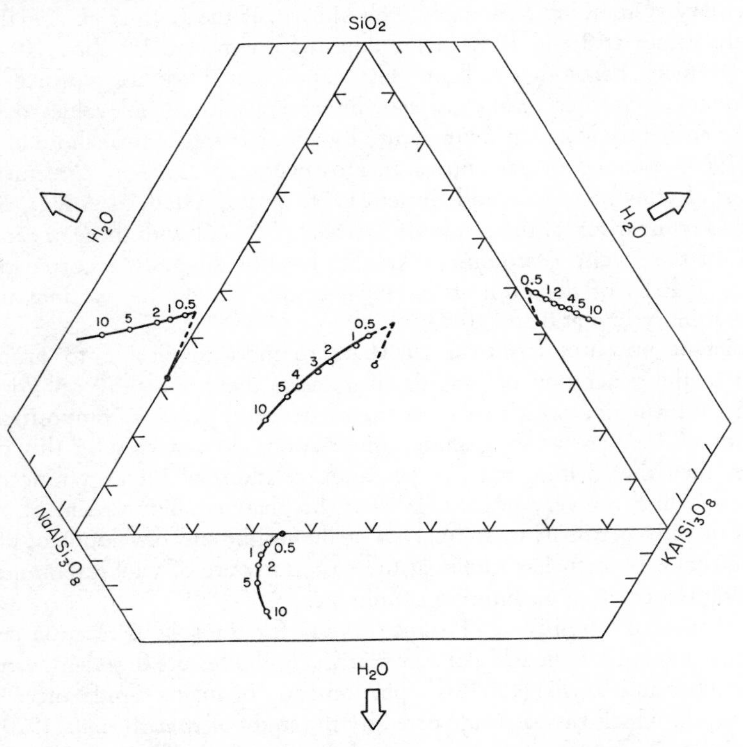

Fig. 5. Compositions of Liquids Participating in the Solidus Reactions as Functions of Pressure. Based on data given in Table 2. Schematic in part, specifically at pressures less than 500 bars. Numerical values refer to pressure in Kbars. Solid circles refer to 1 bar data in the anhydrous systems. H_2O content of quaternary liquid given in Fig. 6. All values in weight fraction, or percent.

of Merrill *et al.* (1970) does not provide data on the composition of the liquid phase. The data on the ternary compositional relations are sparse at pressures greater than 4 Kbar, and non-existent below 500 bars. The solubility of H_2O in silicate liquids pertaining to this quaternary system is one key aspect of the compositional relations. Burnham and Jahns (1962) and Burnham (1967) have

TABLE 2

Composition of the liquid phase participating in the solidus reactions

Pressure (bars)	Temperature (°C)	Composition weight % ab	or	q	H_2O wt %	Reference See Table 1
1	990	33	33	34	0	19
490	770	30	30	40	3.0	1
981	720	33	29	38	4.4	1
1 961	685	39	26	35	6.5	1
2 942	665	42	25	33	8.3	1
3 923	655	—	—	—	9.7	1
4 000	655	47	23	30	9.9	2
5 000	640	50	22	28	11	3
10 000	620	56	21	23	17	3
1	1 063	65	35		0	5
490	925	68	32		2.4	1
981	865	70	30		4.4	1
1 961	800	70	30		6.3	1
2 500	760	70	30		7.7	20
2 942	765	70	30		—	1
5 000	703	72	28		10	7
5 000	700	74	26		—	3
5 000	695	72	28		11	6
10 000	660	72	28		—	3

Pressure (bars)	Temperature (°C)	Composition weight % ab	or	q	H_2O wt %	Reference See Table 1
1	1062	68		32	0	8
490	925	57		43	>2.3	1
981	780	59		41	>4.8	1
1 961	750	61		39	>7	1
2 942	712	65		35	—	1
5 000	680	69		31	—	3
10 000	665	74		26	—	3
1	990		58	42	0	10
490	800		53	47	>2.5	1
500	825		54	46	2.8	11
981	760		55	45	3.8	1
1 000	795		57	43	4.3	11
1 961	755		57	43	5.6	1
2 000	767		58	42	5.1	11
2 942	745		59	41	—	1
4 000	735		59	41	6.3	11
5 000	735		63	37	—	3
10 000	730		63	37	—	3

provided excellent discussions of the problems related to such determinations and have thoroughly reviewed the earlier contributions of Goranson (1931, 1936, 1938) and Tuttle and Bowen (1958).

The curves shown in Figs. 5 and 6, based on the data given in Table 2, represent an attempt to illustrate the variation in composition of the liquid phase participating in the solidus reactions of the quaternary system as a function of pressure. These curves should be treated as approximations in an attempt to

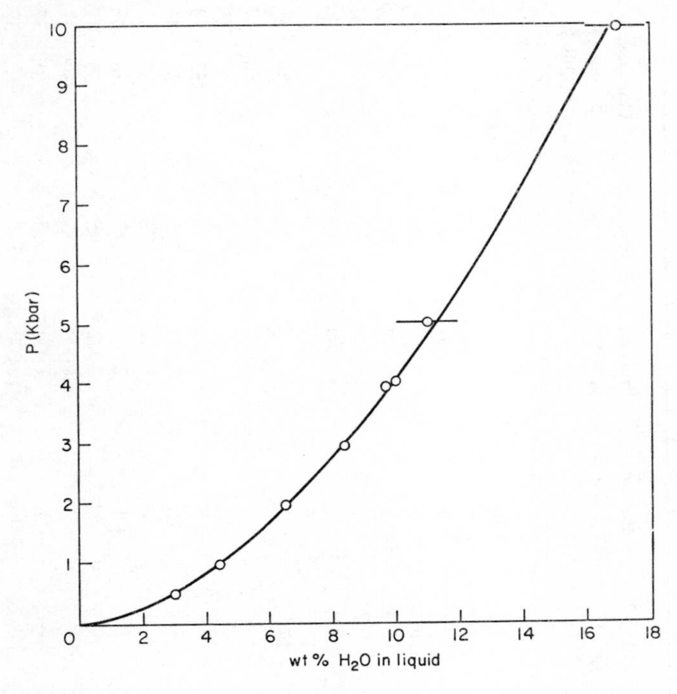

Fig. 6. Solubility of H_2O in the Quaternary Liquid Participating in the Solidus Reaction. Data from Table 2.

obtain an internally consistent presentation. The lack of experimental data at pressures less than 500 bars is particularly troublesome in attempting to relate studies at 1 bar (Schairer and Bowen, 1935, 1955, 1956) to the solidus reactions in the quaternary system. In addition, this pressure range, corresponding to depths of less than 2 km, is of major importance in understanding crystallisation relations of granitic compositions in this near surface regime. Tuttle and Bowen (1958) discussed the vapour-saturated liquidus relations as they related to the polymorphs of silica and the alkali feldspar solvus. In order to demonstrate the vital importance of this P–T regime, diagrams are presented in Fig. 7 which illustrate, in schematic form, the pressure–temperature composition relations for pressures less than 1 Kbar regarding the solidus reactions. It is clear from the data of Schairer and Bowen (1935, 1955, 1956) and Tuttle and Bowen (1958) that when the silica phase participating in the solidus reaction is tridymite the tendency for the composition of the liquids participating in the solidus reactions is toward enrichment in silica with increasing pressure. In contrast when the silica phase is βquartz the tendency is for decreasing amounts of silica with increasing pressure. The set of relations (Fig. 7) is necessary in the analysis of vapour-saturated liquidus relations in this low pressure environment. It

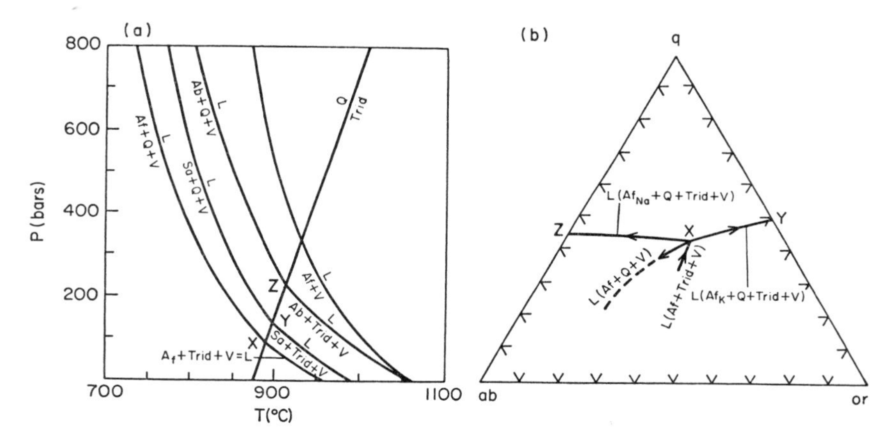

Fig. 7. Low Pressure Equilibria Involving the Tridymite-βQuartz Transformation.
(a) Pressure–Temperature Projection of Solidus Reactions. Symbols as in previous figures. Schematic in the region below 500 bars.
(b) Pressure–Composition Variation of Liquids Participating in the Quaternary Reactions. Temperature, at a given pressure, corresponding to the L(Af_{Na}, Q, Trid, V) and L(Af_K, Q, Trid, V) curves obtained directly from the quartz–tridymite transformation. Arrows indicate increasing pressure. Phase designations given previously. The components ab, or, q, refer to $NaAlSi_3O_8$, $KAlSi_3O_8$, and SiO_2 respectively.

seems very clear to me that one of the greatest needs regarding experimental data in the haplogranite system is for definitive data at pressures below 500 bars.

Vapour Phase. Published compositional data on the aqueous vapour phase which participates in the ternary and quaternary solidus reactions are so sparse as to be essentially non-existent. Luth and Tuttle (1969) *estimated* the composition of the vapour phase participating in the quaternary solidus reaction on the basis of experimental data obtained at 25°C above and below the solidus temperatures at 2, 5 and 10 Kbar with some data at 4 Kbar. This information is subject to considerable uncertainty in view of the indirect means employed in the determinations, and the (necessarily) linear interpolations between the "data" points obtained at each pressure and temperature. Figure 8 (Fig. 18 of Luth and Tuttle,

1969) illustrates the conclusions reached by these authors on the basis of their interpretation of the experimental data. This set of relations may well serve as a guide in lieu of direct determinations. However additional information is

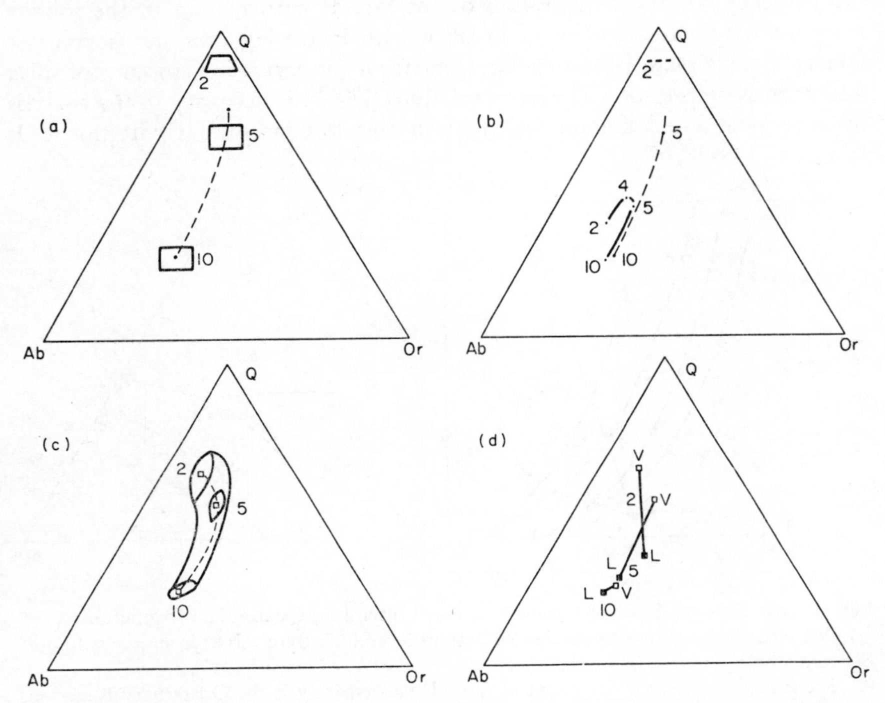

Fig. 8. Vapour Phase Composition Relationships in the Haplogranite System. Figure 18 of Luth and Tuttle (1969). "Projections onto the anhydrous base of the tetrahedron Ab-Or-Q-H_2O illustrating vapour-phase relations. (a) The composition of the vapour phase which coexists with quartz, and albite-rich alkali feldspar, and a potassium-rich feldspar at temperatures approximately 25° below the solidus at 5 and 10 Kbar. No such point exists at 2 Kbar, but the vapour phase which coexists with quartz and an alkali feldspar is quite silica rich as shown. (b) Relationship of subsolidus vapour phase composition (dashed curve), and the vapour phase compositions coexisting with melt (dash–dot curve) at the respective pressures. (c) Approximate mid-points at 2, 5, and 10 Kbar from Fig. 18b, illustrating the inferred composition of the vapour phase which coexists with a liquid at the solidus. The generalised uncertainty in the individual points is shown, as is the uncertainty in the location of the polybaric polythermal curve (dashed). (d) Projection of the L–V tie lines at the solidus temperatures and pressures indicated on the diagram. The points denoted by L are from Tuttle and Bowen (1958) and Luth and others (1964). The points denoted by V are from Fig. 18C."

required. Several tentative conclusions can be drawn on the basis of this study. *First*, the vapour phase coexisting with crystalline alkali feldspar(s) and quartz at pressures less than 5 Kbar and 25°C below the solidus temperature is significantly enriched in normative quartz relative to the composition of the liquid participating in the quaternary solidus reaction. As pressure is increased from 5 to 10 Kbar the composition of this vapour phase approaches that of the liquid participating in the quaternary solidus reaction in terms of normative quartz. *Second*, the vapour phase coexisting with quartz, liquid and potassium feldspar tends to be enriched in both normative quartz and albite relative to the coexisting liquid at pressures less than 5 Kbar and temperatures 25° above the solidus. In a similar fashion the vapour phase which coexists with quartz, liquid and sodium feldspar is enriched in both normative quartz and orthoclase relative to the coexisting liquid. At pressure between 5 and 10 Kbar the normative albite, orthoclase and quartz of the vapour phase coexisting with quartz, sodium (or potassium) feldspar, and liquid approach values similar to those of the liquid. *Third*, at pressures of 5 Kbar and below, the composition of the vapour phase undergoes a pronounced change over the 50°C interval covered by the experiments. The tendency is for enrichment in normative quartz of the vapour phase with decreasing temperature. *Fourth*, inferentially, the composition of the vapour phase participating in the quaternary solidus reaction is enriched in normative quartz relative to the coexisting liquid at pressures of less than 10 Kbar, but is also either enriched in or has essentially the same normative albite at pressures less than 5 Kbar. The Luth and Tuttle (1969) data indicate that the amount of dissolved silicate constituents in the aqueous vapour phase increases from about 5% at 2 Kbar to about 8% at 10 Kbar. Burnham's (unpublished) data on the bounding hydrous binary and ternary systems indicate that these values are somewhat high at low pressure and somewhat low at high pressure. Spengler (1965, and Clark, 1966) has obtained data on the composition of the vapour phase in the system $KAlSiO_4$-SiO_2-H_2O. These data are at temperatures well above the solidus reactions in the system and do not permit ready extrapolation to the composition of the vapour phase participating in the solidus reaction. In addition the compositions used by Spengler (1965) were prepared using the "gel" method and had significant amounts of Na_2O present (see Lambert *et al.* (1970), for a thorough discussion). Luth and Tuttle (unpublished data 1963–1965) studied the composition of the vapour phase participating in the solidus reaction at pressures of 2 to 10 Kbar in the system $NaAlSi_3O_8$-SiO_2-H_2O. The data remain unpublished due to experimental problems in obtaining the equilibrium composition of the vapour phase participating in the solidus reaction. The experimental data obtained in this study indicated that there was an initial high and metastable solubility of silica in the vapour phase participating in the albite + liquid + vapour, liquid + vapour, and quartz + liquid + vapour, "equilibria" at temperatures slightly above the solidus. With increasing duration of the experiments the apparent composition of the vapour phase shifted away from the SiO_2-H_2O system. Experiments of greater than ten days were required to approach the equilibrium vapour phase composition. Although this effect was most pronounced using gel starting materials, it was also found in the case of glass and crystalline starting materials though to a somewhat lesser extent.

Feldspar Phase. The alkali feldspar(s) participating in the quaternary solidus reactions are also solution phases capable of variation in composition along the join $NaAlSi_3O_8$-$KAlSi_3O_8$. When the solidus reaction is of the minimum type the alkali feldspar participating in the solidus reaction has a composition which is coplanar, in the composition tetrahedron, with the coexisting quartz, liquid and vapour. At pressures above the singular point (described previously) where the solidus reaction is $Af_K + Af_{Na} + V = L$ and of the eutectic type, the compositions of the coexisting alkali feldspars are obtained directly from the alkali

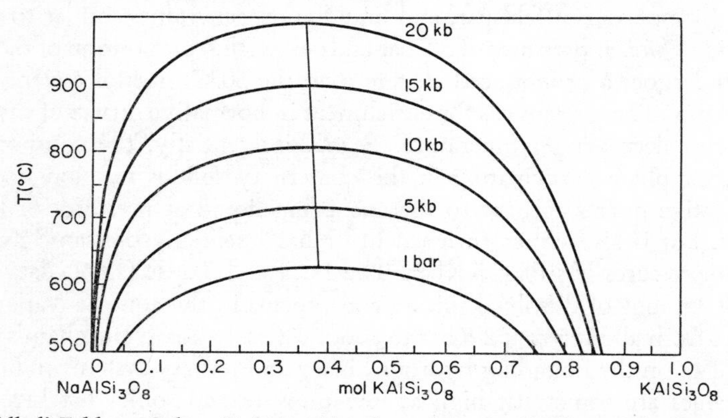

Fig. 9. Alkali Feldspar Solvus Relationships from 1 bar to 20 Kbar. Calculated from the polybaric Margules parameters given by Luth *et al.* (1974, table 3, p. 302). Composition of the critical phase indicated by the nearly vertical curve.

feldspar solvus at the pressure of interest. Neither Luth and Tuttle (1966) nor Morse (1970) found any evidence to indicate that the coexisting alkali feldspars were affected by $SiO_2/(NaAlSiO_4 + KAlSiO_4)$ ratio of the bulk composition. Consequently it is assumed that crystalline solution is restricted to the alkali feldspar join in the system $NaAlSiO_4$-$KAlSiO_4$-SiO_2-H_2O. The alkali feldspar solvus has been studied extensively by a number of workers (see Luth, 1974, for a review). The level of agreement between different workers does not seem to be wholly satisfactory at present, though the form of the solvus and generalised compositions of the coexisting feldspars are reasonably well known. For purposes of this review the solvus relations calculated from the high-temperature data set of Luth *et al.* (1974) will serve as an adequate guide. These solvi are shown in Fig. 9 as calculated for pressures of 1 bar, and at 5 Kbar intervals from 5 Kbar to 20 Kbar. The data of Goldsmith and Newton (1974) suggest that extrapolations of the Luth *et al.* (1974) data to the higher pressure regime are not wholly satisfactory.

In summary, the compositional data for the phases participating in the

TABLE 3

Normative composition of granitic specimens used in experimental studies

	(1)	(2)	(3)	(4)	(5)	(6)	(7)	(8)	(9)	(10)
Ab	36.75	30.31	30.72	36.41	34.13	39.02	39.03	43.11	32.83	28.77
Or	31.79	34.49	24.86	17.06	17.12	12.20	13.05	7.08	20.41	24.03
An	3.30	7.60	15.69	19.91	21.92	29.15	28.72	37.93	16.20	11.17
Q	28.16	27.59	28.73	26.62	26.83	19.63	19.19	11.88	30.56	36.04
DI	89.63	87.86	76.30	70.40	68.40	58.10	54.00	49.10	77.60	84.30

	(11)	(12)	(13)	(14)	(15)	(16)	(17)	(18)	(19)	(20)
Ab	32.64	27.91	33.25	34.28	32.95	49.19	31.00	58.06	42.71	46.16
Or	22.56	29.04	13.48	12.72	20.54	9.76	21.36	34.40	24.64	30.81
An	9.88	5.11	31.99	29.80	10.64	35.79	11.83	3.77	0.00	0.00
Q	34.93	37.93	21.28	23.20	35.87	5.26	35.82	3.77	32.65	23.02
DI	86.70	92.80	54.00	59.60	85.00	51.30	80.50	84.20	97.40	84.70

Numbers in Table refer to samples used in the following references:
(1) Quincy granite; (2) Westerly granite (Tuttle and Bowen, 1958); (3) Granodiorite, 766; (4) Granodiorite, 779; (5) Granodiorite, 768; (6) Tonalite, 520; (7) Tonalite, 1 213; (8) Tonalite, 751 (Piwinskii and Wyllie, 1968); (9) Granodiorite, 678; (10) Quartz monzonite, 685; (11) Quartz monzonite, 774; (12) Granite, 705 (Piwinskii and Wyllie, 1970); (13) Quartz diorite, DR 510; (14) Quartz diorite, DR 126; (15) Quartz monzonite MO 1–B; (16) Quartz diorite, CP 2–1; (17) Granodiorite, JSP 6–2 (Piwinskii, 1973a); (18) Syenite, 26 443; (19) Riebeckite–astrophyllite granite, 26 272; (20) Riebeckite-aegirine microgranite, 58303 (McDowell and Wyllie, 1971).
Sample numbers as used by the individual authors. Normative constituents other than ab, or, an and Q not listed. DI = Differentiation Index = ab + or + q (Thornton and Tuttle, 1960).

solidus reactions of the quaternary system and its bounding hydrous ternary systems are not complete. However, it is possible to construct useful diagrams illustrating the variation in composition of the phases which participate in the solidus reactions as a function of pressure over the range of 0.5 to 10 Kbar. More detailed information, particularly in the low pressure environment and on the composition of the aqueous vapour phase, is necessary before any feeling of complacency is warranted.

Vapour Saturated Liquidus Relations

Tuttle and Bowen (1958), Luth *et al.* (1964), Steiner (1970) and Steiner *et al.* (1974) have presented data and diagrams illustrating the vapour-saturated liquidus relations at pressures between 0.5 and 10 Kbar. Unfortunately no data are available at either higher or lower pressure. Diagrams which illustrate these relations are among the most commonly used by igneous petrologists in attempting to correlate experimental information with field and analytic data on the granitic rocks relative to crystallisation and melting relationships. Consequently, it seems desirable to discuss briefly the level of applicability of such diagrams and provide explicit information on limitations regarding their use.

In terms of direct applicability to specific granitic occurrences the diagrams which illustrate the vapour-saturated liquidus relations are of very restricted utility. This is largely a consequence of two related factors. *First*, changes in liquidus and solidus relations are produced by the presence of other components such as anorthite, excess (relative to the 1:1 Na + K/Al molar ratio) of alumina or alkalis, femic components, and volatile components other than H_2O. *Second*, the relationships indicated on the vapour saturated liquidus diagrams pertain only to liquids which are saturated with respect to an aqueous vapour phase, and are demonstratably (Luth, 1969) inapplicable when considering a liquid which is not saturated with respect to an aqueous vapour phase. The influence of these factors will be explored in subsequent sections of this review and it is sufficient to note here that the modifying effects are not negligible, and the level of predictability to the more complex cases based on these diagrams is rather low.

Vapour-saturated liquidus diagrams in the quaternary system are of greatest utility in discussing generalised processes which are of broad, rather than specific, applicability. Regardless of whether or not equilibrium was actually obtained in the experiments leading to the construction of these diagrams, they are treated as equilibrium diagrams. Thus, even assuming that these diagrams do portray equilibrium relationships it is only possible to use them in comparing different equilibrium states. Information on the process by which a system changes from one equilibrium state to another when intensive parameters are changed is not contained in such diagrams. The familiar treatment of equilibrium and (perfect) fractional crystallisation (Tuttle and Bowen, 1958, p. 63–73) or melting (Presnall, 1969, Presnall and Bateman, 1973) assume that thermodynamic equilibrium is maintained throughout the process, and that a continuum is being considered. Although the idealisation is useful as a conceptual starting point, both the experimental data and natural occurrences suggest that it is an idealisation, not fact, and that kinetic factors related to crystal nucleation, growth, and dissolution cannot be ignored. The reader is referred to the papers of the authors just cited for very useful and thorough treatments of the crystallisation/melting relations in terms of the equilibrium model. Some of the aspects of kinetics bearing on the haplogranite system will be explored in a subsequent section.

Each of the contributions dealing with experimental studies on the vapour saturated liquidus in the haplogranite system relationships has stressed the fact that the liquid compositions were quaternary and that the relationships portrayed in the familiar triangular diagrams with apices at $NaAlSi_3O_8$, $KAlSi_3O_8$, and SiO_2 are projections from the H_2O apex of the composition onto the anhydrous base. The importance of phase relations which involve the composition region below (at lower H_2O content) the vapour–saturated liquidus is sufficient to warrant separate treatment. Indeed it is more convenient to discuss the nature of the vapour-saturated liquidus relations in the context of the overall quaternary system rather than as a specific example of one set of surfaces (Luth *et al.*, 1964) in the system.

Vapour–Undersaturated Liquidus Relations

The principal purpose of this portion of the review is to demonstrate the importance of the pressure–temperature–composition (P–T–X) region in the haplogranite system which involves liquids that are not saturated with respect to an aqueous vapour phase. A secondary purpose of this section is to emphasise the need for direct experimental data on the anhydrous sub-systems of the haplogranite system. This is related to the fact that the P–T–X region involving vapour-undersaturated liquids is the bridge between phase relations in the anhydrous systems and vapour–saturated liquidus relations in their hydrous counterparts. Finally, it seems to the reviewer that a firm understanding of the relationships involving vapour-undersaturated liquids in the haplogranite system is necessary before a meaningful treatment of the more complex relationships in either synthetic systems with more components, or natural granitic rocks, is possible.

Because visualisation of geometric phase relations in quaternary systems is a complex matter much of the following discussion will deal with the ternary systems $NaAlSi_3O_8$-SiO_2-H_2O and $KAlSi_3O_8$-SiO_2-H_2O, where geometric relations are somewhat more conveniently illustrated. The treatment will be largely qualitative in view of the lack of wholly satisfactory experimental data.

As noted previously the solidus reaction for the system $NaAlSi_3O_8$-SiO_2-H_2O

is albite + quartz + vapour = liquid (Ab + Q + V = L) over a large portion of the pressure regime. Considering the liquidus relations in this system at constant pressure this reaction indicates that there are three stable bivariant phase assemblages possible at temperatures immediately above the solidus reaction: Ab + L + V; Q + L + V; and Ab + Q + L. The first two of these bivariant assemblages involve a liquid which coexists with (is saturated with respect to) an aqueous vapour phase and a crystalline phase. The third liquid coexists with, or is saturated with respect to, two crystalline phases, but not with respect to an aqueous vapour phase. The liquids which coexist with albite and quartz, L(Ab, Q), at temperatures above that of the ternary solidus reaction lie on some curve in the composition triangle that has a low temperature termination at the composition and temperature corresponding to the solidus reaction values for the liquid. The high temperature termination of this curve is in the anhydrous system $NaAlSi_3O_8$-SiO_2 and has temperature composition co-ordinates appropriate to the binary solidus reaction, at the same pressure. The bivariant liquids which coexist with quartz and vapour L(Q, V), lie on a polythermal curve extending from the composition of the ternary liquid participating in the ternary solidus reaction to the composition of the binary liquid participating in the solidus reaction in the system SiO_2-H_2O. In a similar sense, the liquid which coexists with albite and vapour L(Ab, V), lies on a curve extending from the composition of the ternary liquid participating on the solidus reaction to the $NaAlSi_3O_8$-H_2O system. These three curves, L(Ab, Q), L(Q, V), and L(Ab, V), are polythermal curves and are conventionally referred to as boundary curves which separate primary phase fields on the liquidus surface. An isotherm, just above the solidus temperature, at this pressure intersects the three curves resulting in points on an isothermal, isobaric ternary diagram.

Still considering the ternary system at constant pressure, there are six possible trivariant phase assemblages stable at temperatures above that of the ternary solidus reaction: L + V; Q + L; Ab + L; Q + V; Ab + V; and Ab + Q. The first three of these two-phase assemblages are of greatest interest in the present context. The liquid of the L + V assemblage, L(V), is saturated with respect to an aqueous vapour phase, but the liquids of the Q + L and Ab + L assemblages, L(Q) and L(Ab) are not saturated with respect to an aqueous vapour phase, rather they are saturated with respect to a crystalline phase. These relations are shown in Fig. 10(a) for a temperature of 720°C and at 5 Kbar. The L(Ab) and L(Q) curves, intersecting at the point L(Ab, Q) of this diagram are the familiar liquidus contour of ternary, isobaric diagrams. Consideration of the phase relations at successively higher temperatures leads to the isobaric, polythermal diagram shown in Fig. 10(b). Of principal concern in this portion of the review is the compositional region which lies below (at lower H_2O content) the curves L(Ab, V) and L(Q, V) of this diagram. Considering the quaternary system for a moment, the curves L(Ab, V), and L(Q, V) represent vapour-saturated liquidus equilibria in one bounding ternary system of the haplogranite system. These curves become surfaces upon addition of the $KAlSi_3O_8$ component in the quaternary system.

Authors concerned with vapour-saturated liquidus relations in systems similar to $NaAlSi_3O_8$-SiO_2-H_2O have developed a convention such that certain aspects

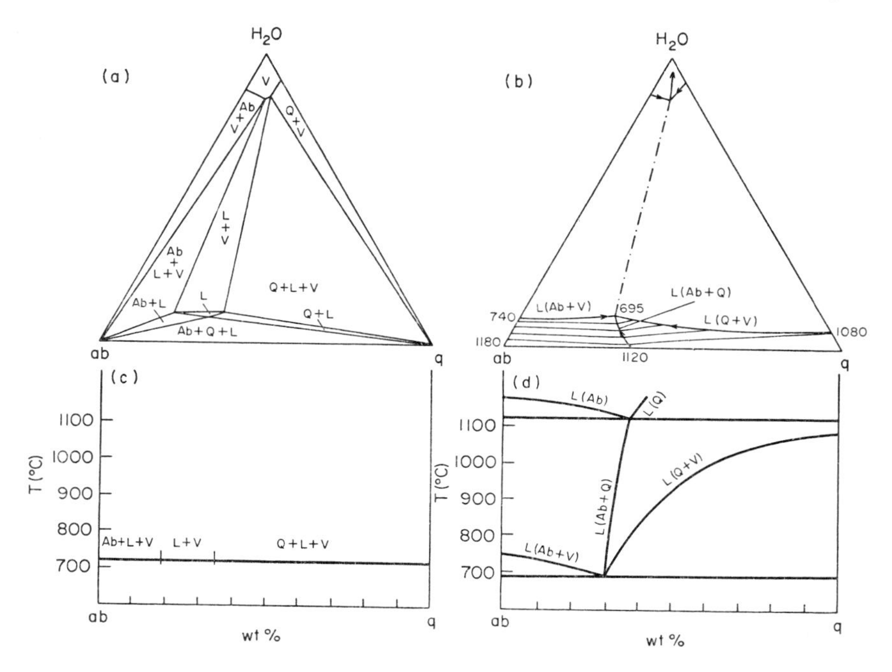

Fig. 10. The $NaAlSi_3O_8$-SiO_2-H_2O System at 5 Kbar.
(a) The 720°C Isotherm. One-, two- and three-phase regions over which the designated phase assemblages are stable. Concentration relations near the H_2O apex are greatly exaggerated to illustrate the difficulty in projecting these relationships. If, and only if, the vapour phase is treated as pure H_2O, then the projected relations shown in (c) are valid and the Ab + V and Q + V regions become coincident with the respective sidelines. Components ab and q refer to $NaAlSi_3O_8$ and SiO_2 respectively. Compositions of vapour saturated liquids which coexist with albite, and quartz, are shown in Fig. 10c, as projected from H_2O.
(b) Polythermal Liquidus Diagram. Temperature values from Luth *et al.* (1964), Stewart (1967), Boyd and England (1963), and Luth (1969). Arrows on the boundary curves indicate the direction of decreasing temperature. Isotherms on the liquidus surface at 100° intervals over the 800–1100°C range are shown schematically. Isotherms on vapour-saturated liquidus, and vapourous surfaces omitted. Solubility of silicates in vapour phase exaggerated. Weight percent frame of reference.
(c) Projection of Ternary 720° Isothermal Equilibria. Only the projected (from H_2O) compositions of vapour-saturated liquids are shown. See discussion under (a) above.
(d) Polythermal Projection of Liquidus Surface. Projection, from the H_2O–T edge, onto the (ab–q)–T face of the (ab–q–H_2O)–T prism, of the liquidus surface. Derived from (b), as described in text.

of these equilibria are shown in a diagram that has the appearance of a binary liquidus diagram at constant pressure. This type of diagram is related in a very straightforward fashion to the ternary diagrams of Figs. 10(a) and 10(b). In Fig. 10(a) construct a line from H_2O through the L(Ab, V) and L(Q, V) points to the $NaAlSi_3O_8$–SiO_2 sideline and locate these points at the appropriate temperature and $NaAlSi_3O_8/(NaAlSi_3O_8 + SiO_2)$ values on a diagram as in Fig. 10(c). If the vapour phase in all cases is assumed to be pure H_2O, then the projection of the L(Ab, V) point on the anhydrous sideline separates a region of Ab + L + V on the left in Fig. 10(c) from a region of L + V on the right. In a similar sense the projection of the L(Q, V) point separates the regions Q + L + V and L + V.

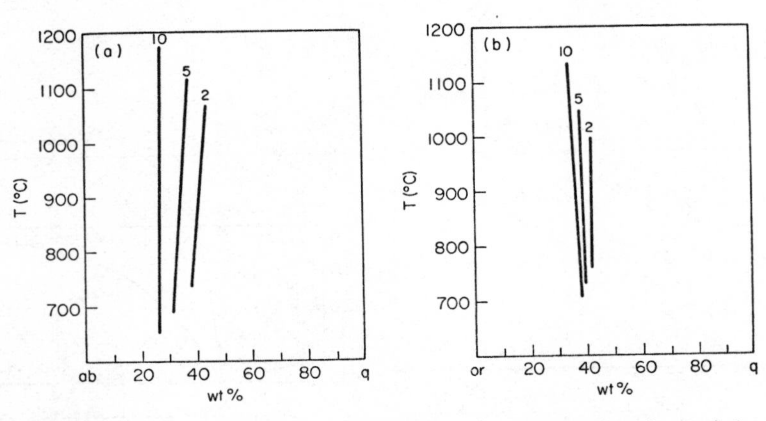

Fig. 11. Vapour Undersaturated, Quartz and Feldspar Saturated, Liquidus Relationships. (a) $NaAlSi_3O_8$(ab)–SiO_2(q)–H_2O System. Schematic illustration of compositional variation of liquids which coexist with albite and quartz at 2, 5 and 10 Kbar. At the lower extremities the liquids (+ab + q) coexist with hydrous vapour.
(b) $KAlSi_3O_8$(or)–SiO_2(q)–H_2O System. As in Fig. 11(a) but for the potassium analogue.

When this procedure is followed at a series of temperature values, at constant pressure, a polythermal diagram of the type shown in Fig. 10(d) is produced. In general, the diagrams similar in form to Fig. 10(d) as presented by Tuttle and Bowen (1958, Figs 17, 20) and Shaw (1963) were not obtained in this manner. Those experiments were performed simply with an excess of H_2O and points on the lines L(Ab, V) and L(Q, V) were obtained by direct determination of the liquidus for albite and quartz, in the presence of excess vapour. Diagrams of the type shown in Fig. 10(d), though very useful in illustrating vapour liquidus relations, do not contain any direct information on the solubility of H_2O in the liquids, or any information on the composition region involving liquids which are undersaturated with respect to an aqueous vapour phase. The nature of the projection technique employed then does not permit illustration of the phase relations in the region: Ab–L(Ab, V)–L(Q, V)–Q–Ab in Fig. 10(a). Yoder *et al.*

(1956) have discussed the analogous set of relations in the system $NaAlSi_3O_8$–$KAlSi_3O_8$–H_2O in considerable detail.

The limitations of the projection technique and the importance of the vapour undersaturated, quartz and feldspar saturated, liquids are further illustrated in Fig. 11. The lines shown connect temperature–composition values in the anhydrous systems to the corresponding vapour saturated liquids in the hydrous systems at pressures of 2, 5 and 10 Kbar, as obtained from values in Table 2, and Luth (1969). It should be clear from Fig. 11(a) that increasing pressure on the system is the dominant factor in shifting the L(Q, Ab) equilibria to more sodic compositions. A similar conclusion holds in the case of the potassic system with regard to the shift of the L(Q, Sa) equilibria to more potassium-rich compositions. However, in the latter case (Fig. 11(b)) the effect of increasing pressure on the anhydrous system is greater than increasing pressure in the hydrous system. The relations shown in Fig. 11 indicate that there is no clear, direct, or simple, method by which one can use vapour-saturated liquidus equilibria as determined at some low pressure, say 2 Kbar, to predict the composition of the vapour-undersaturated liquids at some higher pressure, 5 or 10 Kbar. Indeed it appears that a more satisfactory, in a quantitative sense, approximation is to assume that the L(Ab, Q) or L(Sa, Q) compositions are constant with respect to the $NaAlSi_3O_8/SiO_2$ or $KAlSi_3O_8/SiO_2$ ratios and completely ignore the vapour-saturated liquidus data obtained at lower pressures.

A further illustration of the difficulties involved in extrapolating the vapour-saturated liquidus data at a low pressure to vapour-undersaturated equilibria at higher pressures, can also be developed using Fig. 11. Consider the phase relations along the $(NaAlSi_3O_8)_{70}:(SiO_2)_{30}$–$H_2O$ join at 2 and at 10 Kbar. The vapour-undersaturated liquids on this join at 10 Kbar have the capability of co-existing* with quartz. This is equivalent to saying that along this join quartz is the liquidus phase. At 2 Kbar albite is the liquidus phase on this composition join. Clearly there is no vapour-undersaturated liquid on the composition join at 10 Kbar which even remotely resembles, in terms of crystallisation history, the vapour saturated liquid (which coexists with albite), on the join at 2 Kbar. The same type of argument holds for the $KAlSi_3O_8$–SiO_2–H_2O system on the basis of Fig. 11(b).

Steiner (1970) studied equilibria involving vapour-undersaturated liquids in the halogranite system at 4 Kbar with particular emphasis on the liquids which coexist with quartz and alkali feldspar. These data were used by Steiner *et al.* (1975) in treating crystallisation of a series of hydrous compositions in the haplogranite system as a function of H_2O content at fixed $NaAlSi_3O_8$: $KAlSi_3O_8:SiO_2$ ratios. These results provide a basis for brief discussion of phase relations involving vapour-undersaturated liquids in the haplogranite system.

* Once quartz + liquid is the stable phase assemblage the liquid must be silica-depleted relative to the join, and thus does not lie on the composition join referred to.

The diagram illustrating the phase relations in the haplogranite system at 4 Kbar, in part schematic, is given in Fig. 12. This diagram (Fig. 1 of Steiner *et al.*, 1975) represents an attempt to illustrate the location of the "saturation surface" (Luth *et al.*, 1964 and Luth, 1969) representing the family of vapour-saturated liquids which coexist with quartz and vapour, sodic feldspar and vapour, and potassic feldspar and vapour, relative to the surfaces involving

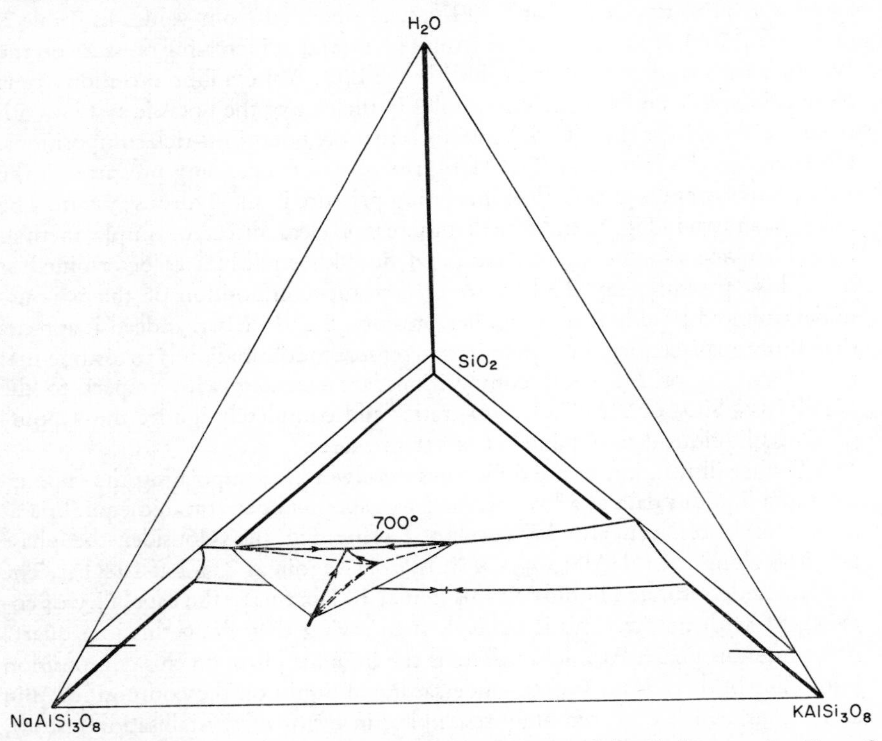

Fig. 12. The Haplogranite System at 4 Kbar. From Steiner *et al.* (1975). "Perspective view of phase relations in a part of the system $NaAlSi_3O_8$-$KAlSi_3O_8$-SiO_2-H_2O at 4 Kbar. Relations involving the compositions of an aqueous fluid (vapour) phase are neglected here, and phase relations in the system $NaAlSi_3O_3$-$KAlSi_3O_3$-H_2O are omitted for the sake of clarity. Diagram is to scale for the compound surface, L(Af_{Na-K}, Q), denoting compositions of liquid that coexists at equilibrium with quartz and alkali feldspars, and the surface L(Af_{Na}, Af_K) denoting compositions of liquid that coexists at equilibrium with two alkali feldspars. Arrows on quaternary boundary curves indicate directions of decreasing temperature."

vapour-undersaturated liquids. The vapour-undersaturated liquids of principal import are those which lie on surfaces beneath (at lower H_2O content) the "saturation surface". These liquids coexist with quartz and alkali feldspar(s) or with two alkali feldspars. The location of the 700°C isotherm on these surfaces is shown to provide a convenient frame of reference. Compositions of the aqueous vapour phase participating in the appropriate equilibria are not shown in Fig. 12. Although this figure provides a convenient means for visualisation of general aspects of the quaternary phase relations, it is a perspective drawing and is not a particularly satisfactory vehicle for discussion of detailed phase relations. In Fig. 13, a projection of the liquidus phase relations onto the anhydrous base, from the H_2O apex, of the composition tetrahedron, the relationships are somewhat more clearly visualised. This diagram, Fig. 3B of Steiner *et al.* (1975), is related to Fig. 12 in the same sense that figures 10(b) and 10(d) are related by the analogous projection technique.

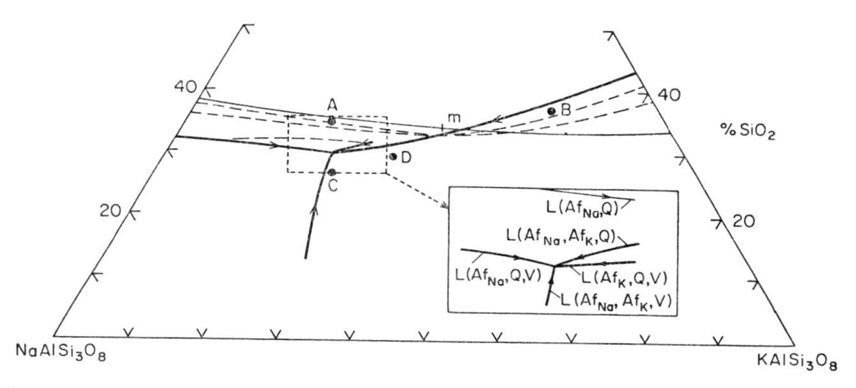

Fig. 13. Projection of Phase Relations in the Haplogranite System at 4 Kbar. From Steiner *et al.* (1975, Fig. 3B). Projection, from the H_2O apex, onto the $NaAlSi_3O_8$-$KAlSi_3O_8$-SiO_2 face of the composition tetrahedron. Quaternary boundary curves are indicated by heavy solid lines. Ternary (anhydrous) boundary curve shown as thin continuous line: the minimum in the anhydrous system is shown by the letter m. Temperature contours are shown by thin broken lines. Expanded view of left-central part of the diagram illustrates the nomenclature applied to boundary curves in terms of phases that coexist with liquid of specific compositions. Af_{Na} = albitic feldspar; Af_K = potassium-rich alkali feldspar; Q = quartz; V = aqueous fluid (vapour) phase. Arrows on boundary curves indicate directions of decreasing temperature. The points A, B, C, D are the compositions featured in Fig. 14.

The orientation of the surfaces in the tetrahedron is particularly important in the analysis of crystallisation relations. In a generalised sense, some of the implications of Figs 12 and 13 regarding crystallisation history can be illustrated by considering stable phase assemblages as functions of temperature and amount of H_2O in the system for a series of selected compositions. Four diagrams showing such relations are given in Fig. 14 and are qualitative in aspects other than as related to the projections of the surfaces shown in Fig. 13. Composition joins A–H_2O, B–H_2O and C–H_2O were selected because they intersect the surfaces (Figs. 12 and 13) defined by liquids which coexist with sodic feldspar and quartz (A–H_2O), potassium feldspar and quartz (B–H_2O) or potassium feldspar and sodium feldspar (C–H_2O), while composition join D–H_2O does not intersect any of these surfaces. The intersection of composition joins A–H_2O, B–H_2O and C–H_2O with the respective surfaces produces the crossing of curves shown in 14(a), 14(b) and 14(c) respectively. Comparison of Figs. 14(a) and 14(b) is particularly noteworthy with respect to the sequence of crystallisation as a function of H_2O content of the bulk composition. The contrast in solidus relations between

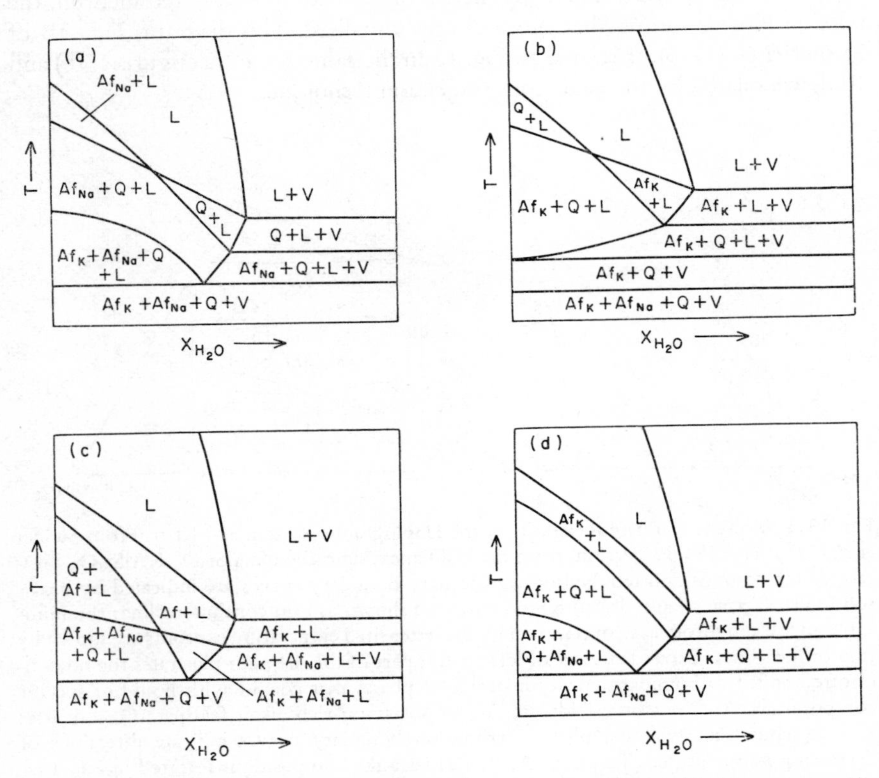

Fig. 14. Schematic Illustration of Variation in Phase Assemblage with Temperature for Silicate-H_2O Joins in the Haplogranite System at 4 Kbar. Phase designations as in the legend to Fig. 1. Anhydrous silicate composition at the left in each diagram. Positions shown in Fig. 13.

(a) Composition A (ab_{45} or ${}_{20}q_{35}$).
(b) Composition B (ab_{15} or ${}_{48}q_{37}$).
(c) Composition C (ab_{50} or ${}_{23}q_{27}$).
(d) Composition D (ab_{40} or ${}_{30}q_{30}$).

those shown in Fig. 14(b) and the other three diagrams is a reflection of the fact that composition B does not lie within the triangle formed by the compositions of the three crystalline phases participating in the quaternary solidus reaction, and compositions A, C, and D do lie within this triangle. Diagrams such as those shown in Fig. 14 are conveniently referred to as phase assemblage diagrams

(Whitney, 1972, 1974), or as pseudobinary diagrams. The term "phase assemblage diagrams" is useful, to distinguish them from other types of pseudobinary diagrams.

If sufficient data were available, diagrams similar to those shown in Fig. 14 could be presented at both higher and lower pressure. Although such data are not available it is possible to utilise the information that is available (Tables 1, 2, Figs 1, 3, 4 and from Luth, 1969) to schematically construct both the 500 and 10,000 bar diagrams analogous to Figs. 13 and 14. These constructions suggest that the analogues to Fig. 14(a) at 500 bars and 10,000 bars have the general appearance shown in Figs. 15(a) and 15(b). Comparison of these two diagrams

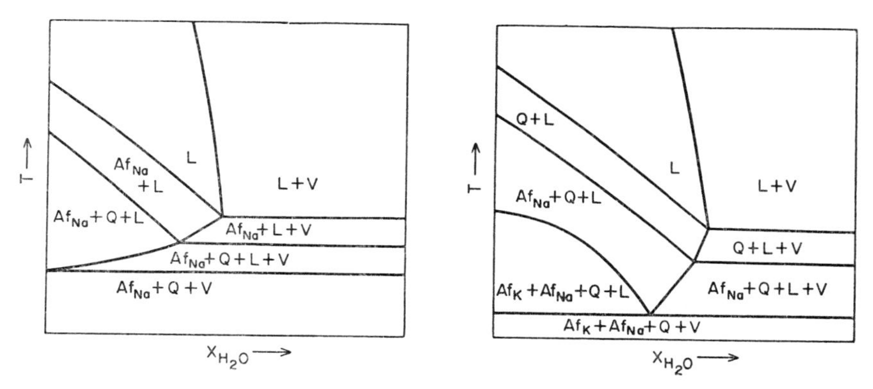

Fig. 15. Pressure Effect on Stable Phase Assemblages for a Specific Silicate–H_2O Join in the Haplogranite System. Completely schematic as no data are available. For comparison with Fig. 14A.

(a) Composition A (ab_{45} or $_{20}q_{35}$) at 500 bars.
(b) Composition A (ab_{45} or $_{20}q_{35}$) at 10 Kbar.

with each other and with Fig. 14(a) provides a graphic illustration of the difficulties inherent in attempting to use either vapour-saturated or vapour-undersaturated liquidus information obtained at a low pressure to predict vapour-undersaturated liquidus equilibria at a higher pressure.

Steiner *et al.* (1975) in their treatment of crystallisation in the haplogranite system at 4 Kbar emphasised the change in chemical composition of the crystal fraction with decreasing temperature in both equilibrium and (perfect) fractional crystallisation. Most conventional treatments of crystallisation concentrate on changes in composition of the silicate liquid, a generally useful approach. However, the haplogranite system is a residua system (Bowen, 1937) and information on the sequence of compositional changes of the crystalline fraction may provide insight into major zoning patterns in intrusions as well as on the compositional contrasts to be expected regarding cogenetic aplites, pegmatites and granites.

The preceeding discussion, treating vapour-undersaturated liquidus relations in the haplogranite system, demonstrates the need for thorough study of the anhydrous equilibria. Information on the boundary curve separating the quartz and alkali feldspar fields on the liquidus in the system $NaAlSi_3O_8$–$KAlSi_3O_8$–SiO_2, particularly with reference to the composition of the ternary minimum liquid, is of key importance. At present no direct data are available. Luth (1969), Steiner (1970), and Steiner *et al.* (1975) all used one method or another of extrapolation techniques to infer phase relations in the anhydrous ternary system. This is clearly not a satisfactory state of affairs, and direct data are essential.

The intent of this portion of the review has been to document and illustrate some of the hazards involved in using vapour-saturated liquidus data (Tuttle and Bowen, 1958, Luth *et al.*, 1965) on the haplogranite system to predict relations involving vapour-undersaturated liquids at some higher pressure. The conclusion of Steiner *et al.* (1975) seems appropriate.

"Known relationships within the haplogranite system also can be of value as indications of what might be expected to occur under given sets of plutonic conditions in nature. Specific petrologic applications, however, should be made by investigators concerned primarily with the rocks. That such applications should be made with caution is dictated by known complexities in the synthetic system, and we suggest that ignoring these complexities in order to accommodate granitic rocks within any simplistic model of genesis can amount to a serious deviation from the route toward petrologic understanding."

(b) THE HAPLOGRANITE SYSTEM IN THE PRESENCE OF ADDITIONAL COMPONENTS

Many authors have discussed the role of additional components on solidus-liquidus equilibria in the haplogranite system. An understanding of the effects of these additional components, considered singly and in combination, is essential in moving toward an understanding of crystallisation and melting relations of the yet more chemically complex natural granitic rocks. Some of the principal additional components of importance relative to the granitic rocks are: (1) those involving departures from the 1:1 alkali/alumina stoichiometry; (2) anorthite; (3) those which result in the capability of forming the common ferromagnesian minerals of the granitic rocks such as biotite, hornblende, and pyroxene; (4) volatile components such as CO_2 and the halogens which may or may not be fractionated in favour of the vapour phase relative to the silicate liquid phase.

For specific petrologic purposes other components will be of particular importance. However, those listed above seem to be of rather general interest and will provide the basis for discussion in this review. When these components are considered together we appear to closely approximate the composition of the granitic rocks. Unfortunately, the data which are available do not permit a wholly satisfactory discussion of the mutual effects of these components. At present it seems that we know little enough, in a quantitative sense, about the effects of each considered individually, as should be made clear in the following discussion.

Departures from the Haplogranite Alkali/Alumina Stoichiometry

The haplogranite system is characterised by a 1:1 alkali (Na + K):Al molar ratio in all of the coexisting phases, when treated as a quaternary system. As shown in Fig. 16, deviations from this stoichiometry may be in the sense of an

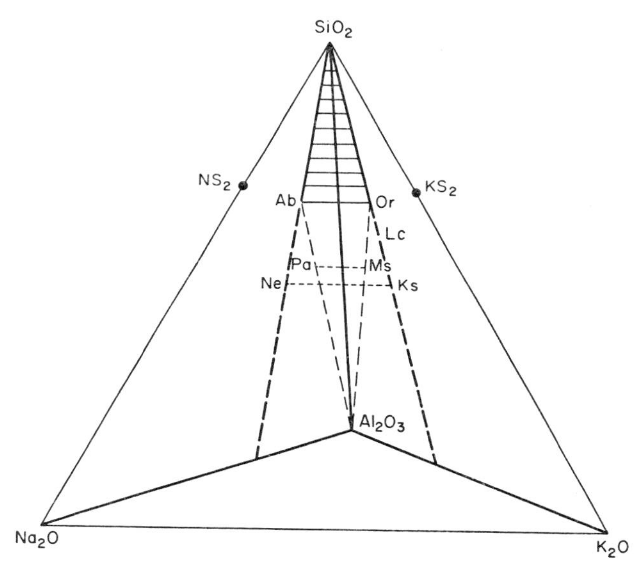

Fig. 16. Composition Tetrahedron Na_2O-K_2O-Al_2O_3-SiO_2. Figure 1 of Luth and Tuttle (1969, p. 517). "Tetrahedron showing the position of the "granite plane", Ab–Or–Q, and related phases in the system Na_2O-K_2O-Al_2O_3-SiO_2. Ne = nepheline, Ks = kalsilite, Ms = muscovite, Pa = paragonite, NS_2 = sodium disilicate, KS_2 = potassium disilicate, Ab = albite, Lc = leucite, Or = orthoclase, Q = quartz."

excess of Al or of alkali. On the one hand the compositions provide possible information bearing on the muscovite–paragonite micas and the alumino-silicates kyanite, andalusite, and sillimanite and on the other hand information on the crystallisation of peralkaline granitic liquids. These two regions of the overall quinary system (K_2O-Na_2O-Al_2O_3-SiO_2-H_2O) will be referred to as the Peraluminous and Peralkaline haplogranite systems for convenient reference.

The Peraluminous Haplogranite System. Phase relationships in the portion of the quinary system involving the coexistence of muscovite–paragonite micas, or alumino-silicates with quartz and/or a silicate liquid have been experimentally studied, or discussed in general terms by a wide variety of workers. This intensive study is a reflection of the importance of muscovite in high grade metamorphism, in melting of pelitic sedimentary rocks, as a potential primary phase in the crystallisation of peraluminous granitic rocks, and as an important phase in pegmatite occurrences.

It would be logical to develop the full set of univariant and invariant equilibria in the K_2O-Al_2O_3-SiO_2-H_2O, Na_2O-Al_2O_3-SiO_2-H_2O and $KAl_2AlSi_3O_{10}(OH)_2$-$NaAl_2AlSi_3O_{10}(OH)_2$-H_2O systems as related to equilibria

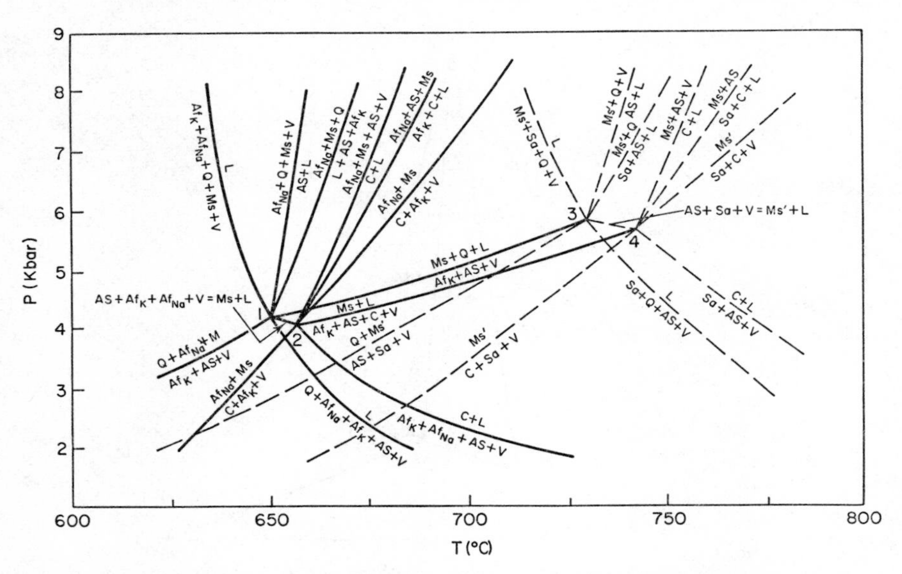

Fig. 17. Pressure–Temperature Projection of Selected Invariant and Univariant Equilibria in the $NaAlSi_3O_8$-$KAlSi_3O_8$-Al_2O_3-SiO_2-H_2O Quinary System and in the $KAlSi_3O_8$-Al_2O_3-SiO_2-H_2O Quaternary System. Quinary reactions are indicated by solid lines with the lettering designating the phases participating in the reactions related to invariant points 1 and 2. Quaternary reactions in the Na-free system are indicated by dashed lines with the lettering indicating phases participating in the reactions related to invariant points 3 (from Stoore and Karotke, 1971) and 4. The location of the Q + Ms′ = AS + Sa + V reaction is from Althus *et al.* 1970). Other reactions from Blencoe's (1974) data and analysis of the published data. AS = Aluminosilicates (kyanite, andalusite and sillimanite); Q = quartz; C = corundum; L = silicate liquid; V = aqueous vapour (fluid) phase; Ms = muscovite-rich muscovite–paragonite crystalline solution; Ms′ = Na-free muscovite; Af_{Na} = Na-rich alkali feldspar; Af_K = K-rich alkali feldspar; Sa = sanidine, Na-free.

involving muscovite, muscovite + quartz, paragonite, paragonite + quartz, and the muscovite–paragonite crystalline solution series before dealing with the phase relations in the quinary system. However, such a process will not be followed here in view of space limitations, the complexity of the relations, and the primary purpose. In addition, Thompson (1974), Blencoe (1974) and Huang and Wyllie

(1974) have recently discussed the available data pertaining to these equilibria. These authors have discussed in detail the geometric relations, in P–T projection, of the invariant and univariant equilibria. In addition Thompson (1974), Blencoe (1974) and Eugster *et al.* (1972) have recently evaluated the muscovite–paragonite crystalline solution stability relations. The experimental data discussed by these authors suggest that the P–T co-ordinates of the various reactions are not in accord at a level such as to warrant extended discussion here. Further, compositional data on the liquid phase participating in the various reactions are essentially non-existent. Of primary concern in the present context are the relations which involve muscovite-rich crystalline solutions that coexist with liquids not far removed from the haplogranite liquids of the 1:1 alkali alumina join in the quinary system.

With these factors in mind certain critical aspects of the quinary equilibria involving muscovite can be discussed in terms of Fig. 17. The P–T projection of univariant and invariant equilibria (slightly modified from Blencoe, 1974) is not an attempt to illustrate completely phase relations in the peraluminous haplogranite system, rather the focus is on the set of reactions important in the latter stages of crystallisation with reference to muscovite and the alumino-silicates.* Of particular importance is the invariant point labelled 1 at which quartz (Q), muscovite (Ms), Na-rich alkali feldspar (Af_{Na}), K-rich alkali feldspar (Af_K), alumino-silicate (AS), liquid (L), and vapour (V) coexist. The P–T co-ordinates for the curves are from Blencoe's (1974) experimental study. As noted previously, liquid compoisitions along the various univariant curves are essentially unknown, other than as indicated by the nature of the reaction. Even casual examination of Fig. 17 indicates that the six-phase univariant reactions provide very little information on liquidus, vapour-saturated or vapour-undersaturated, phase relations. Diagrams of this type are most useful in discussing phase relations which are divariant (five-phase coexistence) rather than liquidus (pentavariant, two-phase) equilibria. The divariant assemblages related to the quinary minimum solidus reactions are of particular importance. At temperatures above the high pressure solidus reaction ($Af_{Na} + Af_K + Ms + Q + V = L$) the possible divariant assemblages are: $L + Af_{Na} + Q + Ms + V$; $L + Af_K + Q + Ms + V$; $L + Af_K + Af_{Na} + Ms + V$; $L + Af_K + Af_{Na} + Q + V$; and $L + Af_K + Af_{Na} + Ms + Q$. These divariant assemblages are stable until other univariant reactions limit them. For example, the geologically important phase assemblage $L + Kf + M + Q + V$ has a wide maximum stability interval limited in part by the quinary reaction: $Ms + Q + L = AS + Af_K + V$ and by the quaternary reaction $Ms' + Sa + Q + V = L$. This stability interval is a maximum in the sense that depending on the specific composition being considered the stability interval may be greatly reduced by passing from the

* With many reservations, I will simplify the alumino-silicate relations relative to the stability of kyanite, andalusite, and sillimanite by referring to them as "alumino-silicates". The reader may overlay his preferred aluminosilicate phase diagram on Fig. 16.

divariant assemblage to any one of the 5 related possible trivariant assemblages: $Af_K + Q + Ms + V$; $L + Q + Ms + V$; $L + Af_K + Ms + V$; $L + Af_K + Q + V$ or $L + Af_K + Q + Ms$. In general, relations of this type are of greatest import when considering specific bulk compositions, rather than the generalised phase relations. Figure 17 may be employed to provide a base for the analysis of the latter stages of crystallisation of peraluminous granitic compositions, but provides negligible information on liquidus relations. It is possible to construct a series of projections of the schematic quinary phase relations to aid in the analysis of possible liquidus relations. In view of the lack of direct experimental data this becomes essentially an academic exercise and will be avoided.

The relationships indicated in Fig. 17 do permit the drawing of several conclusions regarding phase relations in the peraluminous haplogranite system. *First*, the muscovite and muscovite + quartz reactions, involving maximum thermal stability at a given pressure, are of far less importance relative to the haplogranite system than are the set of quinary reactions. *Second*, muscovite as a primary magmatic phase coexisting with a granitic liquid, either vapour saturated or vapour undersaturated, is consistent only with a relatively high (greater than about 4 Kbar) pressure environment. *Third*, muscovite is capable of coexisting with a wide compositional range of vapour-undersaturated and vapour-saturated silicate liquids at high pressure. *Fourth*, the solidus reactions in the peraluminous haplogranite system are only slightly lower in temperature at a given pressure than in the haplogranite system. This suggests that only small amounts of excess Al_2O_3 result in saturation of the haplogranite liquid with respect to an aluminous phase.

The Peralkaline Haplogranite System. An understanding of the effects of excess alkalis on melting and crystallisation relations in the haplogranite system is of major importance in treating the granitic rocks which contain minerals such as acmite, aegirine–acmite, and alkali amphiboles, or which show normative acmite and sodium metasilicate in the chemical analyses.

Tuttle and Bowen (1958) noted that the haplogranite system is not a residua system in the presence of excess alkalis. This conclusion was based on their study of phase relations at 981 bars in the K_2O-Al_2O_3-SiO_2-H_2O system and previous studies. An important aspect of the modifying effects of excess alkalis is the tendency to form silicate liquids which exhibit continuous solubility with respect to vapour phase, as described by Morey and Fenner (1917), as contrasted with the limited solubility relations shown in the haplogranite system described previously. The continuous solubility relations imply a dramatically reduced solidus temperature, far below the magmatic regime, in such systems.

Thompson and MacKenzie (1967) have presented data obtained at 1 Kbar on three sections in the peralkaline haplogranite system. The sections studied were the haplogranite system with 5% Na_2SiO_3, with 5% K_2SiO_3, and 5% $KNaSiO_3$. This study was an elaboration of the previous study on the haplogranite liquidus studies by Carmichael and MacKenzie (1963) where the effects of both acmite ($NaFeSi_2O_6$) and sodium metasilicate together were evaluated. In both cases the experiments were performed in the presence of "excess" H_2O, and no information was presented to indicate the amount of H_2O present or soluble in the liquid. This factor limits the utility of the results, as liquidus temperatures should depend on H_2O content by analogy with known phase relations in the K_2O-Al_2O_3-SiO_2-H_2O and Na_2O-Al_2O_3-SiO_2-H_2O systems. Although the experiments were performed in the presence of "excess" H_2O these authors do not indicate explicitly that an aqueous vapour phase was present. Both studies were primarily concerned with vapour-saturated (presumably) liquidus relations and did not yield information on solidus relations. Bailey and Schairer (1964) stressed the difficultues involved in treating the phase relations in the complex quinary system by geometric means. Particular emphasis was placed on the problems related to projections of the liquidus data on a normative *ab–or–q* triangle. Bailey and Schairer (1964) were particularly concerned about the K/Na ratio in the liquid and in the coexisting alkali feldspar, both as measured and as apparent values in projection. They suggested the term "orthoclase effect" for the separation of feldspars with a higher K/Na ratio than the peralkaline liquid because the separation of such feldspar is fractionating potash from strongly sodic liquids. Thompson and MacKenzie (1967) agreed with the basic conclusions regarding the "orthoclase effect" but suggested that the term was of little value because it links a number of related factors rather than being a single well-defined "effect". While not wishing to be drawn into the controversy, it seems to me that the major problem lies in attempting to portray phase relations in a very complex system, where the data are restricted to a limited suite of compositions, on two-dimensional paper. Considering that the experiments were in five and six component systems it does not seem surprising that difficulties are encountered in projecting results obtained on sections through the appropriate hyperspace. Not intending to minimise the importance of the studies of Carmichael and MacKenzie (1963) and Thompson and MacKenzie (1967), it seems that some of the important aspects of phase relations in the peralkaline haplogranite system can be modelled using somewhat simpler systems.

Mustart (1972) obtained experimental data on liquidus phase relations on the join $NaAlSi_3O_8$-$Na_2Si_2O_5$-H_2O in the system Na_2O-Al_2O_3-SiO_2-H_2O. His figure 12 is reproduced here as Fig. 18. These diagrams, illustrating liquidus relations at essentially 5 Kbar (4.96) and 1.25 Kbar, clearly show the effect of sodium disilicate on increasing the solubility of H_2O in the liquid, and on increasing the amount of dissolved silicates in the vapour phase. For a wide range of peralkaline compositions crystallisation will take place such that there is a continuum between silicate liquid and vapour. Compositions lying above the dotted line in Figs. 18(a) and 18(b) will exhibit the "normal" liquid vapour immiscibility relations on cooling. It is useful to illustrate these relationships in

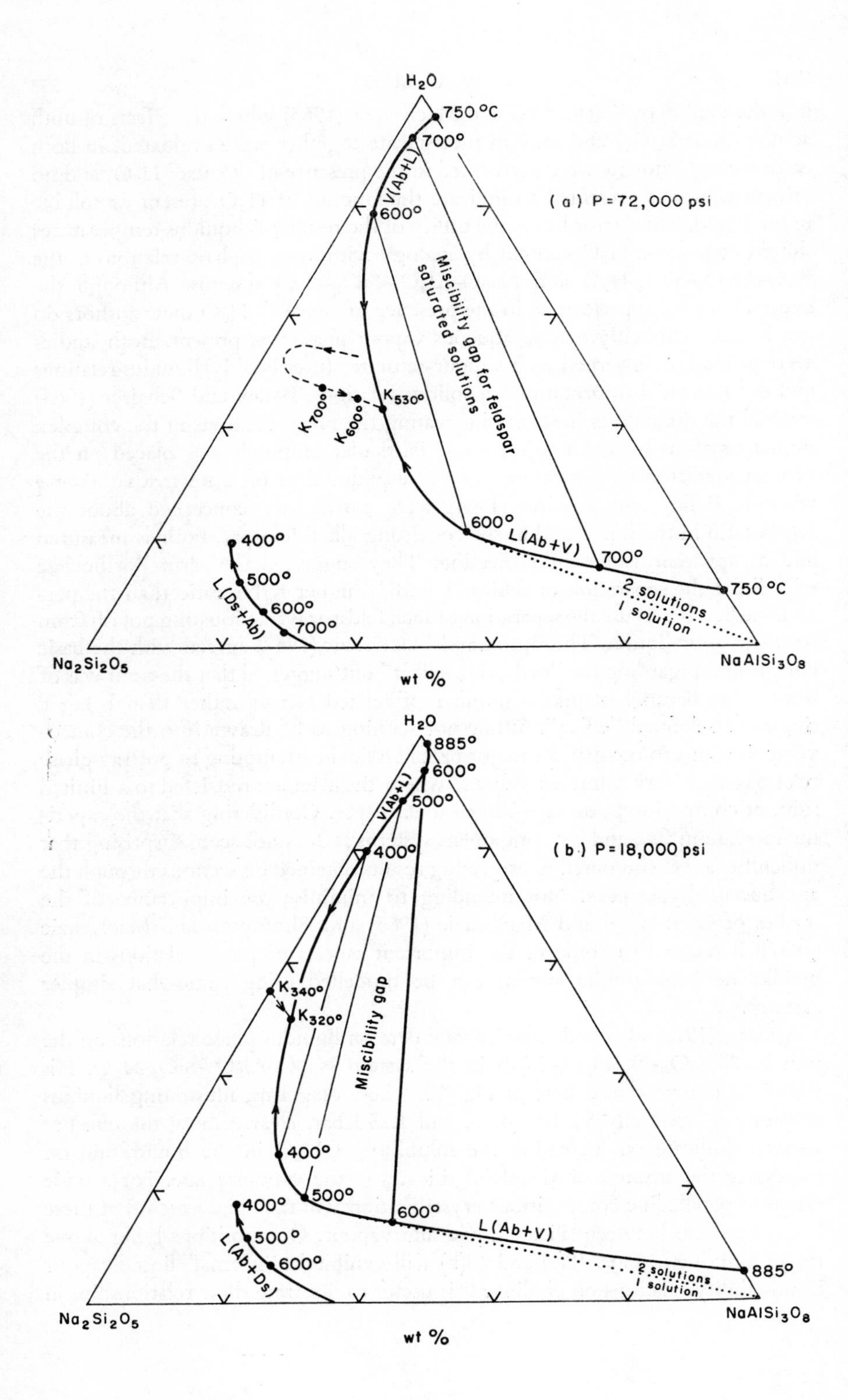

H2O
750 °C
700°
V(Ab+L)
600°
(a) P = 72,000 psi
Miscibility gap for feldspar saturated solutions
K530°
K600°
K700°
600°
L(Ab+V)
700°
750 °C
2 solutions
1 solution
400°
500°
L(Ds+Ab)
600°
700°
Na2Si2O5
NaAlSi3O8
wt %
H2O
885°
600°
V(Ab+L)
500°
400°
(b) P = 18,000 psi
K340°
K320°
Miscibility gap
400°
500°
600°
L(Ab+V)
400°
500°
L(Ab+Ds)
600°
885°
2 solutions
1 solution
Na2Si2O5
NaAlSi3O8
wt %

terms of joins from the $NaAlSi_3O_8$-$Na_2Si_2O_5$ sideline to H_2O as shown in Fig. 19 (from Mustart, 1972, Fig. 13). Compositions X and Y correspond to the two contrasted types of crystallisation history, and are discussed in the caption. Phase relations on this join at 1.25 Kbar and will be generally similar, as can be seen by examination of Fig. 18(b), although the T–X_{H_2O} co-ordinates of the curves will be quite different. Comparison of Fig. 19 with the analogous diagram for the $NaAlSi_3O_8$-H_2O system serves to illustrate the effect of $Na_2Si_2O_5$ on the phase relations. The contrast is most pronounced at low H_2O content. In addition, for a wide range of H_2O content values the three-phase Ab + L + V extends to much lower temperatures than in the $NaAlSi_3O_8$-H_2O system, where it is limited to a unique value.

Because relations similar in form to those of Fig. 18 have been obtained for the joins $Na_2Si_2O_5$-SiO_2-H_2O, $K_2Si_2O_5$-SiO_2-H_2O, and $K_2Si_2O_5$-$KAlSi_3O_8$-H_2O (Carman, Tuttle and Luth, unpublished data) it is believed that Fig. 19 serves as a not unreasonable, though very simplified, model for consideration of the peralkaline haplogranite phase relations. Luth and Tuttle (1969) presented preliminary experimental data on the join $50NaAlSi_3O_8 \cdot 25SiO_2$-$Na_2Si_2O_5$-$H_2O$ at 5 Kbar which indicated enhanced solubility of H_2O in the liquid and enhanced solubility of silicates in the vapour phase relative to the haplogranite system.

In spite of the complexity of phase relations in the peralkaline haplogranite system several conclusions can be drawn with regard to the modifying effects

---

Fig. 18. Isobaric Polythermal Projections for the Ternary System $Na_2Si_2O_5$-$NaAlSi_3O_8$-H_2O at Pressures of 72,000 psi and 18,000 psi. From Mustart (1972). "Solid lines with arrows indicate, with decreasing temperature, the change in composition of crystal-saturated solutions L(Ds, Ab), L(Ab, V), and V(Ab, L). The dashed line with arrows describes the migration of isobaric consolute points, K, which mark for each temperature the point at which crystal-free solutions, L and V, become identical in composition. The dotted line tangent to the path of L(Ab, V) separates the region of compositions which upon isobaric crystallisation will evolve two solution phases, from those compositions which can produce no more than one solution phase. The region extending from the ab–H_2O sideline and denoted as a miscibility gap, is the field in which albite may coexist with the two immiscible solutions L and V. As only ternary relations are presented, no information is given below the 300°C isotherm where the appearance of analcite necessitates a quaternary treatment.

(a) Isobaric Polythermal Projection at a Constant Pressure of 72,000 psi. The paths for L(Ab, V) and V(Ab, L) meet at the ternary critical end point, L = V(Ab), at 530°C which is marked $K_{530°}$. The dashed path of consolute points, also ending at the ternary critical point, does not intersect the ds–w sideline but shows a recurve with decreasing temperature.

(b) Isobaric Polythermal Projection at a Constant Pressure of 18,000 psi. The path of isobaric consolute points is short and extends from the binary critical point on the ds–w sideline at 340°C, to the estimated ternary critical point at 320°C. With decrease in pressure from 72,000 to 18,000 psi, the miscibility gap becomes larger while the path of vapour–undersaturated liquid L(Ab, Ds), remains essentially unchanged."

of excess alkali on liquidus and solidus relations. *First*, both liquidus and solidus temperatures tend to be lowered relative to the analogous (in terms of Na:K:Si ratio) compositions in the haplogranite system. *Second*, the presence of excess alkali tends to enhance the solubility of H_2O in the silicate liquid, even to the extent where there is a continuum between the silicate liquid and the vapour phase. *Third*, the relative depression of the solidus temperatures suggests that coexistence of two alkali feldspars with a silicate liquid is possible at lower pressures in the peralkaline haplogranite system than in the haplogranite system. *Fourth*, comparison of experimental results on the peralkaline haplogranite system with those on the haplogranite system is a difficult problem. However, the "orthoclase effect" seems to be verified, whether or not it should be considered by that name.

In view of the importance of the phase acmite in peralkaline granite rocks it is not surprising that phase relations involving this phase have been studied rather intensively. Of particular importance are the studies of Bailey (1969) and Gilbert (1969) on acmite stability, of Bailey and Schairer (1966) on the Na_2O-Al_2O_3-Fe_2O_3-SiO_2 system at 1 bar, and the previously cited studies by Carmichael and MacKenzie (1963) and Thompson and MacKenzie (1967) on the hydrous systems. Bailey (1969) notes that it is important to distinguish between peralkalinity (molecular excess of alkalis over alumina) and excess alkali silicate (molecular excess of alkalis over alumina and iron oxide) in treating acmite stability and crystallisation relations. He concluded that a prerequisite for acmite crystallisation is that the liquid contain excess sodium silicate.

The upper thermal stability of acmite + H_2O was determined by Bailey at pressures to 5 Kbar using the hematite–magnetite, nickel–nickel oxide and quartz–magnetite–fayalite buffers. Gilbert (1969) determined the analogous acmite = hematite + magnetite + liquid reaction in the anhydrous system at pressures to 40 Kbar. At high oxygen fugacity acmite is stable to about 850°C (5 Kbar), 880°C (1 Kbar), in the presence of excess H_2O according to Bailey's (1969) results. Decreasing oxygen fugacity results in a decrease of the acmite + fluid thermal stability, and under the wüstite–magnetite buffer Bailey observed a reaction involving the stability of an arfvedsonitic amphibole + liquid + vapour at low temperature being replaced by the fayalite + liquid + vapour assemblage at temperatures above 675°C (from 1 to 4 Kbar). The thermal stability of acmite is significantly higher under anhydrous conditions. Gilbert (1969) obtained the equation $T(°C) = 988 + 20.87\,P(Kbar) - 0.155P^2$ describing the stability limit of acmite in the anhydrous system. Bailey (1969) concluded that arfvedsonitic rather than riebeckitic amphiboles are favoured as the primary magmatic amphiboles of the peralkaline granites and the riebeckitic amphiboles are likely to be deuteric or metasomatic phases in such rocks. Carmichael and MacKenzie (1963) and Thompson and MacKenzie (1967) did not observe acmite as a stable phase on the liquidus in their studies of the peralkaline haplogranite system, though it was present as a component. The lack of saturation of the liquids with respect to acmite is presumably a consequence of the particular P–T–composition–oxygen fugacity region studied and does not imply that acmite lacks a stability region on the liquidus.

Bailey and Schairer (1966) have provided a thorough discussion of the liquidus relations in the Na_2O-Al_2O_3-Fe_2O_3-SiO_2 system as determined at 1 bar. This study provides a base for experimental studies in both the hydrous and anhydrous systems at higher pressure. The studies of Bailey (1969) and Gilbert (1969) are examples of such studies and indicate that much remains to be done.

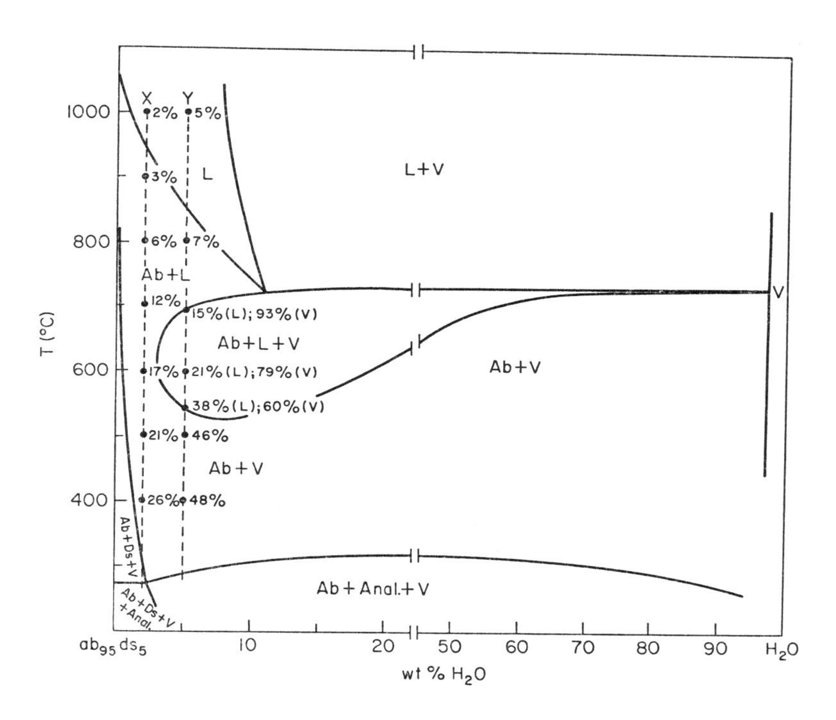

Fig. 19. Isoplethal Temperature–Composition Section at 72,000 psi, through the Ternary Prism with $Na_2Si_2O_5$-$NaAlSi_3O_8$-H_2O as the Base, and Temperature as the Vertical Axis. From Mustart (1972, p. 114). "The figure illustrates phase regions intersected along the join $ab_{95}ds_5$-H_2O. Because relations are not truly binary, the section should be considered as a pseudobinary phase assemblage diagram. The dashed vertical lines indicate two possible crystallisation sequences starting from initial silicate melts, X and Y, with dissolved water contents of 2 and 5 wt % respectively. Percentages refer to the water content of solutions at each point. It will be noted that for path X, no more than a single solution phase is ever present during the cooling history. On the other hand for path Y, a separate vapour phase does evolve at 695°C. This, however, is eventually redissolved in the liquid, and below 540°C feldspar coexists with a single fluid. Finally, near 300°C, analcite begins to form at the expense of albite."

Anorthite

It is my impression that the effects of anorthite on phase relations in the haplogranite system have been the subject of more interest and concern than any other single component. Indeed, the role of anorthite has frequently been assumed to be so important in modifying solidus and liquidus relations that the experimental studies of Tuttle and Bowen (1958) and others concerned primarily with the haplogranite system were of relatively little value. It should be clear,

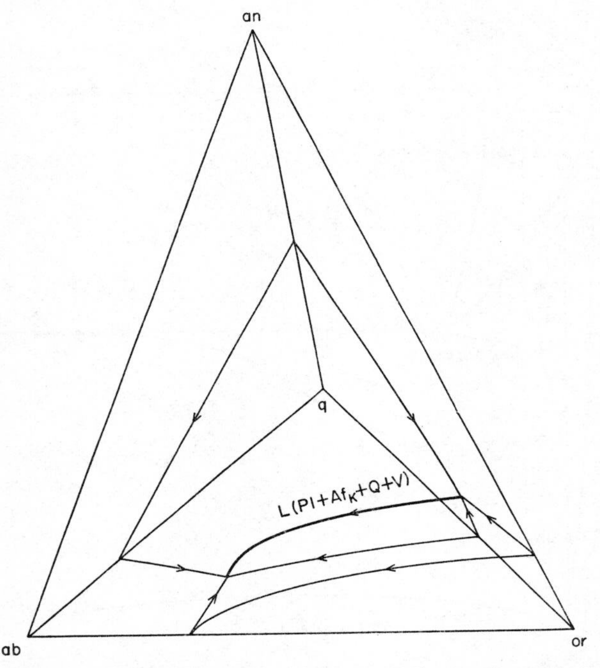

Fig. 20. Schematic Diagram Illustrating Vapour-Saturated Liquidus Relationships in the Haplogranodiorite System at 5 Kbar. The curve L(Pl, Af_K, Q, V), discussed in the text, is constructed in accord with Whitney's (1972, 1975) data. Pl = plagioclase; Af_K = Potassium feldspar; Q = quartz; L = liquid; V = vapour. Component designations are: ab = $NaAlSi_3O_8$; or = $KAlSi_3O_8$; an = $CaAl_2Si_2O_8$; q = SiO_2.

from the sequence of topics in this chapter if nothing else, that I do not subscribe to such a view. Nonetheless, the importance of plagioclase as a phase in the granitic rocks and the normative anorthite content of chemical analyses of the granitic rocks suggest that anorthite is a very important component. I believe that it is sufficiently important in considering problems related to the granodiorites, quartz–monzonites, tonalites and related rocks that it is not unreasonable to refer to the $NaAlSi_3O_8$-$KAlSi_3O_8$-$CaAl_2Si_2O_8$-SiO_2-H_2O system as the "haplogranodiorite" system and will do so in this review. The quaternary haplogranite system is only one of the quaternary systems which bound the quinary system and each quaternary system is of equal importance from the standpoint of phase relationships. Consequently each system merits as full a treatment as was given the haplogranite system. However, the focus in this chapter is on the granitic rocks and the bulk of the experimental evidence indicates that the haplogranite system does represent a residua system for the haplogranodiorite system. With this in mind, somewhat as an apology, the important works of many researchers on the anorthite-bearing subsystem will not be treated other than as they directly relate to the main theme.

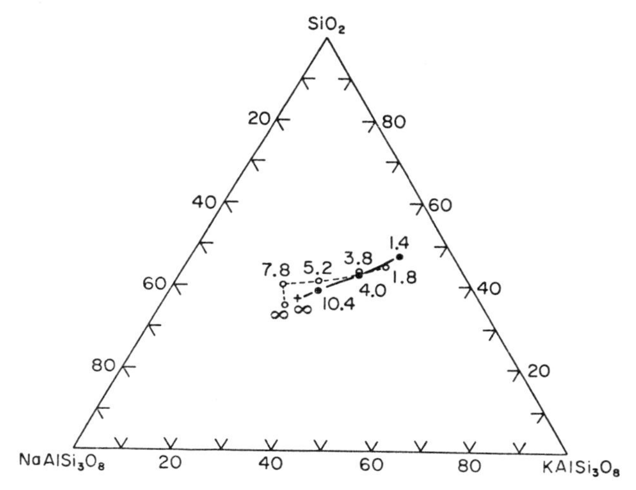

Fig. 21. Projection of the L(Pl, Af_K, Q, V) Curve in the Haplogranodiorite System from $CaAl_2Si_2O_8$ and H_2O. From James and Hamilton (1969, Fig. 9, p. 130). "Comparison of eutectics (v. Platen, 1965) and piercing points (this study) for the system Ab-Or-An-Q-H_2O. Solid circles, this study; open circles, v. Platen (1965); numerals indicate the Ab/An ratios—values of ∞ are for the minimum compositions in the granite system."

Von Platen (1965) and James and Hamilton (1969) have presented experimental data on the vapour-saturated liquidus equilibria in the haplogranodiorite system at 2 Kbar and 1 Kbar respectively. Both workers approached the problem by means of considering the vapour-saturated liquidus relations along non-quaternary joins in the quinary system. Von Platen used a series of compositions which had anhydrous equivalents along planes of constant $NaAlSi_3O_8$/$CaAl_2Si_2O_8$ ratio in the tetrahedron $NaAlSi_3O_8$-$KAlSi_3O_8$-$CaAl_2Si_2O_8$-SiO_2. The ratios chosen correspond to the following plagioclase compositions: $an_{11.4}ab_{88.6}$, $an_{16.1}ab_{83.9}$, $an_{20.8}ab_{79.1}$, $an_{35.7}ab_{64.3}$. James and Hamilton (1969) used a series of compositions along planes parallel to the $NaAlSi_3O_8$-$KAlSi_3O_8$-SiO_2 base of the anhydrous composition tetrahedron, at 3, 5, 7.5 and 10% $CaAl_2Si_2O_8$. Both studies involve the determination of phase relations on sections through the composition tetrahedron (ignoring H_2O). James and Hamilton (1969) clearly recognised the nature of the problem in that the phases observed experimentally do not lie compositionally on the join being studied. They, and Weill and Kudo (1968) have criticised the analysis of the experimental data of von Platen (1965) relative to this problem.

A schematic diagram showing the vapour-saturated liquidus is given in Fig. 20 for purposes of discussion. The three curved surfaces represent the composition of liquids which coexist with: quartz (Q), plagioclase (Pl), and vapour (V); Pl, alkali feldspar (Af_K) and V; and Af_K, Q, V. These surfaces meet along a curve representing the composition of liquids which coexist with Pl, Af_K, Q and V. Figure 21 represents constant pressure equilibria at some (arbitrary) value less than 3.5 Kbar. Little information is available, other than by inference, on the direction in which these surfaces may shift with changing pressure. The location of the curve is shown in projection from H_2O and $CaAl_2Si_2O_8$ in Fig. 21 (Fig. 9 of James and Hamilton, 1969) incorporating the results of both workers. The location of the minima on the alkali feldspar–quartz–liquid–vapour boundary curve from Tuttle and Bowen (1958) at 1 and 2 Kbar are also indicated for reference. At 1 Kbar the curve of James and Hamilton (1969) trends directly toward the minimum in the haplogranite system, but the 2 Kbar curve of von Platen indicates marked deviation from the minimum in the haplo-granite system. The possible complexities of phase relations involving the low temperature termination of this boundary curve are analogous to those discussed by Stewart and Roseboom (1962) in connection with phase relations in the $NaAlSi_3O_8$-$KAlSi_3O_8$-$CaAl_2Si_2O_8$-H_2O system. At pressures above the singular point on the solidus reaction curve for the haplogranite system (page 344) the projection of the quinary boundary curve would be continuous to the projected composition of the haplogranite eutectic liquid, although the minimum temperature on the boundary curve need not be that of the haplogranite system solidus.

Both of the projected curves in Fig. 21 lie well toward the high normative quartz content values of the analysed plutonic and volcanic rocks with greater than 80% normative quartz + orthoclase + albite (Luth *et al.*, 1964) and with greater than 80% normative quartz + orthoclase + albite + anorthite (James and Hamilton, 1969). However, the correspondence of the projected quinary boundary curve with the maxima in the normative compositions of the aplitic rocks (James and Hamilton, 1969, Fig. 12) is excellent. The haplogranodiorite boundary is of primary importance in considering the later stages of crystallisation, and the onset of melting, for granodioritic compositions. Liquids may encounter this boundary curve through crystallisation of Q and Af_K, Pl and Q, or Pl and Af_K in the presence of a vapour phase, or through crystallisation of Pl, Af_K and Q in the absence of a vapour phase. Although liquidus relations are poorly known in the system, attention is drawn to the very large volume of Fig. 20 which involves plagioclase as a liquidus phase.

Whitney (1972, 1975) has approached the problem of phase relations in the haplogranodiorite system from an entirely different standpoint. Stable phase assemblages were determined for four silicate compositions in the $NaAlSi_3O_8$-$KAlSi_3O_8$-$CaAl_2Si_2O_8$-SiO_2 system as a function of temperature and the amount of H_2O added to the silicate composition (X_{H_2O}) at 2 and 8 Kbar. The

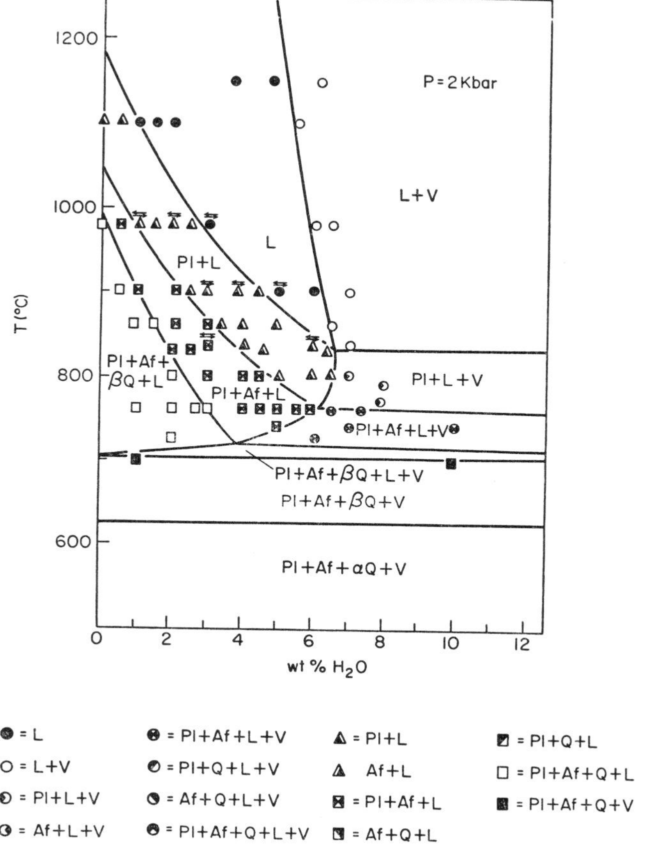

Fig. 22. 2 Kbar Temperature vs. wt % H_2O Diagram for the Synthetic Granite Composition (Rl). From Whitney (1975, Fig. 3). "All experiments are synthesis runs unless otherwise noted. Pl = plagioclase; Af = alkali feldspar; αQ = alpha quartz; βQ = beta quartz; L = silicate liquid; V = hydrous vapour; ⇌ = a pair of experiments, one using crystallised material and one using fired gel."

phase assemblages observed along these four pseudobinary sections in the quinary system illustrate the importance of the vapour-undersaturated liquidus relations, and demonstrate once again the necessity to distinguish clearly between the effects of pressure and of the component H_2O as intensive variables. The bulk compositions selected by Whitney (1972, 1975) were chosen to approximate insofar as possible, the average compositions obtained by Nockolds (1954) for the hornblende–biotite bearing granites (R1), adamellites (R4), granodiorites (R5) and tonalites (R6). The normative constitution of these four compositions are $ab_{32.0}or_{34.0}an_{7.50}q_{26.5}$(R1); $ab_{34.2}or_{30.2}an_{14.1}q_{21.5}$(R4);

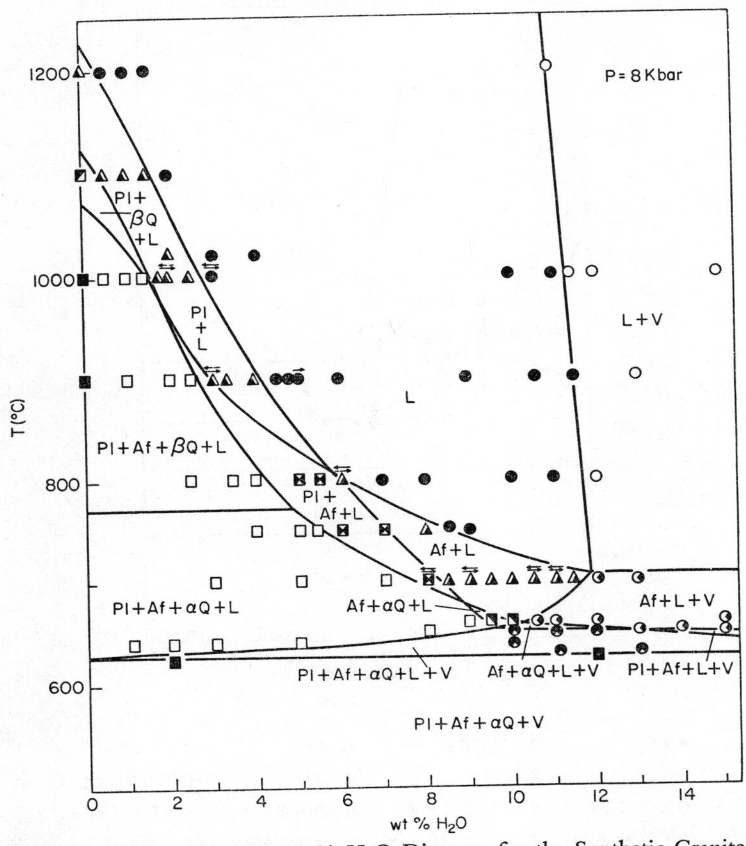

Fig. 23. 8 Kbar Temperature vs. wt % H_2O Diagram for the Synthetic Granite Composition (Rl). From Whitney (1974, Fig. 4). Symbols as in legend to Fig. 22.

$ab_{37.3}or_{19.8}an_{19.8}q_{23.1}$(R5); and $ab_{31.8}or_{11.2}an_{29.9}q_{27.1}$(R6). Although phase relations along each of these pseudo-binary joins are of considerable importance, only one will be discussed in the present context, Rl–H_2O. Diagrams from Whitney (1975) are reproduced here as Figs 22, 23 and 24, showing the stable phase assemblages encountered at 2 and 8 Kbar as functions of temperature and X_{H_2O}, and at 750°C as a function of pressure and X_{H_2O}.

In considering paragenetic sequences obtained through petrographic observations, Fig. 23 is of particular interest. For a series of specific compositions along the Rl–H_2O join we see the pattern given below for the sequence of crystallisation:

←T

	L	Pl+L	Pl+Q+L	Pl+Af+Q+L	Pl+Af+Q+L+V	Pl+Af+Q+V
	L	Pl+L	Pl+Af+L	Pl+Af+Q+L	Pl+Af+Q+L+V	Pl+Af+Q+V
X_{H_2O} ↓	L	Af+L	Pl+Af+L	Pl+Af+Q+L	Pl+Af+Q+L+V	Pl+Af+Q+V
	L	Af+L	Af+Q+L	Af+Q+L+V	Pl+Af+Q+L+V	Pl+Af+Q+V
	L	Af+L	Af+L+V	Af+Q+L+V	Pl+Af+Q+L+V	Pl+Af+Q+V

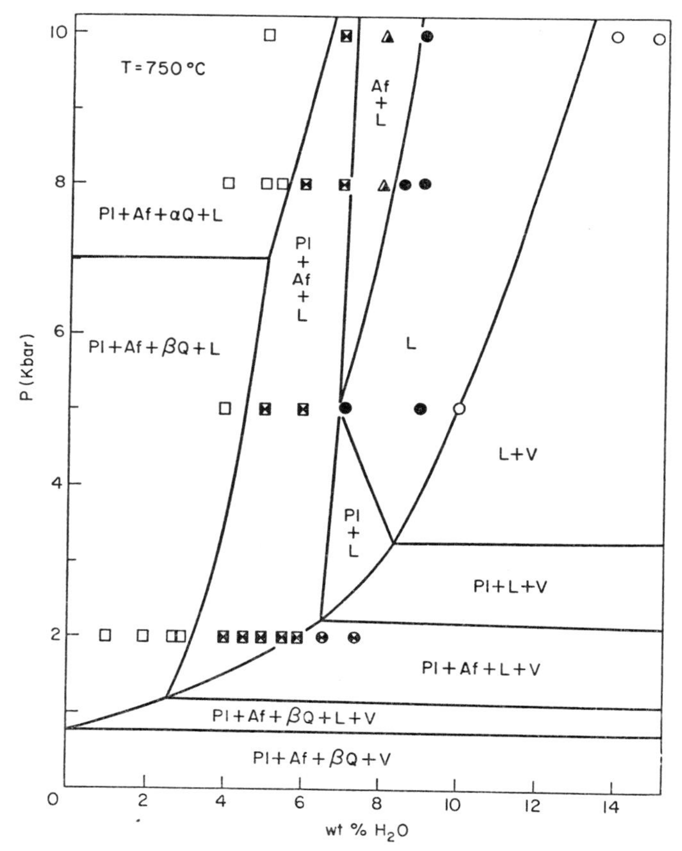

Fig. 24. 750°C Pressure *v.* wt % H_2O diagram for the Synthetic Granite Composition (Rl). From Whitney (1975, Fig. 5). "The data shown at 2 Kbar was actually gathered at 760°C, but is presented here as it was used in construction."

If it is assumed that the specific silicate composition (R1) was originally a liquid composition containing some dissolved H_2O then Fig. 23 demonstrates that, at this pressure, the sequence of crystallisation, and the temperature interval over which crystallisation takes place depends dramatically on the H_2O content of the initial liquid.

Comparison of Figs 22 and 23 shows that the sequence of crystallisation involves plagioclase followed by alkali feldspar over the entire range at 2 Kbar, while at 8 Kbar this sequence exists only over an intermediate ($2 < X_{H_2O} < 6$) range. This provides support for extending the conclusions reached in treating vapour-undersaturated liquidus relations in the haplogranite system to the haplogranodiorite system. Again, it does not seem possible to predict vapour-undersaturated liquidus equilibria at high pressure on the basis of vapour-saturated equilibria relative to a lower pressure.

Diagrams such as those of Figs 22, 23 and 24 provide very little information on the compositions of the coexisting phases. The $T–X_{H_2O}$ regions over which the phase assemblages are stable correspond to intersections of the compositional join, R1–H_2O, with phase regions in the quinary composition hyperprism at the pressure of interest. Relative to the preceding discussion concerning the quinary boundary curve in the haplogranodiorite system, the $T–X_{H_2O}$ region in Figs 22 and 23 over which the stable assemblage is Pl + Af + Q + L + V assumes particular importance. This $T–X_{H_2O}$ region in Figs 22 and 23 involves liquids which lie along the quinary boundary curve, and at constant temperature and the pressure referred to, the compositions of the five coexisting divariant phases have unique values, independent of H_2O content. Depending on the H_2O content, liquids of composition on the Rl–H_2O join may encounter the quinary boundary curve through crystallisation of liquids which coexist with: Pl + Af + Q; Pl + Af; or Pl + Af + V (in Fig 22); Pl + Af + Q; Af + Q; Af + Q + V; Af + V; or Pl + Af + V (in Fig. 23). In both Figs 22 and 23 the silicate liquids participating in the phase assemblage Pl + Af + Q + L + V lie along the quinary boundary curve for only a short temperature interval above the solidus. This was also found to be the case for the other three compositions studied by Whitney at both 2 and 8 Kbar.

Although the compositions of the coexisting phases in the five-phase region Pl + Af + Q + L + V (Figs 22 and 23) are independent of X_{H_2O} and a function of temperature only, at the respective constant pressure, this is not expected to be the general case for regions involving the coexistence of less than five phases. In each of these regions of Figs 22 and 23 the composition of the coexisting phases will depend on *both* temperature and X_{H_2O}, and in Fig. 24 on both pressure and X_{H_2O}. Sufficient data are not available to define details of such dependence.

Whitney (1972, 1975) used the diagram of Fig. 24 to discuss resorption of quartz in granitic magmas which are moving upward in the crust at essentially constant temperature. This figure also provides a graphic demonstration of the interrelated effects of pressure and of H_2O as a component.

The points marking the intersection of curves in Figs 22, 23 and 24 refer to intersections of the particular join, Rl–H_2O, with elements of the phase relations in the quinary isobaric (isothermal for Fig. 24) phase diagram. For example, the point in Fig. 22 at which the Pl + Af + Q + L, Pl + Af + Q + L + V, Pl + Af + L and Pl + Af + L + V fields meet refers to a specific point on the Rl–H_2O join at which this join intersects the Pl + Af + L limit of the Pl + Af + Q + L + V phase (hyper) volume. In general the points involving the intersections of curves in the phase assemblage diagrams obey restrictions similar in origin to the commonly used "Schreinemakers Rules" employed in the pressure–temperature projections of invariant and univariant equilibria. The curves have a discontinuity in slope at the point of intersection, and the relationship of stable and metastable extensions of the curves can easily be derived through consideration of simple ternary and quaternary equilibria.

Intersecting curves, such as shown in Figs 22, 23 and 24, are expected to be characteristic of a wide variety of silicate–H_2O joins in the haplogranodiorite system. Extrapolations of Whitney's (1972, 1975) data indicate that such intersections are likely for the majority of common granitic compositions as given in terms of normative ab:or:q values.

A key feature in comparing the phase relations in the haplogranodiorite system with those of the haplogranite system is related to the importance of $CaAl_2Si_2O_8$ in the feldspar crystalline solutions. Seck (1971) has obtained data in the $NaAlSi_3O_8$-$KAlSi_3O_8$-$CaAl_2Si_2O_8$-H_2O system at 650°C and 1, 5 and 10 Kbar on the compositions of the coexisting ternary feldspars that illustrate the effect of anorthite on the alkali feldspar solvus. The wide compositional range in the ternary system over which two feldspars coexist is in marked contrast to the solvus relations in the alkali feldspar system. Seck's (1971) data provide an experimentally based confirmation for the many previous objections to the use of the "feldspar geothermometer" proposed by Barth (1951, 1962) by demonstrating that the effect of a pressure increase of 10 Kbar is equivalent to a decrease in temperature of 125–150°C on the distribution coefficient.

Presnall and Bateman (1973) have provided a thorough discussion of the fractional fusion process involving vapour-saturated liquids in the haplogranodiorite system. The application to the rocks of the Sierra Nevada Batholith by these authors is on somewhat less firm grounds than is the theoretical treatment. These authors were concerned solely with the vapour-saturated liquidus relations, and it is clear from Whitney's (1972, 1975) study that such relations are appropriate only to compositions which have a large amount of water present, or where temperature is only slightly above the solidus.

In view of the importance of anorthite as a modifying component on phase relations in the haplogranite system it may seem surprising that there are so few experimental data available on composition relations in the haplogranodiorite system. In large part this is a reflection in the difficulties in determining the compositions of the coexisting phases in the experimental studies. In the case of plagioclase, composition determination by indirect means using X-ray diffraction techniques is not feasible due to the small changes in unit cell parameters with composition. Optical techniques are potentially more useful; however, the crystal size of the plagioclase in the experimental products is often less than 5 μm. It is hoped that widespread use of the electron microprobe on both liquid and crystalline phases will provide some of this vitally needed information in the future.

In summary, it is possible to draw several conclusions regarding phase relations in the haplogranodiorite system. *First*, the quinary boundary curve in the system is of importance primarily in the later stages of crystallisation, or the beginning of melting, for typical granite to tonalite compositions. Much of the crystallisation history of such compositions involves liquids which may be far removed, in composition, from such liquids. *Second*, the vapour-under-saturated liquidus relations at high pressure bear no simple, direct relationship to the vapour-saturated liquidus relations based on data obtained at a lower pressure. *Third*, a key feature in contrasting phase relations in the haplogranite and haplogranodiorite systems is the much larger primary field of plagioclase in the latter as contrasted with the field of albite in the former.

Femic Components

Biotite and hornblende are the common ferromagnesian minerals of the granitic rocks and are frequently accompanied by magnetite and ilmenite. Locally other phases including ortho- and clino-pyroxene, fayalitic olivine, alkali amphiboles or pyroxenes may be important. In treating the phase relations involving the iron bearing phases a firm understanding of the role of oxygen fugacity, f_{O_2}, is essential. Primary references regarding the role of f_{O_2} with particular reference to the granitic rocks are Eugster (1957), Eugster and Wones (1962), Wones and Eugster (1965), Eugster and Skippen (1967), Shaw (1967), and Wones and Gilbert (1969) among the many treatments. The cautionary note of Whitney (1972a) regarding the treatment of f_{O_2} when the silicate liquid is not saturated with respect to an aqueous vapour phase is of particular importance.

Each of the ferromagnesian phases listed above is a crystalline solution which may require several components for adequate description of compositional variations. Rarely are these components directly related to the components of the haplogranite or haplogranodiorite systems. This factor precludes treatment of phase relations involving these ferrogmagnesian phases by a method related in a straightforward fashion to the previously discussed liquidus and solidus equilbria. Consequently the approach taken here will emphasise the data on the stability relations of the individual phases, and evaluate, insofar as possible, the potential for existence as a phase participating in liquidus–solidus relations of the haplogranite and haplogranodiorite systems.

It seems worthy of note at the outset that the experimental data are generally consistent with the conclusion that granitic liquids (excepting those that are strongly peralkaline) are capable of dissolving only small amounts of femic constituents at temperatures near the solidus. These (sparse) data support the concept of Bowen (1937), based on equilibria in the anhydrous systems, that

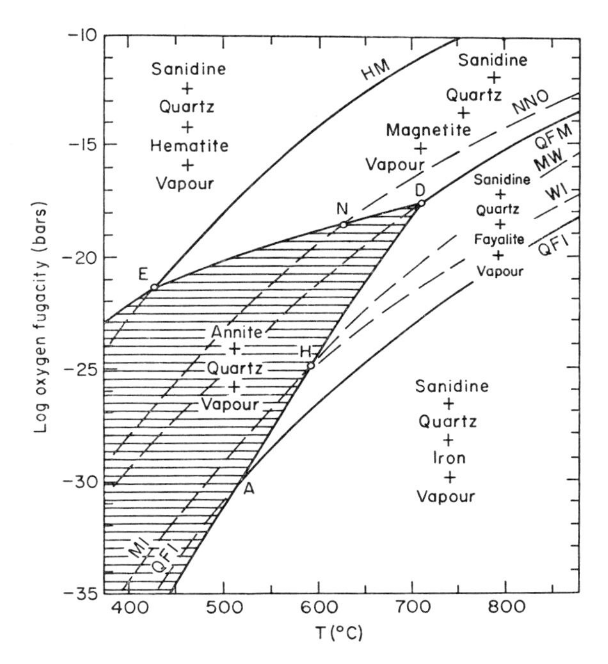

Fig. 25. Annite + Quartz Stability. From Eugster and Wones (1962, Fig. 6, p. 105). "Phase relations of annite + 3 quartz (+ vapour) bulk composition as a function of oxygen fugacity and temperature at constant total pressure of 2070 bars. H, D, N, E are experimentally determined points. QFI, WI, MI, MW, QFM, NNO, HM are oxygen buffer curves."

the system $KAlSiO_4$-$NaAlSiO_4$-SiO_2 is a "residua" system. At present the silica saturated portion ($NaAlSi_3O_8$-$KAlSi_3O_8$-SiO_2) of Petrogeny's Residua System is of principal concern.

Biotite. Eugster and Wones (1962), Wones and Eugster (1965), Luth (1967), Kushiro and Yoder (1969), and Rutherford (1969) have reported on a variety of reactions which limit end-member stability or stability of biotite crystalline solutions. Of principal concern in the present context are data which permit analysis of stability in the presence of quartz, alkali feldspar and plagioclase. Unfortunately such information is not available other than through rather long extrapolations of the experimental data.

Eugster and Wones (1962) determined the stability of annite ($KFe_3AlSi_3O_{10}$-$(OH)_2$) in the presence of quartz as the same as annite alone at oxygen fugacities above the quartz–fayalite–magnetite (QFM) buffer curve at 2 Kbar. However the presence of quartz resulted in a marked reduction in the stability of annite at lower f_{O_2} through formation of fayalite. These relations are indicated in Fig. 25 (Fig. 6 of Eugster and Wones, 1962). The solidus temperature for the haplogranite system at this pressure is approximately 690°C, suggesting that annite is not a likely solidus/liquidus phase in the haplogranite + iron oxide system at this pressure. Wones and Eugster (1965) demonstrated that as the Mg/Fe ratio of the biotite increases in the phlogopite–annite crystalline solutions so does the maximum thermal stability of the biotite. They presented a schematic diagram illustrating the quartz + biotite stability relations but no direct experimental data. Luth (1967) demonstrated the capability of the coexistence of the phlogopite end-member with a silicate liquid and sanidine in the $KAlSiO_4$-Mg_2-SiO_4-SiO_2-H_2O system at pressures greater than 500 bars. Rutherford (1969) examined the annite + quartz (and annite) stability at 2 Kbar in the system $KAlSi_3O_8$-$NaAlSi_3O_8$-Fe-O-H. He found that crystalline solution of the Na analogue of annite in the annite phase reduced the stability of the annite + quartz assemblage. Rutherford (1969) concluded that it was unlikely that the iron–biotites (Mg-free) could coexist with granitic liquids at pressures less than 4 Kbar. The importance of Na–K crystalline solutions in biotites is also suggested by the synthesis of Na–annite (Weidner and Carman, 1968) and by Carman's (1969, 1974) studies on the Na analogue of phlogopite.

Biotites in the granitic rocks typically (for example Dodge *et al.*, 1969) are complex Mg, Fe^{2+}, Fe^{3+}, Al crystalline solutions with $Mg/Mg + Fe^{2+}$ ranging from about 0.3 to 0.6, and with Na/K approaching 0. Considering the effect of Mg/Fe^{2+} ratio on annite + quartz stability, and the effect of Na annite crystalline solution, it seems quite reasonable to expect biotite as a phase participating in solidus–liquidus equilibria in the haplogranite + Fe + Mg system at pressures greater than 1 Kbar. The details of the compositional relations involving coexisting biotite and haplogranite liquids are essentially unknown. However, the experimental studies on natural granites, to be discussed in detail later, demonstrate that biotite does participate in the solidus–liquidus equilibria for a wide range of granitic compositions at pressures greater than about 1 Kbar.

On the basis of an analogy with the results of Luth (1967) on the Na and Fe free system it is expected that biotites coexist with a wide range of vapour saturated, and vapour undersaturated haplogranitic liquids, possibly as a primary phase on the vapour-saturated, or vapour-undersaturated liquidus. In the Na and Fe free system the liquids which coexist with sanidine + quartz + vapour become saturated with respect to the magnesian phases enstatite or phlogopite with only slight amounts of the Mg component present.

Wones and Eugster (1965) noted the importance of reactions of the type

$$\underset{\text{Biotite}}{K(Mg, Fe)AlSi_3O_{10}(OH)_2} + \underset{\text{quartz}}{3SiO_2} = \underset{\text{sanidine}}{KAlSi_3O_8} + \underset{\text{orthopyroxene}}{3(Mg, Fe)SiO_3} + \underset{\text{water}}{H_2O}$$

for the intermediate composition biotites. To a first approximation, such reactions separate assemblages characteristic of the charnockitic rocks from the more familiar biotite bearing equivalents.

Wones and Eugster (1965) provided an analytic expression (equation 6″) for the coexistence of quartz, magnetite, sanidine and biotite, in the presence of an H_2O rich fluid phase. The dependent variable was chosen as fugacity of H_2O (f_{H_2O}) and the independent variables were: 1/T(°K); mol fraction of annite in biotite; f_{O_2}; activity of $KAlSi_3O_8$ in potassium feldspar; and activity of Fe_3O_4 in magnetite. Application of such equations for the estimation of intensive parameters relative to natural occurrences requires a demonstration that the analogous phases present in the rock represent an equilibrium assemblage, and that none of the phases have undergone subsequent (to the time of equilibriation) chemical changes. Petrographic observations on granitic rocks suggest that these conditions are rarely achieved.

Hornblende. Experimental studies on the amphiboles, even more than in the case of the biotites, have been concentrated on stability limits of the end-members. Boyd (1959) and Ernst (1968) have reviewed these data, and several important recent studies have continued the approach. In the context of solidus–liquidus relations pertinent to the granitic compositions these data are of restricted applicability. Of principal importance in the present discussion are the stability relations in the presence of quartz but data are notably lacking.

On the basis of the available analytic data on natural occurrences it seems that the common hornblendes of the granitic rocks can be represented in terms of a range of crystalline solutions with end-members: pargasite–ferropargasite ($NaCa_2(Mg, Fe)_4AlSi_6Al_2O_{22}(OH)_2$); magnesiohastingsite–hastingsite ($NaCa_2(Mg, Fe)_4Fe^3Si_6Al_2O_{22}(OH)_2$); and tremolite–ferrotremolite ($Ca_2(Mg, Fe)_5Si_8O_{22}(OH)_2$ to a first approximation. The pargasite–ferropargasite and hastingsite–magnesiohastingsite end-members are silica deficient, the decomposition products involving phases such as olivine, nepheline and spinel. The tremolite–ferrotremolite series is slightly silica-excess. Indeed CIPW norms calculated from analyses of amphiboles in granitic rocks typically show a deficiency of silica. It may seem that this factor would suggest that such phases are not likely in a high silica environment such as a granitic rock. However, the evidence is overwhelmingly in the opposite sense on the basis of field occurrences. An analogy may be drawn with the Mg end-member of the biotite series, phlogopite. Phlogopite is also a silica-deficient phase with decomposition products involving forsterite, leucite, and kalsilite. The phlogopite + quartz assemblage is stable over a large P–T range, limited (in part) by a reaction involving as the products enstatite and sanidine.

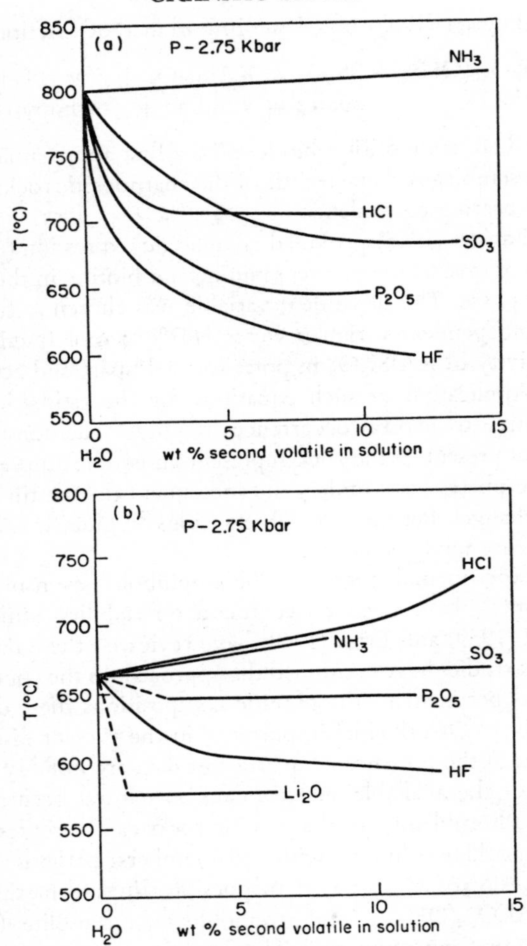

Fig. 26. Effects of a Second Volatile Component on Solidus Temperatures in the Albite–H_2O and Granite–H_2O Systems. (a) Albite–H_2O. (From Wyllie and Tuttle, 1964, Fig. 1, p. 931) "The temperature of beginning of melting in the systems $NaAlSi_3O_8$-H_2O-X, where X is SO_3, P_2O_5, HCl, HF, or NH_3, at a constant pressure of 2750 bars. Mixtures contained 50 wt % of albite and 50 wt % of solution (H_2O + X)." (b) Granite–H_2O. (From Wyllie and Tuttle, 1964, Fig. 3, p. 936). "The temperature of beginning of melting in the "systems" granite-H_2O-X, where X is SO_3, P_2O_5, HCl, Li_2O, HF, or NH_3, at a constant pressure of 2750 bars. Mixtures contained 50 wt % of granite and 50 wt % of solution (H_2O + X)."

Further discussion of hornblende stability in the solidus–liquidus region of the granitic rocks will be postponed to the section dealing with melting/crystallisation behaviour of natural materials. Direct experimental data on "simple" synthetic systems which are appropriate in this context are simply not available.

The relationships involving arfvedsonite–riebeckite amphiboles pertinent to the peralkaline granites have been discussed previously in connection with Bailey's (1969) experimental study of acmite stability. Ernst (1960) demonstrated that magnesioriebeckite was stable under magmatic pressure temperature conditions and noted that because the decomposition products included phases not saturated with respect to silica, the stability of magnesioriebeckite + quartz would be lower than that of riebeckite. He also demonstrated that riebeckite has a significantly lower maximum thermal stability than magnesioriebeckite, and that in the case of riebeckite the decomposition products were silica saturated.

Pyroxenes. In general, the pyroxenes are relatively minor constituents of the granitic rocks. Augite is occasionally found either as discrete grains or as cores within hornblende crystals in the granodiorites, and is a rather common phase in the quartz diorites. Orthopyroxenes are the characteristic ferromagnesian mineral of the charnockitic rocks, and alkali pyroxenes of the acmite, aegirine–augite group are the typical pyroxenes of the peralkaline granites. The experimental data on the pressure–temperature–composition relations involving these phases, as coexisting with granitic liquids are essentially lacking for appropriate synthetic systems.

In general, the experimental studies dealing directly with the stability relations of the ferromagnesian minerals pertinent to solidus–liquidus equilibria involving granitic compositions are of very limited extent. This lack of data is largely a reflection of the dramatically increased chemical complexity of the synthetic system involving these phases, and the extensive crystalline solubility exhibited within each group. At present, the major source of information concerning the mafic minerals as related to granitic compositions is derived from the studies of natural compositions. These results will be discussed in a subsequent section.

Volatile Components in Addition to H_2O

Wyllie and Tuttle (1960) discussed the generalised effects of a second volatile component on solidus and liquidus equilibria using the $NaAlSi_3O_8$-H_2O system as a base. This treatment included a comparison of their previous data (1957) on the effects of CO_2, in addition to H_2O, on the beginning of melting temperatures of feldspars and granite with the generalised model. This (1960) treatment was followed by reports (1961, 1964) dealing with the effects of NH_3, HF, SO_3, P_2O_5, HCl and Li_2O, as second volatile components on solidus reaction temperatures in the albite–H_2O and Westerly Granite–H_2O systems. Their results are summarised in Figs 26(a) and 26(b) (from Wyllie and Tuttle, 1964, Figs 1 and 3). Of principal interest in this suite of studies was the effect of a second volatile component on solidus reaction temperatures and the implications regarding the relative (to H_2O) solubility of the second volatile component in the silicate liquid.

CO_2 seems to be a particularly important volatile component in magmatic processes (White and Waring, 1963), and the following treatment will be primarily concerned with this component. The halogens, particularly Cl and F, are also of considerable importance; however the available data are somewhat less directly applicable in the present context.

Millhollen *et al.* (1971) reported experimental results on the pressure–temperature co-ordinates of the solidus reaction in the $NaAlSi_3O_8$-H_2O-CO_2 system along the albite–1:1CO_2:H_2O join. They suggested that the solubility of CO_2 in the silicate liquids is very low, in accord with the results of Wyllie and Tuttle (1959), and Millhollen (1971). Khitarov and Kadik (1973) reviewed the data on the solubility of CO_2 in basaltic and granitic hydrous liquids and also concluded that the solubility was quite low, an order of magnitude less than H_2O in the 3–5 Kbar range. These results indicate that in silicate–H_2O–CO_2 systems, analogous to granitic compositions, the principal expected effect of CO_2 is to increase the solidus temperature, and affect the liquidus relations only slightly.

Swanson (1974) experimentally evaluated the effect of CO_2, in addition to H_2O, on the solidus–liquidus relations for the same silicate compositions (page 382) used by Whitney in his study of the four silicate–H_2O joins. Swanson reported results at the same pressures, 2 and 8 Kbar, and for the silicate–96% $H_2O \cdot 4\%CO_2$ join. Although the maximum solubility of CO_2 in the silicate liquid was not obtained in this study, solubility of up to 0.5 wt %CO_2 at 8 Kbar, and 0.3% at 2 Kbar, was found for compositions on these joins. The low solubility of CO_2 in the silicate liquids results in the expected slight increase in solidus temperatures relative to the pure H_2O case. However, both the vapour-saturated and the vapour-undersaturated liquidus relations were changed significantly. This results in a marked change of the temperature interval over which specific phase assemblages are stable at a given $H_2O(+CO_2)$ value. This is illustrated by comparing the results obtained for the synthetic granite (R1)–H_2O join at 2 Kbar (Fig. 22) as obtained by Whitney (1972, 1974) with those on the R1–96$H_2O \cdot 4CO_2$ as obtained by Swanson (1974) at the same pressure. Swanson's results are shown in Fig. 27 with the results of Whitney given by the continuous lines (from Fig. 22). As noted previously the solidus temperature is increased only slightly on addition of CO_2, and the vapour-undersaturated liquidus relationships at low $H_2O(+CO_2)$ content are also changed only slightly. However, the vapour-undersaturated liquidus relations at higher $H_2O(+CO_2)$ values and the vapour-saturated liquidus relations are changed dramatically. The liquidus temperatures are increased nearly 100°C in the case involving CO_2. The temperature-composition region over which plagioclase coexists with a vapour-undersaturated liquid is increased in the CO_2 bearing system, largely at the expense of the alkali feldspar + plagioclase + liquid region. Although direct analytical data are lacking, it is expected that the plagioclase-liquid tie lines are somewhat different in the two cases for a given T–$X_{H_2O(+CO_2)}$ value. Swanson's (1974) results on these four composition joins at the two pressures are consistent with low solubility of CO_2 in the hydrous silicate liquid; however, they also demonstrate that the small amounts of CO_2 are not negligible relative to liquidus relationships and the sequence of crystallisation. The low solubility of CO_2 in these liquids is supported by the presence of calcite coexisting with the vapour-undersaturated and vapour-saturated hydrous silicate liquids at temperatures less than 730°C at 8 Kbar on these four joins.

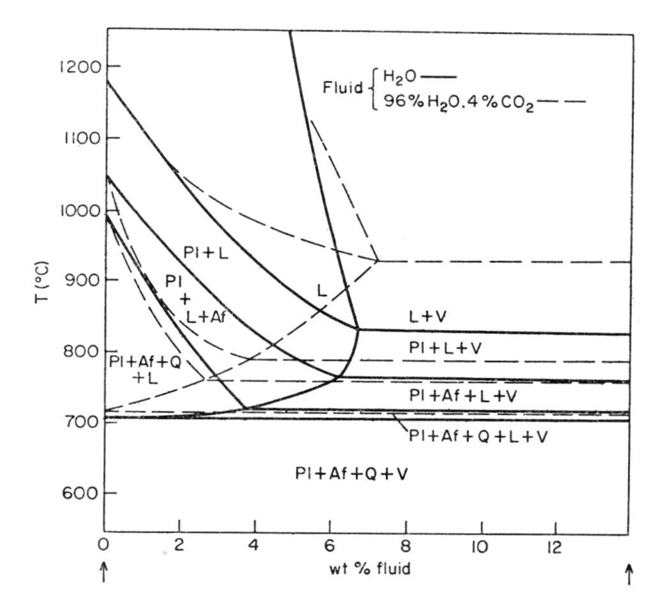

Fig. 27. Comparison of Effects of H_2O and $H_2O + CO_2$ on Liquidus Phase Relations of the Synthetic Granite (RI) Composition at 2 Kbar. Phase assemblages as a function of temperature and wt % fluid (H_2O or $H_2O + CO_2$) at 2 Kbar. Solid lines refer to the H_2O system as do the designated phase assemblages (see Fig. 22) from Whitney (1974). Dashed lines refer to the $H_2O + CO_2$ system from Swanson (1974) and the phase assemblages are in the same sequence as in the case of H_2O.

A large number of other workers have contributed to the data on silicate-H_2O-CO_2 systems, and in general to our knowledge of systems involving a second volatile component. The majority of these studies are somewhat less directly applicable to solidus–liquidus relations appropriate to granitic compositions. Burnham (1967), Burnham and Jahns (1962) and Jahns and Burnham (1969) have provided thorough discussions of the later stages of the crystallisation process, particularly with respect to pegmatite genesis and the role of halogens.

(c) NATURAL SYSTEMS

The early experimental studies (Goranson, 1931, 1932) on natural granitic compositions in the presence of H_2O were primarily concerned with evaluation of the solidus (beginning of melting) reaction co-ordinates in terms of pressure and temperature, and determination of the maximum solubility of H_2O in granitic melts. These data have been reviewed and extended by many workers, in particular Tuttle and Bowen (1958), Jahns and Burnham (1958), Kranck and Oja (1960), Winkler and von Platen (1961a, b), Burnham and Jahns (1962), Brown (1963) and many others more recently. For the most part these studies dealt with discrete specimens and the principal emphasis was on the solidus reactions and, to a lesser extent on the details of the vapour-saturated liquidus relations. Piwinskii and Wyllie (1968) have summarised these results and their implications regarding compositions of the granitic liquids. Experimental studies on individual specimens have been the subject of many additional studies in the past decade. These studies, significant in their own right, are so numerous as to prohibit effective review in this chapter.

Wyllie (1960, 1963) proposed experimental studies of genetically related suites of specimens from specific petrographic associations as a keystone for the further development of petrologic theory. He and his co-workers have been actively engaged in such studies during the past decade. Selected results from these extensive and intensive studies provide the basis for this section of the review.

The experimental studies of suites of genetically related rock specimens have provided much data of importance regarding the *possible* crystallisation history of such specimens. The word "possible" is stressed because direct application of the experimental results to the origin and development of the petrographically related suite of rocks is subject to several limitations. *First*, each specimen is treated as a closed system, other than with respect to H_2O. In evaluating the crystallisation history of the natural material relative to a closed system, equilibrium-based model, direct petrographic observations relative to mineral zoning, reaction relationships, and post-magmatic reactions may be such as to prohibit application of the closed system model. *Second*, considerable caution is required in assuming that a silicate composition, as represented by the rock specimen, was completely liquid at any given stage in its history. If such an assumption is made, it must be consistent with the field and petrographic observational data for the specimen and its relationship to surrounding rocks. *Third*, experiments involving vapour-saturated liquidus relationships may provide little information of use on the crystallisation sequence, or history, if vapour-undersaturated liquidus relations are involved. In this context it is also necessary to consider the effects produced by reducing the activity of H_2O in the liquid through dilution of a coexisting vapour phase with additional volatile components. In general these effects would not be expected to be the same as produced through the absence of a vapour (H_2O + additional components) phase. Textural relationships in granitic rocks of other than the epizone seem to indicate (to me at least) that either the major portion of the crystallisation history has taken place in the absence of a separate vapour phase, or that subsequent post-magmatic recrystallisation has eliminated the evidence for the presence of a vapour phase early in the magmatic history. It is worth remembering that many rhyolitic obsidians, which are samples of granitic *liquids*, have low contents of H_2O (see also Bailey (IIB2), this vol.).

Even with these limitations in mind, the experimental studies on natural granitic rocks have provided much information pertinent to the origin and development of these rocks. It is possible to treat only a few of the many and important studies on this subject in this chapter and the selection process is certain to be unsatisfactory for any specific purpose. The studies selected for review involve only 20 of the large number of compositions studied. For the most part we will be concerned with rocks of the granitic batholiths, ranging from quartz diorite to true granite in composition.

Piwinskii (1968, 1973a, 1973b, 1974) and Wyllie (1968, 1970) are the principal authorities on melting relations of batholithic granites, particularly those of the western United States. These studies have been selected primarily because of the quantity and quality of the experimental results, and because they represent a self-consistent body of experimental data. In addition, data are available on crystallisation sequence, relative to vapour-saturated liquidus relations, over the range from 1 to 10 Kbar for a variety of these samples. The experimental studies of Whitney (1969), Robertson and Wyllie (1971b) and Stern and Wyllie (1973) are of particular importance regarding the vapour-undersaturated melting and crystallisation relations of granitic compositions at low and high pressure. McDowell and Wyllie (1971) have provided experimental data on the vapour-saturated melting relations of the peralkaline granites for contrast with the granites of the batholithic calc-alkaline suite. Huang and Wyllie (1973) have determined the melting relations for a muscovite granite at pressures to 35 Kbar, including data on the vapour-undersaturated liquidus relations at 15 Kbar.

To provide a convenient frame of reference the normative compositions of the specimens to be discussed are given in Table 3 and illustrated in Figs. 28(a) and 28(b). It is clear that a very wide range of compositions is to be considered, ranging from quartz diorite to granite and including both syenites and syeno-diorite.

Piwinskii and Wyllie (1968, 1970) reported on melting relations of nine related granitic compositions from the Wallowa Batholith, Oregon. Five specimens, three granodiorites and two tonalites, were from the main body of the Needle Point Pluton (1968). Four specimens, a granite, two quartz monzonites, and a granodiorite, were from a late stage felsic body intrusive into the main body of the Needle Point Pluton. Data were reported for 1, 2 and 3 Kbar relative to vapour saturated liquidus relations. Although oxygen fugacity was not controlled, magnetite was a phase participating in the entire range of solidus–liquidus relations. Presumably the oxygen fugacity was not far removed from that appropriate to the Ni–NiO buffer curve. The greatest amount of definitive data is available at 2 Kbar and the crystallisation sequences for these nine compositions are shown in Fig. 29 for this pressure. Magnetite is the primary phase on the vapour-saturated liquidus, and was present in all experiments reported. In the progression from granite to quartz monzonite to granodiorite to tonalite the solidus temperature increases by about 70°C. Examination of Fig. 29 suggests that this is largely controlled by K-feldspar participation (or lack thereof) in the beginning of melting reaction. The constancy of temperature at which quartz, K-feldspar, and biotite join the stable phase assemblage on

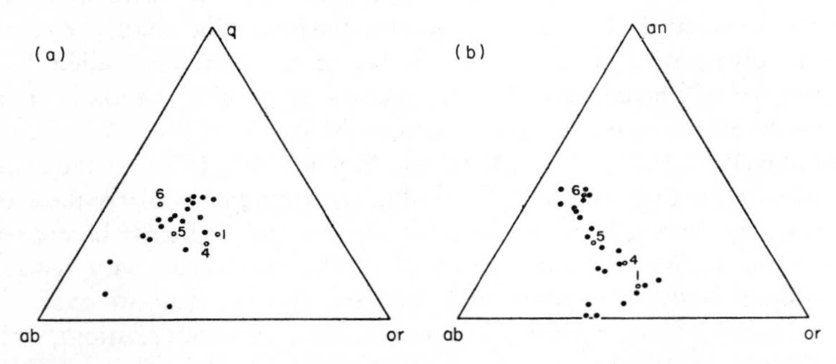

Fig. 28. Normative ab, or, an, q Compositions of a Series of Granitic Rocks which have been Studied Experimentally. Normative compositions given in Table 3 for samples indicated by filled circles. Compositions used by Whitney (1972, 1974) in the Haplogranodiorite system shown as open circles. (a) Normative ab–or–q Triangular Diagram. (b) Normative ab–or–an Triangular Diagram.

crystallisation is remarkable in view of the wide variety of compositions considered (Fig. 28, Table 3). This is contrasted with the behaviour of plagioclase, and to a lesser extent hornblende, which depend to a significant degree on bulk composition. Without direct analytical data or accurate determination of proportions and compositions of the phases it is not possible to evaluate the variation in composition of the liquids which participate in the vapour-saturated liquidus equilibria. Unfortunately, this information may be necessary if the interpretation of the field and petrographic data indicate a need for analysis in terms of a fractional crystallisation or fusion model.

Piwinskii (1973a) reported experimental data of vapour-saturated phase relations at pressures of 1–10 Kbar for three quartz diorites, a granodiorite, and a quartz monzonite from the Central and Southern Coast Ranges of California. As in the previous study magnetite was a phase participating in the equilibria over the entire range studied. The sequence of crystallisation for these compositions is shown in Fig. 30 for both 2 and 10 Kbar, relative to the vapour-saturated liquidus. There is obviously a major pressure effect on the sequence of crystallisation, particularly with respect to the plagioclase, quartz and biotite relationships. This is shown in Fig. 31 (Fig. 2 of Piwinskii, 1973a) for enhanced clarity. The negative slope of the plagioclase, hornblende, and biotite "out" curves and the reversal of slope of the quartz "out" curves are consistent features in these P–T diagrams. With respect to Figs. 30 and 31 it is clear that at low pressure the crystallisation is dominated by plagioclase and hornblende as the liquidus phases, while high pressure crystallisation is dominated by the hornblende and quartz liquidus relations. Examining Fig. 30 relative to the crystallisation sequence at low pressure (2 Kbar), we again note the relative constancy of

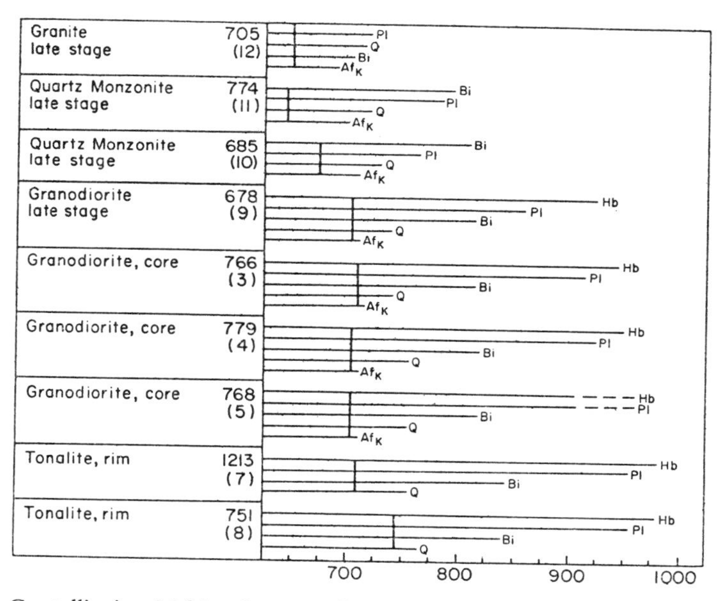

Fig. 29. Crystallisation–Melting Sequence for Rocks of the Needle Point Pluton. Experimental data at 2 Kbar from Piwinskii and Wyllie (1968, 1970). Compositions given in these references and, in part, in Table 3. Numbers in parentheses refer to numbers in heading of Table 3. Magnetite and vapour present in all experiments. Pl = plagioclase, Hb = hornblende, Bi = biotite, Q = quartz, Af_K = K-feldspar. Piwinskii (1973b, p. 202) suggests that the biotite relations may be in considerable error through difficulty in distinguishing between primary and quench biotite. Vertical bar shows the solidus.

the temperature at which quartz, biotite and K-feldspar join the stable phase assemblage on crystallisation (ignoring the results on the Santa Lucia quartz diorite which is characterised by an absence of modal K-feldspar). This pattern, also found at 1 Kbar in both studies, is much less discernible at the higher pressures (Fig. 30).

Piwinskii (1973b, 1968) studied the melting relations involving vapour-saturated liquids of a tonalite, two granodiorites and a biotite granite from the

Sierra Nevada Batholith. The crystallisation sequence for these four compositions at 2 and 10 Kbar are shown in Fig. 30. Magnetite was present in all experiments. As in the previous two cases the relative constancy of temperature at which quartz, biotite and K-feldspar begin to appear in the crystallisation sequence at 2 Kbar is notable. Also similar to the rocks of the Coast Ranges is the dominance of hornblende in the high pressure crystallisation sequence as contrasted with the simultaneous effects of hornblende and plagioclase at low pressure.

To anyone who is familiar with the complexities of crystallisation in relatively simple silicate systems of three and four components, the relative

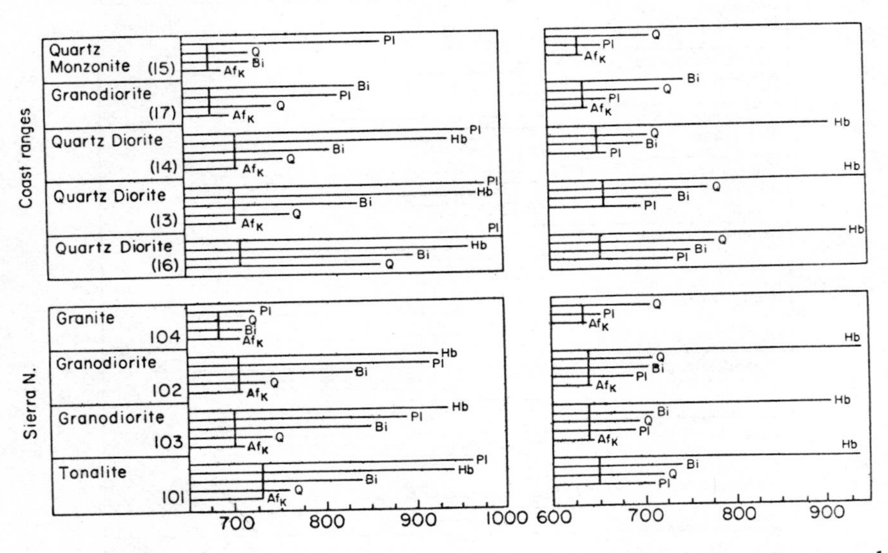

Fig. 30. Crystallisation–Melting Sequence for Rocks of the California Coast Ranges and the Sierra Nevada Batholith. Symbols and nomenclature as in Fig. 29. Results at 2 Kbar shown on the left for both suites and results at 10 Kbar shown on the right. Magnetite and vapour present in all experiments. Experimental data from Piwinskii (1968, 1972a, 1973b). Compositions given in original references and, in part, in Table 3.

constancy of the temperature at which quartz, biotite, and K-feldspar join the phase assemblage for this extremely wide range of compositions in (at least) a seven component system is astonishing. Certainly there is some variation in temperature of appearance with bulk composition; however if random compositions through the seven component system were selected it is a certainty (to me at least) that much wider variation in sequence and temperature of crystallisation would be the case. The only possible explanation that I can see is that the compositions are not random in space, but are systematically arranged such that the analogues in hyperspace to the boundary curves involving liquids which coexist with: plagioclase, hornblende, and biotite (+ vapour); plagioclase, hornblende, biotite and quartz (+ vapour); plagioclase, hornblende, biotite, quartz and K-feldspar (+ vapour) are encountered on crystallisation at very similar temperature composition values for an extremely wide range of "granitic" compositions. For this reason it would be extremely desirable to have direct analytic information on the liquid compositions which coexist with these phases. The preceeding argument suggests even more evidence than the frequently cited abundance plots of normative albite, orthoclase and quartz

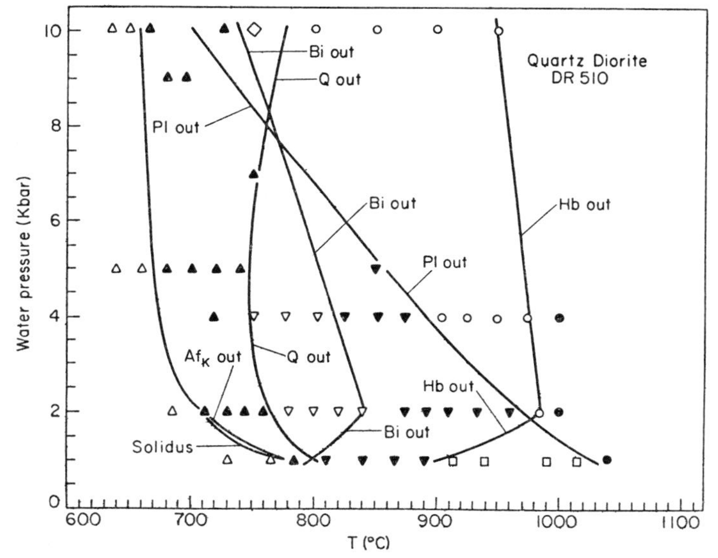

Fig. 31. Vapour-Saturated Liquidus Relations for the Bodega Head Quartz Diorite to 10 Kbar. From Piwinskii (1973a, Fig. 2, p. 114). "Water-pressure-temperature projection of the phase boundaries for Bodega Head quartz diorite DR 510 in the presence of excess water."

for the profound role of crystal-liquid equilibria in controlling granitic compositions. This is also consistent with the relatively small variation in solidus temperature for such a wide range of compositions that involve widespread crystalline solution among all but one (quartz) of the phases participating in the equilibria. Although the argument given above is based on the 2 Kbar vapour-saturated data, further examination of the cited references clearly indicates validity at 1 Kbar and, to a lesser extent, at 10 Kbar.

The experiments referred to in the preceding discussion of natural materials all pertain to phase relations determined in the presence of an aqueous vapour phase. We have seen, in the case of the synthetic haplogranite and haplogranodiorite systems that vapour-undersaturated liquidus relations present a new set of problems and we must inquire as to the analogous effects on the natural systems.

Evolution of thought regarding the need for direct experimental data on phase relations involving vapour-undersaturated liquids in natural systems has been similar to that in synthetic systems. It was originally assumed that knowledge of vapour-saturated liquidus relations as determined at low pressure would be directly applicable to phase relations involving vapour-undersaturated liquids at higher pressures. As time progressed, and the amount of data increased, reliance on this approach has diminished and pleas for more direct data on the vapour-undersaturated liquidus relations have increased. The evolution of thought can be best illustrated by direct quotes from the literature.

Piwinskii and Wyllie (1970) in summarising their results pertaining to the granitic rocks of the Needle Point Pluton (*op cit.*, 1970): "From the excess water results, it is possible to make reasonable estimates of phase relationships under conditions with $P_{H_2O} < P_{total}$, and Fig. 11 can therefore be used as a guide to the phase relationships with $P_{H_2O} = 2$ Kbar, but with load pressure of 3, 5 or even 10 Kbar." Figure 11, referred to in this quotation, is a plot of temperature versus differentiation index (Thornton and Tuttle, 1960) illustrating the sequence of melting, or crystallisation, for granitic rocks of the Needle Point Pluton at 2 Kbar for the vapour-saturated liquidus relations.

McDowell and Wyllie (1971) in comparing their results on the rocks of the Kungnat Syenite Complex, Southwest Greenland, with those of Piwinskii and Wyllie (1968, 1970) on the rocks of the Needle Point Pluton:

"The experimentally determined temperature—D.I. diagrams (water-excess) are consistent with the following statements, although they do not prove them: (1) rocks in the investigated calc-alkaline series are related to each other directly through liquid-crystal reactions, and (2) the Kungnat syenites and granites are not related directly to each other through liquid-crystal reactions.

The magmatic reactions involved in the formation of these rocks would have occurred in the absence of a vapor phase (except in the late stages of crystallization) but the regularity or irregularity of phase relationships with excess water can be extrapolated to vapor-absent conditions. . . . These experimental results can be related to the petrogenesis of the rocks only with due caution and the realization that the experiments represent a limiting situation with water present in excess, and with the oxygen fugacity buffered by the apparatus." (*op. cit.*, pp. 190 and 192.)

Robertson and Wyllie (1971b) in discussing their results on the rocks of the Deboullie Stock, North Maine with particular reference to studies involving vapour-undersaturated liquidus equilibria:

"Work at higher temperatures to bracket additional phase boundaries, particularly for the hydrous minerals, is essential for precise determination of the dependence of crystallisation and fusion sequence on H_2O content as outlined in figure 4. Water-deficient studies for other compositions, representative of both crust and mantle, are needed to enlarge our knowledge of magma generation and crystallization." (*op. cit.*, p. 570.)

Piwinski (1973a) in summarising his results on the granitic rocks of the Coast Ranges, California: "Because the experimental investigation was undertaken in the presence of excess water, no direct link can be established with physico-chemical processes occurring within the earth." (*op cit.*, p. 125.) An essentially identical statement appears in Piwinskii's (1973b) summary of his results on rocks of the Sierra Nevada Batholith.

Stern and Wyllie (1973) in discussion of their extension of Piwinskii's (1968) study of melting relationships of the Dinkey Creek biotite granite from the Sierra Nevada Batholith:

"Only reactions near the vapor-saturated solidus in figure 2 can be considered as directly applicable to problems of magma genesis because of the high water content required to saturate the liquids and thus to approach the near liquidus phase relationships. Therefore, we have studied in addition the melting relationships of granite under water deficient conditions." (*op. cit.*, p. 165. (Figure 2 is a P–T diagram.))

Without belabouring the point further it seems clear that emphasis in recent experiments involves stressing the importance of vapour-undersaturated liquidus equilibria, and lack of direct relationship to vapour-saturated liquidus equilibria obtained at lower pressure, regarding experiments on natural materials. This is in accord with predictions based upon results obtained in the simple haplogranite and haplogranodiorite systems by Luth (1968, 1969), Jahns and Burnham (1969), and Whitney (1972).

Whitney (1969) and Whitney and Luth (1970) reported on melting relations of three non-genetically related granitic rocks from Cape Ann, Massachusetts; Quincy, Rhode Island; and Mount Airy, North Carolina; with data obtained at 2 Kbar involving both vapour-saturated and vapour-undersaturated liquidus relations. These studies involved temperatures of less than 800°C and provided little information on the high temperature vapour-undersaturated liquidus relations. Difficulties in demonstrating the attainment of equilibrium and problems related to definitive determination of the sequence of crystallisation over the rather small temperature interval at which a number of crystalline phases joined the assemblage suggest that extended discussion here is not warranted. However, these data were consistent with the predicted (Luth, 1968, 1969) change in the sequence of quartz, plagioclase and K-feldspar as liquidus phases for the Westerly Granite, similar to those discussed in connection with the haplogranodiorite system.

Robertson and Wyllie (1971a) discussed the pressure–temperature–H_2O content region involving vapour-undersaturated liquidus relations for rock-water systems, paying particular reference to one of the granodiorites from the Needle Point Pluton studied by Piwinskii and Wyllie (1968). The analysis was largely theoretical and treated the schematic phase relationships in considerable detail. They were particularly concerned about the role of hydrous minerals, such as biotite and hornblende, as coexisting with vapour-undersaturated liquids. They distinguished between four types of subsolidus assemblages in silicate-water systems. The definitions given are (*op cit*., p. 253):

Type I: *Water absent*. An assemblage of anhydrous silicate minerals with no vapour phase.

Type II: *Water-deficient and Vapour-absent*. An assemblage of silicate minerals which must include hydrous minerals, but with no vapour phase.

Type III: *Water-deficient and Vapour-present*. An assemblage of silicate minerals with or without hydrous minerals and with vapour phase. There is insufficient water present to saturate the liquid when the crystalline assemblage is completely melted at the existing pressure.

Type IV: *Water-excess*. An assemblage of silicate minerals, with or without hydrous minerals, and with at least enough water to saturate the liquid when the crystalline assemblage is completely melted at the existing pressure. A vapour phase is present.

The emphasis on the component water is perhaps unfortunate in the attempt to relate to natural or synthetic systems involving more than one volatile component, though it may be justified in view of the dominance of water, both quantitatively and in terms of affect on phase relations.

Figure 32 (Fig. 8 of Robertson and Wyllie, 1971b) serves to illustrate the general pattern for these four regions. As these authors were primarily concerned with the possible relationships of biotite and hornblende in the region involving vapour-undersaturated liquid they concentrated on this topic in their extrapolations to higher pressure. It is clear from Fig. 9 of Robertson and Wyllie (1971b) that the sequence of crystallisation from vapour-saturated liquids at high pressure (10 Kbar) was expected to be the same as at low pressure, other than for a reversal in sequence of plagioclase and biotite. Piwinskii (1973a, b) has shown that this is not the case for three granodiorites studied and has suggested (personal communication, 1974) that the behaviour of the granodiorite referred to by Robertson and Wyllie (1973) will be similar to the other granodiorites studied. Piwinskii's (1973a, b) results clearly show (Figs 30, 31) the importance of the decreasing temperature at which plagioclase, and the increasing temperature at which quartz, join the liquidus phase assemblage with increasing pressure. These data suggest that the extrapolations to higher pressure of the Piwinskii and Wyllie (1968) data at 1, 2 and 3 Kbar, as used by Robertson and Wyllie (1971b), are not valid regarding the sequence of crystallisation at high pressure. Analysis of Piwinskii's (1973a, b) experimental results at low and high pressure, and construction of diagrams similar to Fig. 32 at higher pressure, demonstrate that the sequence of crystallisation of plagioclase, quartz, hornblende and biotite from granitic liquids depends strongly on both pressure and water content of the bulk composition. Both hydrous and anhydrous phases are affected dramatically relative to the sequence of crystallisation.

Robertson and Wyllie (1971b) reported experimental results on four rock-water systems from the Deboullie Stock, Maine. Two syenites, a syenodiorite, and a granodiorite were studied at 2 Kbar. In addition data were obtained regarding the region involving vapour-undersaturated liquids for the

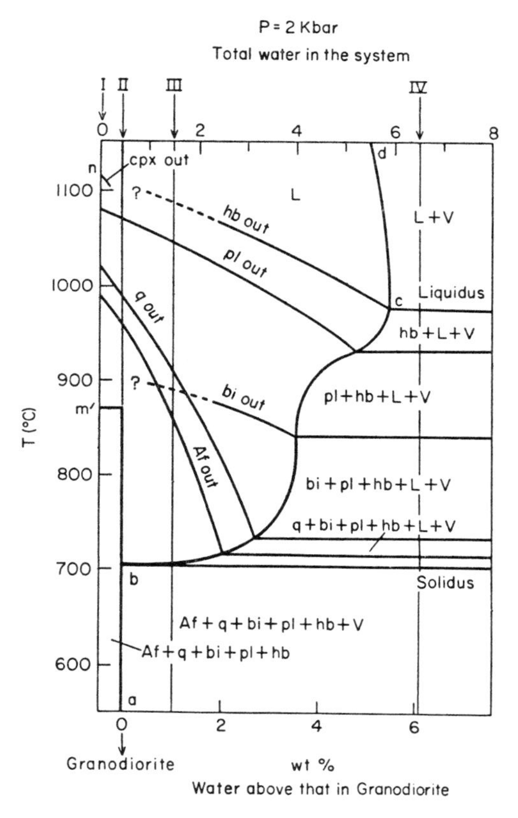

Fig. 32. Vapour-Undersaturated Liquidus Relations for a Granodiorite–H_2O Join. From Robertson and Wyllie (1971a, Fig. 8, p. 265). "Schematic isobaric T–X section for granodiorite 766–H_2O showing the estimated upper stability limits for individual minerals in the vapour-absent region, obtained by interpolation between dry type I and H_2O excess type IV results, or by extrapolation from type IV results. This is a section through Fig. 5. There is not much variation possible in the positions of boundaries for the anhydrous minerals, but there is uncertainty for hornblende and biotite. The temperatures of biotite and hornblende boundaries are sensitive to mineral composition and oxygen fugacity. In other rocks, and possibly in this rock, the liquidus phase in the vapour absent region may change as a function of H_2O content; pyroxene and plagioclase may occur on the liquidus. The temperature m′ where biotite begins to dehydrate in the type II assemblage could be considerably lower in temperature than indicated, relative to the completion of reaction and solution for this bulk composition. See Fig. 9 for the effect of pressure. The abbreviations are: Af, alkali feldspar; q, quartz; bi, biotite; pl, plagioclase; hb, hornblende; cpx, clinopyroxene; L, liquid; V, vapour.

two syenites. The crystallisation sequences for these rock-water systems in the presence of a vapor phase was as follows:

Syenite 34B: biotite + clinopyroxene + magnetite
→ + hornblende → + plagioclase
→ + K-feldspar → + quartz

Syenite 102: hornblende + biotite + clinopyroxene
+ magnetite → + plagioclase
→ + K-feldspar → + quartz

Syenodiorite: hornblende + biotite + clinopyroxene
+ magnetite → + plagioclase
→ + K-feldspar → + quartz

Granodiorite: biotite + hornblende → + magnetite
→ + plagioclase → + clinopyroxene
→ + K-feldspar → + quartz

In the region involving vapour-undersaturated liquids the crystallisation sequence at all H_2O content values studied was the same as in the vapour-saturated case for syenite 34B. However this was the case only at high H_2O content for syenite 102, and the sequence in which quartz and K-feldspar appeared was reversed at lower H_2O contents.

Experimental studies on a granite from the Sierra Nevada by Piwinskii (1968, 1973b) and Stern and Wyllie (1973) also illustrate the hazards involved in the prediction of vapour-undersaturated liquidus relations at high pressure on the basis of vapour-saturated liquidus data obtained at low pressure. For this composition plagioclase is the vapour-saturated liquidus phase at low pressure. At high pressure quartz is the vapour-saturated and vapour-undersaturated liquidus phase. It does not appear that the crystallisation sequence at low pressure in the vapour-saturated liquidus region can be duplicated over any part of the $T\text{–}X_{H_2O}$ range at high pressure in the case of vapour-undersaturated liquids.

3. NON-EQUILIBRIUM STUDIES

Although much of the experimental data which provided the basis for the preceding discussion was quite possibly non-equilibrium data, in a rigorous context (Piwinskii, 1967, Piwinskii and Martin, 1970), it is commonly referred to as equilibrium data in the sense that the direction of equilibrium is obtained, and the results are treated by methods of equilibrium thermodynamics. In this section the topic of primary concern is the experimental data and interpretations which are focussed in the sense of non-equilibrium processes. Of the many possible topics only one will be considered in any detail, crystal growth. Other topics of vital importance in considering non-equilibrium processes are vapour-phase transport, open-system equilibria, and magma mixing. I am not disputing the need for experimental studies of this type, but limitations of space and time permit only a limited treatment. Unfortunately, the theoretical base for the analysis of many of these non-equilibrium processes, including crystal growth, is much less well developed than that used in treating equilibrium conditions.

(a) CRYSTAL GROWTH

By the very nature of things the process of crystal growth is a non-equilibrium process since if equilibrium is attained, quite literally, nothing happens. Crystal growth is a consequence of departure from the equilibrium state and is often considered in terms of a "thermodynamic driving force" (Tiller, 1970). Of principal concern in the present context are data on the kinetics of nucleation and growth of the major phases of the granitic rocks, quartz, plagioclase and K-feldspar.

It is frequently assumed that the presence of large crystals of these phases implies slow growth over long periods of time. Although this may be the case the intent here is to demonstrate that it does not necessarily hold. There is a further implicit assumption employed by many workers that the presence of large crystals of these phases requires the presence of an aqueous fluid phase. Again, though this may be true, it need not be the case.

It is interesting to note that experimental studies of crystal growth in silicate systems of the rock-forming minerals began with the first major paper published from the Geophysical Laboratory, The Carnegie Institution of Washington, "The Isomorphism and Thermal Properties of the Feldspars" by Day, Allen and Iddings (1905). As in so many other areas of what we now call experimental petrology, workers at this institution pioneered and led the way. In their experiments on the plagioclase compositions they observed a pronounced tendency for the formation of spherulitic crystal aggregates from undercooled liquids with compositions more calcic than labradorite, and the great difficulty in nucleation and growth of the albitic feldspars as contrasted to the more calcic varieties.

Although many experimental studies of the "equilibrium" type have provided information on generalised growth rates and nucleation density relations, few studies have been along lines which are directly related to granitic compositions. Jahns and Burnham (1958), as discussed by Wyllie (1963), were among the first to study explicitly the crystal growth problem in granitic compositions. In their experiments they observed many features similar to those observed in the natural pegmatites as reflecting various aspects of the crystal–liquid–vapour coexistence. I should stress that we are still awaiting publication of details, experimental data and conclusions.

Although several workers have reported on aspects of the devitrification process in natural, and synthetic, obsidians the primary emphasis here is on direct crystallisation from the magmatic liquid. Lofgren (1973, 1974) and Fenn (1972, 1973, 1974) have studied critical aspects of the crystallisation process in the plagioclase-water and alkali feldspar-water systems over a relatively wide pressure–temperature–composition range. Swanson (1974) and Swanson *et al.* (1972) have considered the details of the crystallisation process for a series of selected compositions in the synthetic haplogranodiorite system. These few studies provide the basis for the following discussion.

Lofgren (1973, 1974) has been primarily concerned with plagioclase crystallisation in the $NaAlSi_3O_8$-$CaAl_2Si_2O_8$-H_2O system in the 4–6 Kbar range from liquids which were both vapour-saturated and vapour-undersaturated. Average growth rate data (approximate mean of the linear growth rates) varied from 0.3 to 6 microns/hour ($1\ \mu$m/hour $= 2.8 \times 10^{-10}$ cm/sec $\simeq 1$ mm/100 years). Through a series of multi-stage experiments Lofgren was able to produce plagioclase feldspars which showed both a dominant normal zoning and a superimposed reverse zoning within each normal zone. He was also able to characterise a sequence of growth forms as a function of the difference between the growth temperature and the liquidus temperature for the specific composition (ΔT). The sequence observed with increasing ΔT was of the pattern: tabular crystals (equilibrium growth form) → skeletal crystals frequently with entrapped glass → dendritic crystal forms → spherulites. The pattern was well defined for bulk compositions more anorthitic than oligoclase but becomes somewhat less well-defined for the sodic plagioclases. In comparing his results (1974, p. 365) on the $ab_{70}an_{30}$ composition relative to the contrast in the vapour-undersaturated and vapour-saturated relations, Lofgren observed a higher growth rate in the absence of a vapour phase than in the presence of a vapour phase at the same value of ΔT. Lofgren (1973, 1974) also observed significant compositional gradients in Na, Ca, Al and Si in both the crystals and the liquid (quenched as a glass) at the crystal–liquid interface. He proposed (1973, p. 373) a qualitative explanation, based on the vapour-saturated liquidus equilibria, for both the zoning in the crystals, and the concentration gradients at the interface. Smith *et al.* (1955) had previously presented analytic solutions to the differential equations of growth which are partially applicable in this case. The boundary conditions employed by Smith *et al.* (1955), though appropriate for their case are not wholly satisfied in these experiments.

Fenn (1972, 1973) studied the kinetics of crystal nucleation and growth in the $NaAlSi_3O_8$-$KAlSi_3O_8$-H_2O system at 2.5 Kbar involving both vapour-saturated and vapour-undersaturated liquidus relations. The pattern of textural development described by Lofgren (1973, 1974) in the preceeding section was not as well-defined in the alkali feldspar system. In part this was a reflection of a somewhat greater Na/K effect on the crystal shape at low undercooling (ΔT) than was the case for the Ca/Na effect. Fenn (1973) observed linear growth rates of up to 6×10^{-6} cm/sec in the case of vapour-undersaturated growth at ΔT of 200°C. For low ΔT of 25°C or less the maximum linear growth rate still exceeded 1×10^{-6} cm/sec. Of particular importance in Fenn's (1973) study were the relative positions of the nucleation density curves and growth rate curves for a series of compositions of fixed ab/or ratio with increasing H_2O content.

This is shown in Fig. 33. Of particular interest is that the crystals have the potential to grow larger in the presence of vapour-undersaturated liquid, though this requires large ΔT. The maximum growth rate from the vapour-saturated liquid is considerably less; however this maximum growth rate is achieved with low ΔT. It is also worthy of note that the nucleation density is much lower at low ΔT for vapour-undersaturated crystallisation than for vapour-saturated crystallisation. It is tempting to generalise the results to

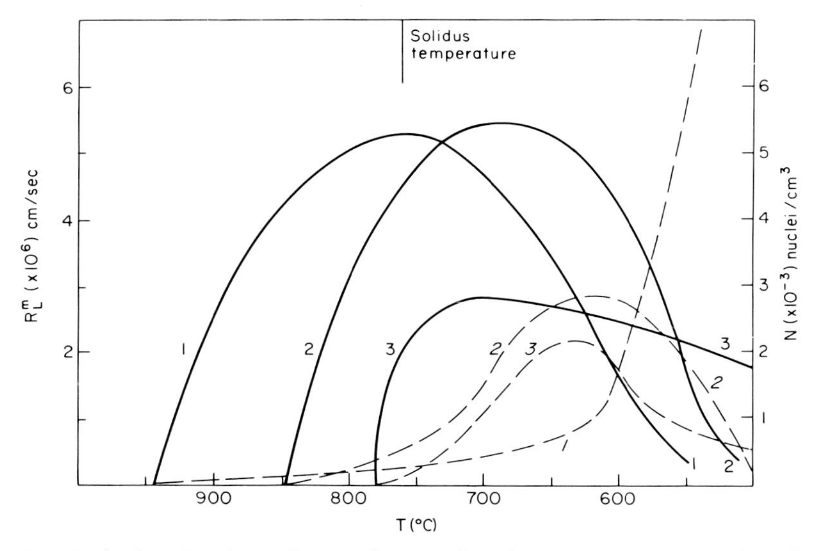

Fig. 33. Nucleation Density and Growth Rate for Three Compositions in the Alkali–Feldspar–H_2O System at 2.5 Kbar. Results, from Fenn (1973), for the three compositions are shown for growth rate (R_L^M) as solid lines with vertical lettering for the composition number, and for nucleation density (N) as dashed curves with slanted lettering. The three compositions, in mole fraction, along the $ab_{.9}or_{.1}$-H_2O join are: (1) $ab_{.639}or_{.071}H_2O_{.29}$; (2) $ab_{.495}or_{.055}H_2O_{.45}$; (3) $ab_{.36}or_{.04}H_2O_{.6}$. Crystal growth occurred in the absence of a vapour phase for composition 1 at $775 \leqslant T(°C) < T_{liquidus}$ and for composition 2 at $780 \leqslant T(°C) < T_{liquidus}$. A vapour phase was present under all conditions for composition 3. Growth rate, R_L^M, refers to the maximum growth dimensions divided by the duration of the experiment. The curves enclose the area defined by a large number of experiments in each case, and refer to an *assumed* (because only initial and final conditions are known) linear growth mode. Nucleation density, N, refers to the numbers of sites per unit volume at the time the experiment was completed as measured micrometrically. See Fenn (1973) for a full discussion.

problems related to the aplites and pegmatites; however, additional data are required. The curves of nucleation density as a function of temperature, or ΔT, are of particular importance in the natural environment, because within a given domain of liquid one would expect that it is the number of nucleation sites rather than the growth rate which determines ultimate grain size.

Fig. 34. Multi-Stage Crystal Growth Experiment in tne Haplogranodiorite System at 8 Kbar. The bulk composition was the synthetic granodiorite composition of Whitney (1975) with 4 wt % H_2O. The initial composition was homogenised at 1100°C, 8 Kbar for 48 hours. Following homogenisation to a hydrous silicate liquid (vapour-undersaturated) temperature was rapidly decreased by 50°C increments; then maintained at the new, lower, temperature for 48 hours. This process was repeated until the final temperature was 700°C and the experiment was quenched.
(a) Photomicrograph. 0.2 mm bar for scale. Plagioclase crystal, with five distinct composition zones, surrounded by two zones of granophyrically intergrown K-feldspar, quartz, and minor plagioclase. The plane of the thin section is sub-parallel to the (010) of the plagioclase crystal.
(b) Electron Microprobe Results. Traverse across the short leg of the "L" in the crystal shown in the photomicrograph, as right to left. Horizontal lines indicate average results of 2–6 analyses in each zone of the plagioclase crystal and in the outer two zones of granophyre. Range of analyses indicates a precision of $\pm$ 1.5 wt %. Dashed lines refer to wt % $KAlSi_3O_8$ and solid lines refer to wt % $CaAl_2Si_2O_8$.
(c) Experimental Results of Whitney (1975) for the R5–H_2O join at 8 Kbar. The specific composition of interest, is indicated by the vertical line at 4% H_2O. Symbols as in Figs. 22 and 23.

(b)

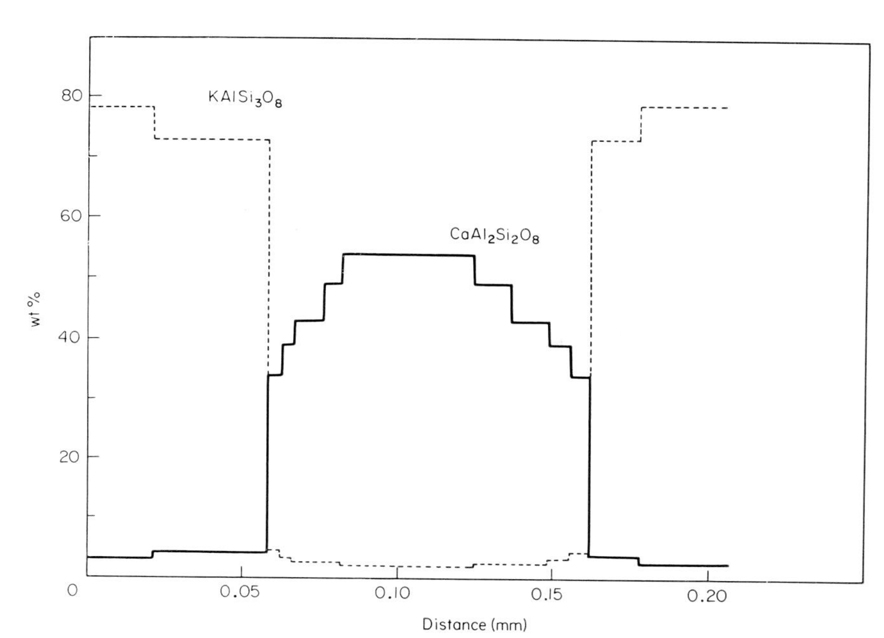

80
60
40
20
0
wt %
$KAlSi_3O_8$
$CaAl_2Si_2O_8$
0.05
0.10
0.15
0.20
Distance (mm)

(c)

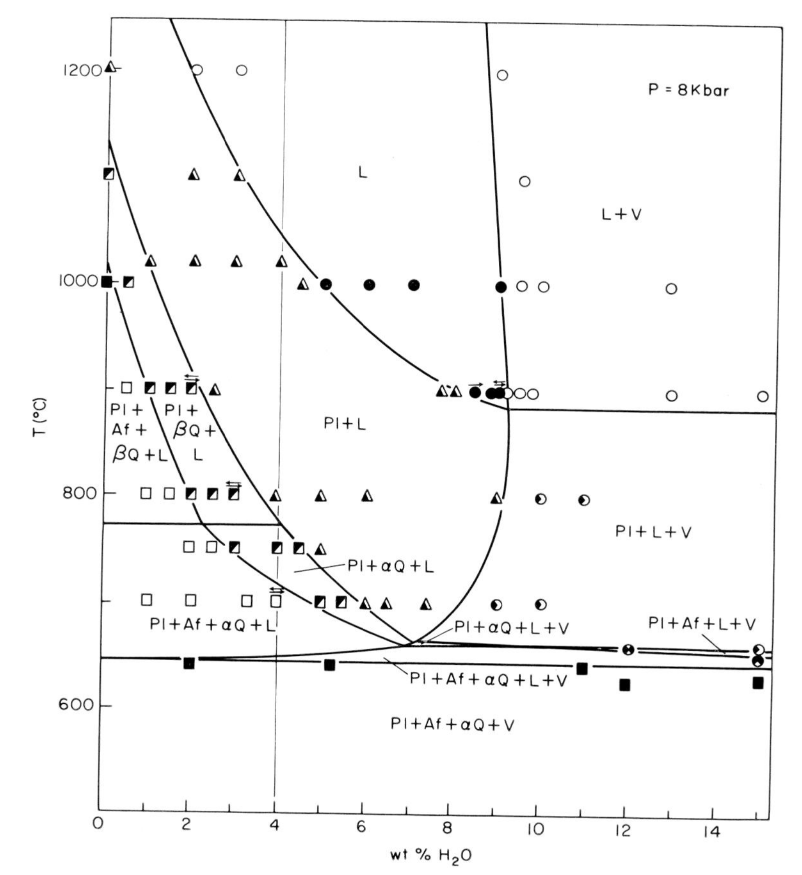

1200
1000
800
600
T (°C)
P = 8 Kbar
L
L + V
Pl + L
Pl + Af + βQ + L
Pl + βQ + L
Pl + L + V
Pl + αQ + L
Pl + Af + αQ + L
Pl + αQ + L + V
Pl + Af + L + V
Pl + Af + αQ + L + V
Pl + Af + αQ + V
0
2
4
6
8
10
12
14
wt % H_2O

Swanson and Whitney (1972) described results of multi-stage crystallisation experiments in four haplogranodiorite experiments which produced zoned plagioclase crystals and granophyric intergrowths of quartz and K-feldspar. The majority of these experiments were relative to vapour-undersaturated crystallisation at 8 Kbar. One of these experiments is discussed in connection with Fig. 34. In this figure both the crystal, the stable phase assemblages for the composition, and the electron microprobe results are shown. The "stratigraphy"

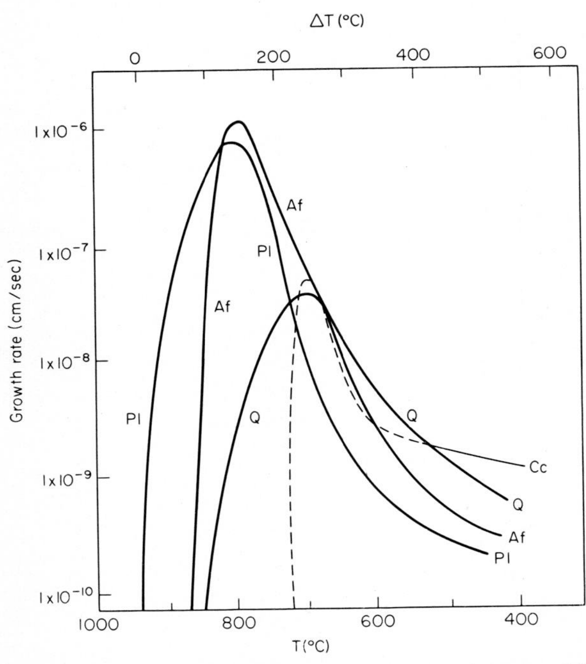

Fig. 35. Relative Growth Rates of Phases in the Haplogranodiorite + CO_2 System. Experiments and interpretation from Swanson (1974). Experiments at 8 Kbar for the synthetic granite (Rl) composition + 5 wt % ($96H_2O . 4CO_2$). Growth rate data obtained in the same manner as described in the legend to Fig. 33.

exhibited by the crystal faithfully reflects the nature of the phase diagram even to the point of being inconsistent with the temperature at which quartz should join the stable phase assemblage. The discrepancy is a consequence of changing the effective bulk composition of the system by thorough extraction of plagioclase as demonstrated by the zoned crystal. This figure not only illustrates the amount of potential information available in such experiments, but also provides a graphic illustration of the effects of fractional crystallisation. The growth rate for each of the zones approximates 1×10^{-9} cm/sec, a not unreasonable geologic value.

Swanson's (1974) results have been discussed previously in connection with the effect of mixed volatiles, specifically CO_2, on equilibria in the haplogranodiorite system. He also studied nucleation density and growth rate for the phases participating in the equilibria. The results are shown in Fig. 35 for one specific composition. The growth rate curves refer to values obtained in discrete experiments at a series of temperatures below the liquidus and must not be interpreted in the sense of a continuous cooling process. They are illustrative of the competitive nature of nucleation and growth in polycomponent–polyphase systems, and provide possible information on the development of granitic textures.

Fenn and Luth (1973) reported on experiments which resulted in the formation of fluid inclusions in feldspar crystals grown from vapour-undersaturated liquids in $NaAlSi_3O_8$-$KAlSi_3O_8$-H_2O. These were suggested to be a result of local saturation of the silicate liquid with respect to an aqueous vapour phase at the crystal–liquid interface produced by relatively rapid crystal growth and low diffusivity of H_2O in the silicate liquid, yielding a boundary region which was markedly enriched in H_2O. Although these results may be of general interest regarding primary fluid inclusions in feldspars, they may have even greater applicability to crystalline inclusions found in minerals of the granitic rocks. If produced by a such a "boundary layer enrichment" mechanism, the paragenetic sequence as obtained by petrographic observations may be in need of revision.

These experimental studies involving crystal growth of minerals of the granitic rocks are not discussed with the idea of presenting definitive information on growth mechanism and kinetics; rather the intent is to suggest that experimental studies on textural development may be just as exciting and informative as the "classical" phase equilibrium studies. It can be argued that we know very little about the origin of textural features of granitic rocks when we cannot quantitatively compare the textures of two different rocks, but must rely on a series of qualitative terms such as those given in a glossary. Studies devoted primarily to the origin and development of textural features, considered in the light of the available "phase equilibrium" data should be one of the major avenues of experimental studies in petrology. Such studies offer one of the principal means for analysis and comparison of magmatic and sub-solidus crystallisation history.

4. CONCLUSIONS

In going back over the contents of this chapter I am astonished at the number of topics and workers that have been omitted. In particular the volcanic rocks of granitic composition have been completely ignored, as have the rocks of pegmatitic and aplitic textures. Vitally important topics related to the ultimate source of the granitic magmas have also not been discussed. The principal thrust of this review has been in the direction of providing a treatment which illustrates the limitations of the experimental approach. Presumably this is a reflection of my belief that in many treatments by the non-experimentalist, the experimental data have been applied without full cognizance of the limitations.

The experimental studies on the relatively simple haplogranite and haplogranodiorite systems have provided considerable information which may be used in a generalised, non-specific way. This is most appropriate in the sense of analysis of possible processes of importance in the formation and subsequent history of the granitic rocks. Crystallisation and melting relationships obtained from studies in the synthetic system are extremely useful in the illustration of possible general trends.

Experimental studies on natural rock–water systems have provided a graphic demonstration of hazards of using results on the synthetic systems to predict specific crystallisation or melting sequences. This "failure" is, of course, a reflection on the role of additional components in the natural systems, primarily anorthite and femic components. The studies on natural granitic rocks have provided dramatic confirmation of the importance of crystal–liquid equilibria in producing the compositions which we see in the great granite batholiths. However, much remains to be done in providing the details of the relationships of one rock to another.

I have stressed the complexities of phase relations involving vapour-undersaturated silicate liquids, largely because there remain attempts (for example, Ragland and Butler, 1972), to utilise the vapour-saturated data obtained at low pressure to predict vapour-undersaturated equilibria at a higher pressure. I believe this problem reflects poorly on the experimentalist in general because we have not presented the data and conclusions with sufficient emphasis on limitations. In a sense we are often guilty of overstating the direct applicability of the experimental studies.

Problems related to kinetics of nucleation and growth of minerals of the granitic rocks must be resolved if we are to understand the contrast in textural development of rocks produced largely through magmatic, as contrasted with post-magmatic processes, as described so elegantly by Tuttle (1952) in his classic paper on this topic.

It should be obvious to the reader that I find this a very exciting time in the application of experimental studies to problems of the granitic rocks. This is not because I believe that we are close to "answers", rather that we are getting closer to the significant questions, as framed on the basis of field and petrographic observations.

ACKNOWLEDGEMENTS

It is a very real pleasure to take this opportunity to express thanks to O. F. Tuttle and R. H. Jahns for their many contributions and discussions. Particular thanks are also due to a large number of graduate students and post-doctoral research colleagues engaged in experimental studies in the School of Earth Sciences, Stanford University. These associates include J. G. Blencoe, J. H. Carman, P. M. Fenn, G. E. Lofgren, R. F. C. Martin, D. A. Mustart, M. T. Naney, A. J. Piwinskii, J. C. Steiner, S. A. Swanson, R. D. Warner, J. A. Whitney, and J. R. Weidner. Financial support by the National Science Foundation and Stanford University, which made possible many of the research efforts referenced here, is also gratefully acknowledged.

REFERENCES

Althus, E., Karotke, E., Nitch, K. H. and Winkler, H. F. G. (1970). *Neues. Jahrb. Min. Mh. Jg.*, 325–336.

Bailey, D. K. (1969). *Am. J. Sci.*, **267a,** 1–16.

Bailey, D. K. and Schairer, J. F. (1964). *Am. J. Sci.*, **262,** 1,198–1,206.

Bailey, D. K. and Schairer, J. F. (1966). *J. Petrol.*, **7,** 114–170.

Barth, T. F. W. (1951). *Neues. Jahrb. Min. ABH*, **82,** 143–154.

Barth, T. F. W. (1962). *Norsk. Geolog. Tidsskr.*, **42,** No. 2 (Feldspar volume), 330–339.

Blencoe, J. G. (1974). "An Experimental Study of Muscovite-Paragenite Stability Relations", Ph.D. thesis, Department of Geology, Stanford University.

Boettcher, A. L. and Wyllie, P. J. (1968). *Geochim. et Cosmochim. Acta*, **32,** p. 999–1,012.

Boettcher, A. L. and Wyllie, P. J. (1969). *Am. J. Sci.*, **267,** 875–909.

Bowen, N. L. (1937). *Am. J. Sci.*, **33,** 1–21.

Boyd, F. R. (1959). Hydrothermal Investigations of Amphiboles *in* "Researches in Geochemistry", Abelson, P. H. (Ed.), John Wiley and Sons, New York.

Boyd, F. R. and England, J. L. (1963). *J. Geophys. Res.*, **68,** 311–323.

Brown, G. M. (1962). *Min. Mag.*, **33,** No. 262, 533–562.

Burnham, C. W. (1967). Hydrothermal Fluids at the Magmatic Stage *in* Barnes, H. L. (ed.) "Geochemistry of Hydrothermal Ore Deposits", Barnes, H. L., Holt, Reinhard and Winston, Inc., New York.

Burnham, C. W. and Jahns, R. H. (1962). *Am. J. Sci.*, **260,** 721–745.

Carman, J. H. (1969). "The Study of the System $NaAlSiO_4$-Mg_2SiO_4-SiO_2-H_2O from 200 to 5,000 bars and 800°C to 1,000°C and its Petrologic Applications", Ph.D. thesis, Department of Geochemistry and Mineralogy, Pennsylvania State University.

Carman, J. H. (1974). *Amer. Miner.*, **59,** 261–273.

Carmichael, I. S. E. and MacKenzie, W. S. (1963). *Am. J. Sci.*, **261,** 382–396.

Clark, S. P. (Ed.) (1966). "Handbook of Physical Constants", Geological Society of America.

Day, A. L., Allen, E. T. and Iddings, J. P. (1905). *Carn. Inst. Wash. Pub.*, **31.**

Dodge, F., Smith, V. and Mays, R. (1969). *J. Petrol.*, **10,** 250–271.

Edgar, A. D. (1973). "Experimental Petrology: Basic Principles and Techniques", Oxford University Press, Oxford.

Ernst, W. G. (1960). *Geochim. et. Cosmochim. Acta*, **19,** 10–40.

Ernst, W. G. (1968). "Amphiboles Crystal Chemistry-Phase Relations and Occurrence", Springer-Verlag, New York.

Eugster, H. P. (1957). *J. Chem. Phys.*, **26,** 1,160.

Eugster, H. P., Albee, A. L., Bence, A. E., Thompson, J. B., Jr. and Waldbaum, D. R. (1972). *J. Petrol.*, **13,** 147–179.

Eugster, H. P. and Skippen, G. B. (1967). Igneous and Metamorphic Reactions Involving Gas Equilibria *in* "Researches in Geochemistry, Vol. 2", Abelson, P. H. (Ed.), John Wiley and Sons, New York.

Eugster, H. P. and Wones, D. R. (1962). *J. Petrol.*, **3,** 82–125.

Fenn, P. M. (1972). *Trans. Am. Geophys. Union*, **53,** p. 1,127.

Fenn, P. M. (1973). "Nucleation and Growth of Alkali Feldspars from Melts in the System $NaAlSi_3O_8$-$KAlSi_3O_8$-H_2O, Ph.D. thesis, Department of Geology, Stanford University.

Fenn, P. M. (1973). Nucleation and Growth of Alkali Feldspars from a Melt, *in* "Proc. NATO Advanced Studies Institute on Feldspars", MacKenzie, W. S. and Zussman, J. (Eds.). Manchester University Press, Manchester.

Fenn, P. M. and Luth, W. C. (1973). *Geol. Soc. Amer., 1973 Ann. Mtg. Program*, 617.

Fyfe, W. S. (1960). *J. Geol.*, **68,** 553–566.

Fyfe, W. S. (1970). Some Thoughts on Granitic Magmas *in* "Mechanism of Igneous Intrusion", Newall, G. and Rast, N. Gallery Press, Liverpool.

Fyfe, W. S. and MacKenzie, W. S. (1969). *Earth Sci. Rev.*, **5,** 185–215.

Gibbs, J. W. (1906). On the Equilibrium of Heterogeneous Substances, *in* "The Scientific Papers of J. Willard Gibbs, Vol. 1: Thermodynamics", Bumstead, H. A. and Van Name, R. G. (Eds.) Longmans, London.

Gilbert, M. C. (1969). *Am. J. Sci.*, **267A,** 145–159.

Goldsmith, J. R. and Newton, R. C. (1974). An Experimental Determination of the Alkali Feldspars Solvus, *in* "Proc. NATO Advanced Studies Institute on Feldspars", MacKenzie, W. S. and Zussman, J. (Eds.). Manchester University Press, Manchester.

Goranson, R. W. (1931). *Am. J. Sci.*, Ser. 5, v. 22, p. 481–502.

Goranson, R. W. (1932). *Am. J. Sci.*, **23,** 227–236.

Goranson, R. W. (1938). *Am. J. Sci.*, **25-A,** 71–91.

Huang, W. L. and Wyllie, P. J. (1973). *Contr. Mineral. Petrol.*, **42,** 1–14.

Jahns, R. H. and Burnham, C. W. (1958). *Geol. Soc. Amer. Bull.*, **69,** 1,592–1,593.

Jahns, R. H. and Burnham, C. W. (1969). *Econ. Geol.*, **64,** 843–864.

James, R. S. and Hamilton, D. L. (1969). *Contr. Mineral. Petrol.*, **21,** 111–141.

Johannes, W., Bell, P. M., Mao, H. K., Boettcher, A. L., Chipman, D. W., Hays, J. F., Newton, R. C. and Seifert, F. *Contr. Mineral. Petrol.*, **32,** 24–38.

Kennedy, G. C., Wasserburg, G. J., Heard, H. C. and Newton, R. C. (1962). *Am. J. Sci.*, **260,** 501–521.

Khitarov, N. I. and Kadik, A. A. (1973). *Contr. Mineral. Petrol.*, **41,** 205–215.

Kim, K. T. and Burley, B. J. (1971). *Canadian J. Earth Sci.*, **8,** 311–337.

Kranck, E. and Oja, R. (1960). *Internat. Geol. Cong.*, 21st, Copenhagen, 1960, rept. 1, pt. 14, 16–29.

Kushiro, I. (1969). *Am. J. Sci.*, **267A,** 269–294.

Kushiro, I. and Yoder, H. S. (1969). *Am. J. Sci.*, **267A,** 558–582.

Lambert, I. B., Robertson, J. K. and Wyllie, P. J. (1969). *Am. J. Sci.*, **267,** 609–626.

Lindsley, D. H. (1967). *Carn. Inst. Wash. Yr. Bk.*, **65,** 244–247.

Lofgren, G. (1974a). Temperature Induces in Synthetic Plagioclase Feldspar *in* "Proc. NATO Advanced Studies Institute on Feldspars", MacKenzie, W. S. and Zussman, J. (Eds.). Manchester University Press, Manchester.

Lofgren, G. (1974b). *Am. J. Sci.*, **274,** 243–273.

Luth, W. C. (1967). *J. Petrol.*, **8,** 372–416.

Luth, W. C. (1968). *Carn. Inst. Wash. Yr. Bk.*, **66,** 480–484.

Luth, W. C. (1969). *Am. J. Sci.*, **267a,** 325–341.

Luth, W. C. (1974). Analysis of Experimental Data on Alkali Feldspars: Unit Cell Parameters and Solvi, *in* "Proc. NATO Advanced Studies Institute on Feldspars", MacKenzie, W. S. and Zussman, J. (Eds.). Manchester University Press, Manchester.
Luth, W. C. and Ingamells, C. O. (1963). *Amer. Miner.*, **48,** 255–258.
Luth, W. C., Jahns, R. H. and Tuttle, O. F. (1964). *J. Geophys. Res.*, **69,** 759–771.
Luth, W. C., Martin, R. F. and Fenn, P. M. (1971). Peralkaline Alkali Feldspar Solvi, *in* "Proc. NATO Advanced Studies Institute on Feldspars", MacKenzie, W. S. and Zussman, J. (Eds.). Manchester University Press, Manchester.
Luth, W. C. and Tuttle, O. F. (0000). *Amer. Miner.*, **48,** p. 1,401–1,403.
Luth, W. C. and Tuttle, O. F. (1966). *Amer. Miner.*, **51,** 1,359–1,373.
Luth, W. C. and Tuttle, O. F. (1969). *Geol. Soc. Amer. Mem.*, **115,** 513–548.
McDowell, S. D. and Wyllie, P. J. (1971). *J. Geol.*, **79,** 173–194.
Merrill, R. B., Robertson, J. K. and Wyllie, P. J. (1970). *J. Geol.*, **78,** 558–569.
Millhollen, G. L. (1971). Melting of nepheline syenite with H_2O and $H_2O + CO_2$. *Am. J. Sci.*, **270,** 244–254.
Millhollen, G. L., Wyllie, P. J. and Burnham, C. W. (1971). *Am. J. Sci.*, **271,** 473–480.
Morey, G. W. and Fenner, C. N. (1917). *J. Amer. Chem. Soc.*, **39,** 1,173–1,229.
Morse, S. A. (1970). *J. Petrol.*, **11,** 221–251.
Mustart, D. A. (1972). "Phase Relations in the Peralkaline Portion of the System Na_2O-Al_2O_3-SiO_2-H_2O", Ph.D. thesis, Department of Geology, Stanford University.
Nockolds, S. R. (1954). Average Chemical Composition of Some Igneous Rocks, *Geol. Soc. Amer. Bull.*, **65,** 1,007–1,032.
Peters, T. J., Luth, W. C. and Tuttle, O. F. (1966). *Amer. Miner.*, **51,** 736–753.
Piwinskii, A. J. (1967). *Earth Planet. Sci. Let.*, **2,** 161–162.
Piwinskii, A. J. (1968). *J. Geol.*, **76,** 548–570.
Piwinskii, A. J. (1973a). *Tschermaks Min. Petr. Mitt.*, **20,** 107–130.
Piwinskii, A. J. (1973b). *N. Jb. Miner. Mh. Jg.*, **5,** 193–215.
Piwinskii, A. J. (1974). *Fortschr. Min.*, 240–255.
Piwinskii, A. J. and Martin, R. F. (1970). *Contr. Mineral. Petrol.*, **29,** 1–10.
Piwinskii, A. J. and Wyllie, P. J. (1968). *J. Geol.*, **76,** 205–234.
Piwinskii, A. J. and Wyllie, P. J. (1970). *J. Geol.*, **78,** 52–76.
Presnall, D. C. (1969). *Am. J. Sci.*, **267,** 1,178–1,194.
Presnall, D. C. and Bateman, P. C. (1973). *Geol. Soc. Amer. Bull.*, **84,** 3,181–3,202.
Ragland, P. C. and Butler, J. R. (1972). *J. Petrol.*, **13,** 381–404.
Ricci, J. E. (1951). "The Phase Rule and Heterogeneous Equilibrium", Van Nostrand, New York.
Robertson, J. K. and Wyllie, P. J. (1971a). *Am. J. Sci.*, **271,** 252–277.
Robertson, J. K. and Wyllie, P. J. (1971b). *J. Geol.*, **79,** 549–571.
Roy, R. (1956). *J. Amer. Ceram. Soc.*, **39,** 145.
Rutherford, M. J. (1969). *J. Petrol.*, **10,** 318–408.
Scarfe, C. M., Luth, W. C. and Tuttle, O. F. (1966). *Amer. Mineral.* **51,** 726–735.
Schairer, J. F. (1950). *J. Geol.*, **58,** 512–517.
Schairer, J. F. and Bowen, N. L. (1935). *Trans. Amer. Geophys. Union*, 325–328.
Schairer, J. F. and Bowen, N. L. (1955). *Am. J. Sci.*, **253,** 681–746.
Schairer, J. F. and Bowen, N. L. (1956). *Am. J. Sci.*, **254,** 129–195.
Schreinemaker, F. A. H. (1915–1925). "Mono-, and Divariant Equilibria". (A series of 49 papers in 29 articles which appeared in the Chemistry Section, *Proc. Konimklijke Akademie Van Wetenschappen Te Amsterdam* from 1915 to 1925, in English); **18,** 116–126, 531–542, 820–828, 1,018–1,025, 1,026–1,037, 1,175–1,190, 1,384–1,398, 1,539–1,552, 1,676–1,691; **19,** 514–527, 713–727, 816–824, 867–880, 927–932, 999–1,006, 1,196–1,205, 1,205–1,217; **20,** 659–667; **22,** 318–322, 542–554; **23,** 1,151–1,160; **25,** 341–353; **26,** 283–296, 719–726; **27,** 57–64, 279–290, 441–450, 800–808, **29,** 252–261.

Seck, H. A. (1971). *Contr. Mineral. Petrol.*, **31,** 67–86.
Seki, Y. and Kennedy, G. C. (1965). *Geochim. et. Cosmochim. Acta*, **29,** 1,077–1,083.
Shaw, H. R. (1963). *Amer. Miner.*, **48,** 883–896.
Shaw, H. R. (1967). Hydrogen osmosis in Hydrothermal Experiments *in* "Researches in Geochemistry," P. H. Abelson (Ed.), Vol. 2. John Wiley and Sons, New York.
Smith, V. G., Tiller, W. A. and Rutter, J. W. (1955). *Can. J. Phys.*, **33,** 723.
Spengler, C. J. (1965). "The Upper Three-Phase Region in a Portion of the System $KAlSi_2O_6$-SiO_2-H_2O at Water Pressures from Two to Seven Kilobars". Ph.D. thesis, Department Geochemistry and Mineralogy, College Mineral Industries, The Pennsylvania State University.
Steiner, J. C. (1970). "An Experimental Study of the Assemblage Alkali Feldspar + Liquid + Quartz in the System $NaAlSi_3O_8$-$KAlSi_3O_8$-SiO_2-H_2O at 4,000 bars". Ph.D. thesis, Department of Geology, Stanford University.
Steiner, J. C., Jahns, R. H. and Luth, W. C. (1975). *Geol. Soc. Amer. Bull.*, **83,** 98.
Stern, C. R. and Wyllie, P. J. (1973a). *Earth and Planet. Sci. Let.*, **18,** 163–167.
Stern, C. R. and Wyllie, P. J. (1973b). *Contr. Mineral. Petrol.*, **42,** 313–323.
Stewart, D. B. (1967). *Schweiz. Mineral. Petro. Mitt.*, **47,** 35–59.
Stewart, D. B. and Roseboom, E. H. (1962). *J. Petrol.*, **3,** 280–315.
Storre, B. and Karotke, E. (1971). *Neues. Jahrb. Min. Mh. Jg.*, **222,** 237–240.
Storre, B. and Karotke, E. (1972). *Contr. Mineral. Petrol.*, **36,** 343–345.
Swanson, S. E. (1974). "Phase Equilibria and Crystal Growth in Granodioritic and Related Systems with H_2O and $H_2O + CO_2$", Ph.D. thesis, Department of Geology, Stanford University.
Swanson, S. E. and Whitney, J. A. (1972). *Trans. Amer. Geophys. Union*, **53,** 1,127.
Thompson, A. B. (1974). *Contr. Mineral. Petrol.*, **44,** 173–194.
Thompson, R. N. and MacKenzie, W. S. (1967). *Am. J. Sci.*, **265,** 714–734.
Thornton, C. P. and Tuttle, O. F. (1960). *Am. J. Sci.*, **258,** 664–684.
Tiller, W. A. (1970). The Use of Phase Diagrams in Solidification *in* "Phase Diagrams, Materials, Science and Technology, Vol. 1", A. M. Alper (Ed.), Academic Press, London and New York.
Tuttle, O. F. (1952). *J. Geol.*, **60,** 107–124.
Tuttle, O. F. and Bowen, N. L. (1958). *Geol. Soc. Amer. Mem.*, **74.**
Von Platen, H. (1965). Experimental Anntexis and Genesis of Migmatites, *in* "Controls of Metamorphism", Pitcher, W. S. and Flinn, G. W. (Eds.), Oliver and Boyd, Edinburgh.
Weidner, J. R. and Carmen, J. H. (1968). *Proc. Mtg. Geol. Soc. Amer.*, 315.
Weill, D. F. and Kudo, A. H. (1968). *Geol. Mag.* [Great Britain], v. 105, p. 325–337.
White, D. E. and Waring, G. A. (1968). *U.S. Geol. Surv. Prof. Pap.*, **440-K.**
Whitney, J. A. (1969). "Partial Melting Relationships of Three Granitic Rocks", B.S. and M.S. thesis, Department of Geology, M.I.T.
Whitney, J. A. (1972a). *Amer. Mineral.*, **57,** 1,902–1,908.
Whitney, J. A. (1972). "History of Granodioritic and Related Magma Systems: an Experimental Study", Ph.D. thesis, Department of Geology, Stanford University.
Whitney, J. A. (1975). *J. Geol.*, **83,** 1–31.
Whitney, J. A. and Luth, W. C. (1970). *Trans. Amer. Geophys. Union*, **51,** 438.
Winkler, H. and Von Platen, H. (1961a). *Geochim. et Cosmochim. Acta*, **24,** 48–69.
Winkler, H. and Von Platen, H. (1961b). *Geochim. et Cosmochim. Acta*, **24,** 250–259.
Wones, D. R. and Eugster, H. P. (1965). *Amer. Mineral.*, **50,** 1,228–1,272.
Wones, D. R. and Gilbert, M. C. (1969). *Am. J. Sci.*, **267A,** 480–488.
Wyllie, P. J. (1960). *Mineral. Mag.*, **32,** 459–470.
Wyllie, P. J. (1963). Applications of High Pressure Studies to the Earth Sciences, *in* "High Pressure in Physics and Chemsitry", Bradley, R. S. (Ed.), Academic Press, London and New York.

Wyllie, P. J. and Tuttle, O. F. (1957). *Trans. Amer. Geophys. Union*, **38,** 413–414.
Wyllie, P. J. and Tuttle, O. F. (1959). *Am. J. Sci.*, **257,** 648–655.
Wyllie, P. J. and Tuttle, O. F. (1960). *Am. J. Sci.*, **258,** 498–517.
Wyllie, P. J. and Tuttle, O. F. (1961). *Am. J. Sci.*, **259,** 128–143.
Wyllie, P. J. and Tuttle, O. F. (1964). *Am. J. Sci.*, **262,** 930–939.
Yoder, H. S. (1950). *Trans. Amer. Geophys. Union*, **31,** 827–835.
Yoder, H. S., Stewart, D. B. and Smith, J. R. (1956). *Carn. Inst. Wash. Yr. Bk.*, **56,** 206–214.

2. Applications of Experiments to Alkaline Rocks

D. K. Bailey

1 Terminology . . . 419
(a) Felsic or Salic rocks . . . 419
(b) Mafic or Femic rocks . . . 420
2 Mineral Ranges and Distribution . . . 421
3 Alkaline Rock Groups . . . 422
(a) Felsic rocks . . . 422
(b) Mafic rocks . . . 449
4 General Factors . . . 460
(a) Alkali transfer . . . 461
(b) The development of peralkalinity . . . 462

1. TERMINOLOGY

Before considering the experimental results that apply to alkaline igneous rocks it is necessary to clarify the meaning and range of the term "alkaline". This expression refers to richness in alkalis and is therefore relative, its meaning varying with context according to the group of rocks under discussion. In another usage, certain magmatic associations or provinces are described as alkaline according to an arbitrarily defined scale (Peacock 1931) and in more recent times even this meaning has lost precision in an ill-defined usage stemming from Peacock's original concept. In this chapter Peacock's connotation is not implied.

(a) FELSIC OR SALIC ROCKS

The most notable disparity in usage is in the application of "alkaline" to rocks rich in alkali feldspars. Those rocks containing feldspathoid are called alkaline, but rocks oversaturated with silica are not unless a non-aluminous alkaline mineral appears in the mode or norm (Table 1). Alkali content is compared with silica in the first instance and with alumina in the second. In recent years authors have avoided this ambivalence by using the term "peralkaline" to describe the condition where alkalis exceed alumina. This term, coined by Shand (1927), is not entirely apt, but at least it allows us to designate accurately those feldspathoidal rocks in which alkalis exceed alumina: the term "sub-aluminous" would ideally describe the condition where alkalis exceed alumina but unfortunately it has been used previously by Shand, with a different meaning.

Three groups of felsic rocks will therefore be considered in this chapter: peralkaline oversaturated; feldspathoidal; peralkaline feldspathoidal.

TABLE 1

Broad groupings of felsic lavas based on alkali:alumina:silica balance. Preferred group name in capitals: alternatives in lower case (highly specific names, e.g. pantellerite, omitted for simplicity). Note that the *RHYOLITE* group is not normally designated as alkaline.

	Oversaturated with silica		Undersaturated with silica
Molecular alkalis less than alumina	*RHYOLITE* *Calc-alkaline rhyolite*	TRACHYTE	PHONOLITE
Molecular alkalis exceed alumina	PERALKALINE RHYOLITE Alkali rhyolite Soda rhyolite		PERALKALINE PHONOLITE Phonolite

(b) MAFIC OR FEMIC ROCKS

"Alkaline", "alkalic", or "alkali" are adjectives commonly used to distinguish basalts that are relatively enriched in alkalis from tholeiites (see Macdonald and Katsura, 1964). The simplest expression of the alkaline character in a basalt is the presence of feldspathoid (or the appearance of *ne* in the CIPW norm). See

TABLE 2

Broad groupings of alkaline mafic lavas based on feldspar: feldspathoid balance. In this Chapter only the feldspar-free group is considered in detail. Note that some nephelinites (and related rocks) may additionally be peralkaline. A large number of names have been given to rocks which are grouped under the heading "feldspar absent". Most types are highly mafic, if not ultramafic, and many of the rocks could as well be described as clino-pyroxenites with appropriate qualification, e.g. leucitite = leucite pyroxenite.

Feldspathoid accessory or occult	Feldspathoid essential (>5%)	Feldspar absent
ALKALI BASALT	BASANITE	NEPHELINITE MELILITITE LEUCITITE

Table 2. With increasing nepheline (or feldspathoid) content basaltic rocks are termed basanites, but a special condition is reached when plagioclase is absent as in the nepheline–pyroxene rocks (nephelinites) and melilite-pyroxene rocks (melilitites). These feldspar-free mafic rocks pose petrogenetic problems distinct from those of the alkali basalts and will therefore receive special consideration: experimental studies pertaining to ordinary basaltic compositions are too extensive to allow coverage within this chapter, and they will consequently be referred to only where they relate directly to the other alkaline rocks under consideration.

Feldspar-free mafic alkaline rocks have additional features that call for

TABLE 3

Experimentally determined stabilities of minerals of special importance in alkaline magmatism. Each reference is intended as a lead to the primary data-sources for each mineral.

Mineral	Key reference
Alkali feldspars	Morse, 1970
Nepheline–kalsilite	Tuttle and Smith, 1958
Leucite	Fudali, 1963
Analcite	Hamilton, 1972
Melilite	Schairer, Yoder and Tilley, 1967
Jadeite	Bell and Roseboom, 1969
Acmite	Bailey, 1969: Gilbert, 1969
Alkali amphiboles	Ernst, 1968
Aenigmatite	Lindsley, 1971
Biotite	Wones and Eugster, 1965
Phlogopite	Yoder, 1970

comment: (a) the alkali ratio ranges from strongly sodic to strongly potassic (most other alkaline rocks are sodic); (b) they are commonly associated with carbonate magmatism; (c) they have characteristics in common with kimberlites; and, therefore, some of the experimental (and other) evidence pertaining to carbonatite and kimberlite is probably relevant.

2. MINERAL RANGES AND DISTRIBUTION

No simple chemical system can encompass the mineralogy of the alkaline rocks, as these range from acid to ultramafic, but the evidence from various systems containing alkaline mineral compositions can elucidate the conditions under which alkaline magmas form and exist. Systems involving alkali feldspars, feldspathoids, alkali pyroxenes and amphiboles, and phlogopite are logical candidates. The phase relations of each separate mineral composition must themselves define some of the limiting conditions of alkaline magmas, but rather than considering mineral compositions separately it is preferable to view them within the context of various groups of alkaline rocks. The more important minerals are listed, with references to relevant experimental systems, in Table 3. Table 4 is a similar listing of systems that describe mineral relations that

TABLE 4

Experimental systems relevant to alkaline rocks.

System	Key reference
Alkali feldspars and feldspathoids	See Table 3
Ne-Ks-SiO_2 (Petrogeny's Residua System)	Bowen, 1937; Schairer, 1950
Ab-Or-SiO_2-H_2O	Luth, 1969
Peralkaline extensions from	Carmichael and MacKenzie, 1963
Ab-Or-Sil-H_2O	Thompson and MacKenzie, 1967
Di-Ne-Ab-H_2O	Edgar, 1964
Di-Ac-Ne-Ab-H_2O	Nolan, 1966
Na_2O-Al_2O_3-Fe_2O_3-SiO_2	Bailey and Schairer, 1966
Ne-Di	Bowen, 1922; Schairer, Yagi and Yoder, 1962
Expanded basalt tetrahedron: Ca_2SiO_4-Ne-Fo-SiO_2	Schairer and Yoder, 1970
Phlogopite-bearing systems	Modreski and Boettcher, 1972
Jadeitic pyroxenes	Bell and Davis, 1969
Acmitic pyroxenes	Cassie, 1971
Amphibole-bearing systems	Kushiro, 1970

are important in alkaline rocks: some of these will be discussed in subsequent sections where they relate directly to the rock group under consideration.

3. ALKALINE ROCK GROUPS

In the following sections each of the alkaline rock groups outlined in Tables 1 and 2 will be examined, to identify any special petrologic problems within a group, and between groups, and to consider how experiments have eliminated or modified these problems. The ultimate aim of the examination is to highlight the major unsolved problems, and to ask what new experiments are needed.

(a) FELSIC ROCKS

These are usefully considered first in three categories, silica-undersaturated melts, peralkaline silica-undersaturated melts, and peralkaline melts oversaturated in silica.

In order to see the effects and influences of experimental petrology on our understanding of the alkaline felsic rocks it is best to trace briefly the growth of ideas. In the early part of this century, when systematic experimental petrology was still in its infancy, most of the hypotheses current today were already long-established (see Iddings, 1909). Some of these, such as molecular diffusion along a thermal gradient in a magma, which was favoured by Iddings himself, and liquid immiscibility (favoured by Daly, 1914) fell into disfavour, partly through experimental results which seemed directly opposed, but more through the influence of experimental results in support of the dominating hypothesis of our own time, magmatic evolution by fractional crystallisation. This form of differentiation proposed by Darwin (1844), had found few supporters, but from 1915 onwards the early results from the Geophysical Laboratory were marshalled and presented by Bowen (1928) in a brilliant advocacy of the process. Although many important results were contributed by others, especially his colleagues at the Laboratory, it was Bowen who gave the lead. His perception of the significance of the synthetic systems, his understanding of how to apply them to rocks, his awareness of petrologic problems, and, above all, his lucid presentation, made him a formidable champion of his cause. The consequent transformation in petrologic thinking is best revealed in a selection of later igneous texts. By 1933 Daly (p. 486) was able to compile a list of eminent petrologists who "believe crystal fractionation to be one of the principal processes involved in the development of alkali-rich rocks". Daly included himself in the list, although he was unable to concede an exclusive role to fractional crystallisation, being still committed to limestone syntexis as an overriding factor in the genesis of feldspathoidal rocks (and also conceding that juvenile volatiles might accomplish the same ends). By the 1950's and 1960's (Turner and Verhoogen, 1951 and 1960; Tilley, 1958) the processes of assimilation, melting, magma mixing, gaseous transfer, etc., had all been relegated to distinctly subordinate roles to fractional crystallisation in explaining the genesis of alkaline rocks. Apart from a tendency to pay lip-service to partial melting (largely seen as the direct converse to fractional crystallisation) the same general situation prevails today, although there is growing support for liquid immiscibility (Roedder and Weiblen, 1970; Ferguson and Currie, 1972) and the controlling influence of volatiles, (Bailey 1966, and 1970; Macdonald *et al.*, 1970).

It is a remarkable fact that the early results from systematic phase equilibria studies largely confirmed what was already evident (and well-known) in the petrology of igneous rocks which had crystallised at low pressures, e.g. that cafemic minerals crystallised early and felsic minerals late, so it is interesting to ask why these results should have had such a profound effect. What Bowen did, more than anything else, was to convey to petrologists the perception which comes from the making and description of an experiment. When crystals separate from, or react with, a melt, the necessity to understand and *depict* the result (either diagrammatically or algebraically) brings with it a new insight into the possible consequences of continuing or interrupting the process. Furthermore, quantifying a process enables a more rigorous and critical test of alternative hypotheses, as witnessed by Bowen's devastating critique of magma differentiation by liquid unmixing (1928, p. 7). His arguments cast a shadow of doubt over silicate liquid immiscibility that still persists, and is indeed difficult to dispel with the evidence available in *crystalline* rocks. But perhaps more than anything the new experiments with melts and crystals stimulated a new mode of thinking about magma differentiation, and promoted speculation about crystal ↔ liquid interactions beyond the range of experimental observation, e.g. Bowen's speculation that amphibole concentration in a basaltic melt might yield a new melt of nephelinitic character (1928, p. 270).

As far as the felsic alkaline rocks are concerned, Bowen's contribution, both experimental and conceptual, is immense, starting with his classic study of the plagioclase system (1913) and carrying through to the equally celebrated work on the "Granite System" (Tuttle and Bowen, 1958). Although neither of these landmarks pertains directly to alkaline rocks, they have illuminated our understanding of alkali feldspar relations, which are central to the issue. In the plagioclase system Bowen was able to quantify the nature and form of the solid solution loop, to demonstrate that Na Si-rich liquids were in equilibrium with Ca Al-rich crystals, that there would be continuous reaction between liquid and crystals with falling temperature, or that if there were continuous separation of crystals the residual liquid would be strongly enriched in Na Si. These and subseqent studies of systems such as plagioclase–diopside (Bowen, 1915) confirmed and supported his growing conception of primitive basaltic magma evolving by fractional crystallisation towards a residuum enriched in alkalis and silica, a concept which subsequently developed into the undergraduate's life-line—Bowen's Reaction Series (1928). In the early days there was great difficulty explaining how trachytes and phonolites could be derived by fractional crystallisation, partly because Bowen himself insisted that rhyolite was the normal end-product (all the more surprising when it is remembered that there was abundant evidence of differentiation in sills of feldspathoidal basalt). This intellectual impasse was not shared by non-experimentalists, who saw no need to explain all magmas by one mechanism and who were severely critical of some of Bowen's tortuous solutions to his own riddle of the trachytes and phonolites (Shand, 1923). Subsequent schemes depending on the incongruent melting of orthoclase (Bowen, 1928) were no more satisfactory but meanwhile important fundamental data were accruing on systems involving feldspathoids (Greig and Barth, 1938) and Bowen had entered into fruitful collaboration with Schairer, whose prodigious efforts resulted in the monumental description of the 1 atm equilibria in nepheline–kalsilite–silica, which Bowen dubbed "Petrogeny's Residua System" (see Fig. 1). This provided the first over-view of the relations between rhyolites, trachytes and phonolites, and Bowen was quick to demonstrate how well natural rock compositions coincided with the lower temperature regions in the phase diagram (1937: before Schairer's phase diagram was completed for publication in 1950). From this flowered the full concept of these felsic liquids being the residua from primitive basalt magma, by variations in the path of fractionation and through differences inherited from the parental basalt. So powerful was Bowen's argument that his conclusion, that felsic magmas are the end-products of low-pressure fractionation of various basalts, is now generally treated as virtually axiomatic (in my experience, dissenters are left in no doubt about their basic eccentricity!).

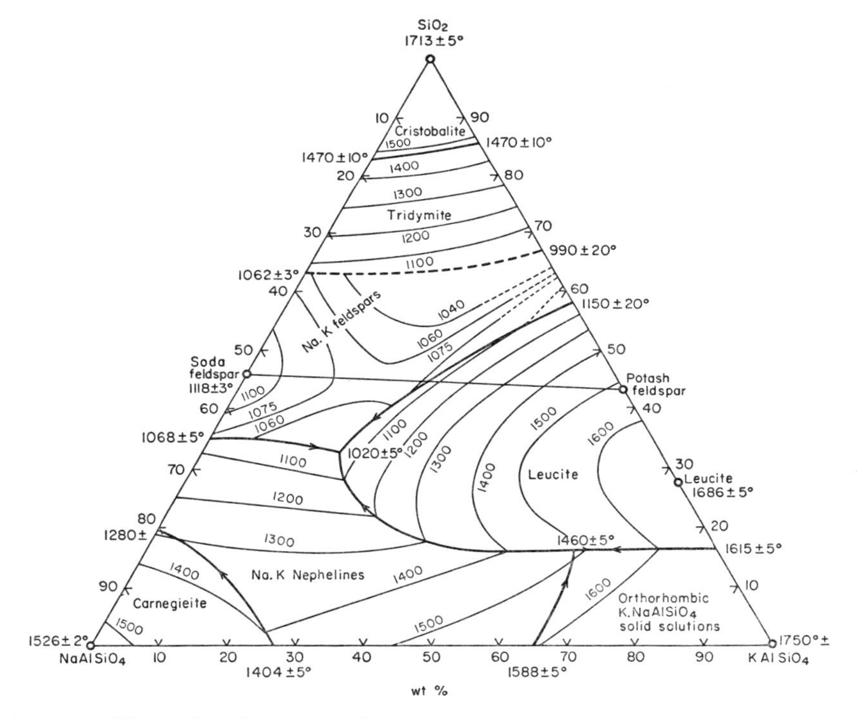

Fig. 1. Equilibrium liquidus diagram for the system $NaAlSiO_4$-$KAlSiO_4$-SiO_2 (nepheline–kalsilite–silica) at 1 atm, after Schairer (1950). The broken phase-boundary, alkali feldspar–tridymite, indicates the region where crystallisation (and quenching experiments) were impossible.

During the same period, the other vital aspect of the alkaline rocks was also being considered in the pioneer studies of the synthesis and stability of acmite pyroxene ($NaFeSi_2O_6$), the normative and modal hallmark of peralkalinity in rocks (Washington and Merwin, 1927; Bowen, Schairer and Willems, 1930). The incongruent melting of this pyroxene and its low temperature of formation were established. It was pointed out that early fractionation of iron oxide from acmitic melts would yield silica-oversaturated residual liquids, thus offering a possible route from an undersaturated peralkaline trachyte to a peralkaline rhyolite residuum (Bowen, *et al.*, ibid.). The question of how liquids achieved the condition of peralkalinity appears to have had no immediacy for petrologists up to the 1930's and one senses from the discussions that this was regarded as a vagary of magmatism. It was Bowen (1945) who highlighted the problem and saw a solution in the "plagioclase effect", in which he skilfully called attention to the fact that it would be impossible to crystallise pure albite from a melt containing calcium. If, therefore, molecular $CaO + Na_2O$ exceeds Al_2O_3 in a melt, the precipitation of plagioclase would normally be expected to yield a residual liquid with excess Na_2O and SiO_2. The new dilemma posed by this discovery,

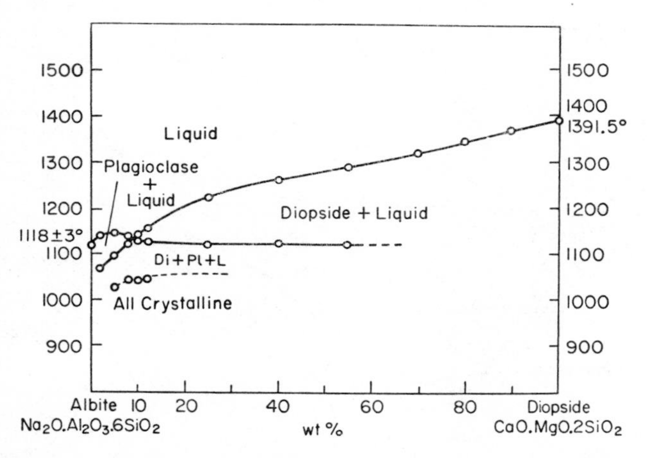

Fig. 2. Pseudo-binary diagram of the join albite–diopside at one atmosphere, after Schairer and Yoder (1960). The hump in the plagioclase liquidus, the three phase region diopside + plagioclase + liquid, and the curve in the solidus, all indicate that the feldspar is not pure albite and that when plagioclase crystallises neither its composition nor the resulting liquid lie in the plane of the diagram. The presence of the anorthite molecule in the plagioclase means that the coexisting liquid is depleted in Al_2O_3 with respect to Na_2O and SiO_2 and is effectively peralkaline.

which is never commented upon, is to explain why tholeiitic residua are not peralkaline, but we shall return to this later. The discovery of the "plagioclase effect" is a good example of how precise phase equilibria studies give an extra cutting edge to petrologic thinking. Subsequently the actual working of the "effect" was examined directly in the system diopside–albite, by Schairer and Yoder (1960). See Fig. 2.

By the 1950's the basic 1 atm data on the system Na_2O-K_2O-Al_2O_3-SiO_2 had been established (Schairer and Bowen, 1955 and 1956) including the portion in which molecular alkalis exceeded alumina, i.e. peralkaline compositions, showing that eutectics involving alkali silicates were at considerably lower temperatures, and were much more highly alkaline than comparable melts in "Petrogeny's Residua System". The time was ripe for the study of a related system involving the acmite composition, Na_2O-Al_2O_3-Fe_2O_3-SiO_2, to ascertain the behaviour of melts crystallising alkali feldspar, acmite, $\pm$ nepheline $\mp$ quartz, and, ultimately, sodium silicates (Bailey and Schairer, 1966). This system constitutes a simplified "Peralkaline Residua System", although the authors stressed that the same peralkaline "residua" could equally well be generated by partial melting of the appropriate bulk composition. Low temperature crystallisation paths were found which provide analogies with peralkaline rhyolites and phonolites, and also malignite or acmite–ijolite compositions, which will be discussed in later sections. In each of these sections, devoted to a particular group of rocks, the more recent experimental studies will be examined against the background of the earlier, and mostly low-pressure, results that have been sketched in above (and listed in Tables 3 and 4).

(1) *Alkali Feldspars*

Before discussing the various groups of felsic rocks a few words on the uses and significance of the experimental data pertaining to the alkali feldspars are in order, because these are the major minerals in rocks of this composition range.

A general description of the alkali feldspar liquidus, solidus, and solvus relations is given in the first part of this book (Figs 17 and 20). In the same place there is a discussion of alkali feldspar crystallisation and melting, when extended into the "granite" system and more complex systems. The most important characteristic of the alkali feldspar join is the temperature minimum in the solid solution loop at any given pressure. Effectively, two solid solution loops, the sanidine series and the anorthoclase series, meet at the minimum, which has a composition Or_{30-40} depending on total pressure. Alkali feldspar crystals at the composition of the minimum will melt directly to a liquid of the same composition. Fractional crystallisation of alkali feldspar liquids of all other compositions will cause any residual liquids to converge on the minimum composition. Thus there is a preferred composition (or restricted range of composition) around which natural hypersolvus alkali feldspars may be expected to cluster, if felsic rocks are products of low-temperature melts. Many analysed alkali feldspar phenocrysts of felsic lavas do, in fact, cluster around the feldspar minimum composition range. Somewhat surprisingly, the relationship persists even into strongly peralkaline lavas, in which the alkali ratio of the liquid may differ strongly from the phenocrysts (Bailey and Schairer, 1964). Feldspar in the minimum range obviously represents a special structural state because it is at Or_{37} that the boundary between anorthoclase (triclinic) and sanidine (monoclinic) appears at lower temperatures. In felsic lavas the common occurrence of phenocrysts in the minimum range almost suggests that the feldspar is behaving as a phase of fixed composition. In a cycle of cooling and crystallisation, this could be envisaged by saying that once the feldspar has moved to the minimum temperature composition it cannot shift, unless the temperature rises and melting ensues.

In experimental studies of felsic systems, wide ranges of feldspar compositions are reported as crystallising from a wide range of liquids (for instance, in the "granite" system, Tuttle and Bowen, 1958), and most of these feldspars change

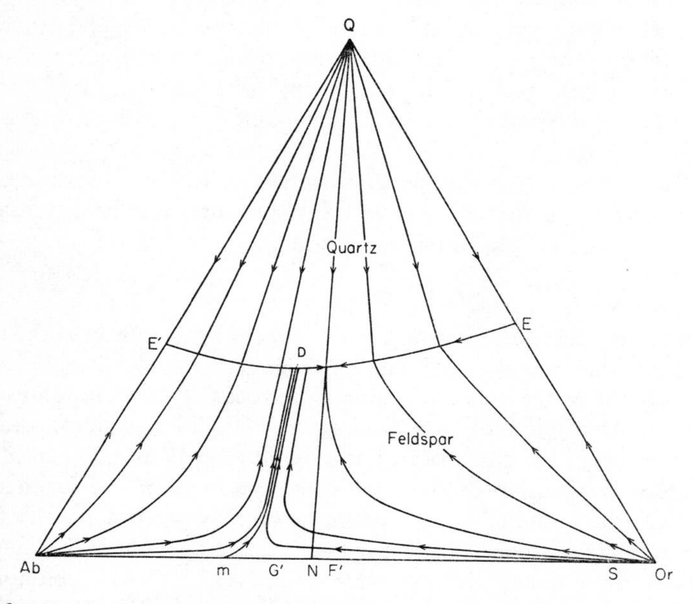

Fig. 3. After Tuttle and Bowen (1958). (a) Isobaric fractionation curves in the "granite" system at $P_{H_2O} = 1000$ kg/cm². The curves indicate the evolutionary paths of liquids when there is no equilibration with precipitated crystals, e.g. if the crystals were separated from the liquid as fast as they form. The relationships in (a) and (b) are projections onto the anhydrous base of the quaternary system $NaAlSi_3O_8$-$KAlSi_3O_8$-SiO_2-H_2O. The line mD is the axis of the "thermal valley" in the "granite" system at $P_{H_2O} = 1000$ kg/cm².

composition strongly during the course of crystallisation. These represent cases of crystallisation on the solid solution loops well away from the minimum—such results are invaluable for helping to define those ranges of composition which do precipitate feldspar of minimum characteristics, but may not themselves be easily applied directly to rocks. Even in the relatively simple systems that have been studied experimentally there is a deficiency of data on the paths of feldspar crystallisation away from the minimum zone. Tuttle and Bowen (1958) provided equilibrium and fractional crystallisation diagrams for the "granite" system at $P_{H_2O} = P_{total} = 1000$ kg/cm², which are reproduced here in Fig. 3. It needs to be pointed out that even with Tuttle and Bowen's published data, it is not easy to reconstruct these diagrams. These data are tabulated in the

Appendix, and the interested reader should first try to reconstruct the original phase diagram (with temperature contours), noting how closely he can define the position of the quartz–feldspar minimum, and the axis of the "thermal valley" in the feldspar liquidus surface. Construction of equilibrium and fractional crystallisation diagrams from the few quoted feldspar determinations is

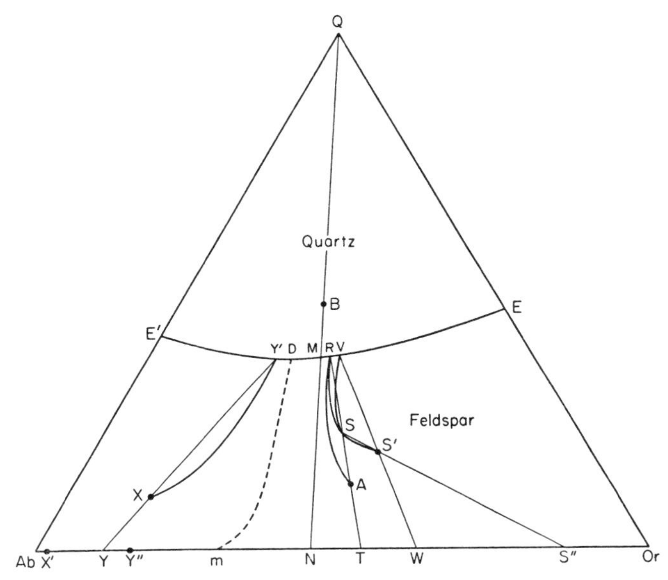

Fig. 3(b). Isobaric equilibrium curves in the "granite" system at $P_{H_2O} = 1000$ kg/cm^2. The curves S′SV, AR and XY′ are evolutionary liquid paths when the feldspar crystals are in continuous equilibrium with liquid. Interruption of equilibrium, e.g by removal of crystals at point S results in a new liquid path SR when crystallisation resumes. The straight lines, such as YXY′, are tie-lines between coexisting crystals (Y) and liquid (Y′) for the bulk composition X at a specific equilibrium.

correspondingly harder. Of course, Tuttle and Bowen had a big advantage since they constructed their diagrams with the benefit of years of experience in making the experiments, and probably more determinations of feldspar compositions than were eventually published. It is, none-the-less, a salutary experience to try to reconstruct what is justifiably regarded as a cornerstone of the experimental study of felsic rocks. Feldspar determinations have been made in subsequent studies (Carmichael and MacKenzie, 1963, and Thompson and MacKenzie, 1967; for oversaturated peralkaline systems: Hamilton and MacKenzie, 1965, and Morse, 1969 and 1970; for undersaturated compositions) but only Hamilton and MacKenzie (1965) provide an equilibrium and fractional crystallisation diagram. All the experiments were made with $P_{H_2O} = P_{total} = 1$ Kbar, except those of Morse ($P_{H_2O} = P_{total} = 5$ Kbar). Data for other pressures, more complex systems, and H_2O deficient conditions are sorely lacking. Where rocks are found with hypersolvus alkali feldspars of widely different composition from Or_{30-40} they presumably contain valuable information about their mode of formation but at the moment there is not much experimental information to use to interpret these cases. The experimental information is too sparse, and is carrying more interpretative weight than it should reasonably be expected to bear. More petrography and analyses of natural feldspars are probably the required stimulus for more searching experimental studies on alkali feldspar crystallisation.

Two warning notes are needed when considering alkali feldspar and liquid relations. The first concerns the use of normative expressions of the liquids from which the feldspars separate. The method has been popular since Bowen (1937) used it to compare rock compositions with those in Petrogeny's Residua System (see Figs 1 and 4). It is valid as long as the rock compositions reasonably approximate to the plane of the System, i.e. the alkali:alumina ratio is roughly unity after removal of anorthite (*an*). For peralkaline compositions the simple normative projection is biased because it neglects excess alkalis, which in the CIPW norm are calculated only as sodium silicates. In the norm projection, the important relationship of alkali ratios in feldspar crystals and their conjugate liquids thus becomes distorted, and may even become seemingly reversed (Bailey and Schairer, 1964). The second area of possible ambiguity lies in the fact that all reported experiments have been made under conditions of excess H_2O ($P_{H_2O} = P_{total}$). Little information is available on the actual compositions of the liquids (usually described in terms of *anhydrous* bulk composition) and none at all on any coexisting hydrous fluid. Alkalis, alumina and silica must be partitioned between liquid and fluid, and the extent of partitioning will be a function of the *amount* and solubility of H_2O in each experiment. No data on these variables are reported in most experiments. Until more data are forthcoming the results of water saturated alkaline systems must be treated with reserve, because the real compositions of the liquids that coexist with alkali feldspars are not precisely known.

Experimental studies of the systematic effects of adding anorthite to felsic alkaline systems have not been reported, although the foundations for such studies are being laid in the "granite" system, as described by Luth (this volume).

In summary, therefore, it can be said that there is still much to be done before the experiments give us an adequate method for interpreting feldspar relations in felsic alkaline rocks. Some noteworthy attempts have been made to use the scarce data which are available (e.g. Nash *et al.* (1969) and Presnall and Bateman (1973)) but the resulting models do not lend themselves to quantitative application. Whilst they have undoubted value for obtaining a "feel" for possible natural processes, extension of the models to detailed comparisons of natural and synthetic compositions, and any resulting conclusions, should be viewed against the background of available data.

(2) *Non-Peralkaline Feldspathoidal Rocks*

Trachyte and phonolite (syenite and nepheline syenite) are the typical and most abundant rocks of this type. As Bowen (1937) pointed out the composition range correlates well with the low-temperature region in the nepheline–kalsilite–alkali feldspar phase diagram, and most of the non-peralkaline types, with an alkali/alumina ratio close to unity, are genuinely analogous to compositions in this system. Examination of the same system at $P_{H_2O} = 1$ Kbar reveals a similar distribution of phase fields and temperature contours, except that temperatures are 200°C–300°C lower, and the stability of leucite is much reduced (Hamilton and MacKenzie, 1965). At 1 Kbar pressure there is only a small shift in the nepheline–feldspar minimum and the authors confirm Bowen's finding and demonstrate a strong concentration of natural rock compositions in the one atmosphere minimum region (see Fig. 4).

At $P_{H_2O} = 5$ Kbar (Morse, 1969) the feldspar–nepheline eutectic temperature has fallen to 635°C, but its composition has shifted only to a slightly higher nepheline content compared with the 1 atm and 1 Kbar determinations. New features under these conditions are the liquidus crystallisation of two alkali feldspars, and the intervention of a narrow stability field for analcite between sodic felspar and nepheline. The field of analcite appears to be expanded even further by increasing P_{H_2O} (Hamilton, 1972). Morse (1969) says that the common occurrence of nepheline syenites with coexisting nepheline and albite shows that such rocks formed under a water pressure less than 5 Kbar. If this is so then any such nepheline syenites in which there is clear evidence of primary *magmatic* crystallisation of two alkali feldspars (i.e. K-rich feldspar in addition to albite) were subject to a further constraint, i.e. the P_{H_2O} must have been sufficiently high for the feldspar solvus to intersect the solidus (between 2 Kbar and 5 Kbar). Most syenites and nepheline syenites, however, have hypersolvus alkali feldspar, and their mafic mineralogy usually suggests that they crystallised under water-deficient conditions. This conclusion is strongly supported by the high temperatures and low H_2O contents of trachyte and phonolite lavas. In this context, it is worth remembering that it would be virtually impossible to erupt water-saturated melts from depth because on rising they would pass below the solidus as soon as the pressure diminished (see Luth, 1969; and Harris *et al.*, 1970): for high level eruption to be possible the melts would have to have positive PT melting curves in the higher part of the pressure range. In this case it might be impossible for the alkali feldspar solvus (with only a small positive PT slope) to intersect the solidus at any reasonable pressure; under these conditions hypersolvus feldspar would be common, as seems to be the case. The main value of the hydrothermal results to date is, therefore, to indicate that increasing P_{H_2O} does not greatly change the composition range of trachytic–phonolitic liquids, and to suggest limits for the P_{H_2O} that may apply to magmas (see Fig. 5).

One of the problems raised by the experimental results is that analcite does not appear on the liquidus of ab–ne–H_2O until $P_{H_2O} \sim 5$ Kbar. This result is seriously at odds with the many descriptions of primary analcite in volcanic rocks, and especially with the oft-recorded analcite syenite streaks in shallow teschenite intrusions (e.g. Tyrrell, 1928). These streaks are generally attributed

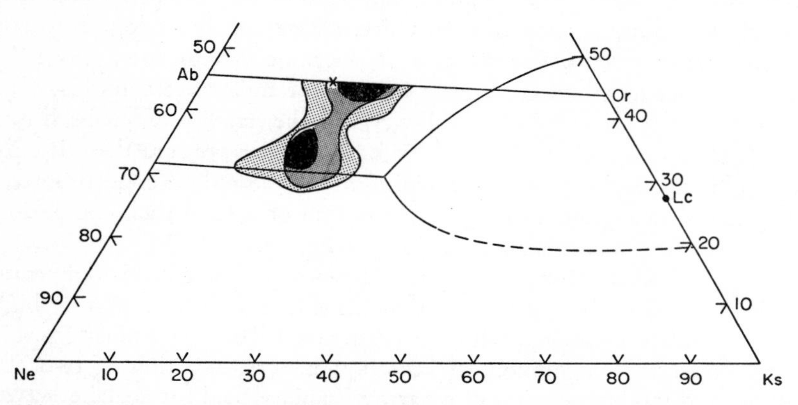

Fig. 4(a). Contour diagram illustrating the distribution of analyses of 102 plutonic rocks in Washington's tables (1917) that carry 80% or more of normative *ab* + *or* + *ne*. (After Hamilton and MacKenzie 1965.)

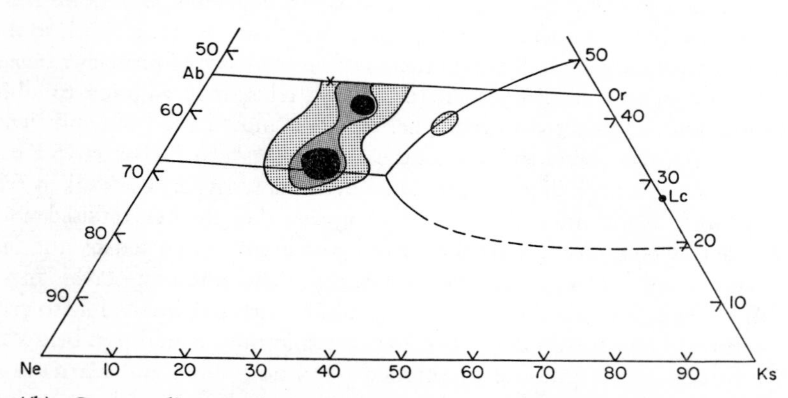

Fig. 4(b). Contour diagram illustrating the distribution of analyses of 122 extrusive rocks in Washington's tables (1917) that carry 80% or more of normative *ab* + *or* + *ne*. (After Hamilton and MacKenzie 1965.)

to the crystallisation of a liquid residuum from the nearly solidified teschenite. It is perhaps worth noting that these analcite syenite streaks are commonly peralkaline or have peralkaline tendencies, and there is a possibility, worth testing by experiment in the system analcite–sodium silicate–H_2O, that the stability field of analcite is extended to lower pressures by the presence of sodium silicate in the melts. Ford (1972) has recently shown that oversaturated hydrous peralkaline liquids, which may be analogous to the analcite-bearing cases, can exist to temperatures below 400°C. The work of Tuttle and Bowen (1958) showed that the addition of excess alkalis to melts in the system Q–Or–Ab–H_2O produced a compositional continuum from hydrous magmatic melts to hydrothermal fluids.

In his discussion of syenites and nepheline syenites, Morse (1969) makes the point that a group of 81 syenite analyses from Greenland tend to cluster around

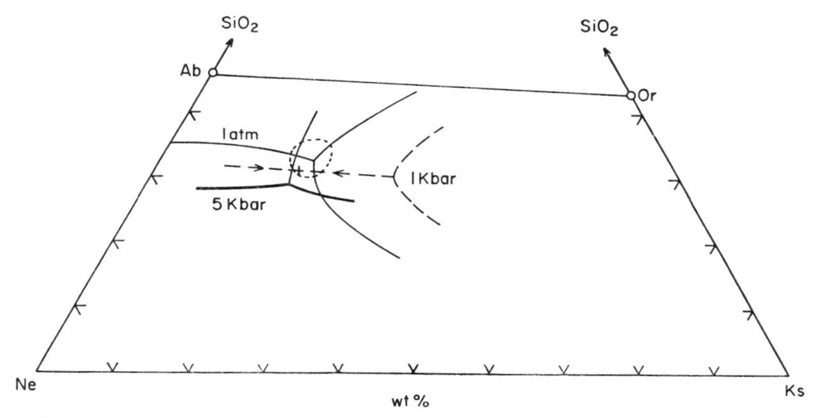

Fig. 4(c). Positions of nepheline syenite minima in petrogeny's residua system at 1 atm and 1 and 5 Kbar P_{H_2O}. The 1 atm boundaries (fine line) are from Schairer (1950); the 1 Kbar boundaries (long dashes) are from Hamilton and MacKenzie (1965); and the 5 Kbar boundaries (heavy line) are from Morse (1970). The 1 atm minimum lies very near the leucite reaction point. The dashed circle is the contour around the statistical maximum of nepheline syenite analyses in (b). (After Morse, 1970.)

Or_{40} when projected into Petrogeny's Residua System, whereas the experimentally determined minimum on the alkali feldspar loop ranges from Or_{35-30} (0–5kb P_{H_2O}). Analyses from Washington's tables (1917) of plutonic and extrusive rocks (Hamilton and MacKenzie, 1965) show similar maxima around Or_{40}, and a subsequent test of trachytes from the Atlantic islands (MacKenzie, 1972) shows the same effect. Both Morse and MacKenzie suggest that the control on this clustering is exerted by the crystallisation of the mafic phases, implying that the clustering of rock analyses is around a natural feldspar composition which is effectively controlled by this cotectic crystallisation. But many analyses of trachyte anorthoclase, and its plutonic perthitic equivalents are nearer to Or_{30} than Or_{40}, and as Morse himself states elsewhere, most of the feldspar in the Labrador and Greenland syenites is a hypersolvus member of the plagioclase–anorthoclase series: he goes on to add, however, that a *second, K-rich* feldspar may appear in the very late stages (my emphasis). This latter observation offers one reason for the Or_{40} clustering of the whole rock analyses—many syenites finished their crystallisation in the presence of a K-enriched residual fluid which is not modelled by the simple experiments on the alkali feldspar join. A further reason for the apparent *or* enrichment when syenite and trachyte compositions are projected into Petrogeny's Residua System is that many of these rocks contain biotite, which can only be expressed as *or* in the CIPW norm. The reader is recommended to try to identify for himself one major area of uncertainty in the experiments which may also be contributing to this "dilemma".

A comment is needed on leucite. It may be seen from Fig. 4 that most felsic undersaturated rocks plot outside the leucite stability field, except at 1 atm pressure. This observation has sometimes been coupled with the absence of leucite in plutonic rocks, and the suggestion made that leucite crystallises from liquids only at low pressures. This is true when $P_{H_2O} = P_{total}$ (the large contraction of the leucite field at $P_{H_2O} = 1$ Kbar being seen in Fig. 4) but as Lindsley (1967) has shown, leucite is stable to 20 Kbar under anhydrous conditions, and the incongruent melting of potash feldspar (to leucite + liquid) persists to 19 ± 1 Kbar. Many felsic magmas are not saturated with water, and may not achieve such a condition even in the late stages of crystallisation, so the absence of plutonic leucite cannot be simply attributed to high P_{H_2O}. Part of the answer must lie in the instability of leucite in the sub-solidus. Scarfe, Luth and Tuttle (1965) have shown that in the pressure range 0–10 Kbar, leucite breaks down to form kalsilite + K feldspar, at temperatures 500°C–800°C, respectively. Their breakdown curve was corroborated by Lindsley (1967) at higher pressures. This breakdown is called the "pseudoleucite reaction" in reference to the observed breakdown of some natural sodic leucites in shallow intrusions to nepheline + orthoclase. It is clear that under slow plutonic cooling there will be ample opportunity for thorough recrystallisation of any earlier formed leucite, whilst the intrusion is in the range 500°C–800°C. Furthermore, it should be remembered that, in general, highly potassic magmas are basic (rather than felsic) and the plutonic equivalents of leucite-bearing lavas typically contain biotite as the potassium-rich phase.

All the experiments so far described indicate that trachyte–phonolite compositions are consistent with liquids that have achieved equilibrium with alkali feldspar + feldspathoid at moderate to low pressures. There is surprisingly little discussion of the possibility that the same situation might extend to mantle pressures: yet there is nothing in the high pressure experiments to rule this out. Jadeite composition ($NaAlSi_2O_6$) is equivalent to albite + nepheline, i.e. a simplified nepheline syenite composition, and its melting relations are of obvious interest. Bell and Roseboom (1969) have shown that the most important effect of increasing pressure up to 25 Kbar is to shift the composition of the nepheline–albite eutectic towards higher nepheline contents, whilst Bell and Davis (1969) have shown that the first melt in jadeite–diopside is jadeite-rich.

Thus, the low-temperature melts from mantle pyroxenes at depths well into the upper mantle may be expected to be rich in the components of nepheline and alkali feldspar. A more realistic attack on this question would bring in potash and a start has been made by Yoder and Upton (1971) on diopside-sanidine at $P_{H_2O} = 10$ Kbar (Fig. 6). The results show that initial melts in the system are rich in sanidine (and poor in diopside) but this work needs extending to sodic compositions. Jadeite–phlogopite would be a system of obvious interest because of the recent experimental confirmation that phlogopite is stable over a wide

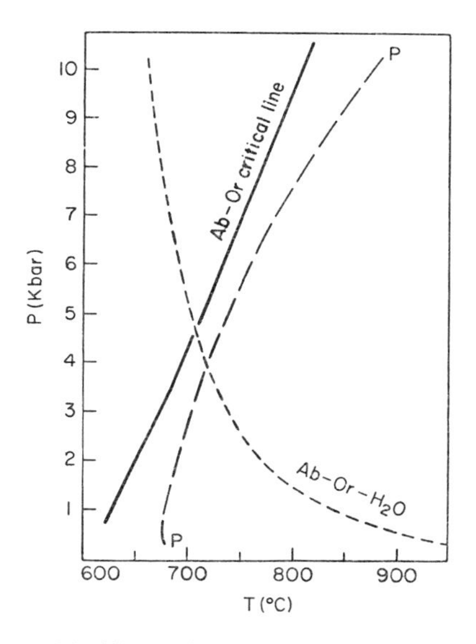

Fig. 5. The alkali feldspar critical line (solvus crest) in PT space (from Luth, this vol.) compared with the synthetic alkali feldspar solidus minimum ($P_{H_2O} = P_{total}$) (Luth, this vol) and the *anhydrous* solidus of a natural glassy trachytic pantellerite, P–P (specimen number KE4, with 18.5% *q* and 53% *or* + *ab*). P–P was determined experimentally by J. P. Cooper (personal communication, 1975) and represents the *minimum* position of the solidus (the specimen was essentially crystalline at higher temperatures). It may be seen from this that dry trachytes (and rhyolites) may be completely solidified above the solvus, at all depths in the crust.

range of upper mantle conditions so it is only a matter of time before systems like jadeite–phlogopite will be examined. An interesting equivalence appears if carbonates are added to the system:

High Pressure		*Low Pressure*
jadeite + phlogopite + calcite + CO_2	=	albite + nepheline + orthoclase + dolomite + H_2O
		(nepheline syenite + carbonatite + H_2O)

Systems of this complexity are of course difficult to study at the required temperatures and pressures, and may be expected to pose severe quenching problems; this example is probably too artificial anyway, but it may stand as a warning against complacency, if only because there are some reported occurrences of trachyte and phonolite containing peridotite (?mantle) nodules (Wright, 1969). There is therefore no experimental evidence directly opposed to an origin for some nepheline syenitic melts in the mantle to depths of 70–80 km, although the implications of the data of Bell and Roseboom (1969) and Kushiro (1968) are that these would be more undersaturated (and presumably more basic) than those rocks that cluster so notably around the minimum in Petrogeny's Residua System (Fig. 4). High pressure melts might be expected to be more akin to the nephelinitic phonolites to be discussed in a later section.

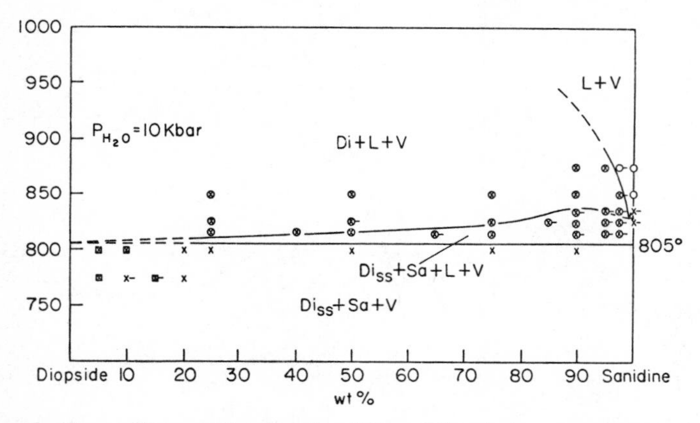

Fig. 6. Pseudo-binary diagram for diopside (Di)-sanidine (Sa) at P_{H_2O} = 10 Kbar. Note that the region of extensive low-temperature melting is for sanidine-rich compositions.

(i) *Corundum-bearing Nepheline Syenites*. Separate consideration is needed for the special problem of the peraluminous character of this group of rocks. Examination of Fig. 7, the phase diagrams of Na_2O-Al_2O_3-SiO_2 and K_2O-Al_2O_3-SiO_2 (Schairer and Bowen, 1955 and 1956) reveals that the peraluminous eutectics in each ternary are close to the lines of Al_2O_3/alkali = 1, and there is nothing to indicate that the situation will be radically different in the quaternary, nor, indeed, at higher pressures. Most CIPW norms of lavas are corundum-free, and those which are not contain such small amounts of *c* that it can be attributed to analytical error or alteration in the sample (see Chayes, 1970). Fresh trachytes and phonolites do not contain alumina-rich phases, such as spinel and white mica, so that some special explanation seems necessary for syenites with modal corundum, or high normative corundum. One possibility is that these syenites represent melts that have differentially lost alkalis (and silica), but there is good evidence from the Bancroft–Haliburton area (which contains some of the best examples of corundum syenites) of the involvement of metasediment in their formation. Tilley (1958) consolidated the findings, and the concept of nephelinisation proposed by Gummer and Burr (1943) for the origin of these rocks. Most of them retain a metasedimentary fabric, but some nepheline syenites are intrusive, and the description of corundum-bearing pegmatites suggests that low-temperature volatile-rich fluids were available. Such fluids would not normally be capable of reaching the Earth's surface as silicate melts, and the absence of equivalent lavas is not surprising. All the evidence conspires to suggest that these corundum-rich melts are the products of solutions reacting with aluminous country rocks. Tilley (1958) favours the view that the metasomatising solutions were themselves nepheline-bearing, possibly derived from nepheline syenite magma. Other workers have favoured granitic emanations (e.g. Moyd, 1949) that were de-silicated by passage through limestone, but even this would be unnecessary if the parent granite were peralkaline and emitted residual fluids such as those in the system Na_2O-Al_2O_3-Fe_2O_3-SiO_2: these are sufficiently low in silica to produce nepheline by reaction with aluminous material (Bailey, 1963). The mixed calcareous–pelitic assemblage of Bancroft–Haliburton would be ideal for the process.

(ii) *Non-peralkaline Nepheline Syenites.* Petrographic text-books tend to favour in their descriptions of nepheline–syenites the peralkaline varieties, containing sodic amphiboles and pyroxenes. This is partly because the non-peralkaline types, containing biotite and hornblende, tend to be more basic and have affinities with nepheline monzonite, nepheline diorite and nepheline gabbro, with the result that they may appear in different parts of the classification. Indisputable nepheline syenites of non-peralkaline character may be less abundant than peralkaline types, but they are not uncommon: the two types constitute the "miaskitic" and "agpaitic" nepheline syenites of Ussing (1911) and they exhibit other distinctive differences, apart from alkali/alumina ratio (see Heinrich, 1966; and Sørensen, 1974 for summaries). In terms of the fractional crystallisation model of magma evolution, miaskitic (alkali/alumina < 1) derivatives from a basaltic parent raise a difficult question, because unless the parental melt were corundum–normative (an extremely unlikely case: Chayes, 1970, p. 180) it might be expected that the operation of the plagioclase effect must lead inescapably to peralkalinity in the felsic residua. The other chemical differences between miaskitic and agpaitic compositions are equally difficult to explain by crystal fractionation. Assimilation or metasomatism offer no way out of the dilemma because miaskitic trachytes and phonolites are not restricted to the continents, any more than are non-peralkaline rhyolites. We do not yet know enough about magma processes or chemistry to answer this question, but it might have been asked sooner!

Another problem emerges when the natural rock compositions are plotted in the experimental system nepheline–kalsilite–silica (Fig. 4). It is true that rhyolites, trachytes and phonolites (and their plutonic equivalents) are noticeably concentrated in the low-temperature region as shown by Bowen (1937), but the widely-accepted conclusion that rhyolites and phonolites are derived through trachytic stems needs closer examination. The distribution of analyses is strongly tri-modal around phonolite, trachyte and rhyolite, which is not to be expected from any smooth evolutionary production of phonolite (nor rhyolite) from trachyte. If these residua are derived by fractional crystallisation from a basaltic parent there is a strong suggestion that many phonolites and rhyolites have not come through a trachytic stage, i.e. either the basaltic parental material may be variable or some other process strongly modifies the fractionation. In the rhyolite case the former alternative may apply because many rhyolites could have come from a tholeiite parent and Chayes (1971, 1973) has demonstrated a marked bimodality in basalt compositions with strong maxima

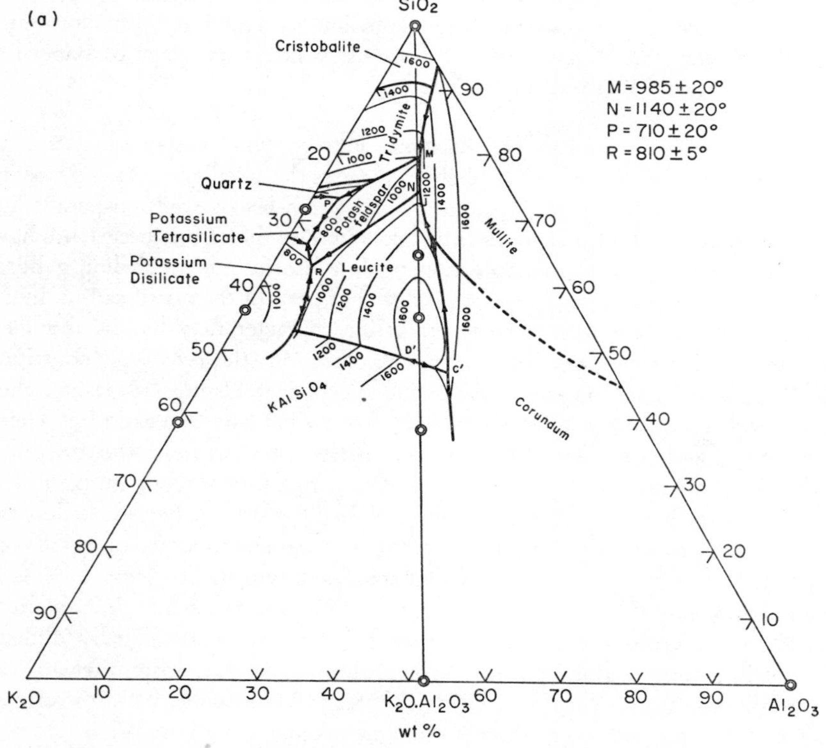

Fig. 7. Ternary liquidus diagrams for the systems (a) K_2O-Al_2O_3-SiO_2, and (b) Na_2O-Al_2O_3-SiO_2, after Schairer and Bowen (1955 and 1956). Only a few of the invariant temperatures are shown, to illustrate the marked differences between peraluminous and peralkaline conditions (full details are given in the sources). Double circles indicate compositions (note that orthoclase composition lies within the leucite field).

around tholeiite and alkali olivine basalt: but this explanation of the rhyolites only makes the trachyte–phonolite relationship even more enigmatic. If the trachyte–phonolite bimodality results from fractional crystallisation of alkali olivine basalts, in which no bimodal distribution is discernible, the explanation for the divergent trends will come only when we can discern the point of separation and identity the cause—until we can do this other origins for phonolite and trachyte must be sought. Coombs and Wilkinson (1969) have pointed out that in some provinces, such as E. Otago, there is no evidence for an alkali basalt–trachyte–phonolite line, but that the association basanite (nephelinite?)–phonolite is common, whatever the connecting links may be. Another simple explanation for the trachyte and phonolite bimodality is that some phonolites are products of partial melting, and their contribution to the sample population is causing the second maximum around the phonolite composition: this possibility is strengthened by the occurrence in East Africa of widespread plateau phonolites, which it is hardly plausible to suppose are products of protracted fractional crystallisation (Bailey, 1974).

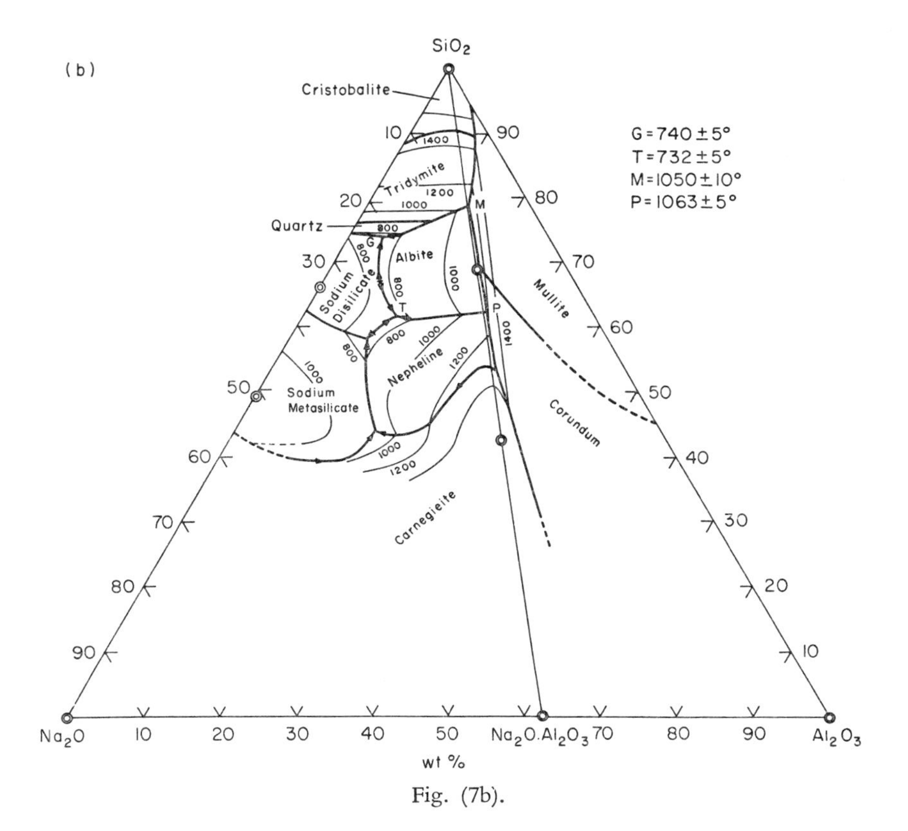

Fig. (7b).

The main contributions to be sought immediately in experimental petrology are better limits on total pressures and volatile pressures involved in nepheline syenite formation. Work on the "granite" system under volatile deficient conditions shows significant differences from the water saturated results (Luth, this volume): similar studies are needed in the silica-undersaturated part of "Petrogeny's Residua System". With improved techniques some exploration of the causes of the differences between miaskitic and agpaitic melts is to be hoped for, and with increasing use of the electron- and ion-probes the detailed unravelling of residual liquid and partial melting trends.

(3) *Peralkaline Feldspathoidal Rocks*

In the previous section it was possible to discuss the non-peralkaline nepheline syenites and phonolites in terms of the system nepheline–kalsilite–silica, because the known eutectics with alumina in excess of alkalis are not far removed from the nepheline–kalsilite–silica plane. Such is not the case for the peralkaline eutectics, as examination of the phase diagrams for Na_2O-Al_2O_3-SiO_2 and K_2O-Al_2O_3-SiO_2 quickly reveals (Fig. 7) (Schairer and Bowen, 1955 and 1956). Addition of excess alkalis to a composition in nepheline–kalsilite–silica will considerably extend the path of crystallisation, both in terms of temperature and composition, and similar indications were found during some experiments in the peralkaline region of K_2O-Al_2O_3-SiO_2-H_2O by Tuttle and Bowen (1958). Clearly, more experiments were needed in peralkaline systems and some progress has been made by the study of sodic compositions involving especially the sodic pyroxene acmite ($NaFeSi_2O_6$). Univariant equilibria involving nepheline + albite + acmite + liquid were found in the appropriate portion of Na_2O-Al_2O_3-Fe_2O_3-SiO_2 at 1 atm (Bailey and Schairer, 1966), where this univariant crystallisation path was found to link an invariant reaction point (nepheline + albite + acmite + hematite + liquid) to a eutectic (nepheline + albite + acmite + sodium disilicate + liquid): Fig. 8. The liquid composition at the reaction point is rich in nepheline and acmite and closely resembles the natural composition of malignite (or late stage liquids in nephelinite or ijolite). For this reason the natural analogues of the univariant path, nepheline + albite + acmite + liquid, are probably those nepheline (and acmite)-rich phonolites found associated with nephelinites, and referred to variously as nephelinitic phonolite and phonolitic nephelinite. Petrographically these phonolites are distinguished by being more strongly undersaturated than the trachytic phonolites, and having nepheline rather than feldspar phenocrysts. The final eutectic involving sodium disilicate must be fairly close in composition to the ternary eutectic (nepheline + albite + sodium disilicate + liquid) in Na_2O-Al_2O_3-SiO_2 (see Fig. 7), i.e. poor in iron and rich in sodium silicate. It probably has no natural rock analogue, because in natural liquids additional components would be present, and other phases would almost inevitably intervene before such an alkali silicate-rich liquid was achieved. For instance, it might be expected that with such an enrichment of alkalis there would be concomitant enrichment in water and other volatiles, with the consequent intervention of a hydrous fluid phase. The chief value in identifying this eutectic is that it indicates the direction in which peralkaline nepheline syenites would trend during crystallisation—most of the rocks that we see represent stages along the way (albeit by more complex paths).

The melting relations of nepheline–albite–acmite in the presence of excess

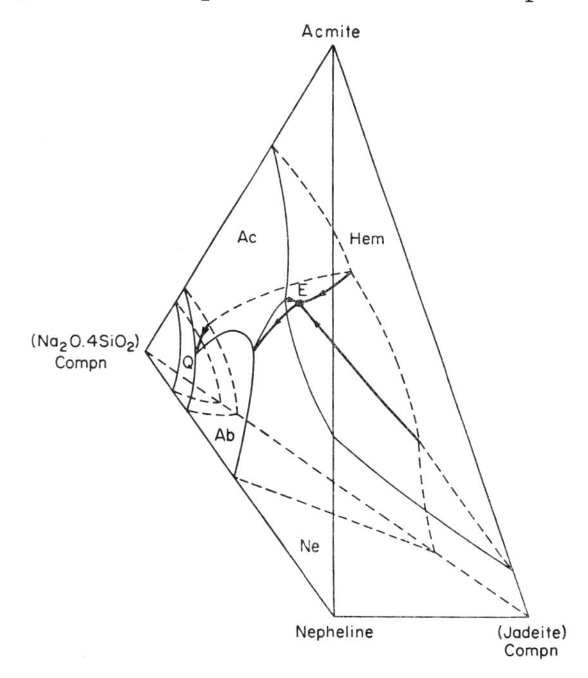

Fig. 8. Illustration of the phase relations in the volume Acmite–Nepheline–(Jadeite)–$(Na_2O.4SiO_2)$, showing the proximity of the quaternary reaction point, E (Ac + Hem + Ne + Ab + liquid) to the ternary reaction point, Ac + Hem + Ne + liquid. Phase boundaries on the faces of the volume—thin lines: univariant curves bridging the faces—heavy lines. Arrows show falling temperatures.

H_2O at 1 Kbar have been described by Nolan (1966) who located a liquid composition (projected into the anhydrous plane of the system) in equilibrium with nepheline + albite + acmite + magnetite (+ fluid). (See Fig. 9.) This liquid is not an isobaric invariant composition, but as projected it is analogous to the reaction point in Na_2O-Al_2O_3-Fe_2O_3-SiO_2, and hence to the natural malignite composition, and it confirms that liquids rich in the constituents of nepheline and acmite are probably relatively low temperature "residua". By adding diopside Nolan was able to produce dramatic increases of albite and nepheline content in the liquid composition in equilibrium with nepheline + plagioclase + pyroxene. Thus the liquid composition moved towards a composition more akin to normal nepheline syenite (with only a small pyroxene content) as the acmite content diminished. Nolan points out, too, that natural pyroxenes in nepheline syenites normally contain low to moderate amounts of the acmite molecule, which strengthens the analogy with the synthetic results. The series of liquids identified by Nolan represent the extension from nepheline–albite–diopside into the peralkaline region, with increasing amounts

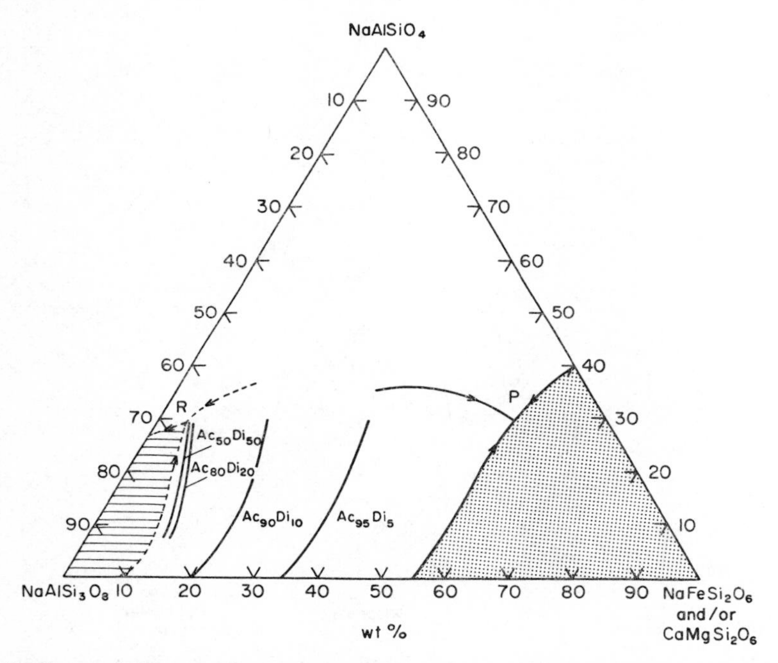

Fig. 9. Composite diagram for a portion of the system $NaAlSi_3O_8$-$NaAlSiO_4$-$NaFeSiO_6$-$CaMgSi_2O_6$-H_2O at $P_{H_2O} = 1000$ kg/cm², after Nolan (1966). Ruled area denotes the plagioclase feldspar phase volume determined in the system $NaAlSi_3O_8$-$NaAlSiO_4$-$CaMgSi_2O_6$-H_2O. Stipple denotes the pyroxene phase volume determined in the system $NaAlSi_3O_8$-$NaAlSiO_4$-$NaFeSi_2O_6$-H_2O. The shift in the trace of the feldspar-pyroxene boundary surface with changing pyroxene composition is also shown. This diagram is a projection of data (for comparative purposes) from a series of composition planes (expressed in anhydrous terms) between *ne–ab–di* and *ne–ab–ac*. In all phase fields a hydrous vapour (fluid) is also present, and in acmite-rich compositions magnetite is a further additional phase. Hence the assemblage at R is plagioclase + nepheline + pyroxene + liquid + fluid, and at P there are six phases, albite + nepheline + acmite + magnetite + liquid + fluid.

of acmite, but the relationships between the liquids in the series, and the ultimate residua, are unknown. Nolan stresses the concentration of rock analyses around the diopsidic end of the series, and doubts whether the more acmitic liquids are petrologically significant (1966, p. 157). However, special factors may be contributing to the paucity of analyses in the acmite-rich region. (1) Increasing peralkalinity will probably be accompanied by increasing volatile enrichment, with rapidly-falling viscosity and the consequent higher chances of residual liquid losses during crystallisation. (2) If acmite-rich fluids are low-temperature "residua" (as much petrographic evidence, and many experiments suggest) then the chances of finding such highly reactive residua isolated or uncontaminated must be correspondingly less. It seems premature therefore to dismiss the finding of acmite-rich liquids as inconsequential, partly because rocks of appropriate composition do exist, but mostly because these liquids indicate the direction in which peralkaline compositions trend—while this may seem quantitatively trivial in terms of magmatic rocks formed from residual liquids, it may be vitally important in understanding partial melting processes, alkali–iron metasomatism, or the concentration of rare elements in agpaitic plutons.

When all this has been said, however, the fact remains that very strongly peralkaline undersaturated lavas are relatively scarce (Macdonald, 1974). The different levels of peralkalinity may be seen in Nolan's separate plots of plutonic and volcanic rocks, in which strongly peralkaline compositions are much more evident among the plutonic samples. Furthermore, it appears that phonolitic lavas seldom achieve the high degrees of peralkalinity found in the oversaturated pantellerites (Macdonald, 1974): the levels of peralkalinity in phonolites are more comparable with those in the oversaturated comendites, and it may be the high peralkalinity of the pantellerites that calls for special explanation.

If many phonolites originate by fractional crystallisation (or partial melting) of alkali basalt, it would be expected from the operation of the plagioclase effect that they would be peralkaline. As noted in the previous section it is the non-peralkaline phonolites that present a problem in this context. None-the-less, the problems of the agpaitic rocks are by no means resolved by the invocation of the plagioclase effect, at best it can be only a partial answer to the problem and the broader question of the development of peralkaline potential will be discussed separately in a later section. In addition, we are still far from knowing why peralkaline rocks have such a distinctive minor and volatile element content. Much more subtle experiments will be needed to research this, and they await a more detailed petrographic and chemical understanding of the natural materials. The experiments to date, at most can give only a general view of the melting conditions of peralkaline rocks. The composition range studied is narrow, e.g. no K_2O; the volatile systems have involved only H_2O which may not be apt; and the total pressure range has extended only to 1 Kbar. Obviously much needs to be done and possibly one of the most promising ways of extending our knowledge in the immediate future will be to use natural peralkaline phonolites, of various types, as experimental starting materials. A valuable start has been made by Piotrowski and Edgar (1970) and Sood and Edgar (1970) (summarised by Edgar (1974)) which indicates the importance of iron and volatile contents in determining the different crystallisation behaviour of miaskitic and agpaitic melts.

(4) *Peralkaline Oversaturated Rocks*

The more common rhyolites and granites contain molecular Al_2O_3 equal to or in excess of alkalis. Much experimental study has been devoted to compositions in which the alkali–alumina ratio is unity, and this forms the topic of another section of this book (Luth: pp. 335–417). The present discussion is limited to those rhyolites and granites which are frequently called "alkaline" but which are more accurately described as peralkaline. It is clear that crystallisation of these peralkaline liquids produces drastic changes in the chemical composition—mainly loss of Na_2O—so that in the discussion which follows the lava terminology will be used exclusively because the plutonic names (no matter how accurately defined) cannot describe natural liquid compositions (see Macdonald and Bailey, 1973, for a review). Among the peralkaline rhyolites it is useful to make the distinction between comendite and pantellerite. Lacroix suggested that the term pantellerite be employed when the normative femics exceeded 12.5%, and this apparently arbitrary subdivision seems to separate two groups of peralkaline rhyolites that have other distinctive chemical attributes (Macdonald and Bailey, 1973) and possibily different origins.

The quartz + feldspar + alkali silicate + liquid eutectics in Na_2O-Al_2O_3-SiO_2 and K_2O-Al_2O_3-SiO_2 (Schairer and Bowen, 1955 and 1956) and the quartz + feldspar + liquid cotectics leading to them, constitute the simplest analogies to peralkaline rhyolites (Fig. 7). A closer analogy was found in the quartz + albite + acmite + liquid path in Na_2O-Al_2O_3-Fe_2O_3-SiO_2 (Bailey and Schairer, 1966): at its low temperature end this path terminates in the quaternary eutectic quartz + albite + acmite + sodium silicate ($3Na_2O \cdot 8SiO_2$) + liquid, which is compositionally close to the ternary eutectic in Na_2O-Al_2O_3-SiO_2. All three eutectics are rich in alkali silicate, and low in temperature, and it is doubtful if natural silicate melts could approach this condition without the intervention of other phase configurations. As with the undersaturated melts, these eutectics serve to indicate the direction of the liquid paths leading to them, and in this respect the paths involving quartz + feldspar + liquid have been shown to provide good analogies for rocks, because most comendites and many pantellerites plot in the experimental cotectic zone (shown in Figs 10 and 11). For a detailed account, see Macdonald and Bailey (1973).

Other synthetic analogues have been provided by Carmichael and MacKenzie (1963) and Thompson and MacKenzie (1967) in which they explore selected joins in the system Na–K–Al–Fe–Si–O–H at 1 Kbar pressure. These experiments have the advantage that the variations in alkali feldspar compositions can be studied, but, in common with all water-saturated experiments, the precise liquid and fluid compositions are unknown, a factor that may be crucial in peralkaline systems. Nevertheless, the quartz + feldspar + "liquid" minima expressed in terms of their anhydrous bulk compositions, show a remarkable correspondence with the 1 atm cotectic region (Figs 10 and 11). The reasons for this are unknown, but it suggests that there may be no marked incongruency between liquid and fluid phases.

A good deal of effort has been expended in the synthetic studies on locating the "thermal valleys" (Carmichael and MacKenzie, 1963) or the "low temperature zone" (Thompson and MacKenzie, 1967) in the alkali feldspar primary phase region in the peralkaline system. The concept of the thermal valley is derived from the "granite system" in which Tuttle and Bowen (1958) depict a

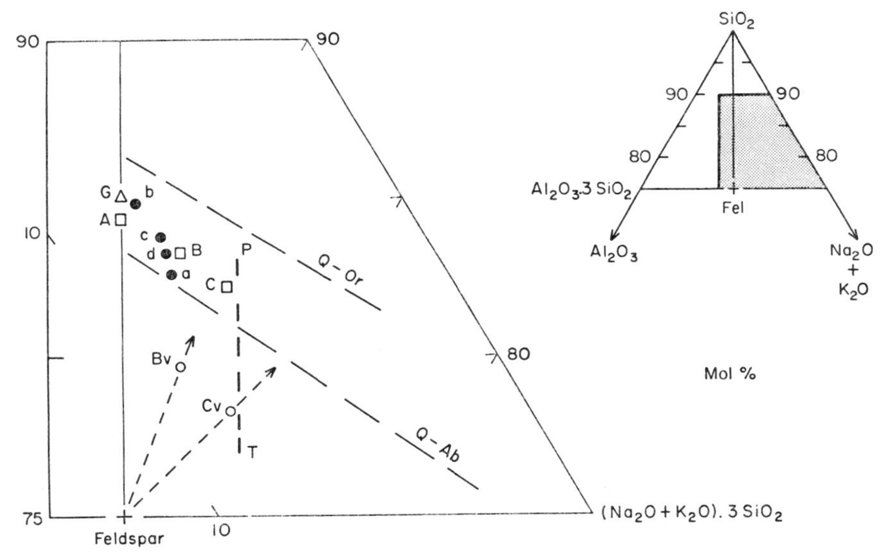

Fig. 10. Enlarged portion of the peralkaline region in alkali–alumina–silica (Bailey and Macdonald, 1969) showing the 1 atm cotectics (Q–Or and Q–Ab, Schairer and Bowen, 1955 and 1956) and the P_{H_2O} = 1 Kbar determinations of quartz–feldspar minima A, B, C from Carmichael and MacKenzie (1963): G from Tuttle and Bowen (1958): a, b, c, d from Thompson and MacKenzie (1967). Compositions in the "thermal valleys", B_v and C_v, from Carmichael and MacKenzie (1963). The arrows through B_v and C_v indicate the liquid trends generated by feldspar separation. The heavy line T–P is the distribution axis of glassy lavas in the series pantelleritic trachyte-pantellerite.

valley in the temperature contours on the feldspar liquidus surface. The axis of the valley was also indicated by the convergence of feldspar–liquid tielines. The "thermal valley" axis in the "granite" system is shown in Figs 18 (Part I) and Fig. 3. The head of this valley is the thermal minimum in the feldspar join where the sanidine and anorthoclase solid solution loops meet. Liquids along the axis of the valley precipitate a narrow range of feldspar composition until the quartz–feldspar boundary is reached. The implication is that the axis of the valley is a locus of liquids in the "granite system" that are in equilibrium with feldspar at the minimum between the alkali feldspar solid solution loops (with the proviso that the feldspar at this minimum may itself vary slightly with bulk liquid composition and changes of pressure). Carmichael and MacKenzie (*ibid.*) depicted the results from their two peralkaline joins by projecting them into the normative triangle *q–or–ab*, in an effort to make a direct comparison with the "granite system". In each of the joins they were able to locate a low temperature region on the feldspar liquidus, which they termed a "thermal valley",

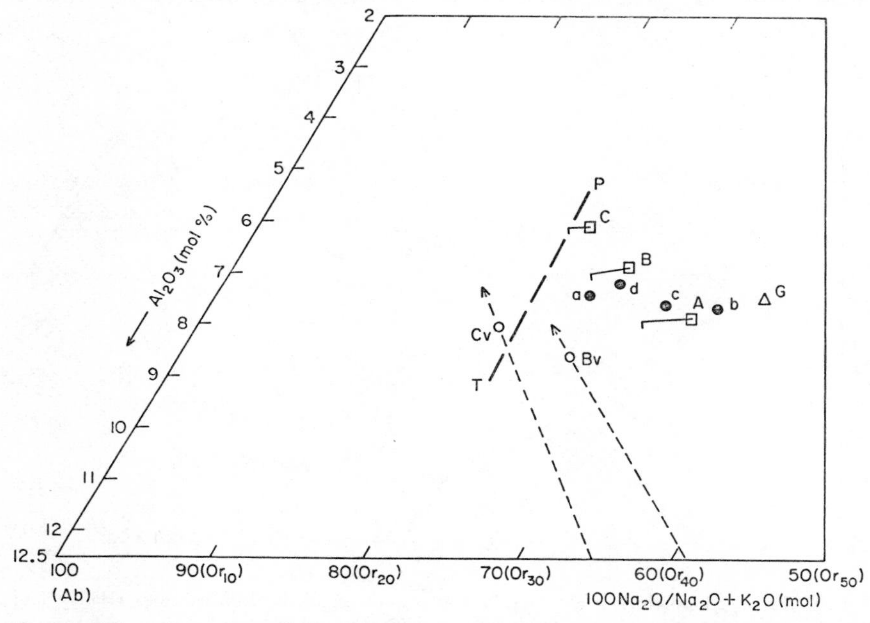

Fig. 11. Enlarged portion of the alkali-ratio *v*. alumina diagram (Bailey and Macdonald, 1969) showing the compositions in Fig. 10. The line and tick attached to A, B, and C show the quartz-feldspar boundary and the intersection of the "thermal valley" in each case. The tielines from B_v and C_v are to the conjugate feldspar compositions: the extensions (with arrows) indicate the liquid trends generated by precipitation of feldspar of constant composition (as found in the experiments: Carmichael and MacKenzie, 1963). It is apparent that this process alone cannot lead to the "thermal valley" intersections with the quartz–feldspar boundaries in B and C.

the axis of which was again indicated by the convergence of feldspar–liquid tielines across the join. In each join they identified one composition on the axis which precipitated feldspar of constant composition over a temperature interval of 30°C–40°C. The tieline between this feldspar and the liquid was used to define the axis of the "thermal valley", which was considered to be straight over most of its length (except close to the feldspar join). This idea was subsequently extended by Thompson and MacKenzie (*ibid.*) into a "low temperature zone" which could be envisaged as a surface in the peralkaline system, with its origin in the thermal valley of the "granite system". This concept is essentially the same as that depicted by Bailey and Schairer (1964, Fig. 3). Unfortunately none of the compositions in the additional peralkaline joins studied by Thompson and MacKenzie appear to lie in the low temperature zone (i.e. they do not report any compositions precipitating constant feldspar compositions). The two such compositions recorded by Carmichael and MacKenzie precipitated alkali feldspar compositions of Or_{41} and Or_{35} which is in the same range as most natural feldspar phenocrysts in peralkaline lavas, and the experimental feldspars recorded by Bailey *et al.* (1974b). There is therefore an impressive amount of data to suggest that in the peralkaline volume the alkali feldspars that are in equilibrium with natural liquids, and with low temperature experimental liquids, are generally $Or_{35\pm5}$. This is in the same range as the minima on the alkali feldspar join itself and suggests that these peralkaline liquids are in fact in equilibrium with feldspar of "minimum" characteristics, i.e. it cannot become more sodic because it is already at the bottom of the anorthoclase loop. The low temperature zone, therefore, should be envisaged as the locus of liquids in the peralkaline system which precipitate feldspar at the minimum between the alkali feldspar solid solution loops.

In Figs 10 and 11 it may be clearly seen that the precipitation of feldspar of constant composition from the two points in the axes of Carmichael and MacKenzie's thermal valleys cannot, by itself, produce residual liquids near the quartz feldspar minima. The situation is analogous to the distribution of natural pantelleritic liquids in the Kenya province (Macdonald *et al.*, 1970, and many unpublished analyses). The natural feldspar tielines show the same oblique arrangement to the liquid composition trend in the rocks (T–P in Figs 10 and 11). Although in both cases (experimental and natural) the liquids may lie in a low temperature zone and precipitate constant feldspar composition, the generation of liquids in this part of the low temperature zone *cannot* be *solely* a function of feldspar fractionation or crystallisation. At the present time, there is no demonstrable case of pantellerite derivation by feldspar fractionation in the range T–P (Figs 10 and 11). Furthermore the experiments so far do not suggest a mechanism to relate this range of compositions.

The addition of excess H_2O to peralkaline compositions causes a big lowering of liquidus and solidus temperatures, well below those known or deduced for natural rhyolites, which when fresh also have very low H_2O contents. Consequently, Bailey *et al.* (1974b) have studied the melting and crystallisation of anhydrous peralkaline obsidians, over a range of pressures. Non-peralkaline obsidians are notoriously resistant to attempts to crystallise them in the anhydrous state, but the peralkaline ones, the pantellerites especially, crystallise in only a few hours. The liquidus curves so far determined are shown in Fig. 12; they exhibit several features of interest. (1) Each curve has an inflection where the available volatiles become completely dissolved in the melt, so that from this point increasing pressure raises the liquidus temperature. (2) The liquidus temperatures fall steadily from trachyte to pantellerite, but the slope of the liquidus surface is considerably less than that found in the hydrous synthetic studies. (3) The temperature ranges are in good agreement with known thermometric determinations on rhyolites (Carmichael, 1967). Points (1) and (3) combine to suggest that the volatile contents of these melts were never vastly greater than

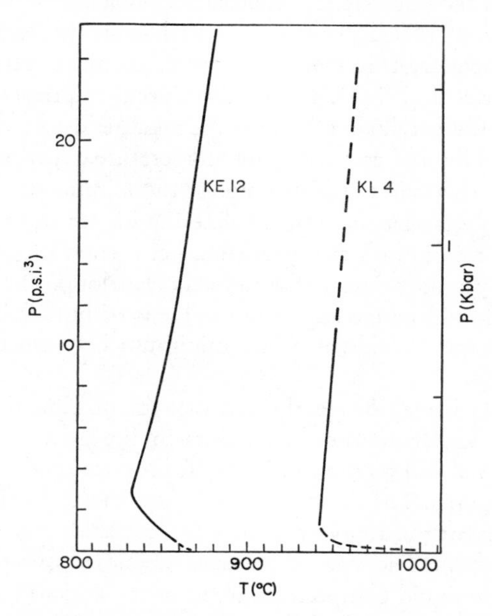

Fig. 12. Anhydrous PT liquidus curves for natural pantellerite (KE 12) and pantelleritic trachyte (KL 4) glasses, after Bailey *et al.* (1974b).

preserved in the obsidians, because had they been much richer in volatiles the melts must either have been at temperatures well above the liquidus for a long period during ascent (which is denied by the phenocrysts), or they were at much lower temperatures at depth (which seems improbable). The fact that a range of obsidians from different volcanoes have similar volatile contents and similar inflection pressures in their liquidus curves, conspires with other evidence (Macdonald *et al.*, 1970) to suggest that the volatile contents are approaching maximum levels for this range of compositions. These experiments illustrate the value of using natural glasses as starting materials, and highlight the possible dangers of relying on synthetic results from systems that depart widely from the natural *liquid* compositions.

Another series of experiments pertaining directly to rock compositions has been carried out by Martin (1974). In these he has heated basaltic compositions in a long temperature gradient in the presence of excess H_2O, and reports that peralkaline compositions develop in the cooler part of the gradient. He finds support in these results for an origin for peralkaline melts by volatile transfer in magma chambers (Smyth, 1927) but there must be some reservations about this conclusion, if only because natural rhyolite glasses (of all types) are notably deficient in H_2O.

As with the peralkaline phonolites, peralkaline rhyolites are developed in both the major divisions of the lithosphere, oceanic and continental, although they are conspicuously lacking in orogenic magmatism. So far, most of our headway has been in defining the problems of this group of rocks more precisely; experiments have helped us to see more clearly the way in which peralkaline melts may be related in physical terms but we are far from being able to explain their origins and derivation, and hence their distribution. Further experiments over a much wider pressure range, and with controlled atmospheres, are needed on the natural obsidians; more attention must be directed to experimentation relevant to the natural volatiles associated with these melts; and, to help with testing hypotheses on natural rocks, we need much more information on the behaviour of the highly characteristic trace elements, such as zirconium, in peralkaline systems. The high degree of consistency in the composition of the peralkaline obsidians (Macdonald and Bailey, 1973) gives cause for optimism that the problems of the physical chemistry may be soluble by well-designed experiments.

(b) MAFIC ROCKS

As was stated in the introduction, a consideration of the experimental petrology of feldspar-bearing mafic rocks is outside the scope of this chapter; instead it was proposed to examine the experimental contributions to the special problems of the feldspar-free alkaline rocks, of which the most abundant are the nephelinites, with their common associates, the melilitites. The additional problems of potash-enrichment encountered in some provinces, expressed in the appearance of leucite and kalsilite, will be considered separately at the end.

For these mafic rocks the historical aspect calls for no separate treatment because they were never the subject of sustained or intensive experimental study until recent years.

(1) *Nephelinites and Melilitites*

The close relationship between these two groups of rocks was perhaps first highlighted by the experiments of Bowen (1922) on nepheline–diopside. This system should be expected to provide a good analogy with nephelinite, which consists chiefly of augite and nepheline, and Bowen found that a wide range of compositions in the middle of the join had olivine on the liquidus, extending the likeness to the olivine nephelinites. But he found, too, that melilite appeared as a later-crystallising phase right across the join and that liquid persisted to temperatures below 1000°C, with no clear indication that diopside and nepheline would dominate the sub-solidus assemblage. Experimentally, the matter rested for forty years, until Schairer, Yagi and Yoder (1962) revised nepheline–diopside, and extended the observations to the sub-solidus (Fig. 13). Bowen (1922 and 1928) had interpreted the diagram, and related it to petrological observations on lamprophyres with his usual skill, but the difficulties for nephelinites are many, and serious. Natural nephelinites *do* consist largely of clinopyroxene with lesser amounts of nepheline, and variable olivine contents, whereas melilite is not an invariable nor even a common major constituent, quite the reverse. In continental volcanic provinces, olivine-poor and olivine-free nephelinites are common (Bailey, 1974); such compositions are even more

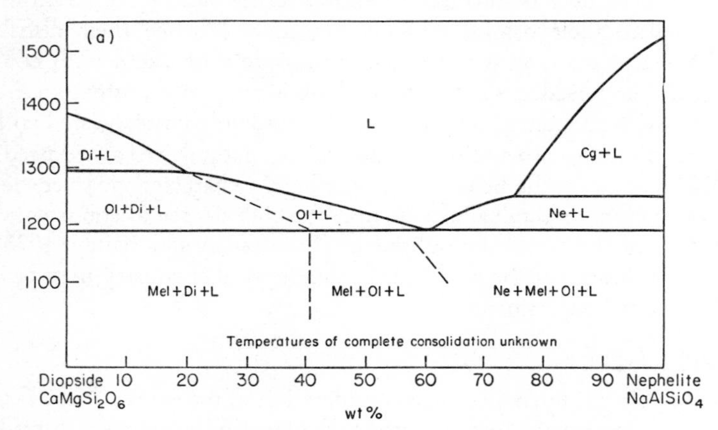

Fig. 13. Pseudo-binary diagram of equilibrium in mixtures of nepheline and diopside. (a) after Bowen (1922); and (b) after Schairer *et al.* (1962). Note that the biggest changes in the form of the diagram are below 1200°C, when several phases are crystallising. The lower temperature data, and the identification of solid solutions, were made possible by powder X-ray diffraction studies.

difficult to account for in experimental terms. Nephelinite is, moreover, a high temperature lava and yet in the synthetic system diopside and nepheline do not crystallise together at high temperatures. But perhaps the most striking difference from the felsic lavas is that the experiments indicated no invariant conditions (such as the low temperature minima in Petrogeny's Residua System) to correspond with the natural liquid compositions. The more recent experiments, referred to below, should be viewed against this background.

Nepheline–diopside, although commonly figured in binary form (Fig. 13) is, of course, a join in the quinary system Na_2O-CaO-MgO-Al_2O_3-SiO_2. Considerable experimental studies at atmospheric pressure have been devoted to this system in recent years (Schairer and Yoder, 1964; Schairer *et al.*, 1968; Onuma and Yagi, 1967) in an effort especially to elucidate the crystallisation relations involving melilite. Most of this work has followed from earlier studies on basalts (Yoder and Tilley, 1962) defined in terms of a composition tetrahedron, forsterite–diopside–nepheline–silica (the "basalt tetrahedron"). This model was found inadequate to describe melilite-bearing lavas, which must be referred to an enlarged system forsterite–larnite–nepheline–silica, which takes account of the calcium-rich composition of most melilites. This is the "expanded basalt tetrahedron". A crystallisation "flow-diagram" was deduced from the studies

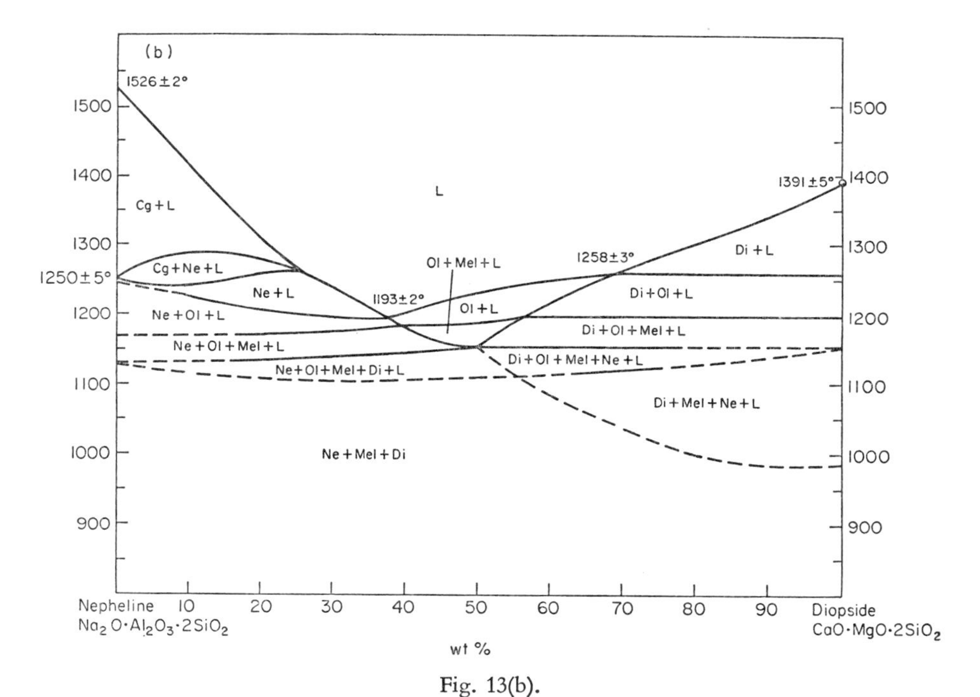

Fig. 13(b).

but this has been modified by later experiments. Differences were found when anorthite was added to the system (Schairer, Tilley and Brown, 1968) where it seemed that one important crystallisation pattern led to a single eutectic (L) at which nepheline–plagioclase–diopside–melilite crystallise simultaneously (see Fig. 14). The crystallisation sequences are such that diopside and nepheline are never seen to crystallise together without either melilite or plagioclase. The same result was found by Onuma and Yagi (1967) on *di–aker–ne*, and by Schairer and Yoder (1960) on *fo–ne–di*. In fact, the assemblage forsterite + diopside + nepheline + liquid, which would approximate to olivine nephelinite, has not been found in any of the new joins. This assemblage has been found only in the join diopside–nepheline–SiO_2 (Schairer and Yoder, 1960a). Forsterite composition does not lie in this join and olivine should be expected to react out under equilibrium crystallisation. Strangely enough, the assemblage forsterite + diopside + nepheline + liquid does not appear in the composition plane forsterite–diopside–nepheline (Schairer and Yoder, 1960). Thus, the "olivine–nephelinite" which figures in the recent flow diagrams can be regarded only as a device to fit the proposed model of crystallisation. Olivine-free nephelinites (without melilite) cannot be fitted to the experimental model at all.

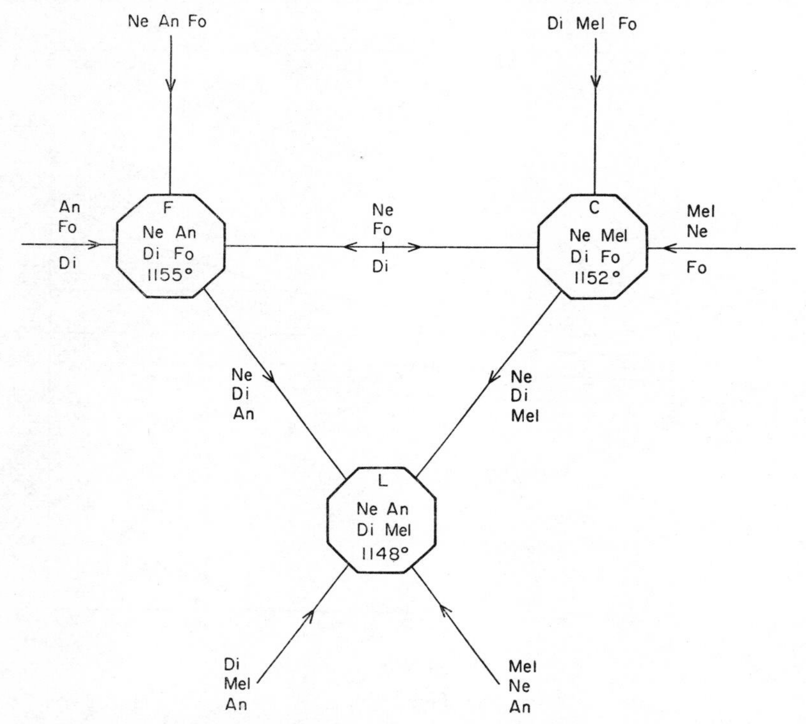

Fig. 14. Part of the flow diagram of Schairer *et al.* (1968). Octagons are "invariant" points (four solids + liquid) from each of which four "univariant" lines (three solids + liquid) arise. The temperatures at the "invariant" points are within ±3°C. F and C are reaction points at which olivine reacts out: L is a "eutectic". Nepheline (Ne), diopside (Di), plagioclase (An) and melilite (Mel) are all solid solutions. This diagram describes only the relations between the low variance phase assemblages, and the temperatures (where known). There is no compositional dimension in a flow diagram; Schairer *et al.* (1968) were able to affirm, however, that these "invariant points" are close in composition.

Some of the apparent conflicts between the experimental products and the rocks have arisen from the use of a quaternary model (flow sheet) to describe what is essentially a quinary system. Distinct quinary behaviour in the expanded basalt tetrahedron has since been recorded by Schairer and Yoder (1969 and 1970) and forced them to resort to experiments in the system CaO-MgO-Al_2O_3-SiO_2

as a truly quaternary model for the melilitite–nephelinite association. O'Hara and Biggar (1969) also base their hypotheses about these rocks on the same quaternary system. Relationships in the system CaO-MgO-Al_2O_3-SiO_2 at 1 atm pressure are complex, but are still far removed from a natural system in which the additional components Na_2O, K_2O, Fe, O, CO_2 and H_2O can be seen in nature to be important. The analogies drawn from CaO-MgO-Al_2O_3-SiO_2 lack conviction and do not seem justified. Indeed, Thompson (1972) has rightly pointed out the dangers of erecting models of *basalt* crystallisation in this system. Nephelinites obviously belong in a much more complex chemical system, and it is probable also that the interrelations between the magmas of the nephelinite association require explanation in terms of pressure variation. This is seen, too, when the melting relations of nephelinitic lavas are examined.

Compared with basalts the nephelinitic lavas tend to have high liquidus temperatures (Tilley, Yoder and Schairer, 1965) but one of the most notable features in their crystallisation, compared with basalts, is the large range of temperature between the crystallisation of the major phases. An olivine–nephelinite from Hawaii has olivine on the liquidus at 1305°C but nepheline does not start to crystallise until 1085°C. If the narrow crystallisation temperature span shown by the basalts is taken to indicate their low variance condition (Yoder and Tilley, 1962) then one must infer that nephelinites are considerably removed from this condition at atmospheric pressure. Such a wide temperature span is also inconsistent with the 1 atm synthetic results (Schairer, Tilley and Brown, 1968) which seem to show the major "nephelinite" invariant points close together in composition and temperature (see Fig. 14).

All these results are pointing up the fact that the one atmosphere experiments fail to model the natural nephelinite system. The outstanding question is, how is this magma formed, or what special conditions are needed to form nephelinitic rather than basaltic magma? The low pressure experiments suggest no differentiation scheme by which nephelinite could be derived from basalt. Obviously the answer lies partly in higher pressure ranges, but before turning to these the effects of adding iron to nepheline rich compositions should be briefly examined.

One atmosphere experiments in the system Na_2O-Al_2O_3-Fe_2O_3-SiO_2 (Bailey and Schairer, 1966) have shed some light on the lower-temperature, end-stages of nephelinite crystallisation. A quaternary reaction point was located, acmite + hematite + nepheline + albite + liquid, which was close in composition to the join nepheline–acmite, i.e. the liquid at the invariant point was relatively rich in the acmite and nepheline components (Fig. 8). This liquid is similar in its composition to the type-malignite (Johannsen, 1938). It is interesting because the final stages of nephelinite crystallisation commonly involve aegirine and nepheline and, in many instances, alkali feldspar (or aegirine + analcite). The synthetic invariant point is thus the analogue of the nephelinite residuum and, of course, malignite. Furthermore, it gives rise to a univariant crystallisation path, acmite + nepheline + albite + liquid, which is analogous to peralkaline nepheline syenite or phonolite. It illustrates a possible lineage from nephelinite to phonolite, and malignite to nepheline syenite. A curious consequence of the proximity of the quaternary reaction point to the join nepheline–acmite is that small fluctuations in conditions could alter the subsequent path of crystallisation, so that it follows a trend of increasing acmite content in the liquid. In the natural environment, fluctuations in the volatile phase might well initiate such a trend and give rise to the aegirine-rich "melteigites" described from many localities. As was pointed out in an earlier section, it is possible to extrapolate these findings to consideration of the subvolcanic conditions, because Nolan (1966) located a similar point in the system acmite–nepheline–albite–H_2O at 1 Kbar pressure.

Before considering the synthetic studies on nephelinite compositions at higher pressures, reference must be made to two plutonic heteromorphs and some relevant experiments.

Following the discovery of pargasitic hornblende xenoliths, of nephelinitic composition, in the nephelinites of the Moroto volcano, Varne (1968) has proposed that the nephelinite lava series derives from the incongruent melting of pargasite, as determined by Boyd (1959). Varne suggests that the two series of lavas represented at Moroto, nephelinite and alkali basalt, are both derived by partial melting of hydrated peridotite: the nephelinites from the hornblendic portion and the olivine basalts from the volatile-poor fraction. It must be noted that the specific application of this idea has been questioned by Wood (1968) because of the incompatibility between the trace elements in the hornblende and those in the nephelinites at Moroto. Furthermore, reference to Fig. 15

Fig. 15. PT diagram showing melting curves for various minerals and assemblages that might be present in the mantle. Ll is the melting curve for spinel lherzolite in the presence of excess H_2O, with pargasitic amphibole stable up to melting for pressures below 17 Kbar. Above 17 Kbar the amphibole decomposes along the broken line Lh. (Kushiro, 1970).

Phl + H_2O (phlogopite + excess H_2O: Yoder and Kushiro, 1969)
Phl − H_2O (phlogopite (water deficient): Yoder and Kushiro, 1969)
Phl = H_2O (phlogopite: Yoder and Kushiro, 1969)
Phl + En (phlogopite + enstatite: Modreski and Boettcher, 1972)
Phl + En + H_2O (with excess H_2O: Modreski and Boettcher, 1972)
Ac(h) (Acmite at high P_{O_2}: Gilbert, 1967)
Ac(l) (Acmite at low P_{O_2}: Gilbert, 1967)
$Ac_{67}Di_{33}$(h) (at high P_{O_2}: Cassie, 1971)
Jd (Jadeite: Bell, 1964)
Di (Diopside: Boyd and England, 1963)
En (Enstatite: Boyd *et al.*, 1964)
Fo (Forsterite: Davis and England, 1964)
Fa (Fayalite: Hsu, 1967)
Parg (Pargasite: Gilbert, 1970)

The long-dash lines labelled 32 and 50 are geothermal gradients at 32°C/Kbar (continental) and 50°C/Kbar (oceanic) respectively.

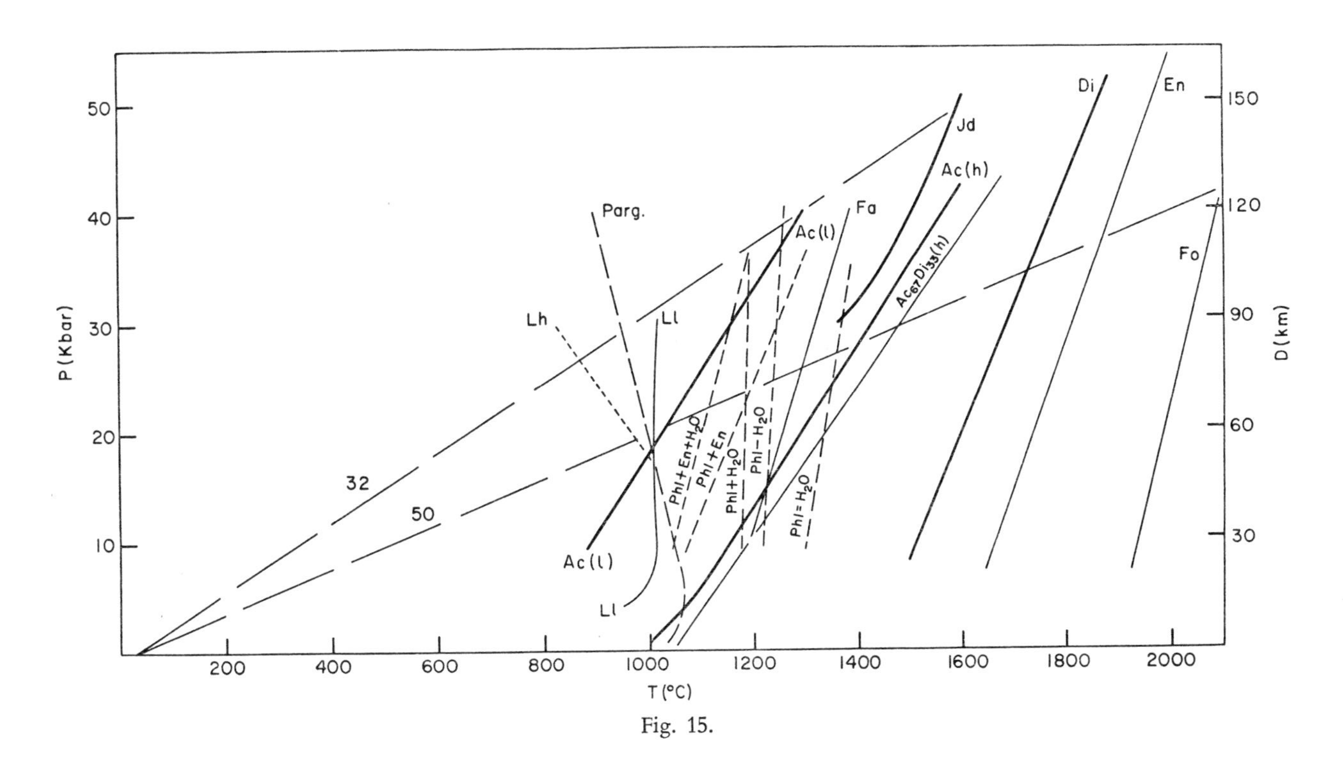
P (Kbar)
50
40
30
20
10
200
400
600
800
1000
1200
1400
1600
1800
2000
T (°C)
D (km)
150
120
90
60
30
32
50
Lh
Parg.
Ll
Ac(l)
Ac(l)
Ll
Phl+En+H_2O
Phl+En
Phl+H_2O
Phl−H_2O
Phl=H_2O
Fa
Jd
Ac(h)
$Ac_{67}Di_{33}(h)$
Di
En
Fo

Fig. 15.

shows that at pressures greater than 20 Kbar the amphibole in a lherzolite (Kushiro, 1970) breaks down at sub-solidus temperatures, and therefore cannot melt directly to provide a nephelinitic liquid.

Tilley and Yoder (1968) in a series of experiments on natural nephelinites have shown that under pressure these are converted to an essentially pyroxenite composition, with accessory biotite, hornblende, olivine and magnetite. In the presence of excess water, *especially at lower pressures*, there is a prominent development of hornblende along with clinopyroxene. These results, and the natural pargasite from Moroto, show the possibility of *nephelinite heteromorphs* at depth, but they do not bring us much nearer to the problem of how the *nephelinite composition* originates as a discrete phase (or assemblage) in the Earth. Although the authors themselves do not comment on it, their experimental results have possibly a direct bearing on the occurrence of melteigite as the plutonic equivalent of nephelinite. Experimentally, it seems that pressure, with or without added volatiles, is sufficient to suppress the stability of olivine in nephelinite compositions. Experiments carried out at different times, on different nephelinitic starting materials and in different laboratories, all show pressure ranges over which no olivine appears below the solidus (Yoder and Tilley, 1962; Tilley and Yoder, 1968; Bultitude and Green, 1970). This must be a significant factor in nephelinite petrogenesis.

In line with the growth of the concept of the formation of basic magmas in the mantle some authors have used high pressure experiments to devise or support possible hypotheses of nephelinite genesis. The most favoured mechanisms and their chief protagonists are: (a) anhydrous crystal fractionation at high pressures (O'Hara and Yoder, 1967; O'Hara, 1968); (b) crystal fractionation or partial melting at high pressures of H_2O (Bultitude and Green, 1968) and (c) partial melting of mantle peridotite at high pressures (Kushiro and Kuno, 1963; Kushiro, 1968, 1972a and 1972b).

(i) *Anhydrous High Pressure Fractionation.* A scheme for nephelinite genesis was developed by O'Hara (1968) in an extension of an earlier hypothesis by O'Hara and Yoder (1967) for the formation of kimberlite. The latter hypothesis was based on a limited number of anhydrous experiments in and around the join diopside–pyrope (in the system CaO-MgO-Al_2O_3-SiO_2) and on mixtures of separated minerals from natural eclogite and garnet peridotite, at 30 Kbar pressure. The reasoning is complex, and there are serious gaps in the evidence, but O'Hara and Yoder propose that the chief melt-product from a garnet peridotite mantle is a picritic liquid, from which garnet and clinopyroxene (bimineralic "eclogite") crystals may separate at high pressure. Protracted "eclogite" fractionation is claimed to produce highly undersaturated liquids resembling kimberlite. Subsequently O'Hara (1968) suggested that the nepheline normative composition of the residual liquid may be enhanced (to give nephelinites) by the separation of a spinel-lherzolite assemblage (spinel + olivine + clinopyroxene + orthopyroxene) at intermediate pressures.

Eclogite fractionation is, as the authors themselves point out (O'Hara and Yoder, 1967, p. 105), an hypothesis and must be tested, not merely adopted as if it were proven as some authors seem inclined to believe. It is impossible, for instance, to find an analysed eclogite that meets the needs of the hypothesis when applied to the potash-rich lavas of S.W. Uganda (Hough, 1972), in spite of the fact that Holmes and Harwood (1932) first proposed eclogite fractionation for this province, and O'Hara and Yoder (1967) believed such lavas to fit their more refined hypothesis. (It must be mentioned, too, that there are plenty of ultramafic and mafic nodules from this province, but no eclogites). There are, then, serious difficulties for the hypothesis even when applied to potash-rich magmas, but when the second stage of lherzolite fractionation is added to produce nephelinites (O'Hara, 1968) the difficulties are stupendous. The fractionation requirements now become exceedingly finicky and even more protracted, and if followed through would produce minute quantities of residual nephelinite liquid. Such a scheme is totally out of keeping with the observed facts of nephelinite volcanoes in the ocean basins (but no kimberlites reported), early nephelinites in major continental magmatic episodes, and the far greater abundance of nephelinite than potash-rich magma. Clearly, a mechanism of more general applicability is needed to account for the majority of nephelinites.

(ii) *Melting or Crystallisation at High* P_{H_2O}. Bultitude and Green (1968) made a series of experiments on olivine nephelinite composition at high pressures of H_2O, and reported orthopyroxene as the dominant liquidus phase. As a result, they claimed that orthopyroxene fractionation would drive the residual liquid towards olivine melilite nephelinite. They also proposed that the initial melts from mantle peridotite at high pressures of H_2O would be nephelinitic in character. For these experiments they had used unsealed capsules, and Kushiro (1969, 1970) obtained conflicting results with sealed capsules, reaching the conclusion that the unsealed experiments must have lost water and alkalis during the runs. Subsequent experiments by Kushiro (1972a and 1972b) have served only to confirm that the residual liquids from hydrous experiments are always more siliceous than those produced by anhydrous melting at high pressures.

It should be added that Bultitude's and Green's hypothesis of orthopyroxene fractionation lacks conviction when real nephelinites are considered. Nephelinite volcanism is characteristically explosive and some nephelinites at least would be expected to contain orthopyroxene phenocrysts, but these have yet to be reported.

(iii) *High Pressure Melting*. In 1963, Kushiro and Kuno deduced that strongly undersaturated mafic magma would be the first product of partial melting in the mantle at high pressure. Subsequent experimental studies by Kushiro (1968, 1969, 1972a and b) on selected joins in the system Na_2O-CaO-MgO-Al_2O_3-SiO_2, and on natural minerals and rocks, all indicate that increasing pressure (up to 60 Kbar (1972)) shifts the olivine–orthopyroxene liquidus boundary, and its connected phase geometry, towards increasingly silica-poor compositions. These results indicate that with increasing depth in the mantle the *initial* melt will be increasingly undersaturated. In his latest experiments on natural lherzolite, Kushiro (1972b) has analysed the melt products by electron microprobe and shows that with increasing pressure these become nepheline normative. The degree of melting in his experiments is relatively large ($\frac{1}{4}$ to $\frac{1}{3}$) and there is the distinct possibility that small degrees of melting would produce even more undersaturated liquids, with higher concentrations of residual elements and volatiles, and thus more akin to nephelinites.

Addition of H_2O, Kushiro discovered (1972a), produces more siliceous liquids, such that only at pressures around 20 Kbar do nepheline-normative liquids become a possibility. Even then the clinopyroxenes in the mantle peridotite must have low $CaAl_2SiO_6/NaAlSi_2O_6$ ratios, and at higher temperatures tholeiitic liquids would develop. In keeping with the observed characteristics of natural nephelinites, these experiments suggest that high P_{H_2O} is not normal in nephelinite genesis.

(2) *Potash-rich Mafic Rocks*

When considering nephelinites it soon became clear that the complexities of the natural system cannot be successfully modelled by the synthetic studies, although Kushiro's experiments seem to offer useful guide-lines. How much more intractable, then, are the problems of those mafic rocks involving potassic phases such as leucite, kalsilite, potash feldspar and biotite. The difficulties of modelling such a system are well stated by Luth (1967) in his description of the system $KAlSiO_4$-Mg_2SiO_4-SiO_2-H_2O in the pressure range 0–3 Kbar. This system is still far-removed from the natural compositions, so that it is difficult to draw direct analogies, but Luth's thoughtful appraisal of the possible low-pressure applications to K-rich lavas and lamprophyres will repay study by petrologists working on these rocks. Some of the more important results of this study were the demonstration of the strongly incongruent melting of phlogopite, and the detailed phase relations about the phlogopite composition. Luth also noted that the stability of phlogopite-bearing assemblages was enhanced in the vapour-absent region, and he stressed the far-reaching effects on phase changes of vapour-present or vapour-absent conditions and the consequent care that is needed in interpreting rocks. The fundamental problem of the potassic rocks, how they achieve their peculiar chemistry, could be only touched on within the confines of Luth's study, but it was during this period that experimentalists first began to appreciate the significance of primary phlogopite in rocks such as kimberlite, and the possibility that K_2O and H_2O might be stored in this mineral in the mantle. The stage was set for an examination of the stability of phlogopite at higher pressures.

Yoder and Kushiro (1969) found that, in the presence of excess H_2O, phlogopite melted at 1180°C at 10 Kbar and 1195°C at 30 Kbar, but that its thermal stability was higher (e.g. 1220°C at 10 Kbar) in the H_2O deficient region. Subsequently Yoder (1970) reported that the thermal stability of phlogopite was raised in the excess H_2O region by addition of CO_2 to the gas phase. His isothermal, isobaric diagram at 1225°C and 10 Kbar, shows an extensive region of phlogopite stability with increasing CO_2, which is thus nearly 50°C above the hydrous melting temperature. He seems inclined to attribute this largely to the relative insolubility of CO_2 in the melt, but part of the explanation must surely lie in the inhibiting by CO_2 of the incongruent solubility of phlogopite in a hydrous gas phase. More recently Modreski (1972) has begun a study of the stability of phlogopite-assemblages, involving enstatite, diopside, spinel, corundum and pyrope, in an effort to approach possible mantle mineralogics. He finds that enstatite causes the greatest lowering of the melting range (see Fig. 15), and, because orthopyroxene is a presumed major constituent of the upper mantle, he suggests that its effect on phlogopite melting in the mantle will be the most significant. If this were so the consequences are intriguing because the melting of the enstatite produces quartz normative liquids in the lower pressure range, such that in the presence of free H_2O the experimental liquids were strongly quartz normative at 25 Kbar (equivalent to 80 km depth). Under anhydrous conditions the melt from phlogopite–enstatite loses its quartz normative character between 10 and 15 Kbar and becomes leucite normative (between 35 and 60 km depth). The effects of other constituents such as Na and Fe, and, especially in view of Yoder's results, CO_2, are as yet unknown. Modreski's preliminary results on the quartz normative character of some of these melts offer a tantalizing hint to the possible formation of the *hy* normative lavas of the volcano Sabinyo, in the midst of the otherwise potassic feldspathoidal province of Bufumbira, in S.W. Uganda, and to the even more enigmatic highly potassic rhyolites of the South Kivu province, further south. These apparently anomalous lavas become explicable in terms of higher levels of melting in a phlogopite bearing mantle, or in terms of increasing P_{H_2O} in the region of magma generation.

All the high pressure experiments on phlogopite support the view that this phase will be stable to considerable depths (depending on the thermal gradient and the associated mineralogy) in the mantle. Modreski (1972) for instance, estimates that mica should persist to depths of 175 km below continental cratons. Many petrologists accept phlogopite as the probable form in which K_2O and H_2O are stored in the mantle, and that phlogopite melting provides these constituents in mantle-derived magmas. But there is still a powerful preference for K-rich magmas to be derived by secondary processes, such as crystal fractionation (O'Hara and Yoder, 1967) and zone-refining (Harris, 1957;

Kushiro, 1968), when it would seem that the obvious first choice is direct melting of phlogopite-rich mantle. The need for secondary concentration only arises if the phlogopite-content of the mantle is always low and the magma system is closed. But once you admit phlogopite as a mantle phase you can only assume that the concentration is everywhere uniformly low if you also assume that the mantle is homogeneous. What is more, the constant association of alkaline mafic magmas with carbonate magmatism and CO_2 enrichment demands a concept of open system melting. This in turn opens up the possibility of alkali metasomatism of the mantle, producing especially phlogopite and amphibole, as a precursor of melting. The tectonic and magmatological developments of these concepts were given by Bailey (1972) and the petrographic and chemical evidence has been provided by Lloyd and Bailey (1975). Significant experimental support is provided by the experiments of Luth (1967) and Yoder (1970) where, in both cases, the gas phase in equilibrium with phlogopite (+ forsterite) must, as Luth points out, contain significant amounts of ($KAlSiO_4$ + SiO_2). This is a very simplified case but the potentialities of such a gas phase as a metasomatising agent under mantle conditions cannot be ignored, especially as Yoder's experiments involved CO_2 as well as H_2O.

In summary, it has not been possible to perceive in the low pressure experiments any satisfactory mechanism for deriving the alkaline mafic rocks from more common magmas. The suggested fractional crystallisation schemes at high pressures are unconvincing in their requirements, and in their applications to real rocks, and they also fail conspicuously to relate to other features of the association; for instance, it becomes very difficult to see how sodic and potassic magmas could be coeval in these schemes. Fractional melting at high pressures suffers from fewer draw-backs, but the abundance of volatiles as well as alkalis in these associations really demands that the region of magma generation was open to hyperfusible components, either during or prior to melting.

In a review of nephelinites and ijolites (Bailey, 1974) it was concluded that "CO_2 and carbonates have an important (if not paramount) role". Recent experiments by Eggler (1974) show that CO_2 is readily soluble in ultramafic melts at mantle pressures and that such melts are more silica-undersaturated than those produced under hydrous conditions, *or* in the absence of volatiles. Eggler believes this "to explain the origin of primary nephelinites and melilite nephelinites". Experiments by Brey and Green (1975) indicate that the production of olivine melilitite will be favoured by the presence of CO_2 in the source region in the mantle.

These new experiments confirm the importance of CO_2 in the formation of alkaline *silicate* melts, and point the way to a new (but difficult) experimental approach to these complex magmas.

4. GENERAL FACTORS

In addition to experiments pertaining to specific groups of alkaline rocks, there are some broader aspects of alkaline petrogenesis that may be common to several rock groups. The two most important of these are alkali transfer, and the development of the condition of peralkalinity in magmas: the salient features of experiments bearing these two questions are outlined in the following sections.

(a) ALKALI TRANSFER

The concept of diffusion or volatile-transport of alkalis within a body of magma has been current for many years, and an hypothesis for the origin of alkaline rocks by such mechanisms was propounded by Smyth (1927). This hypothesis is frequently dismissed, with the inference that it is not susceptible to test, but in truth there can be no question that alkali transfer is possible and the real problem for supporters of alternative hypotheses is to explain why it has not happened in their examples! The natural evidence is abundant: in the alkaline exhalations and springs of active volcanic regions; in the alkali losses of obsidians on devitrification and crystallisation; but most of all in the extensive aureoles of alkali metasomatism that are so characteristic of alkaline intrusions. In this last instance the evidence of alkali transfer is locked in the solid structure of the country rocks. The mobile magma cannot hold a clear imprint of such a process, but is it rational to assume that it has not operated?

Most experimentalists working in alkaline systems can testify to alkali losses, usually to a coexisting vapour phase (e.g. Bailey, 1963), and clear-cut cases were noted in the previous discussion of phlogopite stability, but little systematic work has been done. Probably the most directly applicable study is that of Orville (1963), who also summarises previous studies of reactions between alkali feldspars and alkaline fluids. Orville studied the equilibria between alkali feldspars and alkali chloride-water solutions at 2 Kbar in the temperature range 350°C–700°C, and even made direct studies of alkali exchanges between feldspars (through the vapour) along a temperature gradient. He was able to demonstrate that the feldspar was Na enriched in the hot zone and more K enriched in the cooler part of the system. These were equilibrium experiments in a closed system and must be applied with care to the natural situation in which a relatively constant concentration (as well as temperature) gradient may apply across a system that is effectively open to alkalis at both ends. In some fenite aureoles, for instance, K rich feldspars have formed near the contact with more sodic feldspars further out, in apparent contradiction of Orville's results, but this arrangement is readily explicable in terms of a steady outflow from the intrusion of metasomatising fluids having a high K activity. It might be questioned that experiments which involve no melting have any relevance to problems of alkaline magmatism but this must be firmly rejected. In the first place, many petrologists have asserted that certain alkaline magmas are the products of melting of metasomatised rocks (e.g. von Eckermann, 1948) and this method of concentrating the alkalis (which is the crucial problem of alkaline magmatism) cannot be ignored. And, in the second place, if alkali transfer can be demonstrated in a solid-vapour system it would seem even more feasible at higher temperature when liquid is also present. Ultimately, we may expect to see more sophisticated experiments in an open system, but to make this really worthwhile we must have a clearer knowledge of the volatiles involved.

(b) THE DEVELOPMENT OF PERALKALINITY

From the outset of this chapter I have tried to preserve the distinction between alkali-enrichment with respect to silica and alkali-enrichment with respect to alumina. This is necessary because, firstly, the term "alkaline" is applied to rocks of both conditions, and secondly, there is a real danger that an important petrochemical distinction may become blurred. General discussions of the genesis of alkaline rocks seem prone to concentrate largely on the alkali–silica part of the question (Tilley, 1958) and where the development of peralkalinity is mentioned it is almost invariably ascribed to the "plagioclase effect" (Bowen, 1945). The classic controversy between Bowen (1928) and Fenner (1929) about whether fractional crystallisation would yield alkali-rich or iron-rich residual liquids was never concerned with the alkali–alumina balance. Later commentators (e.g. Lindsley, 1963) have suggested that the relative enrichments of alkalis and iron in a residual liquid would be dependent on the precipitation of iron oxide during the fractionation sequence (and hence on the P_{O_2}). But iron oxide (or iron silicate) precipitation cannot directly influence the alkali–alumina ratio. The great complexity of the problem becomes more evident when we remember that peralkalinity tends to correlate with increasing iron, i.e. when alumina is deficient alkalis and iron increase together! It is essential therefore to consider the available experimental evidence bearing on the causes of peralkalinity.

(1) *Low Pressure Fractionation: Plagioclase Effect*

Ever since Bowen suggested it (1945) the "plagioclase effect", operating during fractional crystallisation, has had great appeal for petrologists seeking to explain peralkaline felsic magmas. The mechanism is simply explained: a liquid containing calcium in addition to the albite components ($NaAlSi_3O_8$) cannot be expected to precipitate crystals of pure albite because some calcium must enter the plagioclase structure. The calcium in the plagioclase requires aluminium in the ratio $CaAl_2$, with the net result that Na and Si are concentrated in the remaining liquid. The process was clearly shown experimentally by Schairer and Yoder (1960a) in the non-binary character of the system albite–diopside ($NaAlSi_3O_8$–$CaMgSi_2O_6$) especially in the low-melting part of the system (close to the albite composition) where some Ca enters the plagioclase that crystallises with pyroxene (see Fig. 2). It is important to remember that operation of the plagioclase effect *requires* that CaO be present in the melt in excess of the Al_2O_3 necessary to form the anorthite molecule, which may be stated as: molecular $Na_2O + K_2O + CaO > Al_2O_3$. In terms of the CIPW norm, the condition is that in any melt containing normative diopside or wollastonite (*di* or *wo*) the plagioclase effect is possible (Bailey and Schairer, 1964). The *requirement* during fractionation is that the amount of anorthite precipitated in plagioclase must exceed the normative anorthite content of the parent melt. This last point must not be overlooked, because virtually all basalts satisfy the chemical condition, but many are associated with non-peralkaline felsic magmas. Even among the alkali basalts the levels of peralkalinity in their associated felsic magmas are highly variable, but possibly the most striking difference is between alkali basalts and tholeiites. It is hard to see any systematic differences in the Ca or Al levels in these two different types (e.g. Macdonald and Katsura, 1964) which are effectively separated by their alkali–silica ratio (Macdonald and Katsura, 1964) which expressed in normative terms yields two maxima in the distribution, alkali basalts with *ne*, and tholeiites with *q* (Chayes, 1971). The latter group of rocks are not associated with peralkaline felsic magmas, and yet on the face of it, there is no reason why the plagioclase effect should fail in this case and not in the alkali basalts. Perhaps the first question that should be raised by this dichotomy is whether felsic rocks are derived from basaltic parents *solely* by fractional crystallisation; in so many instances the process seems unable to account for the distribution and behaviour of alkalis in igneous suites. But, recognising the assumption implicit in the "plagioclase effect", let us consider the felsic members of the alkali-basalt and tholeiite associations as the end-products of fractional crystallisation, and ask how the "effect" could have been suppressed in the tholeiite case. Mineralogically alkali basalt is distinguished by Ca-rich augite and strongly zoned plagioclase (with An-rich cores), from tholeiite, with subcalcic augite, or orthopyroxene, and feebly-zoned plagioclase (with composition approaching the normative value). Although the contents of Ca and Al are similar in the two kinds of basalt, the Ca activity appears to be significantly lower in the early stages of crystallisation of the tholeiite, with the result that potential *an* survives in the lower temperature liquid and effectively prevents the achievement of the peralkaline condition. Another way of stating this is to say that during tholeiite crystallisation the *an* content of the crystal fraction never exceeds the *an* content of the original liquid. The reasons are unknown, and may be subtle: the relatively higher silica and lower alkalis in the tholeiites are factors that spring to mind. But is this not another way of saying that only melts with a high alkali activity develop peralkaline tendencies, and, if so, perhaps the "plagioclase effect" is irrelevant. How do alkali basalts attain the prerequisite of high alkali activity, would seem to be a much more compelling question.

The last question becomes even more compelling when we remember that plagioclase separation can be invoked only for the pressure range within which plagioclase is stable. The upper pressure limit is probably in the range 15–25 Kbar in mixed assemblages (Bell and Roseboom, 1969), while the demonstration by Emslie and Lindsley (1969) that clinopyroxene $\pm$ garnet replaces plagioclase as the liquidus phase, even in alumina-rich basic rocks, between 12–20 Kbar indicates that the influence of plagioclase separation on residual liquids will be considerably muted in the upper stability range. Many basaltic magmas must originate in (or at least equilibrate with) a peridotitic mantle at pressures beyond the stability of plagioclase, and thus derive their inherent peralkaline tendencies without the influence of plagioclase melting or crystallisation. Furthermore, some nephelinites are intrinsically peralkaline and it is virtually impossible to relate these to a basaltic parentage by low-pressure fractionation. Clearly the "plagioclase effect" can never provide more than a partial solution to the problem of peralkalinity.

(2) *High Pressure and the Peralkaline Condition*

Under high pressure conditions, beyond the stability of plagioclase, the alkalis may be located partly in phlogopite as noted in the previous discussion of potash-rich magmas (Fig. 15) but beyond 40 Kbar even this is in doubt (Kushiro *et al.*, 1967). Clinopyroxene must be the favourite choice as the source of alkalis among the solid phases throughout most of the upper mantle, because plagioclase, amphibole and mica have only limited stabilities. Therefore, it is to the stabilities of alkali pyroxenes at high pressures that we should look for possible causes of the peralkaline condition in a melt originating in the mantle. Reference to Fig. 15, shows that the jadeite and acmite melting curves are at much lower temperatures than diopside, enstatite or forsterite. It is to be expected that clinopyroxenes in the mantle would be complex solid solutions containing the jadeite and acmite molecules. Calculated acmite contents are less commonly reported than jadeite in mantle pyroxenes (perhaps because most modern analyses are by electron probe and iron is quoted as "FeO") but these give 5–10% "$FeSiO_3$" in ultramafic specimens (see Boyd, 1974). As "FeO" increases so, generally, does Na_2O. In eclogites of presumed mantle origin the clinopyroxenes may contain enough soda to calculate as 50% "jadeite" (part of which should undoubtedly be reckoned as acmite) (see Eggler and McCallum, 1974). The limited experimental evidence on the melting behaviour of clinopyroxene solid solutions indicates quite clearly that the liquid from the equilibrium melting of jadeite–diopside solid solutions will be as much as 20–30% richer in the jadeite component (Bell and Davis, 1969). Fractional crystallisation, or disequilibrium melting, may produce highly sodic liquids. The low-melting potential of acmite is even greater. Cassie (1971) has reported on acmite–diopside and shows the minimum temperature on the solidus is at acmite composition. He states that his experiments indicate a narrowing of the incongruent melting range with increasing pressure, and he suggests "that the effectiveness of partial melting of rocks containing acmite–omphacite in generating soda- and iron-enriched magmas may . . . be truly effective only at relatively shallow depths in the mantle". In the absence of any evidence of a minimum on the *solidus* between acmite and diopside, Cassie's suggestion requires that the mantle pyroxenes must be low in acmite, homogeneous and subject only to equilibrium melting. In any case, the melt would still be acmite normative, or potentially peralkaline, unless counteracted by melting of spinel, pyrope or aluminous enstatite, which seems unlikely in the temperature range. But much more important than this is that Cassie's experiments were deliberately made at very high P_{O_2} (by adding excess oxygen to the charge)—hardly a likely condition in the mantle. In his report on the pure acmite composition at high pressures Gilbert (1968) reported a lowering of the melting curve by 250°C–300°C at low P_{O_2} throughout the range to 45 Kbar total pressure. The position of such a low P_{O_2} melting curve for acmite is sketched on Fig. 15, and it transforms the whole melting curve pattern. Within the mantle it is to be expected that oxidation conditions would approximate more closely to this curve: it is difficult, therefore, to see how early low-melting liquids could fail to be acmite normative, nor how subsequent (more voluminous) melts at higher temperatures could fail to inherit this potential, albeit overprinted by an anorthite normative characteristic. It is worth pointing out, too, that the strong P_{O_2} dependence of acmite melting means that low-melting liquids enriched in soda and iron could be generated, at constant P and T, solely by reduction of the P_{O_2} (Bailey, 1969).

In conclusion it may be said that the presently known melting relations of alkali and iron silicates make the generation of peralkaline melts in the upper mantle an ever-present possibility. More extensive melting may be expected to contribute more Ca and Al to the liquids such that they become *an* normative, but it is clear that these liquids rarely, *if ever*, attain the condition of molecular $Na_2O + K_2O + CaO < Al_2O_3$. In other words, melts from the mantle, if not actually *ac* normative, are *di* or *wo* normative, and hence subject to the possible operation of the plagioclase effect during any low-pressure fractionation. Not only may some mafic melts be peralkaline directly from the mantle, e.g. certain nephelinites, but even the basalts derive the potential from this environment. The real problems, as suggested in earlier discussions, would be in accounting for alkaline magmas with excess alumina, without invoking crustal contamination, and in explaining why only alkali and transitional basalts are normally associated with peralkaline, low temperature magmas.

REFERENCES

Bailey, D. K. (1963). *Carn. Inst. Wash. Yr. Bk.*, **62,** 131–133.
Bailey, D. K. (1966). *In* "The Carbonatites" (O. F. Tuttle and J. Gittens (Eds.)). John Wiley, New York.
Bailey, D. K. (1969). *Am. J. Sci.*, **267A,** 1–16.
Bailey, D. K. (1970). *Geol. J. Special Issue*, **2,** 177–186.
Bailey, D. K. (1972). *J. Earth Sci. (Leeds)*, **8,** 231–245.
Bailey, D. K. (1974). *In* "The Alkaline Rocks", H. Sørenson (Ed.). John Wiley, New York.
Bailey, D. K. (1974a). *In* "The Alkaline Rocks", H. Sørensen (Ed.). John Wiley, New York.
Bailey, D. K., Cooper, J. P. and Knight, J. L. (1974b). *In* "Oversaturated Peralkaline Volcanic Rocks". Special Vol. *Bull. Volc.*
Bailey, D. K. and Macdonald, R. (1969). *Am. J. Sci.*, **267**, 242–248.
Bailey, D. K. and Schairer, J. F. (1963). *Carn. Inst. Wash. Yr. Bk.*, **62,** 124–131.
Bailey, D. K. and Schairer, J. F. (1964). *Am. J. Sci.*, **262,** 1,198–1,206.
Bailey, D. K. and Schairer, J. F. (1966). *J. Petrol.*, **7,** 114–170.
Bell, P. M. (1964). *Carn. Inst. Yr. Bk.*, **63,** 171–174.
Bell, P. M. and Davis, T. C. (1969). *Am. J. Sci.*, **267-A,** 17–32.
Bell, P. M. and Roseboom, Jr., E. H. (1969). *Mineral. Soc. Amer.* Special Publication No. 2.
Bowen, N. L. (1913). *Am. J. Sci.*, **35,** 577–599.
Bowen, N. L. (1915). *Am. J. Sci.*, **40,** 161–185.
Bowen, N. L. (1922). *Am. J. Sci.*, 1–34.
Bowen, N. L. (1928). ' The Evolution of the Igneous Rocks". Princeton University Press, Princeton.
Bowen, N. L. (1937). *Am, J. Sci.*, **33,** 1–21.
Bowen, N. L. (1945). *Am. J. Sci.*, **243A,** 75–89.
Bowen, N. L., Schairer, J. F. and Willems, H. W. V. (1930). *Am. J. Sci.*, **20.**
Boyd, F. R. (1959). *In* "Researches in Geochemistry, Vol. 1", P. H. Abelson (Ed.), John Wiley, New York, 377–396.
Boyd, F. R. (1974). *Carn. Inst. Wash. Yr. Bk.*, **73,** 285–293.
Boyd, F. R. and England, J. L. (1963). *J. Geophys. Res.*, **68,** 311–323.
Boyd, F. R., England, J. L. and Davis, B. T. C. (1964). *J. Geophys. Res.*, **69,** 2101–2109.
Brey, G. and Green, D. H. (1975). *Contr. Mineral. Petrol.*, **49,** 93–103.
Bultitude, R. J. and Green, D. H. (1968). *Earth Planet. Sci. Lett.*, **3,** 325–337.
Bultitude, R. J. and Green, D. H. (1970). *Nature*, **226,** 748–749.
Carmichael, I. S. E. (1967). *Contr. Mineral. Petrol.*, **14,** 36–64.
Carmichael, I. S. E. and MacKenzie, W. S. (1963). *Am. J. Sci.*, **261,** 382–396.
Cassie, R. M. (1971). *Carn. Inst. Wash. Yr. Bk.*, **69,** 170–175.
Chayes, F. (1970). *Carn. Inst. Wash. Yr. Bk.*, **68,** 179–182.
Chayes, F. (1971). *Carn. Inst. Wash. Yr. Bk.*, **69,** 295–299.
Chayes, F. (1973). *Carn. Inst. Wash. Yr. Bk.*, **72,** 673–676.
Coombs, D. S. and Wilkinson, J. F. G. (1969). *J. Petrol.*, **10,** 440–501.
Daly, R. A. (1914). "Igneous Rocks and their Origin". McGraw-Hill, New York.
Daly, R. A. (1933). "Igneous Rocks and the Depths of the Earth". McGraw-Hill, New York.
Darwin, C. (1844). "Volcanic Islands". Smith, Elder, and Co., London.
Davis, B. T. C. and England, J. L. (1964). *J. Geophys. Res.*, **69,** 1,113–1,116.
Eckermann, H. von. (1948). *Sver. Geol. Undersök. Ser. Ca.*, **36.**
Eckermann, H. von. (1961). *Bull. Geol. Inst. Univ. Upsala*, **40,** 25–36.
Edgar, A. D. (1964). *Amer. Miner.*, **49,** 573–585.

Edgar, A. D. (1974). *In* "The Alkaline Rocks". H. Sørensen (Ed.). John Wiley, New York, 355–389.
Eggler, D. H. (1974). *Carn. Inst. Wash. Yr. Bk.*, **73,** 215–224.
Eggler, D. H. and McCallum, M. E. (1974). *Carn. Inst. Wash. Yr. Bk.*, **73,** 294–300.
Emslie, R. F. and Lindsley, D. H. (1969). *Carn. Inst. Wash. Yr. Bk.*, **67,** 108–112.
Ernst, W. G. (1968). "Amphiboles". Springer-Verlag, Berlin.
Fenner, C. N. (1929). *Am. J. Sci.*, **18,** 225–253.
Ferguson, J. and Currie, K. L. (1972). *Nature (Phys. Sci.)*, **235,** 86–89.
Ford, C. E. (1972). *Nat. Envir. Res. Council Pub. Series D*, **2,** 161–164.
Fudali, R. F. (1963). *Bull. Geol. Soc. Amer.*, **74,** 1,101.
Gilbert, M. C. (1967). *Carn. Inst. Wash. Yr. Bk.*, **65,** 241–244.
Gilbert, M. C. (1969). *Am. J. Sci.*, **267-A,** 145–159.
Gilbert, M. C. (1969a). *Carn. Inst. Yr. Bk.*, **67,** 167–170.
Greig, J. W. and Barth, T. F. W. (1938). *Am. J. Sci.*, **35A,** 94–112.
Gummer, W. K. and Burr, S. V. (1943). *Science*, **97,** 286–287.
Hamilton, D. L. (1972). *Nat. Envir. Res. Council Pub. Series D*, **2,** 24–27.
Hamilton, D. L. and MacKenzie, W. S. (1965). *Min. Mag.*, **34,** No. 268, 214–231.
Harris, P. G. (1957). *Geochim. Cosmochim. Acta*, **12,** 195–208.
Harris, P. G., Kennedy, W. Q. and Scarfe, C. M. (1970). *J. Geol. Special Issue*, **2,** 187–200.
Heinrich, E. Wm. (1966). "The Geology of Carbonatites". Rand McNally and Co., Chicago, U.S.A.
Holmes, A. and Harwood, H. F. (1932). *Q. J. Geol. Soc. Lond.*, **88,** 370.
Hough, F. E. (1972). Ph.D. thesis, University of Reading.
Hsu, L. C. (1967). *J. Geophys. Res.*, **72,** 4,235–4,244.
Iddings, J. P. (1909). "Igneous Rocks". John Wiley, New York.
Johannsen, A. (1938). "A Descriptive Petrography of the Igneous Rocks". University of Chicago Press, Chicago.
Kushiro, I. (1968). *Jour. Geophys. Res.*, **73,** 619–634.
Kushiro, I. (1969). *Tectonophys.*, **7,** 427–436.
Kushiro, I. (1970). *Carn. Inst. Yr. Bk.*, **68,** 245–247.
Kushiro, I. (1970). *Carn. Inst. Wash. Yr. Bk.*, **68,** 240–247.
Kushiro, I. (1972). *J. Petrol.*, **13,** 311–334.
Kushiro, I. (1972). *Carn. Inst. Wash. Yr. Bk.*, **71,** 357–362.
Kushiro, I. and Kuno, H. (1963). *J. Petrol.*, **4,** 75–89.
Kushiro, I., Syono, Y. and Akimoto, S. (1967). *Earth Planet Sci. Lett.*, **3,** 197–203.
Lindsley, D. H. (1963). *Carn. Inst. Wash. Yr. Bk.*, **62,** 60–66.
Lindsley, D. H. (1967). *Carn. Inst. Wash. Yr. Bk.*, **65,** 244–247.
Lindsley, D. H. (1971). *Carn. Inst. Wash. Yr. Bk.*, **69,** 188–190.
Lloyd, F. E. and Bailey, D. K. (1975). "Physics and Chemistry of the Earth, Vol. 9". (Proceedings of the First International Conference on Kimberlites), Pergamon, Oxford.
Luth, W. C. (1967). *J. Petrol.*, **8,** 372–416.
Luth, W. C. (1969). *Am. J. Sci.*, **267A,** 325–341.
Macdonald, R. (1974). *In* "The Alkaline Rocks". H. Sørenson (Ed.) John Wiley, New York.
Macdonald, R. and Bailey, D. K. (1973). *U.S. Geol. Survey Prof. Paper*, **440-N-1.**
Macdonald, R., Bailey, D. K. and Sutherland, D. S. (1970). *J. Petrol.*, **11,** 507–517.
Macdonald, G. A. and Katsura, T. (1964). *J. Petrol.*, **5,** 82–133.
MacKenzie, W. S. (1972). *Nat. Envir. Res. Council Pub. Series D*, **2,** 46–50.
Martin, R. F. (in press). *In* "Peralkaline Acid Volcanic Rocks". (D. K. Bailey, F. Barberi and R. Macdonald, Eds.). *Special Issue: Bull. Volc.*
Modreski, P. J. (1972). *Carn. Inst. Wash. Yr. Bk.*, **71,** 392–396.
Modreski, P. J. and Boettcher, A. L. (1972). *Am. J. Sci.*, **272,** 852–869.

Morse, S. A. (1969). *Carn. Inst. Wash. Yr. Bk.*, **67,** 112–126.
Morse, S. A. (1970). *J. Petrol.*, **11,** No. 2, 221–251.
Moyd, L. (1949). *Amer. Miner.*, **34,** 736–751.
Nash, W. P., Carmichael, I. S. E. and Johnson, R. W. (1969). *J. Petrol.*, **10,** 409–439.
Nolan, J. (1966). *Quart. J. Geol. Soc.*, **122,** 119–158.
O'Hara, M. J. (1968). *Earth Sci. Rev.*, **4,** 69–133.
O'Hara, M. J. and Biggar, G. M. (1969). *Am. J. Sci.*, **267A,** 364–390.
O'Hara, M. J. and Yoder, H. S. Jr. (1967). *Scott. J. Geol.*, **3,** 67–117.
Onuma, K. and Yagi, K. (1967). *Amer. Miner.*, **52,** 227–243.
Orville, P. M. (1963). *Am. J. Sci.*, **261,** 201–237.
Peacock, M. A. (1931). *J. Geol.*, **39,** 54–67.
Piotrowski, J. M. and Edgar, A. D. (1970). *Meddr. Gronland, BD.*, **181,** No. 9, 1–62.
Presnall, D. C. and Bateman, P. C. (1973). *Geol. Soc. Amer. Bull.*, **84,** 3,181–3,202.
Roedder, E. and Weiblen, P. W. (1970). *Science*, **167,** 641–643.
Scarfe, C. M., Luth, W. C. and Tuttle, O. F. (1965). *Geol. Soc. Amer.* (Annual Meeting Program), 144–145.
Schairer, J. F. (1950). *J. Geol.*, **58,** 512–517.
Schairer, J. F. and Bowen, N. L. (1955). *Am. J. Sci.*, **253,** 681–746.
Schairer, J. F. and Bowen, N. L. (1956). *Am. J. Sci.*, **254,** 129–195.
Schairer, J. F., Tilley, C. E. and Brown, M. A. (1968). *Carn. Inst. Wash. Yr. Bk.*, **66,** 467–471.
Schairer, J. F., Yagi, K. and Yoder, H. S. Jr. (1962). *Carn. Inst. Yr. Bk.*, **61,** 96–98.
Schairer, J. F. and Yoder, H. S. Jr. (1960). *Am. J. Sci.*, **258-A,** 273–283.
Schairer, J. F. and Yoder, H. S. Jr. (1960). *Carn. Inst. Wash. Yr. Bk.*, **59,** 70–71.
Schairer, J. F. and Yoder, H. S. Jr. (1964). *Carn. Inst. Wash. Yr. Bk.*, **63,** 65–74.
Schairer, J. F. and Yoder, H. S. Jr. (1969). *Carn. Inst. Wash. Yr. Bk.*, **67,** 104–105.
Schairer, J. F. and Yoder, H. S. Jr. (1970). *Carn. Inst. Wash. Yr. Bk.*, **68,** 202–214.
Schairer, J. F., Yoder, H. S. Jr. and Tilley, C. E. (1967). *Carn. Inst Wash. Yr. Bk.*, **65,** 217–226.
Scrope, G. P. (1825). "Volcanoes". W. Phillips, London.
Shand, S. J. (1923). *Trans. Geol. Soc. S. A.*, **25,** 16–32.
Shand, S. J. (1927). "Eruptive Rocks". Thomas Murby and Co., London.
Shand, S. J. (1930). *Geol. Mag.*, **67,** 415–426.
Smyth, C. H. Jr. (1927). *Proc. Amer. Phil. Soc.*, **66,** 535–80.
Sood, M. K. and Edgar, A. D. (1970). *Medd. Grønland*, **181,** 41.
Sørensen, H. (1974). (Ed.) "The Alkaline Rocks", John Wiley, New York.
Thompson, R. N. (1972). *Am. J. Sci.*, **272,** 901–932.
Thompson, R. N. and MacKenzie, W. S. (1967). *Am. J. Sci.*, **265,** 714–734.
Tilley, C. E. (1958). *Q. J. Geol. Soc.*, **113,** 323–360.
Tilley, C. E. and Yoder, H. S. Jr. (1968). *Carn. Inst. Wash. Yr. Bk.*, **66,** 457–460.
Tilley, C. E., Yoder, H. S. Jr. and Schairer, J. F. (1965). *Carn. Inst. Wash. Yr. Bk.*, **64,** 69–82.
Turner, F. J. and Verhoogen, J. (1951). "Igneous and Metamorphic Petrology". McGraw-Hill, New York.
Turner, F. J. and Verhoogen, J. (1960). "Igneous and Metamorphic Petrology". McGraw-Hill, New York.
Tuttle, O. F. and Bowen, N. L. (1958). *Geol. Soc. Amer. Mem.*, **74.**
Tuttle, O. F. and Smith, J. V. (1958). *Am. J. Sci.*, **256,** 571–589.
Tyrell, G. W. (1928). *Quart. J. Geol. Soc.*, **84,** 540–569.
Ussing, N. V. (1911). *Medd. om Grønland*, **38.**
Varne, R. (1968). *J. Petrol.*, **9,** 169–190.
Washington, H. S. (1917). *U.S. Geol. Surv. Prof. Paper*, **99.**
Washington, H. S. and Merwin, H. E. (1927). *Amer. Miner.*, **12,** 233–252.
Wones, D. R. and Eugster, H. P. (1965). *Amer. Miner.*, **50,** 1,228–1,272.

Wood, C. P. (1968). Unpubl. Ph.D. thesis. University of Leeds.
Wright, J. B. (1969). *Min. Mag.*, **37,** 370–374.
Yoder, H. S. Jr. (1970). *Carn. Inst. Wash. Yr. Bk.*, **68,** 236–240.
Yoder, H. S. Jr. and Kushiro, I. (1969). *Am. J. Sci.*, **267A,** 558–582.
Yoder, H. S. Jr. and Tilley, C. E. (1962). *J. Petrol.*, **3,** 342–532.
Yoder, H. S. Jr. and Upton, B. G. J. (1971). *Carn. Inst. Wash. Yr. Bk.*, **70,** 108–112.

APPENDIX I

A. Abbreviations for phases in figures

PHASE	ABBREVIATION	PHASE	ABBREVIATION
Acmite	Ac	Kyanite	Ky
Aegirine	Aeg	Larnite	Lar
Akermanite	Ak	Leucite	Lc
Albite	Ab	Liquid	L
Alkali feldspar	Af	Magnesite	M
Almandine	Alm	Magnetite	Mt
Aluminosilicates (Al_2SiO_5)	AS	Melilite	Mel
Amphibole	Am	Merwinite	Me
Analcite	Anal.	Monticellite	Mo
Andalusite	A	Mullite	Mull
Anorthite	An	Muscovite	Ms
Anthophyllite	Anth	Na-rich feldspar	Af_{Na}
Augite	Aug	Nepheline	Ne
Biotite	Bi	Olivine	Ol
Brucite	B	Omphacite	Omph
Calcite	Cc	Orthoamphibole	Oam
Carnegieite	Cg	Orthoclase	Or
Celadonite	Ce	Orthopyroxene	Opx
Chlorite	Chl	Paragonite	Pa
Chloritoid	Ctd	Pargasite	Parg
Clinopyroxene	Cpx	Periclase	P
Cordierite	Cd	Phlogopite	Phl
Corundum	C	Plagioclase	Pl
Diopside	Di	Pyrope	Py
Dolomite	Dol	Pyroxene	Px
Enstatite	En	Quartz	Q
Epidote	Ep	Rankinite	Ra
Fayalite	Fa	Sanidine	Sa
Feldspar	F	Sapphirine	Saph
Forsterite	Fo	Serpentine	Serp
Garnet	Gt	Siderite	Sid
Gehlenite	Ge	Silica minerals	SiO_2
Grossularite	Gr	Sillimanite	Sill
Haematite	Hem	Spessartine	Spess
Hercynite	Hc	Sphalerite	Sph
Hornblende	Hb	Spinel	Spl
Ilmenite	Ilm	Spurrite	Sp
Jadeite	Jd	Staurolite	St
Kalsilite	Ks	Talc	Tc
K-rich feldspar	Af_K	Tilleyite	Ti
Kornerupine	Korn	Tremolite	Tr

APPENDIX I (cont.)

Tridymite	Trid	Wollastonite	Wo
Ulvöspinel	Usp	Zoisite	Zo
Vapour (fluid)	V		

B. Abbreviations for oxygen buffers (note that excess H_2O is present in all buffer assemblages)

BUFFER	ABBREVIATION	BUFFER	ABBREVIATION
Quartz + fayalite + iron	QFI	Haematite + magnetite	HM
Wustite + iron	WI	Bixbyite + pyrolusite	BP
Magnetite + wustite	MW	Hausmannite + bixbyite	HsB
Quartz + fayalite + magnetite	QFM	Manganosite + hausmannite	MaHs
Nickel + nickel oxide	NNO	Magnetite + iron	MI
		Manganese + manganosite	MnMa

APPENDIX II

Common logarithms of oxygen fugacities in equilibrium with selected redox reactions

Buffer	Abbrev.	$\log_{10} f_{O_2} = -\frac{A}{T} + B + C\frac{(P-1)}{T}$ T in °K P in bars			Reference
		A	B	C	
Fe_2O_3–Fe_3O_4	HM	24912	14.41	0·019	Eugster & Wones 1962
Fe_3O_4–'FeO'	MW	32730	13·12	0·083	Eugster & Wones 1962
'FeO'–Fe	WI	27215	6·57	0·055	Eugster & Wones 1962
Fe_3O_4–Fe	MI	29260	8·99	0·061	Eugster & Wones 1962
NiO–Ni	NNO	24709	8·94	0·046	Eugster & Wones 1692
SiO_2–Fe_2SiO_4–Fe_3O_4	QFM	25738	9·00	0·092	Wones & Gilbert 1968
Mn_2O_3–MnO_2	BP	8850	10·97	0·0192	Huebner 1969
Mn_3O_4–Mn_2O_3	HsB	9848	7·896	0·0052	Huebner 1969
$Mn_{1-x}O$–Mn_3O_4	MaHs	25715	13·42	0·0818	Huebner 1969
Mn–$Mn_{1-x}O$	MnMa	40057	7·431	0·0630	Huebner 1969

APPENDIX III

Experimental petrology at very high pressures

Mao and Bell (1975) have developed a diamond-windowed high pressure cell capable of experiments up to 500 Kbars, in which the sample (alone) may be heated to 2000°C by a high-power laser beam. Temperature is measured by an optical pyrometer: pressure by the shift of the strong line in the fluorescence spectrum of ruby. One beauty of the method is that the pressure cell is mounted in a microscope system and the specimen can be viewed *during* the experiment. This opens up the prospect of systematic studies under conditions equivalent to a depth of 1000 km in the Earth's mantle.

Reference: Mao, H.K., and Bell, P.M. (1975). *Carn. Inst. Wash. Yr. Bk.*, **74,** 402–405.

APPENDIX IV

The table opposite gives the results of Tuttle and Bowen (1958) on the system $NaAlSi_3O_8$–$KAlSi_3O_8$–SiO_2–H_2O, at P_{H_2O} = 1000 kg/cm^2.

Plot the compositions in a triangular diagram, indicating the liquidus phase and the liquidus temperature in every case. Then draw in the phase boundaries and the liquidus contours.

Additional data required (from the bounding "binary" systems at P_{H_2O} = 1000 kg/cm^2):

1. Quartz-albite eutectic: Temperature 780°C; Composition $Q_{42}Ab_{58}$.
2. Quartz-orthoclase eutectic: Temperature 760°C; Composition $Q_{47}Or_{53}$.
3. Minimum on Ab–Or liquidus: Temperature 860°C; Composition $Or_{30}Ab_{70}$.

(Note: in the following experiments the starting-material was stated as "partially crystalline", except for the experiment * which was "all crystalline" and the experiment + which was "glass").

Initial composition wt%			Temperature °C	Time	Result	Composition of crystals by X-ray
SiO_2	Or	Ab				
				$P_{H_2O} = 1000$ kg/cm²		
9.1	26.4	64.5	830	7 days	Crystals + glass	Or_{26}
9.1	26.4	64.5	870	7 days	All glass	—
9.1	26.4	64.5	850	7 days	Feldspar + glass	—
9.1	29.4	61.5	850	4 days	Feldspar + glass	Or_{23}
9.5	35.2	55.3	850	5 days	Feldspar + glass	Or_{59}
9.5	35.2	55.3	860	7 days	All glass	—
10.4	52.9	36.7	870	7 days	Feldspar + glass	Or_{81}
10.4	52.9	36.7	900	7 days	All glass	—
17.7	8.8	73.5	850	9 days	All glass	—
17.7	8.8	73.5	830	9 days	Feldspar + glass	Or_{5}
27.1	17.7	55.2	820	7 days	All glass	—
27.1	17.7	55.2	800	8 days	Feldspar + glass	—
18.4	26.4	55.2	810	7 days	Feldspar + glass	Or_{25}
18.4	26.4	55.2	830	6 days	All glass	—
18.9	35.1	46.0	830	4 days	All glass	—
18.9	35.1	46.0	800	3 days	Feldspar + glass	Or_{50}
18.9	35.1	46.0	760	8 days	Feldspar + glass	Or_{48}
28.1	35.2	36.7	800	5 days	Feldspar + glass	—
28.1	35.2	36.7	820	4 days	All glass	—
27.7	26.4	46.0	800	5 days	All glass	—
+27.7	26.4	46.0	790	1 day	Feldspar + glass	—
36.0	8.8	55.2	790–800	11 days	All glass	—
36.0	8.8	55.2	780	7 days	Feldspar + glass	—
36.4	17.7	46.0	790	3 days	All glass	—
36.4	17.7	46.0	735	3 days	Quartz + feldspar + glass	Or_{11}
*36.9	26.4	36.7	725	41 days	Quartz + feldspar + glass	Or_{39}
36.9	26.4	36.7	740	7 days	All glass	—
36.9	26.4	36.7	730	1 day	Quartz + glass	—
37.1	35.2	27.7	765	6 days	All glass	—
37.1	35.2	27.7	745	6 days	Feldspar + glass	—
37.1	35.2	27.7	725	3 days	Feldspar + quartz + glass	Or_{80}
37.6	44.0	18.4	790	7 days	All glass	—
37.6	44.0	18.4	770	6 days	Feldspar + glass	Or_{93}
45.6	17.7	36.7	840	7 days	All glass	—
45.6	17.7	36.7	820	7 days	Quartz + glass	—
46.4	35.2	18.4	820	7 days	All glass	—
46.4	35.2	18.4	800	4 days	Quartz + glass	—
28.3	44.0	27.7	815	6 days	All glass	—
28.3	44.0	27.7	795	6 days	Feldspar + glass	—
28.3	44.0	27.7	755	3 days	Feldspar + glass	Or_{83}

* Three beginning-of-melting experiments on this composition are recorded by Tuttle and Bowen, 1958 (Table 10). These show that when completely crystalline this composition starts to melt between 715–720°C at this pressure.

Subject Index

Accuracy, 3
ACF diagram, 325
Acmite (Ac), 37, 162, 195, 376, 425, 454, 464
Actinolite, 123, 127, 276
Activity coefficients, 201
Activity diagram, 87
Activity (FeS), 215
Adirondack, 230
AFM diagrams, 325
Agpaitic, 437, 443, 444
Akermanite, 238, 239, 272, 289
AKF diagrams, 311, 325
Albite, 67, 103, 156, 159, 195, 225, 267, 279
Albite-epidote-hornfels facies, 325
Algebraic methods, 52.
Alkali-alumina-silica diagram, 445.
Alkali basalt, 420, 463
Alkali exchanges, 461
Alkali feldspar, 279, 427, 445, 461
Alkali feldspar critical curve, 345, 435
Alkali feldspar solvus, 48, 58, 311, 352, 356, 431
Alkali pyroxenes, 464
Alkali-ratio *v* alumina diagram, 446
Alkali transfer, 419, 461
Alkaline rocks, 268, 419
Almandine, 130, 195, 223, 225, 231, 265, 314, 319, 320
Almandine-spessartine garnet, 266
Alps, 157
Aluminium silicates (Al_2SiO_5), 195, 217, 225, 268, 314
Aluminium silicates (+ Ti, V, Cr and Fe), 218
Amesite, 138
Amphibole group, 123, 125, 276, 460
Amphibolite, 171, 173, 176, 177
Amphibolite facies, 264
Analcite, 109, 154, 155, 164, 341, 431
Analcite syenite, 432
Andalusite, 195, 216, 217, 264, 266, 267, 268, 270, 271, 278, 305, 306, 312
Andradite (Melanite), 267, 268
Annite, 195, 207, 278, 388
Annite + quartz, 228
Anorthite, 247, 250, 280, 298, 300
Anthophyllite, 195, 216, 221, 239, 240, 266, 275, 277, 283, 284, 286, 287, 305, 306, 321
Antigorite, 239
Application of experimental results, 187
Aragonite, 115
Arfvedsonite-riebeckite amphiboles, 376, 391

Barrovian zones, 93
Basalt-eclogite transition, 175, 177
Basalt tetrahedron, 67, 451
Basalt tetrahedron (expanded), 451
Basaltic composition, 276, 325
Basanite, 420
"Beginning of melting" curve, 85, 312
Binary compatibility figures, 65
Binary system, 75
Binary system (isobaric), 24
Biotite, 195, 208, 216, 229, 231, 266, 278, 282, 311, 387, 458
Blueschists, 102
Bodega Head Quartz Diorite, 399
Boehmite, 216, 220
Brucite, 107, 240, 266, 275, 282, 284, 305, 306
Buchites, 261, 262, 264, 266, 268, 278, 279, 281, 282
Buffer curves (and technique), 15, 92
Buffering and gas equilibria, 187, 203
Bulk composition, 26, 58

Calcareous schists, 272
Calcite, 237, 238, 248, 249, 250, 266, 267, 282, 289, 293, 300
Calcite-aragonite inversion, 116
Calcite-siderite solvus, 234
Calibration, 3, 6, 16
California Coast Ranges, 396, 398, 401
Calorimeter, 158, 161
Carbon Dioxide (CO_2), 200, 201
Carbonates, 187, 233, 234, 282, 324
Carbonates in the presence of water, 187
Carbonatite, 435
Celebes, 150, 157
Charnockites, 176

Chemical equilibrium during rock crystallisation, 4
Chlorite, 138, 139, 195, 216, 221, 222, 231, 266, 267, 278, 304, 305, 306, 311, 319, 320
Chlorite-quartz, 309
Chloritoid, 195, 216, 223, 231, 266, 271, 314, 319
Chrysotile, 239
Clausius-Clapeyron equation, 103
Clinochlore, 139, 278
Clinopyroxenes, 266, 274
Clinozoisite-epidote, 271
Coincidence rule, 81
Coincident phases, 81
Cold seal, 6, 9
Cold seal pressure vessel, 8, 20
Colinear phases, 81
Comendite, 444
Compatibility relations, 283
Compatibility triangles, 65, 66, 71, 72
Component, 21
Component reduction, 48
Composition barrier, 29
Compositions of fluid, 288
Compressibility (H_2O), 89
Compression reactions, 310, 315, 324
Congruent melting, 28, 63
Constructing a phase diagram, 3
Contact aureoles, 263, 269
Contact marbles, 272
Contact metamorphism, 261, 264, 265, 318
Control of Temperature and Pressure, 3
Controlled atmosphere, 6
Cordierite, 195, 216, 221, 222, 231, 264, 266, 277, 280, 282, 304, 305, 306, 311, 314, 319, 324
Cordierite-corundum, 309
Cordierite + forsterite, 310
Cordierite-gedrite-garnet rocks, 322
Cordierite-K-feldspar, 265, 311
Cordierite-K-feldspar hornfels facies, 325
Cordierite-muscovite, 264
Cordierite + phlogopite, 311
Cordierite + spinel, 310
Corundum, 195, 208, 225, 266, 267, 283, 298, 300, 305, 306, 319
Cotectic, 45
Cotectic region, 445
Cotectic surface, 45
Cotectic zone, 444
Cristobalite, 279
Crystal growth, 404, 405, 408
Crystallisation, 26, 131
Crystallisation-melting sequence, 397, 398
Cummingtonite, 266
Cummingtonite-grunerite, 277

Deboullie Stock, North Maine, 400, 403
Decarbonation and dehydration reactions in H_2O-CO_2, 187
Decarbonation reactions in the absence of water, 187, 290
Deformational strain-energy, 122
Degenerate invariant point, 80
Degenerate system, 79, 82
Degenerate univariant lines, 80
Degrees of freedom, 21, 83, 190
Dehydration boundary, 87
Dehydration reactions, 83, 285, 309, 313
Dehydration reactions in the absence of CO_2, 187
Development of peralkalinity, 462
Diaspore, 216, 220
Diatomites, 279
Differential thermal analysis (D.T.A.), 59
Dinkey Creek, 401
Diopside, 105, 162, 195, 237, 238, 240, 247, 266, 267, 274, 289, 293, 454, 464
Diopside-forsterite, 296
Direction of reaction, 5
Disequilibria, 263
Distribution coefficient, 320
Divariant area, 65, 75
Divariant equilibrium, 22, 30
Divariant field, 26, 73
Divariant surface, 33
Dolomite, 238, 249, 266, 282, 289, 293

Eclogite, 88, 102, 112, 129, 169, 171, 172, 173, 176, 456, 464
Eclogite fractionation, 456, 457
Electron microprobe, 53, 408, 458
Enstatite, 37, 195, 221, 238, 240, 267, 274, 275, 283, 284, 286, 287, 305, 306, 454, 464
Epidote-group, 266, 271
Equilibration, 263
Equilibrium constant, 196
Equilibrium crystallisation diagrams, 429
Equilibrium (dynamic), 63
Equilibrium failure, 40
Equilibrium (static), 63
Equipment, 3, 5
Eulites, 274

Eutectic, 26
Examining the charge, 16
Experiment Time, 3, 15
Experimental charge, 19
Experimental Method, Limitations, 18
Exsolution, 47
Extensive variables, 22
Externally heated pressure vessel (*see* Cold Seal) 6

Fayalite, 195, 207, 266, 319, 454
Fe-chloritoid, 231
Fe-cordierite, 208, 223, 271
Fe-Mg exchange reactions, 322
Fe-staurolite, 231
Feldspar, 279, 282, 311
Feldspar-liquid relationships, 45, 427
Feldspar-nepheline eutectic, 431.
Feldspathoidal Rocks, 431
Felsic rocks, 419, 424
Ferriannite, 229
Ferro-anthophyllite, 319
Ferropargasite, 276
Ferrosilite, 274, 320
Ferruginous hornfels, 322, 324
Ferruginous pelites, 314
Fibrolite, 269
Flow-diagram, 35, 36, 452
Fluid inclusions, 198
Fluid phase in metamorphic systems, 187, 198
Fluid pressure, 174, 178, 263
Fluorine, 206, 227
Forsterite, 195, 221, 238, 240, 247, 266, 275, 284, 286, 288, 289, 293, 297, 305, 306, 308, 454, 464
Forsterite + calcite, 292, 296
Franciscan Formation, 115, 118, 120, 124, 129, 138, 140, 146, 149, 150, 154, 157, 158, 168
Fractional crystallisation, 423, 424, 438, 456, 459, 460, 462, 463
Fractionation curves, 428

Gabbro contact aureoles, 322
Garnet, 129, 195, 216, 221, 265, 315, 319, 321
Gas Equilibria, 203
Gases (C-O-H), 204
Gedrite, 195, 216, 221, 223
Gehlenite, 247, 272, 298, 300
Gels, 11, 12, 13, 14, 18
Geothermal gradients, 102, 118, 119, 120, 148, 161, 166, 167, 454
Geothermometry, 287
Gibbs' phase rule, 21, 190
Gibbsite, 216, 220
Glasses, 11, 12, 18, 448
Glaucophane, 139
Glaucophane I and II, 126
Glaucophane schist, 112, 123, 124, 133, 166, 167
Global tectonics, 262
Gneisses, 264, 277, 278
Grandites, 268
Granite (melting), 89
"Granite" system, 44, 45, 48, 392, 424, 428
"Granite" system (isothermals), 61
Granulite, 88, 102, 171, 176, 177
Granulite facies, 273
Graphite, 204, 205
Grossularite, 133, 247, 248, 267, 298, 300
Grossularite-andradite, 266
Grossularite + quartz, 300
Growth rate, 407, 410
Grunerite, 266

Haplobasalt, 55
Haplogranite system 338, 364, 365, 366, 367, 368, 370
Haplogranite system (+ anorthite), 378
Haplogranite system (peralkaline), 372
Haplogranite system (peraluminous), 369
Haplogranite system (vapour phase), 354
Haplogranodiorite system, 378, 379, 385, 408, 410
Hausmannite, 236
HCl, 206
Heat change, 59
Heat input, 55
Heating curve, 59, 63
Hedenbergite, 195
Hemimorphite, 240
Hercynite, 195, 207, 208, 216, 271, 314, 319, 321, 324
Heterogeneous equilibria, 21
H_2O, 199, 201
H_2O (excess), 86
Homogeneity of the sample, 12
Hornblende, 170, 266, 276, 389
Hornfels facies, 264
Hornblende hornfels facies, 325
Hornfelses, 264, 266, 274, 277, 278, 323
Humite-group minerals, 266

Hydrate stability, 87, 108
Hydrogen (H_2), 200, 201
Hydrogen fugacity, 92
Hydrogrossular, 133
Hydrothermal activity, 287
Hyperfusible components, 460
Hypersolvus feldspars, 427, 431
Hypersthene, 195, 322, 324

Idocrase, 272
Ijolite, 440
Ilmenite ($FeTiO_3$), 209
Ilvaite, 266
Incongruent melting, 38, 63, 424
Incongruent solubility, 459
Indifferent phases, 82
Influence of other components, 187
Initial liquid, 85
Intensive variables, 22
Internally Heated Pressure Vessel, 9, 115, 116, 119, 151, 158
Interpretation of Run Products, 3, 17
Invariant equilibrium, 22, 23
Invariant point, 23, 69, 73, 75, 78, 79, 95
Ion microprobe, 53
Iron formations, 266
Isobaric univariant, 24, 26
Isothermal sections, 54, 56, 57
Isothermal sections (solid solution), 59
Isotope fractionation, 103

Jadeite, 67, 103, 149, 150, 154, 162, 339, 340, 454, 464

K-feldspar (*see also* Orthoclase), 195, 225, 249, 250, 266, 267, 458
Kalsilite, 458
Kanto Mountains, Japan, 121, 157
Kaolinite, 195, 216, 220
Kilchoanite, 273
Kimberlite, 456, 457, 458
Kinetics (*see* Nucleation and Reaction)
Kornerupine (boron-free), 304, 307, 310
Kungnat Syenite, Greenland, 400
Kyanite, 216, 217, 264, 267, 305, 306, 312

Lamprophyres, 458
Larnite, 238, 239, 273, 289
Latent heat of fusion, 55
Laumontite, 145
Lawsonite, 142
Leucite, 431, 434, 458
Leucitite, 420
Lherzolite, 454, 456, 457, 458
Lime-aluminium silicates, 144
Lime silicate, 272, 276, 298
Lime silicate hornfelses, 266
Limestones, 289
Liquid immiscibility, 423, 424
Liquidus, 24, 40
Liquidus contour, 56
Low-high albite transition, 160
Low temperature zone, 445, 447
Low variance conditions (equilibria), 28, 53

Mafic rocks, 419, 449
Magma evolution, 61
Magnesioriebeckite, 123, 124
Magnesite, 238, 266, 283, 286, 287
Magnetite, 195, 208, 216, 315, 319, 321
Magnetite-hercynite solvus, 210, 211
Malignite, 440, 441, 453
Manure (horse), 18
Marbles, 266
Margarite, 278, 298
Marls, 298
Materials and Methods, 11
Melange, 168
Melilite group, 272
Melilitite, 420, 449, 451
Melteigites, 454, 456
Melting, 83
Melting (fractional), 59.
Melting (incongruent compound), 37
Melting (isobaric), 37
Melting (solid solution), 40
Melting at an invariant point, 27
Melting (controlled by hydrate), 91
Melting curve (anhydrous solid), 88
Melting curve (negative), 89
Melting curve ($P_{H_2O} = P_{total}$), 88
Melting curve (P_{H_2O}/P_{total}), 90
Melting curve (P_{H_2O} limited), 91
Melting curve (positive), 89
Merwinite, 266, 273, 289
Metabasic rocks, 266, 271, 284, 325
Metamorphic facies, 83, 93, 102, 104, 187, 189, 325, 326
Metamorphic grade, 80, 94
Metamorphic iron formations, 314
Metamorphic Rocks, 99
Metamorphism—retrograde, 108, 109
Metasomatism, 169, 174, 187, 263, 287, 437, 443, 460, 461

Metastability, 161, 169
Metastable equilibrium, 19
Metastable phases, 14, 263, 281, 282
Metastable steps, 114
Mg-cordierite, 222, 231, 267
Mg-cummingtonite, 216
Miaskitic, 437, 444
Mica group, 227, 266, 278
Mica Stabilities, 137, 224
Migmatites, 313
Mineral assemblages, 283
Mineral facies, 83, 187, 189
Mineral powders, 11
Mineralising fluid, 131, 156
Minima, 445
Minnesotaite, 221
Mixed gas system, 3, 91
Mn-chlorite, 233
Mont Blanc, 198
Monticellite, 238, 239, 266, 272, 273, 289
Morey-Schreinemakers expression, 78
Morey-Schreinemakers rule, 75
Mullite, 217, 266, 268, 270, 308
Multicomponent rock systems, 325
Muscovite, 195, 216, 224, 225, 231, 248, 250, 266, 267, 278, 282, 311
Myrmekitic intergrowths, 323, 324

Na- feldspar (*see also* Albite), 195
Na-K iron biotites, 232
Natural systems (granitic), 394
Needle Point Pluton, 395, 397, 400
Nepheline, 195, 273, 431
Nepheline-albite eutectic, 434.
Nepheline syenites, 431, 435, 442, 454
Nepheline Syenites (corundum-bearing), 436, 437
Nephelinisation, 437
Nephelinite, 420, 440, 449, 464, 465
Non-congruently melting compounds, 36
Non-Equilibrium Studies, 404
Normative projections, 396, 446
Nucleation density, 338, 406, 407
Nucleation kinetics, 336, 405, 406

Obsidians, 447, 449
Olivine, 216, 266, 283, 320
Omphacite, 104, 149
One atmosphere experiments, 6
Open system, 462
Ophicalcite, 292, 296
Orthoamphibole, 266, 277, 321
Orthoclase, 37, 195, 207, 216, 250
Orthopyroxene, 216, 266, 274, 320
Orthopyroxene fractionation, 457
Ostwald Steps, 114, 140, 166
Osumilite, 266, 282
Osumilite-type phase, 314
Overpressure, 114, 120, 140, 158
Oxide equilibria, 187
Oxygen buffers, 92
Oxygen fugacity, 15, 208, 266, 314
Oxygen isotopes, 113, 138

Pacheco Pass, 157, 158, 162, 165
Pantellerite, 444, 448
Paragonite, 195, 224, 225, 266, 267, 278
Paragonite-muscovite solvus, 226
Paralavas, 261, 281
Pargasite, 195, 276, 454
Partial melting, 58, 59, 178, 265, 313, 456
Partial pressure (carbon dioxide: P_{CO_2}), 92
Partial pressure (oxygen: P_{O_2}), 92, 465
Partial pressure(water: P_{H_2O}), 86, 87, 156, 175
Pelites, 300, 304, 319, 324
Pennine Alps, 121
Penninite, 139
Peralkaline, 419, 420
Peralkaline feldspathoidal rocks, 420, 440
Peralkaline oversaturated rocks, 420, 444
Peralkaline residua system, 427
Peralkalinity, 426, 462
Periclase, 240, 266, 275, 282, 284, 288, 289, 305, 306
Peridotite, 58, 300, 454, 456, 457
Perthite, 279
Petrogenetic grid, 80, 83, 166, 190
Petrogeny's Residua System, 13, 424, 430, 433, 436
Petrogeny's Residua System (Rock compositions), 432
Phase, 21
Phase assemblage diagrams, 367
Phase change steps, 63
Phase diagram construction, 21
Phase relations, interpretation, 19
Phases of siliceous carbonate rocks, 233
Phengite, 134, 137, 227, 228
Phlogopite, 195, 249, 250, 278, 311, 454, 458, 459, 460
Phonolite, 420, 424, 436, 438, 440, 454
Piercing point, 52.

Piston-cylinder apparatus, 10, 151, 156, 159, 160, 162
Plagioclase, 40, 41, 195, 266, 279
"Plagioclase effect", 426, 462, 464
Plastic strain energy, 103, 115, 168
Poly-component systems, 52
Polymorphic inversion, 63, 81
Porcellanites, 279
Pore space, 88
Porphyroblasts, 264
Potash-rich lavas, 457
Potash-rich Mafic Rocks, 458
Potassic rhyolites, 459
Predazzite, 292
Prehnite, 147, 148, 298, 300
Pressure (fluid or gas. *see also* Partial pressure), 106, 115, 156, 174
Pressure (total or lithostatic), 86, 106
Pressure Control, 16
Pressure medium, 8
Precision, 3
Primary phase areas, 30
Primary phase regions, 32
Primary phase volumes, 33
Prograde reactions, 26, 32
Prograde thinking, 26
Progressive melting, 57
Progressive metamorphism, 93
Projected phase diagrams, 48
Projected phase diagrams (normative), 373, 430, 432, 433, 434, 446
Pseudo-binary, 49, 50, 426
Pseudo brookite (Fe_2TiO_5–$FeTi_2O_5$), 209
Pseudo-quaternary, 52
Pseudo-system diagrams, 50
Pseudo-ternary, 52, 53
"Pseudoleucite reaction", 434
PT constructions, 73
PT diagrams, 3, 22, 64, 65, 66, 69, 73
PT diagrams (liquids), 83
PT diagrams (Metamorphism), 92, 93
PT diagrams (multi-system), 82
PT liquidus curve, 85
PT melting curves, 454
PT slopes, 85
Pumpellyite, 146, 148
PX diagrams, 3, 61
PX diagram for nepheline-silica, 70
Pyrite, 215, 267, 283
Pyrometamorphism, 264
Pyrope, 130, 132, 136, 267, 320, 322
Pyrophyllite, 195, 216, 220
Pyroxene Group, 274, 391
Pyroxene-hornfels facies, 325
Pyroxenite, 456
Pyrrhotite, 214, 215, 267, 283

Quaternary eutectic (isobaric), 33
Quaternary "flow-diagram", 52
Quaternary invariant point, 70
Quaternary system (isobaric), 33, 34
Quaternary univariant curve, 52
Quartz, 195, 207, 216, 217, 238, 248, 249, 250, 279, 345
Quartz-feldspar cotectic, 45
Quenching, 4, 8, 10, 11, 19
Quenching furnace, 6, 7
Quinary invariant points, 72
Quinary system, 36

Range, Precision and Accuracy, 11
Ranges of Physical Conditions, 3, 5
Rankinite, 238, 266, 273, 289
Reaction kinetics, 116, 174, 263
"Reaction Point", 38
Reaction rates, 4, 5
Reaction series, 37, 424
Reaction series (continuous), 41
Reaction series (discontinuous), 40
"Reduced" components, 50
Regional metamorphism, 262, 264
Retrograde reactions, 26
Reversal, 5, 13, 337
Rhodochrosite, 236
Rhyolite, 13, 420, 424, 438, 449
Riebeckite, 123, 195, 376
Rutile (TiO_2), 209, 237, 238

Sagvandites, 287
Sample compositions, 3, 11
Sample containers, 3, 14
Sample size, 3, 14
Sanbagawa terrain, 124
Sanidine, 195, 266
Sanidinite facies, 263, 264, 266, 273, 276, 282, 289, 296, 325
Sanidinites, 264, 274, 279
Sapphirine, 221, 266, 273, 304, 305, 306
Sauconite, 240
Schists, 266
Schreinemakers analysis, 73, 385
Sericite, 134
Serpentine, 138, 141, 195, 216, 221, 239, 240
Serpentinisation, 122, 128, 133, 139, 140

Serpentinites, 300
(Si Al) disordering, 160
Siderite, 236
Sierra Nevada Batholith, 398
Siliceous carbonate rocks, 233, 279, 289, 293
Sillimanite, 195, 216, 217, 266, 267, 268, 270, 305, 306, 312, 315
Simple squeezer, 116, 117, 143, 155
Single-phase area, 55
Singular phases, 82
SiO_2-Polymorphs, 279, 345
Skarns, 266, 271, 276
Soda Metasomatism, 113, 129
Solid media pressure apparatus, (*see* Piston-cylinder), 10
Solid solution, 40, 83, 154
Solid solution loops, 43, 427
Solid solution melting, 63
Solid solution series, 105
Solids (equilibria), 3, 61, 63, 263
Solidus, 26, 40, 48
Solidus reactions, 350
Solubility (CO_2), 460
Solubility (H_2O silicate liquids), 89, 350, 352
Solution calorimetry, 113, 115, 150
Solvus, 43, 210, 211, 226, 234
Solvus crest, 435
Spessartite, 133, 233, 267
Sphene, 237, 238
Spinel, 266, 283, 305, 306, 315, 320, 321, 324
Spotted slates, 264, 266, 278
Spurrite, 237, 238, 239, 266, 273, 289
Starting materials (and limitations), 5, 18
Staurolite, 195, 216, 223, 231, 266, 269, 314
Staurolite + quartz, 271, 318
Stress, 102
Stress gradients, 88
Subduction zones, 120
Subfacies, 325
Subsolidus equilibrium, 35, 61
Sulphide equilibria, 55, 187, 214, 283
Syenites, 431
System, 21
Systems (*see* Systems Index, following)

Tactites, 266
Talc, 195, 216, 221, 240, 247, 266, 275, 279, 283, 284, 286, 288, 293, 297, 305, 306, 311, 321
Talc-calcite, 296
Talc schists, 284
Technological problems, 5
Tectonic blocks, 113, 124, 130
Temperature control, 16
Temperature gradient, 449, 461
Temperature minimum, 43, 45, 47
Ternary eutectic (isobaric), 32
Ternary invariant point (isobaric), 32
Ternary liquidus (isobaric), 30
Ternary system (isobaric), 30
Ternary univariant (isobaric), 32
Thermal decomposition curves, 128
"Thermal divides", 29
Thermal sink, 27
Thermal stability, 67, 87
Thermal valley, 45, 445, 446
Thermocouple, 6
Tholeiite solidus, 195
Tholeiites, 463
Three-phase area, 57
Ti-andradites, 268
Tie-lines, 58, 59, 67, 70, 72, 429, 446
Tilleyite, 238, 239, 266, 273, 289
Time scale, 5
Topological rules, 73
Trachytes, 424, 436, 438, 448
Tremolite, 127, 128, 195, 240, 247, 249, 250, 266, 267, 276, 293
Tremolite-calcite, 296
Tridymite, 266, 267, 269, 279
Tridymite-Quartz, 345, 353
Two-mica assemblage, 311
Two-mica reaction, 313
Two-phase area, 57
TX diagrams, 3, 24, 241, 243, 244, 246, 248
TX diagrams involving liquids, 28

Ultrabasic rocks, 266, 279, 284, 287, 288
Ulvospinel (Fe_2TiO_4), 209
Univariant curves (liquid), 83
Univariant line (isobaric), 33
Univariant lines (reaction boundaries), 23, 65, 66, 73, 75, 80
Univariant reaction, 65
Unmixing of solutions, 47
Upper Mantle, 179

Vapour-absent condition, 458
Vapour (hydrous), 87
Vapour phase, 20, 353
Vapour-present, 458
Vapour pressure curves, 109
Vapour saturated liquidus relations, 358

Vapour-undersaturated liquidus relations, 359
Variance, 21
Variations in P_{O_2}, 3
Very high pressure, 11
Vesuvianite, 266, 272
Volatiles, 448, 449
Volatile transfer, 449
Volatile transport, 461
Volcanic rocks, 282

Wallowa Batholith, Oregon, 395
Water-Carbon Dioxide, 201
Water vapour pressure (*see* P_{H_2O}), 156, 172, 178
Water (*see* H_2O)
Western Alps, 132
Western Pennine, 157
Willemite, 240
Wollastonite, 238, 247, 266, 267, 273, 276, 289, 297, 298, 300

Xonotlite, 239

Yoderite, 221

Zeolite, 109
Zoisite, 247, 271, 298, 300
Zone-refining, 459
Zoning (plagioclase), 41

I. Systems defined by ideal mineral components

Acmite-albite (Ac-ab), 163
Acmite-albite-diopside-nepheline(Ac-ab-di-ne), 442
Acmite-albite-nepheline (Ac-ab-ne), 441, 454
Acmite-augite-jadeite (Ac-aug-jd), 153
Acmite-diopside (Ac-di), 464
Albite-anorthite (Ab-an: Plagioclase), 424
Albite-anorthite-diopside (Ab-an-di), 42
Albite-auorthite-diopside-forsterite (Ab-an-di-fo), 42
Albite-anorthite-orthoclase-quartz (Ab-an-or-q: *see also* Haplogranodiorite), 381, 311 and 378 (with H_2O)
Albite-diopside (Ab-di), 105, 426, 462
Albite-orthoclase (Ab-or), 43, 47
Albite-orthoclase-quartz-water (Ab-or-q-H_2O: *see* Granite and Haplogranite), 44, 338, 339, 344
Albite-orthoclase-SiO_2 (Ab-or-SiO_2), 387
Albite-orthoclase-water (Ab-or-H_2O), 338, 339, 346
Albite-quartz-water (Ab-q-H_2O), 89, 338, 339, 347, 361, 362
Albite-silica (Ab-SiO_2), 24
Albite-sodium disilicate-water (Ab-$Na_2Si_2O_5$-H_2O), 373, 375, 377
Albite-water (Ab-H_2O), 110, 339, 348, 392
Anorthite-diopside-forsterite (An-di-fo), 54, 55, 56

Diopside-jadeite (Di-jd), 434, 464
Diopside-nepheline (Di-ne), 49, 50, 449
Diopside-pyrope (Di-py), 50, 456
Diopside-sanidine (Di-sa), 435, 436

Enstatite-phlogopite (En-phl), 459

Forsterite-kalsilite-silica-water (Fo-ks-SiO_2-H_2O), 458
Forsterite-nepheline-silica (Fo-ne-SiO_2), 67, 71
Forsterite-silica (Fo-SiO_2), 38

Kalsilite-nepheline-silica (Ks-ne-SiO_2), 425 (*see also* "Petrogeny's Residua System").

Nepheline-silica (Ne-SiO_2), 29
Nepheline-silica-water (Ne-SiO_2-H_2O), 340

Orthoclase-water (Or-H_2O), 339, 348
Orthoclase-quartz-water (Or-q-H_2O), 338, 339, 347, 362

II. Systems defined by chemical formulae. Arranged in cation alphabetical order with CO_2 and H_2O always following SiO_2.

Al_2O_3-CaO-K_2O-SiO_2-CO_2-H_2O, 187, 247, 248
Al_2O_3-CaO-MgO-SiO_2, 452
Al_2O_3-CaO-MgO-SiO_2-H_2O, 147, 148
Al_2O_3-CaO-MgO-Na_2O-SiO_2, 450, 458
Al_2O_3-CaO-SiO_2-CO_2-H_2O, 300
Al_2O_3-CaO-SiO_2-H_2O, 298
Al-Fe-O, 209
Al_2O_3-FeO-K_2O-MgO-SiO_2-H_2O, 187, 216, 231
Al-Fe-K-Na-Si-O-H, 444
Al_2O_3-Fe-O-MgO-SiO_2-H_2O, 319
Al_2O_3-Fe_2O_3-Na_2O-SiO_2, 53, 427, 440, 444, 453
Al_2O_3-Fe-O-SiO_2-H_2O, 223, 314
Al_2O_3-K_2O-MgO-SiO_2-H_2O, 310
Al_2O_3-K_2O-Na_2O-SiO_2, 426
Al_2O_3-K_2O-SiO_2, 436, 444
Al_2O_3-MgO-SiO_2-H_2O, 221, 304
Al_2O_3-Na_2O-SiO_2, 436, 444

Al_2O_3-Na_2O-SiO_2-H_2O, 373
Al_2O_3-SiO_2, 217
Al_2O_3-SiO_2-H_2O, 220

CaO-MgO-SiO_2-CO_2, 289
CaO-MgO-SiO_2-CO_2-H_2O, 187, 244, 246 293
CaO-SiO_2-H_2O, 239
$CaCO_3$, 116, 117
$CaCO_3$-$FeCO_3$, 234
$CaCO_3$-$MgCO_3$, 234
$CaCO_3$-$MnCO_3$, 234

Fe-S, 214
Fe-S-Zn, 216
FeO-SiO_2-H_2O, 221

MgO-SiO_2-CO_2-H_2O, 244, 285
MgO-SiO_2-H_2O, 220, 242, 243, 284

SiO_2-H_2O, 348

ZnO-SiO_2-H_2O, 239